Second International E-Conference on Physics of Materials and Nanotechnology (ICPN 2021)

IOP Conference Series: Materials Science and Engineering
Volume 1221

Online
27 - 30 October 2021

Editors:

H Devendrappa
Sangaraju Shanmugam
Robah Boukherroub
Praveen C Ramamurthy
Ramakrishna Matte H S S
Rajashekhar Bhajantri

ISBN: 978-1-7138-5435-7
ISSN: 1757-8981

Printed from e-media with permission by:

Curran Associates, Inc.
57 Morehouse Lane
Red Hook, NY 12571

Some format issues inherent in the e-media version may also appear in this print version.

Institute of Physics
Dirac House, Temple Back
Bristol BS1 6BE UK

Phone: 44 1 17 929 7481
Fax: 44 1 17 920 0979

techtracking@iop.org

Additional copies of this publication are available from:

Curran Associates, Inc.
57 Morehouse Lane
Red Hook, NY 12571 USA
Phone: 845-758-0400
Fax: 845-758-2633
Email: curran@proceedings.com
Web: www.proceedings.com

TABLE OF CONTENTS

doi:10.1088/1757-899X/1221/1/011001

IOP Material Science & Engineering Conference Proceeding

Second International E-Conference on Physics of Materials & Nanotechnology (ICPN-2021) held during October 28-30, 2021

Chief Editor

Dr.Devendrappa H
Professor

ICPN 2021

MANGALORE UNIVERSITY

Department of Physics
Mangalore University, Mangalagangothri, India

IOP Conf. Series: Materials Science and Engineering 1221 (2022) 011001 doi:10.1088/1757-899X/1221/1/011001

Preface

Second International E-Conference on Physics of Materials and Nanotechnology (ICPN-2021)

On behalf of Organizing committee, it's my great pleasure to bring this beautiful collection of research article which were presented in the Second International E-Conference on Physics of Materials and Nanotechnology (ICPN-2021) which was organized by Department of Physics, Mangalore University, Mangalagangothri, Mangalore (Karnataka), India during October 28 -30th, 2021. This conference will serve as an ideal platform for young researchers, academician and students from all over the country as well as from the globe to share their ideas and experiences in Physics, Nanotechnology and Nano science research. This conference provided meaningful platform for young researchers to update and upgrade themselves. The main themes of the conferences were Nano materials, Chemistry of materials, Functional, and smart materials, Polymers bulk and nano composite, Dielectrics and ferroelectrics, Soft materials & Magnetic materials, Crystal growth & its Nonlinear Optics, Bio-materials and Biopolymers, Optical materials, Semiconductor physics, Gas Sensors, Photovoltaic devices, thin films, Radiation processing of materials etc. I am glad that experts in the various field of Nanotechnology are here to present and discussed their ideas and latest development in the field. ICPN 2021 has received a great response throughout the country and globe. It includes 11 invited talk, 34 oral presentation and more than 170 poster presentations. We have received more than 78 papers for the publications in this issue, after reviewing and editing processes, 66 papers were accepted for the proceedings. We thank the authors for their contribution and the reviewers for their comments to improve the quality of article. I would like to express my gratitude to all members of organizing committee,

IOP Conf. Series: Materials Science and Engineering 1221 (2022) 011001 doi:10.1088/1757-899X/1221/1/011001

staffs, research scholars and postgraduate students of the Institute for their willingness, commitments, dedication and relentless efforts in ensuring that this conference is a success.

Editorial Board;

Devendrappa H, Sangaraju Shanmugam, Robah Boukherroub, Praveen C Ramamurthy, Ramakrishna Matte H S S, and Rajashekhar Bhajantri,

IOP Conf. Series: Materials Science and Engineering 1221 (2022) 011002 doi:10.1088/1757-899X/1221/1/011002

Peer Review Statement

All papers published in this volume have been reviewed through processes administered by the Editors. Reviews were conducted by expert referees to the professional and scientific standards expected of a proceedings journal published by IOP Publishing.

- **Type of peer review:** Undefined
- **Conference submission management system:** Morressier
- **Number of submissions received:** 78
- **Number of submissions sent for review:** 75
- **Number of submissions accepted:** 66
- **Acceptance Rate (Submissions Accepted / Submissions Received × 100):** 84.6
- **Average number of reviews per paper:** 3
- **Total number of reviewers involved:** 8
- **Contact person for queries:**
  Name: Devendrappa H
  Email: dehu2010@gmail.com
  Affiliation: Mangalore University - Physics

Synthesis, characterisation and AC conductivity study of PANI/α-Fe_2O_3 nanocomposites

Rahul Jabnoor[1] , Manjunatha B[2], M V N AmbikaPrasad[3], Arunkumar Lagashetty[4], and Vijaykumar Basavarajappa[5], Sangshetty Kalyane [6]
[1&6]Department of Physics Bheemanna Khandre Institute of Technology, Bhalki –585328 Karnataka., India
[2]Department of Physics Gulbarga university kalaburagi PG center Raichur –584133 Karnataka.,India
[3&5]Department of Physics Gulbarga university kalaburagi –585106 Karnataka.,India
[4]Department of Studies in Chemistry, Vijayanagara Sri Krishnadevaraya University, Ballari, Karnataka, 583105, India

ABSTRACT

The disperse α-Fe_2O_3in polyaniline matrix to form its nanocomposite enhances the structure and properties of PANI for wider applications. Present study reports the preparation of alfa iron oxide dispersed PANI nanocomposite (α-Fe_2O_3/PANI NC) with the help of oxidize agents APS by technique insitue method. t. Composite of PANI with 10%, 20 %, 30%, 40% & 50% of α-Fe_2O_3 has been synthesized by adding iron oxide powder gradually in the reaction mixture during oxidative polymerisation with APS. The composites were well characterized like as XRD,SEM,FTI. AC conductivity of the pure polyaniline & α-Fe_2O_3/PANI NC with 10%, 20%, 30%, 40% & 50% is studied as function of frequency. The value of ac conductivity increases as frequency increased from 1 KHz – 1MHz.

Keywords: PANI, (α-Fe_2O_3, SEM, XRD, AC conductivity, FT-IR

---

*Corresponding author e-mail: sangshetty2016@gmail.com

1. Introduction

IOP Conf. Series: Materials Science and Engineering 1221 (2022) 012001 doi:10.1088/1757-899X/1221/1/012001

Polymers material has developed into areas of rising attention in investigate since the information of this material contains huge potential the state of solid device [1].

The polymeric composites material studying in the field of technological and application oriented polymer material this polymeric material obtaining the oxides of the material.consititute these composite polymeric matrials are built for the further study of characteristics and applications. [2-4].

The polymeric conducting, materials contain a multiplicity of application in the Scientifics, Medicals, Industrial applications similar to anticorrosion, coatings the statid, EMI etc come beneath primary generations,. next Generations of electrically polymer have application such as, solar cells transistors batteries LEDs, etc. Control conductivities, high temperatures resistances, cheap and easiness of bulkiness research make this material good-looking in the engineer and Scientifics globe [2-6].

The important applications of conducting polymers such as reversibility, which can be ease to used for layering, shape, high-quality environment constancy improve their latent use for practically application. The mainly study the conduct polymer; PANI obtain chemically or electrochemically direct. Polymers have turn out to be an area of rising attention in investigate because of the detail that these material have huge possible for solid states device [4-6].

Polyaniline has been attacked much more attention because of only its unique application that is reversible proton doping, highly conductivity for electrical properties, moreover, it is ease to preparation with cost effective. Now-a-days it is highly alarming and demand of high quality materials because of electromagnetic compatibility [7, 8].

Metals oxide distributed polymers composite have involved a huge contract of attention starting researcher, since they regularly exhibits unforeseen cross property synergistically derives from both mechanism. One of the best examples for oxide materials is the iron oxide; this is well known for useful resources of oxide with important futuristic application [8-11].

The conductive polyaniline contain the property likes metallic oxides resources well-known as composites polyaniline also different composition might lead to desirable's property. particularly those polyaniline materials were important to owing their bridging role between the worlds of polyaniline conductivity of polymeric [12-15].

The polyaniline exhibiting variety of forms with differ in the physical and chemical properties [16-18]. The yellow color is found by the reduce the lecumerldine,browinian colour is prooemeralidine's and violet is nigrniline's . Various oxidizing agent/monomer ratios have been used. Ammonium peroxydisufate is commonly used as oxidizing agent. There are benzenoid and quinonoid resonances according to amine and imine structures. Altering the ratio's of amine's to imine's, a variety of formation are formed [19,28].

Emeraldine, nigraniline and pernigrandline in highly crystalline forms were produced by oxidation with hydrogen peroxide, perchloric acid and potassium dichromate respectively [20]. Polyaniline has good environmental stability [21]. Forming the base of emeraldines has alternates amines and imines repeats unit [22-23]. PANI have chemical flexibly –NH grouping backbones which is response for motivating chemical and physical aspects [24]. It was found that IR spectrum of emeraldine contained three Gaussian bands, nigraniline spectrum showed two Gaussian band and pernigraniline spectrum had one Gaussian band. This was consistent with stacking of benzenoid and quinonoid rings along one direction and with coupling of electrons with ring puckering vibrations of benzenoid rings [25]. The current survey focusing the fabricates compositing polyaniline materials these composites have contains large property of the applications. The α-Fe_2O_3 may be an add-on advantage in the fields of polymer; in exacting, it may supply high-quality applications while dopes the PANI. The property of composites similar to higher stiffness and lofty melt points can be use the intractable agent. Little previous information explains the sensitivities and selectiveness in research of nanostructure PANI with the adding of dissimilar oxides of the metals [26,28].

2. Experimental

2.1. Materials and Method:

Chemicals used in the present investigation are of analytical grade and in the preparation of the solutions. Chemical oxidative polymerization of aniline with iron oxide method is used for the synthesis of PANI/α-Fe_2O_3 NC.

Synthesis of α-Fe_2O_3

Synthesis of PANI/α-Fe_2O_3 NC

The prepared solution of 0.1M of aniline in volumetric flask is dissolved in a beaker along with prepared solution of 0.1N of hydrochloric acid and this mixer is stirred for one hour by magnetic stirrer at room temperature. After one hour of constant stirring, the prepared solution of 0.1M of ammonium persulphate was adding with solutions of anilines hydrochloric acid using pipette. As prepared α-Fe_2O_3 nanomaterial was this mixer is stirred for 10 hours for the completion of reaction, and then it left for 24 hours to settle down. Later the prepared solution is filtered using vacuum pump and Buchner funnel. At last, yield was obtained from the hot air oven twenty four hours at sixty degree temperature.The said composite were prepared for 10, 20, 30, 40 and 50 wt.% of iron oxide is prepared to study the mechanistic variation of the composite. The complete synthesises method is shown in scheme-1.

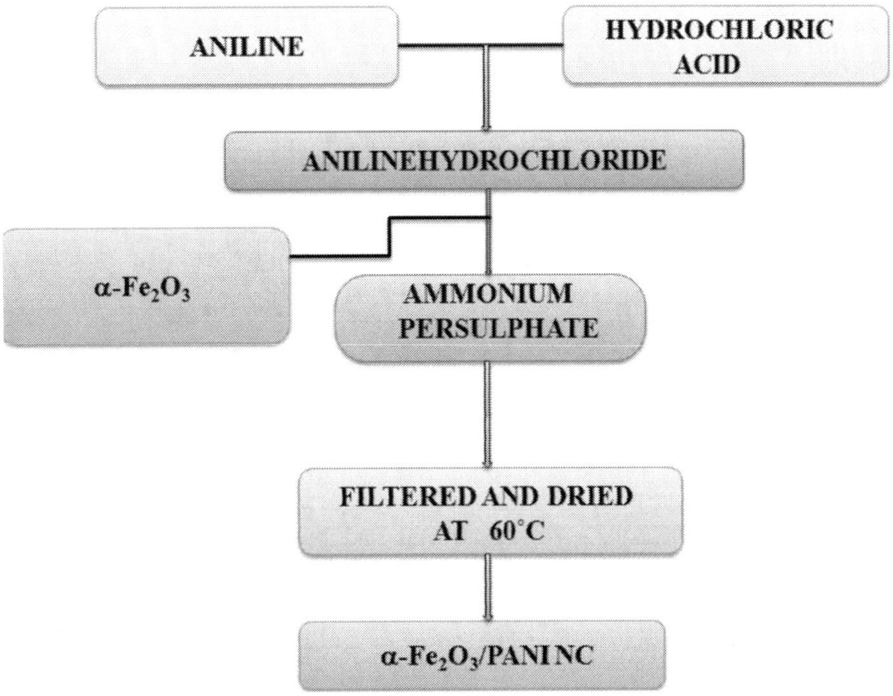

Scheme-1: Synthesis method of PANI/α-Fe_2O_3 NC

3. Characterization Techniques

Structural characterization of the prepared sample was well studied by powders XRD the peaks was record at thirty kilo vat the XRD uses the radiations of CuKα (1.54 Å). Using the MODEL NO. THOUSAND Perkin-Elmer SHOWS THE Bonding nature of the sample Morphological study of the sample was carried out by bright field TEM picture and get from 190 kilo electron volt with the model of JOEL 100 CX operated at 190 KeV. JEOL JSM-6380 LA .

4. Result and Discussion:

4.1 XRD

The solid nature also sequence pack of the synthesize polymeric nanocomposites was well studied by XRD. Figure 1 shows indexed XRD pattern of as prepared chemically oxidative derived PANI/ α-Fe_2O_3 NC. It is reported that, plain PANI shows partially crystalline nature and shows its characteristic peak at $2\theta = 28^0$ [27]. In addition to this, XRD pattern of PANI/ α-Fe_2O_3 nanocomposite shows number of Bragg reflection patterns because crystallinity natures of the sample. Further, the development of crystalline nature in the PANI is due to matrixes of alfa iron oxide nanomaterial and is indexed in the composite pattern in accordance with the its JCPDS file 33-0664. The presence of α-Fe_2O_3 reflections in the PANI/ α-Fe_2O_3 sample confirms the formation of complete PANI/ α-Fe_2O_3 NC.[28]

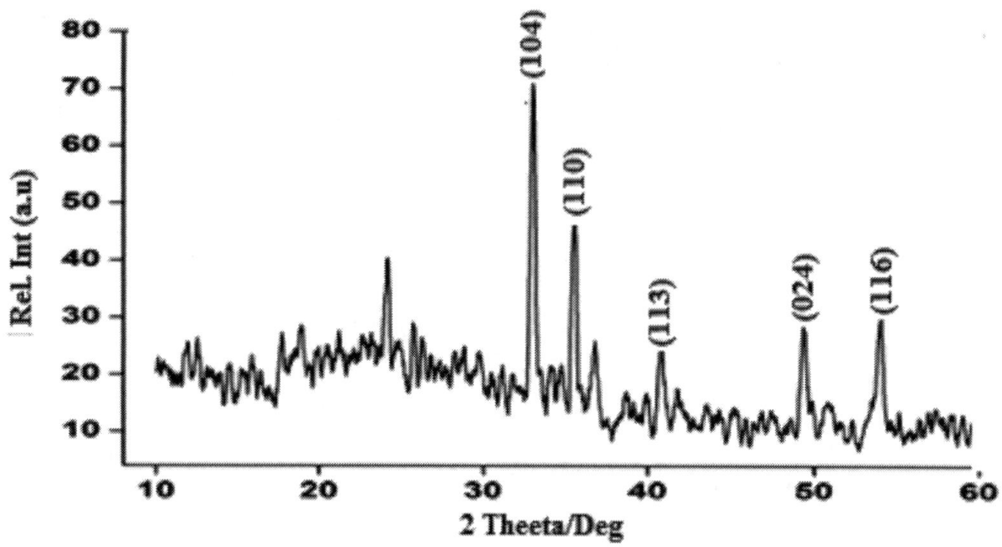

Figure-1: XRD pattern of PANI/ α-Fe_2O_3 nanocomposite

4.2 SEM study

Figure 2(a-b) shows SEM image made α-Fe_2O_3-PANI nanocomposites of sample at low and high resolution respectively. It is observed in the image that, the cluster of spherically shape particle with elongates sequence patterns. Fine particles iron oxides are well dispersed in the PANI matrix shows compact structure. It is also seen clearly from the SEM image these particle are nonporouses, shape of the garnulars agglomerates aloso these particles are organised or interconnecting. However in large magnifications it reveal homogeneous distribution of fine crystalline particles throughout the polymer sample, which enhances the crystalline nature of the sample. In addition to this, the sample image also indicate numerous aggregate particle and pore that have been reduces due to it homogeneities. The contrasting in the image is due to the difference in spreading from variety surfaces area as a effect of geometrically difference b/w polyaniline and dispersed oxide sample [28,29].

Figure 2(a-b): Shows SEM photograph of Polyaniline- Fe_2O_3 composites

4.3 FTIR Studies

Table-1 shows vibrational bands of as prepare of the nanocomposites sample PANI/ α-Fe_2O_3. These band at 1550 and 1478 centimeter are attributing to C=N and C=C stretch modes of the quinonoid's and benzenoid's unit of respective polyaniline. Patterns at 1290 and 1230 cm^{-1} are assign to C–N stretch modes of benzenoid rings of the nanocomposite sample. This type of typical pattern at 1245 centimeter because this polyaniline have conductivity by protanating. In addition to this, peak at the centimeter of 1110 is considered to be the planes bend vibrational of C– H modes. Similarly the band at 810 cm^{-1} originating out of planes C– H bends vibrational of the aromatics polymer. Further, the sample spectrum showed 2 pattern at 645 and 650 cm^{-1}; which are assigned due to the occurrence of iron oxides (M-O). Presence these metal oxides pattern in the composites spectrum substantiate the structure of PANI/ α-Fe_2O_3 nanocomposite.[28,30]

Table-1

Vibrational bands of PANI/ α-Fe_2O_3 nanocomposite

Sample	PANI/ α-Fe_2O_3 nanocomposite
Frequencies	1550, 1478, 1290, 1230, 1245, 1110, 810, 645, 650

5. A C Conductivity Study

Figure-3 tells the variation of AC conductivities of pure polyaniline & polyaniline/ α-Fe_2O_3 nanocomposite like 10%, 20%, 30%, 40% & 50% as a function of frequency. The value of AC conductivity increases as frequency increased from 1 KHz – 1MHz. The frequencies depends conductivities for disorder material such as polymeric material can be due to interracially polarizations at contact and grains boundary of the samples. The value of ac conductivity was found to be $2.91x10^{-5}$ S/m at 1KHz frequency and which increases as frequency increased & reaches the value 0.0025 S/m at 1MHz frequency. Previously reported that, the ac conductivity of pure polyaniline was found to be approximately $3x10^{-5}$ S/m at 1 KHz and it increases to 2.5x 10^{-5} S/m at 1 MHz In all the cases ac conductivity increases as frequency increases and also increases as content of iron oxide increases in the polyaniline matrix upto 10%. It is due to the extending chains length of polyaniline's which facilitates the hop of charging carrier when the contents of iron oxide is up to 50 wt% [31].

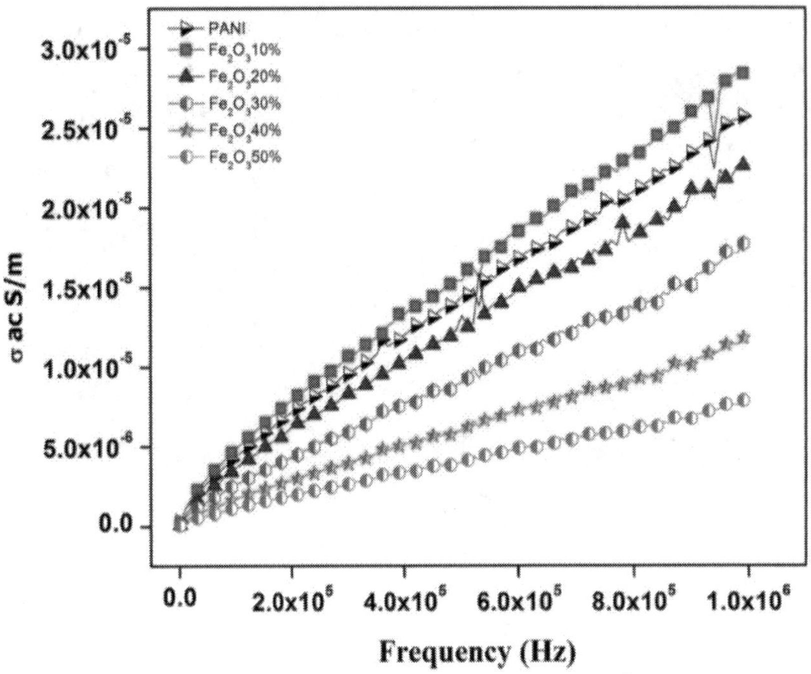

Figure-3: AC conductivity of PANI& PANI- Fe_2O_3composite

Conclusion:

In organic polymer, the polyaniline is one of among them and which is considered as oldest well-known organically polymers which consist of an ill-defines classes of material obtain by the chemically polymerizations of aniline. PANI place very important role in the field of the application of sensors, electronic devices and rechargeable batteries due to its novel properties such as the property of stability in the environment, lowered price of the PANI and its higher electrical conductivity. Thus these materials are popular in the scientific interest and also in much commercial application. PANI is a one of the type of polymer of conducting state which can be prepared by the oxidation aniline under polymerization reaction. This can do by many different methods of synthesis of polyaniline. The property of the PANI mainly depends on the method of synthesis used for the development of PANI. PANI attracted interest due the variation of electrical property by the different methods. Synthesis method of the PANI can affected to its surface morphology, structural property and electrical property. The electrical conductivity of the polyaniline can be tailored by the process of doping. When polyaniline was doped with many different fillers can doping agent, the electrical conductivity of the polyaniline changes. Also the electrical conductivity cases ac conductivity increases as frequency increases and also increases as content of iron oxide increases in the polyaniline matrix upto 10%. This may be due to the extended chain length of polyaniline which facilitate the hopping of charge carriers when the content of iron oxide is up to 50 wt%

Acknowledgement:

The authors gratefully acknowledge, and this work was supported by VGST (Vision Group on Science and Technology), Department of Physics, Bheemanna Khandre Institute of Technology, Bhalki, Bidar.

References :

1. H. Devendrappa,U.V. Subba Rao, M.V.N. Ambika Prasad Study of dc conductivity and battery application of polyethylene oxide/polyaniline and its composites,Journal of Power Sources, Vol. 155(2), 2006, 368-374,
2. Morawetz H Encyclopaedia of Polymer Science and Engineering, 2nd edn. vol. 7, pp. 722-745 (Wiley, New York) (1987).
3. Angew.chem.Int. NO.19.3456 WEBSITE http://www.ch.ic.ac.uk/local/ organic/ tutorial/steinke/4yr PolyConduct (2003).
4. Bovey F. A., Academic Press, New York, (1982).
5. Mac Gregor E. A. and Greenwood, C.,*Polymers in Nature* (Wiley, Chichester, (1980).
6. Epstein A.J., Ginder J.M., Zuo F., Bigelow R.W., Woo H., Tanner D.B., Richter A.F., Huang W. and MacDiarmid G.A., *Synth. Met.,* Vol. 18 *pp* 303-309. (1987).
7. Shirakawa H., Lewis E.J., MacDiramid A.G., Chaing C.K. and Heeger A.J J. Chem. Soc. Chem. Commun. pp 578. (1977).
8. MacDiarmid A. G., Angew, Chem. Int. Ed. Vol. 40, 2581 (2001).
9. Sailor M. J. and Curts C. L.. Adv. Mter, Vol. 6, 688 (1994).
10. Gangopadhyay R., H. S.Nalwa (Ed.) pp. 105-131, American Scientific Publishers, Los Angeles (2004).
11. Martin C. R., Acc. Chem. Res. Vol. 28, 61 (1995).
12. Huang W. S., Humphery B. D., MacDiarmid A. G., J. Chem. Soc., Faraday Trans. Vol. 1 82, 2385 (1986).
13. MacDiarmid A. G., Chiang J. C., Halpern M., Huang W. S., Mu S. L., Somasiri N.

    L. D., Wu W.Q., Yaniger S. I. Mol. Cryst. Liq. Cryst. Vol. 121, 173 (1985).
14. N. Zhang, R. Wu, Q. Li, K. Pakbaz, C. O. Yoon, F. Wudl Chem. Mater., Vol. 5 (11), pp 1598–1599 (1993).
15. Suarrez M. F. Herreara-conducting polymers –Electrochemistry Encyclopedia of life support system.
16. Bakshi A.K. and Geetika Bhalla journal of scientific and industrial research vol. 63, pp 715-728. (2004).
17. W. R. Salaneck, I. Lundström, and B. Rånby Nobel Symposium in Chemistry Ed"s (Oxford Sci., Oxford, (1993)).
18. H. Shirakawa, E.J. Louis, A.G. MacDiarmid, C.K. Chiang and A.J. Heeger, J Chem Soc Chem Comm. Vol. 579 (1977).
19. Chiang C.K., Druy M.A., Gau S.C., Heeger A.J., Louis E.J., MacDiarmid A.G., Park Y.W. and Shirakawa *H., J. Amer. Chem. Soc.,* Vol. 100,*pp* 1013-1015. (1978).
20. Kaveeta P.J., Prasad R.G., Venkata S.J., Aparna R.S. and Phani A.R., *Nano Biomed. Eng.,* Vol. 4 144-149. (2012).
21. Rahman M., Kumar A., Su-Park D. and Shim Y., *Sens.,*Vol. *8 pp* 118-141 (2008).
22. Heeger Alan J. reviews of modern physics, vol. 73 (3) (2001).
23. Shirakawa H. Heeger Alan J., MacDiarmid A. G. Chem. Commun.1-4 (2003).
24. Ramkrishnan S. RESONANCE. Vol.16 (12) December (2011).
25. De Albuquerque J. E., Mattoso L.H.C., Faria R.M., Masters J. G. and MacDiarmid A.G. , *Synth. Met.,* Vol. *146,* 1-10, (2004).

26. Chiang, C. K., Druy, M. A., Gau, S. C., Heeger, A. J., Louis, E. J., MacDiarmid, A. G., Park, Y. W., & Shirakawa, H. (1978). Synthesis of highly conducting films of derivatives of polyacetylene,(CH) x. *Journal of the American Chemical Society, 100*(3), 1013–1015
27. Arunkumar Lagashetty,Amruta Pattar, Sangappa K. Ganiger Synthesis, characterization and antibacterial study of Ag doped magnesium ferrite nanocomposite Heliyon. Vol. 5 (5) e01760
28. B. Manjunatha & Arjun N. Shetty & S. Kaveri & Sundar S. Mety & K. C. Anjaneya & Ramakrishna Reddy & Sangshetty Kalyane 5 Chemical Mediated Synthesis of Polyaniline/Tungstenoxide (PANI/WO3) Nanocomposites and Their Antibacterial Activity Against Clinical Pathogenic Bacteria BioNanoScience BioNanoScience 2020, Vol. 10(1), 73-80
29. Stejskal J.,Sapurina I.,Trchov´aM and Konyushenko EN, Macromolecules, 41, pp 3530-3536 (2008).
30. Manjunatha B, Mahantappa Limbitot, Manjula c Sangshetty Kalyane Studies on structural, surface morphology and transport properties of PANI-WO3 JEITR, 2018, 5(9) 564-570
31. Tseng R. J., Huang J., Ouyang J., Kaner R. B., Yang, Y., Nano Lett. Vol. 5(6), pp 1077. (2005).

IOP Conf. Series: Materials Science and Engineering 1221 (2022) 012002 doi:10.1088/1757-899X/1221/1/012002

Gamma-Ray Interaction of Selected Inorganic Scintillators Used in HEP Experiments

Jasjot Singh Dhillon [a] and Yogesh K. Vermani[b]
Department of Physics, Sri Guru Granth Sahib World University, Fatehgarh Sahib-140407, India.
[a] Email ID: jasjot.jsd@gmail.com,
[b] Corresponding author's Email ID: yugs80@gmail.com

Abstract. We attempted to study gamma-ray attenuation and sensing properties of conventional and modern inorganic scintillators being employed in the high energy physics (HEP) experiments. The mass attenuation coefficient (μ_m), effective atomic number (Z_{eff}) and mean free path (mfp)were theoretically evaluated for the conventional scintillators materials such as CsI and NaI (Tl) and compared with advanced scintillator materials: PWO, PbF_2, and BGO along with rare earth elements based scintillators such as LYSO:Ce, LuAG:Ce, BaF_2:Y which have been proposed for applications in the future HEP experiments. Thegamma-ray attenuation parameters are analyzed within the framework of online software toolkit `py-MULBF' over wide photon energy range from 0.015 MeV to 15 MeV.Variationof μ_m (and, Z_{eff}) with photon energy follows a trend similar for most of the inorganic scintillator materialsinvestigated here.CsI, however, maintained almost same effective atomic number value with respect to photon energy which signifies that CsI may be suitable for specific gamma-ray detection and sensing applications. Lead-based scintillator materials such as PbF_2, PWO along with $high-Z$ BGO are observed to exhibit better radiation attenuation capabilities.

1. Introduction

Rontgen and Becquerel discovered X-rays along with α-, β-, andγ-rays in 1895 and 1896, respectively. One of the important characteristic properties of these radiations was to excite phosphors such as those used in photographic plates ($K_2Pt(CN)_4$) that led to their discovery. These photographic plates, however, are not ideal for detection of highly penetrating radiations. This was the time when scintillating materials came into the picture. These are the luminescent materials capable of absorbing radiations effectively and converting absorbed energy into electromagnetic radiations in the UV-vis spectra[1]. Scintillation is a general property of matter and can be exhibited by materials in gas, liquid and solid phases. An efficient scintillator material is capable of detecting incident radiation and converting the signal into a flash of light[2].A basic scintillator is required to possess high density (ρ) and atomic number (Z) so as to increase the stopping power of the material. A scintillation material with good light output (measured in photons/MeV) is very crucial in those applications, whereparticlesinvolved are of relatively lower energy but this property hasn't much significance in High Energy Physics (HEP) experiments. Decay time(τ) of shorter duration is an amiable factor for significant timing resolution and operations involving high counting rates.

For most of applicationsa perfect scintillator can be a combination of high stopping power, higher light output,and shorter decay timeHigh energy physics (HEP) experiments, in general, involve machinery operating in a very high radiation environment and, thus, radiation sensing and damage inflicted on the detector components due to inadequate *radiation hardness* becomes a matter of concern for scintillator materials used in electromagnetic calorimeters for observing particles with extreme momenta. Although, radiation hardness may be irrelevant for most of medical applications, but is of extreme importance for radiation detectors employed in the HEP experiments [3, 4,5].$CaWO_4$ and ZnSbased powders were one of the conventional materialsused in earlier times for conversion of invisible radiation intovisible light spectrum.However, detection based upon bulk single-crystalline materials actually began in the late 1940s with the advent of scintillation detector materials like NaI (Tl) and CsI (Tl). These halide scintillatorscontributed extensively for detection of gamma-rays. Besides halide scintillators, first oxide-based $CdWO_4$scintillator and subsequently introduced$Bi_4Ge_3O_{12}$ (BGO) scintillator have been widely used as standardised material toevaluate performance of modern detector materials[2].

One of the widespread utilization of scintillators was explored in HEP experiments conducted at CERN,Geneva in 1989 in the form of electromagnetic radiation calorimeter (ECAL) incorporating 12,000 BGO crystals (24 *cm* in length occupying a total volume of 1.2 m^3). However, despite being dense and a rugged material, the next generation of high-precision ECALs requireda faster detection in theproton collider built at CERN. The lead tungstate ($PbWO_4$) provided for cheaper solution in view of shorter radiation length and higher density[6].This led to reduced amount of crystal required in bulk form for the detection purpose. About 75,000 crystals of $PbWO_4$ (PWO)are employed in electromagnetic calorimeter of Compact Muon Solenoid (CMS)in Large Hadron Collider (LHC) at CERN. This calorimeter has a total crystal volume of about 11 cubic meters which is the largest volume among all the existing calorimeters.The lead tungstate (PWO) based calorimeter has played a major role in the discovery of Higgs bosonin the CMS experiments [7]. Apart from these inorganic scintillators, CsI(doped and undoped) crystal has been widely used as crystalline scintillation detector in FNAL, SLAC, and KEK accelerators in high-energy physicsexperiments [8].The electromagnetic calorimeters of future generation face a serious challenge as significant loss of light yield has been observed in the CMS PWO crystals employed in the LHC experiments.For experimentsperformed at high luminosity, LHC crystals should survive high doses of radiation.$BaF_2$and Ce^{3+} dopedLuAG and LYSOinorganic scintillators are found to be radiation *hard*for high luminosity LHC experiments. Thus aLYSO MIPtiming detector has been proposed for the CMS upgrade at LHC [5,7]. In future, one needs to develop inorganic scintillators with strong *radiation hardness* and minimum *pile-upeffect*. Yttrium doping in BaF_2 crystals was found to increase the ratio of light outputs of the two components corresponding to decay times of 600 ns and 0.5 ns, in favour of the faster component [5, 9].

In view of these interesting developments, it would be of interest and relevance to undertake gamma-ray interaction studies for these inorganic scintillators. We shall employ `py-MLBUF' software toolkit [10]to evaluate important gamma-ray shielding and sensing parameters.

2. Computational Approach

Like other shielding materials (absorber), scintillator materials interact with gamma-rays either viacomplete photon absorption or photon scattering. Photoelectric effect and pair production phenomena. The phenomenon of photon scattering can either be coherent, called as Rayleigh scattering, or it could be incoherent which is usually known as the Compton scattering.Theoretically, mass attenuation coefficient (μ_m) can be determined using the relation given by Jackson and Hawkes in 1981 [11]popularly known as *mixture rule*:

$$\mu_m = \frac{\mu}{\rho} = \sum_i w_i (\mu_m)_i \text{ Eq. (1)}$$

Here, w_i is the weight fraction of the ith constituent element present in the compound or mixture. The mass attenuation coefficient value encompasses the probability of interaction of gamma - ray photons with mass per unit area of the material and represents the resultant attenuation via all interaction mechanisms.

Next, we define the effective atomic number for shielding material computed over wide range of photon energy.Firstly, we calculate the effective atomic cross-section per atom(σ_a) that can be obtained using the formula [12]:

$$\sigma_a = \frac{M}{N_A \sum_i n_i} \mu_m . \qquad \text{Eq. (2)}$$

Here, M== $\sum_i A_i n_i$ is the molecular weight of the material. A_i, n_i, and N_A denotes, respectively, the atomic weight, number of atoms of the ith constituent element and the Avogadro's number.Total (effective) electronic cross-section per electron, σ_e can be written as:

$$\sigma_e = \frac{1}{N_A} \sum_i \frac{f_i A_i}{Z_i} (\mu_m)_i, \text{Eq. (3)}$$

where, $f_i = \frac{n_i}{n}$ is the fractional abundance of ith element, Z_i and $(\mu_m)_i$ are the atomic number and mass attenuation coefficient of the ith constituent element, respectively. The effective atomic numberZ_{eff} of shielding material can then be defined as follows:

$$Z_{eff} = \frac{\sigma_a}{\sigma_e} \quad \text{Eq. (4)}$$

Effective atomic number is anuseful photon attenuation parameter that signifies the stopping power of the scintillator material against gamma-ray photons. Larger value of Z_{eff} is highly desirable for better interaction with material and provides effective attenuation of gamma radiations.Another parameter of immense importance is the mean free path (mfp).It is defined as the scintillator material thickness(in cm) that shall be sufficient to reduce the intensity of the beam to 37% of its original value at a given photon energy [10, 13].It is obtained from the attenuation coefficient using the following equation:

$$mfp = \frac{1}{\mu} \quad \text{Eq. (5)}$$

For present work, we shall study and compare various radiation attenuation parameters such as mass attenuation coefficient (μ_m), effective atomic number (Z_{eff}) and mean free path (mfp) *etc.*for selected inorganic scintillators computed using the recently developed online platform namely `py-MLBUF' [10].We have reported in **table 1** some of the recently developed inorganic crystals being used in HEP experiments along with their important physical and optical properties.

3. Results and Discussion

Mass attenuation coefficient (μ_m) determines the probability of interaction of a photon with absorber material via different interaction mechanisms that depends upon interaction mechanism at particularphoton energy. We display in **Fig. 1**(top panel),the mass attenuation coefficient (μ_m) values computed for selected inorganic scintillator materials as a function of photon energy. In photoelectric (PE) absorption energy domain, μ_m values are quite high for all the scintillation materials due to complete photon energy absorption. Beyond 100 keV photon energy, absorption process in the absorbing medium doesn't take place as often and μ_m value drops off quickly for higher photon energies. At lower energies, lead containing inorganic scintillators such as PWO and PbF_2 along with bismuth based `BGO' scintillators are observed to have best radiation attenuation efficiencies up till 400 keV. There are observed sharp jumps in the mass attenuation coefficient values in the PE energy domain owing to presence of absorption edges of constituent elements in the scintillator materials Beyond $E_\gamma > 400\ keV$, there is observed decline in the μ_m values and all curves converging to similar values. This happens due to moderate dependence of photon absorption cross-section (σ_{CE}) on the atomic number (Z) and inverse dependence on the photon energy in the Compton scattering (CE) energy regime. As seen in **Fig. 1(a)**,CsI bears the highest value of the mass attenuation coefficient value in the photon energy range 0.04 MeV to 0.1 MeV. For the energies greater than 100 keV and in the Compton scattering energy regime, scintillators such as PbF_2, PWO and BGO still exhibit better photon attenuation capabilities.Fig.1(b)displays the effective atomic number (Z_{eff}) varyingwith incident photon energy for theselected HEP scintillator materials. Z_{eff}parameter is dependent uponthe elemental composition of the scintillator material. Consequently, lead based scintillators *i.e.*PWO and PbF_2 possess the higher effective atomic number values throughout the photon energy range studied. For every scintillator material chosen other than CsI, there occurs exaggerated spikes inthe Z_{eff} curves in the PE energy domain that corresponds to the absorption edges of the constituent

TABLE 1. Important properties of inorganic scintillation crystals being used in conventional, recent and proposed High Energy Physics (HEP) experiments.

Scintillation crystal	Elemental composition (wt.%)	Light yield (photons/ MeV)	Decay time, τ (ns)	Radiation length (cm)	Density (g/cm^3)	Ref.(s)
PWO ($PbWO_4$)	45.5%Pb; 40.5%W 14.0 % O	250	5-15	0.89	8.28	[7, 14, 15]
PbF_2	84.5%Pb; 15.5% F	<1000	30, 6	0.93	7.77	[16, 17]
CsI	51.1%Cs; 48.8%I	16,800	10	1.86	4.51	[17, 18]
LYSO:Ce (0.05 mol.%)	57.5%Lu; 3.24%Y 24.63%Si; 14.6%O 0.01% Ce	30,000	40	1.14	7.11	[2, 5, 8, 15, 19]
LuAG:Ce (0.1 mol.%)	61.6%Lu; 15.8%Al 22.5%O; 0.02%Ce	25,000	820, 50	1.45	6.76	[5, 20]
BaF_2:Y (5 mol.%)	76.29%Ba; 21.1%F 2.6%Y	2,000	600, 0.5	2.03	4.89	[5, 9]
BGO ($Bi_4Ge_3O_{12}$)	67.1 %Bi; 17.5%Ge 15.4%O	8200	300	1.12	7.13	[18, 20]
NaI:Tl (0.3 mol.%)	15.27%Na; 84.31%I 0.41%Tl	40,000	230	2.59	3.67	[1, 20]

elements present in the scintillation materials. Except for CsI scintillator, the Z_{eff} values for all other materials tend to decrease with increase in photon energy in similar fashion to that for mass attenuation coefficient values.The CsI almost shows nearlyconstant value (≈ 54) which lies in between the atomic number of Cesium (Z = 55) and for Iodine (Z= 53) [21]. PbF_2, PWO and BGO scintillators show absorption edges around 100 keV, while absorption edges for Lutetium based LuAG:Ce and LYSO:Ce scintillators occurs around ≈ 60 keV photonenergy. Clearly one can infer that lead based PWO, PbF_2 andBGO scintillators possess higher Z_{eff} values than conventional NaI (Tl) scintillator.

Next, we study that *mean free path*(mfp) computed as function of photon energy for selected inorganic scintillators. A lower value of mfp is vital factor and highly desirable characteristic of the gamma-ray sensing device. The knowledge of *mfp* values across wide range of gamma-ray energies can guide us to employ crystal based scintillation detectors of required thickness in HEP experiments.The variation of mfp with respect to the photon energy is depicted in **Fig. 1(c)**.The mean free path (mfp) being inversely proportional to linear attenuation coefficient tends to increase with increase in the photon energy. All scintillators arefound to show smaller mfp valueswith exaggerated spikesin the lower energy regimethat corresponds to occurrence of absorption edges of the constituent elements. Afterwards, mfp values continue to rise smoothly with incident photon energy. For gamma-ray energies $E_\gamma \geq 1$ MeV, there is a gradual decrease in the curve slopes and the mfp values get saturated due to dominance of Compton scattering phenomenon. As expected, lead containingscintillators (PWO and PbF_2) and BGO possess lowest of mfp values, while conventional NaI (Tl) and CsI scintillators possess moderately higher mfp values over whole energy domain. Lutetium based LYSO:Ce and LuAG:Ce scintillators exhibit nearly the same mfp values as illustrated in **Fig. 1(c).** From this study, one can conclude that lead based scintillators (*i.e.* PWO and PbF_2) and bismuth based (BGO) inorganic scintillators are found to possess higher mass attenuation coefficient and lowermfp values, and thus may be regarded as the better attenuators of highly penetrating gamma-rays.

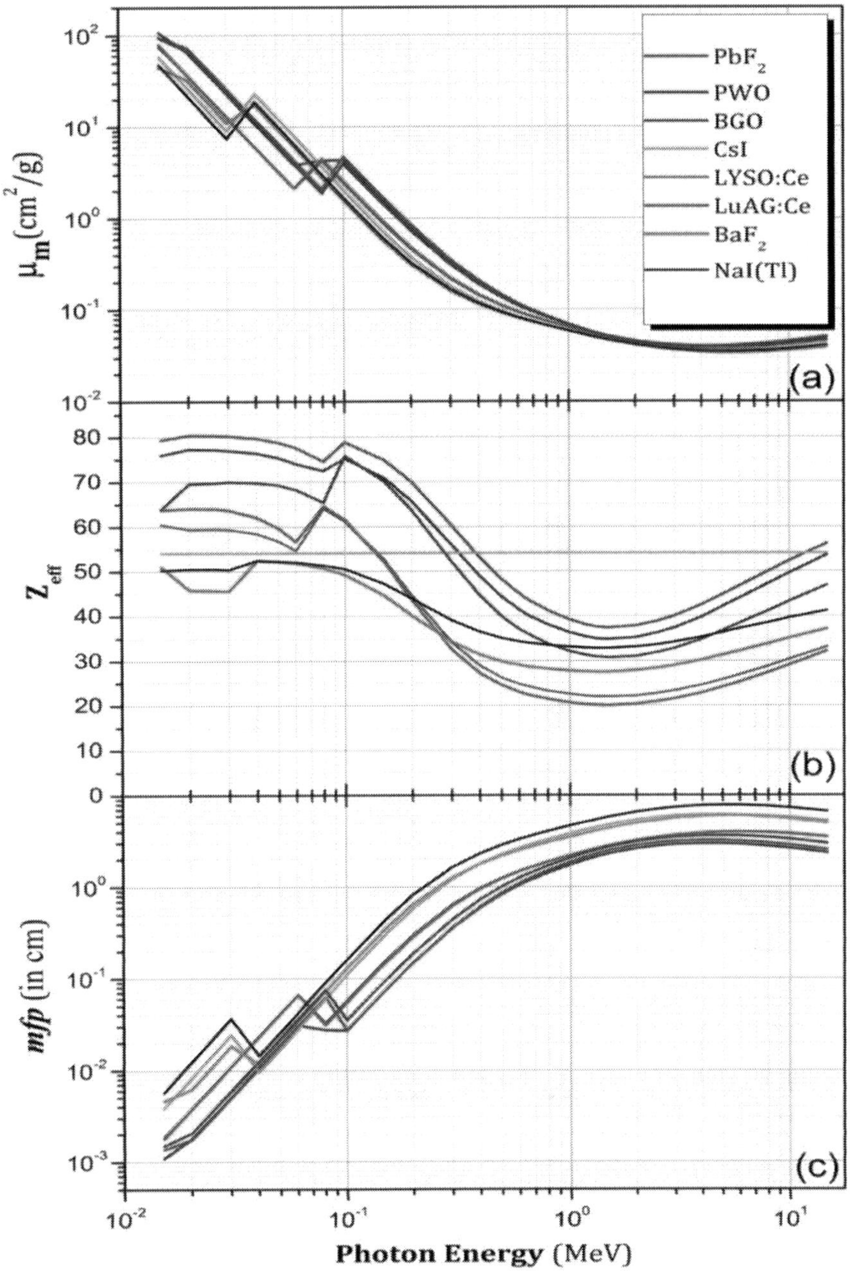

FIGURE 1. (a) Mass attenuation coefficient (μ_m); (b) Effective atomicnumber (Z_{eff}); and (c) mean free path (mfp) values computed asfunction of photon energy for inorganic scintillatormaterials.

4. Conclusion

Present study summarizes the gamma-ray interaction behavior analyzed forselected inorganic scintillators being used in recent HEP experiments.The lead-based scintillators (PbF_2 and PWO) are found to possess superior radiation attenuation capabilities than the conventional scintillation materials. Even though showing lesser Z_{eff}values, newly developed scintillators containing rare earth elements such as LYSO:Ceand LuAG:Cealso perform relatively better than traditional NaI(Tl)scintillator as can be seen in the calculations of $\mu_m$and mfpvalues.Future studies would be focused onthe extensive and in-depth analysisof gamma radiation shieldingand sensing properties of such advanced scintillator materials for exploring application in HEP experiments.

References

[1]Blasse G, 1994, Scintillator materials*Chem. Mater.***6**1465-1475.

[2]Martin N and Yoshikawa A 2015, Recent R&D trends in inorganic single crystal scintillator materials forradiation detection Adv. *Optical Mater.***3**463-481.

[3] DerenzoS E,WeberMJ,Bourret-Courchesne E,and KlintenbergMK, 2003, The quest for the idealinorganic scintillator*Nucl. Instrum. Methods Phys. Res.A***505**111-117.

[4] MelcherCL2005Perspectives on the future development of new scintillators *Nucl. Instrum. MethodsPhys. Res. A***537**6-14.

[5] HuC*et.,* 2018, Ultrafast inorganic scintillators forGHz hard X-Ray imaging*IEEE Trans. Nucl. Sci.***65(8)**2097-2104.

[6] Moses William W, 2002, Current trends in scintillator detectors and materials*Nucl. Instrum. MethodsPhys. Res. A***487**123-128.

[7] DujardinC,AuffrayE,BourretE,DorenbosP,LecoqP,NiklM,Vasil'EvAN,YoshikawaA,and ZhuR-Y, 2018, Needs, trends and advances in inorganic scintillators*IEEE Trans. Nucl. Sci.***65(9)** 1977-1997.

[8] ZhuR-Y, 2018, The next generation of crystal detectors*Radiat. Detect. Technol. Methods***2**2.

[9]HuC,XuC,Zhang L,ZhangQ,and ZhuR- Y 2019, Development of Yttrium-doped BaF_2crystals forfuture HEP experiments*IEEE Trans. Nucl. Sci.***66 (7)**1854-1860.

[10]MannK S and MannS S, 2021, Py-MLBUF: Development of an online-platform for gamma-ray shielding calculations and investigations*Ann. Nucl. Energy***150**107845.

[11]JacksonDFand HawkesDJ1981X-ray attenuation coefficients of elements and mixtures*Phys. Rep.***70(3)**169-233.

[12] Singh K,SinghH,SharmaV,NathuramR,KhannaA,KumarR, BhattiSS, and SahotaHS 2002, Gamma-ray attenuation coefficients in bismuth borate glasses*Nucl. Instrum. Methods Phys. Res. B***194**1-6.

[13]ŞakarE,ÖzpolatÖzgür F,AlımBünyamin,SayyedMI,andKurudirekM 2020, Phy-X / PSD:Development of a user friendly online software for calculation of parameters relevant to radiation shielding and dosimetry*Radiat. Phys. Chem.***166**108496.

[14] LecoqP*et al* 1995, Lead tungstate ($PbWO_4$) scintillators for LHC EM calorimetry*Nucl. Instrum. Methods Phys. Res. A***365**291-298.

[15] ZhuR-Y, 2019, Ultrafast and Radiation Hard Inorganic Scintillators for future HEP Experiments*J. Phys.: Conf. Ser.***1162**012022.

[16] AndersonDF,KobayashiM,WoodyCLandYoshimuraY, 1990, Lead Fluoride: An ultra-compactCherenkov radiator for em calorimetry*Nucl. Instrum. Methods Phys. Res. A***290**385-389.

[17] MaoR,ZhangL,and ZhuR-Y2011Crystals for the HHCAL detector concept*IEEE Nuclear Science Symposium Conference Record*p1911-1917;doi:10.1109/NSSMIC.2011.6154384.

[18] AnnenkovAA,KorzhikMV,and LecoqP2002Lead tungstate scintillation material*Nucl. Instrum. Methods Phys. Res. A***490**30-50.

[19] CookeDW,McClellanKJ,BennettBL,RoperJM,WhittakerMT,and MuenchausenRE, 2000, Crystal growth and optical characterization of cerium-doped $Lu_{1.8}Y_{0.2}SiO_5$*J. Appl. Phys.* **88**7360.

[20] HuC,LiJ,YangF,JiangB,ZhangL,andZhuR-Y, 2020, LuAG ceramic scintillators for future HEP experiments*Nucl. Instrum. Methods Phys. Res. A***954**161723.

[21] KumarTKand ReddyKV1997Effective atomic numbers for materials of dosimetric interest *Rad. Phys. Chem.***50(6)**545-553.

IOP Conf. Series: Materials Science and Engineering 1221 (2022) 012003 doi:10.1088/1757-899X/1221/1/012003

Evaluating Shielding Materials for High Energy Space Radiation

Ankit Gohel[1,2*] **,Rajnikant Makwana**[2,3] **and Bhargav Soni**[2]

[1]Department of Physics, Government Arts and Science College, Patdi, Saurashtra University, Rajkot, 382765, India
[2]Department of Physics, Faculty of Science, The M. S. University of Baroda, 390002, India
[3]Departmental of Fundamental Physics, University of Salamanca, 37007, Spain
E-mail address: ankitphysics007@yahoo.com

Abstract. Space radiation is very hazardously harmful to human health and physically damaging to electronics equipment. The major radiation sources are Solar Particle Events (SPE) and Galactic Cosmic Rays/Radiation (GCR). GCR composed of high energy protons, alpha particles and electrons travelling from outside solar system are especially difficult to shield. Previous study has been shown that aluminum and higher atomic number structural alloys, provide fairly little shielding which may increase secondary radiation. In this work, we evaluate the range of materials for GCR shielding. Materials are aluminum, for its low atomic number, Liquid Hydrogen, Polyethylene,Water, Polybenzoxazine, Lithium hydride, Liquid Methane and Hydrogenated Graphite Nano fiber (HGNF) materials which containing high proportions of hydrogen which effectively slowing secondary neutrons and GCR radiation. Boron nitride storing hydrogen absorbs the secondary neutrons, GCR, and SPE effectively. Liquid Methane performance is also noticeable. The dose equivalent for each shielding materials are calculated using On-line Tool for the Assessment of Radiation In Space (OLTARIS) and High Charge and Energy TRaNsport (HZETRN) and compared for 2010 Solar minimum GCR. The results show that Liquid hydrogen is the good space radiation shielding material for a given thickness, followed by Lithium Hydride, Liquid Methane, Boron nitride storing hydrogen, Polyethylene, Polybenzoxazine, HGNF, and then aluminum. In addition, we plotted dose equivalent for Liquid hydrogen and Polyethylene at different densities.

1. Introduction

The high energy from cosmic environment can create biological hazards to humans and physical damages to electronics. Solar particle events (SPE) and Galactic cosmic radiation (GCR) are the main source of such radiation. The SPE, Solar flares, and Coronal Mass Ejection (CME) are having constitutes of electrons, protons and heavy nuclei. GCR contains high energetic protons, alpha particles, electrons and nuclei of atomic number ($Z>2$) which travels nearly speed of light from outside the solar system with energies ranging up to 10 GeV/nucleon or more [1]. In Solar minimum, GCR particles accelerated by shock waves, which penetrates deeply in the solar system due to less solar activity. Additionally when these radiations pass through spacecraft material it creates secondary radiation. The GCR ions and secondary neutrons interact with molecules in cells, alter DNA structure, and can cause severe damage to the human body.The National Council on Radiation Protection and Measurements (the NCRP) has thus recommended the career dose limits forastronauts. Without improved shielding materials, there are severe limitations forspace exploration into deeper space and for longer time periods [1].

The important reason for considering GCR simulation is to sending astronauts on long duration space missions. In previous studies, Aluminum was widely used in spacecraft for its relatively low atomic number, light weight and structural stability [1].Hydrogen is the lightest element and has no neutrons. Hydrogen is effective at (1) fragmenting heavy ions such as are found in GCR, (2) stopping protons such are found in SPE, and (3) slowing down neutrons such as are formed as secondary when the GCR and SPE interact with matter [2]. Polyethylene, Water, Liquid Hydride contains hydrogen concentration. Boron and Nitrogen have largest neutron absorption cross section and hydrogen stored boron nitride absorbs the secondary neutrons.Hydrogen-rich benzoxazine material showed shielding efficiency comparable to Polyethylene [3].There has been numerous dose reduction studies performed over years for types of space missions using transport codes [1-7].

For this study, we used OLTARIS [8] and HZETRAN 2015 [9-16] to simulate dose equivalent in mSv/year for various materials with various thickness in g/cm^2. The simulation was performed using 2010 solar minimum GCR space radiation environment [16].

2. Materials and Methods

2.1. Materials

Dose equivalent was calculated using various shielding materials with thickness distributions ranging from 5 to 100 g/cm^2. The materials analysed included aluminium (Al, 2.7 g/cm^3), Liquid Hydrogen (LH_2, 0.0708 g/cm^3), Polyethylene (CH_2, 1 g/cm^3), Water (H_2O, 1 g/cm^3), Polybenzoxazine ($C_{36}H_{56}N_2O_4$, 1.07 g/cm^3), Lithium hydride (LiH 0.82 g/cm^3), Liquid Methane (CH_4. 0.423 g/cm^3), HGNF (2.25 g/cm^3), and Boron Nitride (BN, 2.1 g/cm^3) respectively.

2.2. GCR Models

The GCR spectrum at one astronomical unit (AU; where 1 AU = 93 million miles) from the sun was applied as the boundary condition in the simulations which simulates the GCR radiation environment in the vicinity of the earth but outside the protective magnetic field comprising the Van Allen belts [17]. The GCR model used as inputs to OLTARIS and HZETRN during 2010 solar minimum in units of flux (MeV^{-1} cm^{-2} s^{-1}). We used web-based tool of OLTARIS to perform the dose equivalent analysis. OLTARIS consists of two features: a web-based interface and an execution environment. It allows the geometry configurations to be specified as slabs, spheres, or complicated spacecraft, habitat, exploration vehicles and space suits [18]. HZETRN 2015 is the deterministic transport code developed at NASA Langley Research Center [9-16]. The HZETRN code is he engine behind Tool for the Assessment of Radiation in Space (TARIS) which is a deterministic space radiation analysis tool.

2.3. Simulation Geometry

In this study, various shielding materials are simulated using slabs mode with the material thickness of the total shielding ranging from 5 to 100 g/cm^2 for 2010 solar minimum GCR environment. In analysis, each slab is 5 g/cm^2 in both OLTARIS and HZETRN transport codes. Practical parameter of physical thickness of the shielding materials; e.g., 50 g/cm^2 of aluminum will be about 18.5 cm thick, whereas an equal thickness of polyethylene will be about 50 cm thick. Aluminum, Liquid Hydrogen, PE, Boron Nitride, hydrogen stored boron nitride, Polybenzoxazine, water, lithium hydride, liquid methane, and HGNF are analyzed to estimate dose equivalent. In addition, comparisons of dose equivalent results between OLTARIS and HZETRN.

3. Results and Discussion

Characteristics of the effective shielding materials are slow down the incident space radiation; reduce the reproduction of secondary radiation; and light weight respectively.

3.1. Dose Equivalent Analysis

The results simulated by OLTARIS to estimate dose equivalent for various shielding materials using slab mode are presented in Figure 1.When comparing shielding materials, a more effectivedose equivalent reduction explains that greater energy loss of the incident radiation. In Figure 1, aluminium drops slowly, whereas Liquid hydrogen drops effectively faster, followed by Lithium hydride, Liquid methane, BN+20 wt% H_2, BN+15 wt% H_2, Polyethylene, BN+10 wt% H_2, Polybenzoxazine, Water, HGNF, BN+5 wt% H_2, Boron nitride (BN), and aluminium. Therefore, Liquid hydrogen is more effective shielding material against GCR. Lithium hydride and liquid methane are the next suitable shielding material. Polyethylene, water, and hydrogen stored BN provide better dose equivalent reduction than aluminium, but are less effective than liquid hydrogen against GCR. In order to absorb the secondary neutron radiation, hydrogen stored boron nitride is selected for shielding. By increasing the hydrogen content to boron nitride, produced secondary neutron radiation decreases.Shielding capacity of hydrogen-rich benzoxazine (Polybenzoxazine) and Polyethyleneare comparable to each other. HGNF is significantly better than boron nitride plus 5% by weight, boron nitride and aluminium.

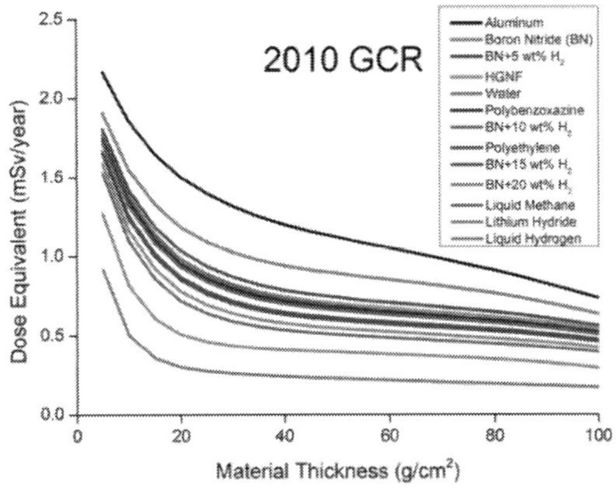

Figure 1. Annual dose equivalent for different shielding materials of varying thickness against 2010 solar minimum GCR using OLTARIS

Figure 2 present inter-code comparisons between OLTARIS and HZETRN for dose equivalent at all depths using various shielding materials. Both the OLTARIS and HZETRN transport codes use the ICRP 60 [19].The very short lived particles, secondary neutrons and protons, are most penetrating particles and hence the dose contributions from these are more noticeable at deeper depths. The dose equivalent values calculated from HZETRN codes are higher for all materials when compared to OLTARIS. The contribution of low energy neutrons to dose equivalent is surpassed only by the primary protons. In Figure 2, there is a good qualitative agreement between the transport codes at low thickness regions, whereas there are significant differences between the codes at high thickness regions. In Figure 2, both transport codes dose equivalent of Aluminum, Lithium hydride, Liquid Methane, and water are comparatively matched up to 20 g/cm^2 thickness. HZETRN estimated dose equivalent of Liquid hydrogen, Polyethylene, Polybenzoxazine, BN+20 wt% H_2, and HGNF are higher than OLTARIS.

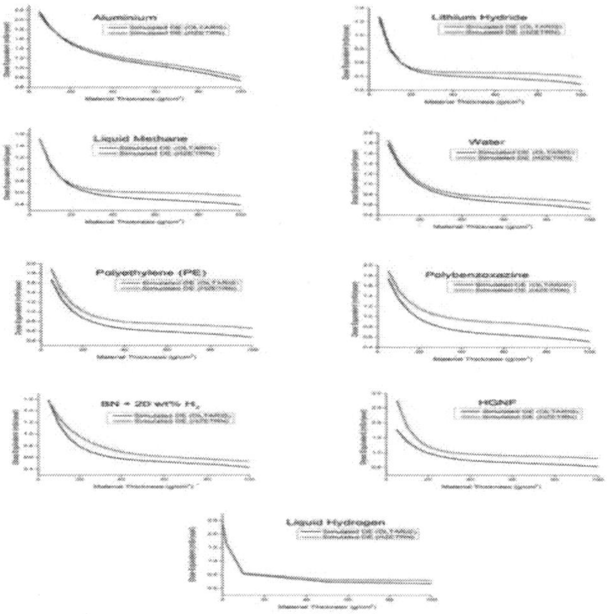

Figure 2. Comparisons of dose equivalent between OLTARIS and HZETRN against 2010 solar minimum GCR

Table 1 and Figure 3 provide the comprehensive densities data and dose equivalent for Liquid hydrogen and Polyethylene against 2010 solar minimum GCR from HZETRN.Various densities of liquid hydrogen at different temperatures (K) are given in Table 1. Low Density Polyethylene (LDPE, 0.91 g/cm^3), Polyethylene (0.92 g/cm^3, 0.941 g/cm^3, 1 g/cm^3), High Density Polyethylene (HDPE, 0.95 g/cm^3), and Ultra High Molecular Weight Polyethylene (UHMWPE, 0.97 g/cm^3) are also simulated. For Liquid hydrogen, dose equivalent is effectively reduced at density of 0.069 g/cm^3. In addition, dose equivalent drops faster at density of 0.91 g/cm^3 for Polyethylene.This analysis concludes that materials optimized at low density reduced the GCR exposures and slow down the secondary neutrons.

Table 1. List of the different densities of the Liquid hydrogen and Polyethylene

Liquid Hydrogen Density (g/cm^3)	Polyethylene (PE) Density (g/cm^3)
0.06900	0.9100
0.07088	0.9200
0.07099	0.9410
0.07100	0.9500
0.07114	0.9700
0.07229	1.0000
0.07300	--
0.07647	--
0.07687	--
0.07691	--
0.07700	--
1.00000	--

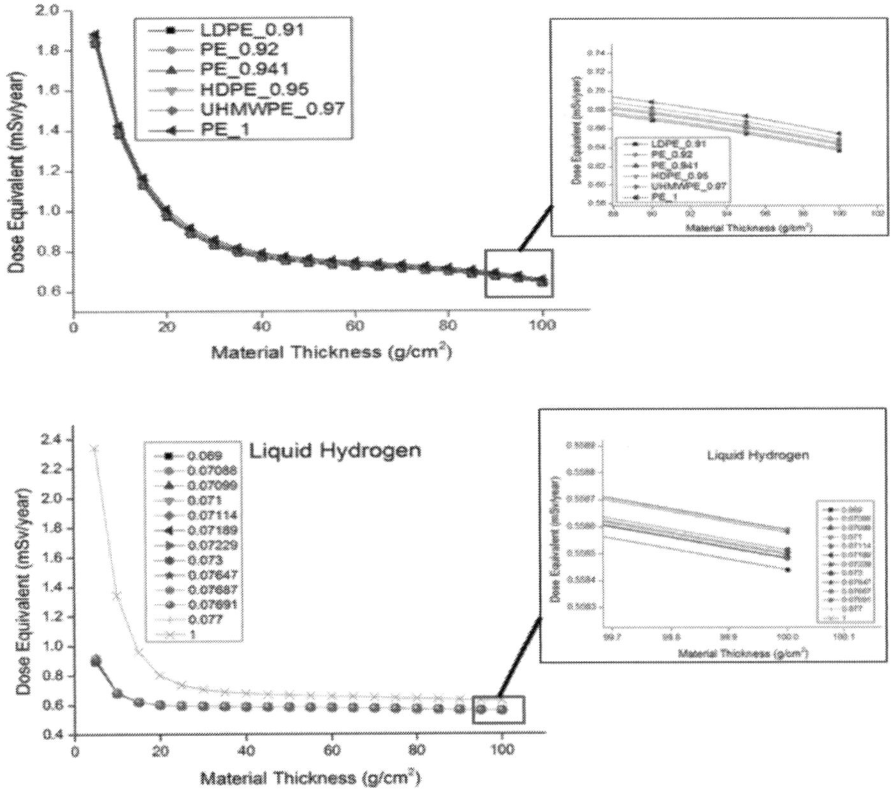

Figure 3. Dose equivalent of Polyethylene and Liquid Hydrogen at different densities using HZETRN for 2010 solar minimum GCR

4. Conclusions

In this study, we have used OLTARIS and HZETRN transport codes to simulate the dose equivalent of various shielding materials using slab mode for 2010 solar minimum GCR. Liquid hydrogen was found to be the effectively good space radiation material for a given thickness, followed by Lithium hydride, Liquid methane, hydrogen stored boron nitride, Polyethylene, Polybenzoxzine, water, HGNF, Boron nitride, and aluminium. Comparison of dose equivalent from OLTARIS and HZETRN hasshown that there are many regions where the codes are comparatively matched, and some regions where the results diverge greatly. The optimized densities effectively reduced dose equivalent at 0.069 g/cm^3 for liquid hydrogen and at 0.91 g/cm^3 for LDPE respectively. In future study, multi-layered materials may provide better radiation protection against GCR.

Acknowledgments

The authors are thankful to OLTARIS and HZETRN transport codes which provided by NASA.

References

[1] Xindi Li 2017 *Simulation of Shielding Materials against Galactic Cosmic Rays* Rice University

[2] Thibeault S A, Fay C C, Lowther S E, Earle K D, Sauti G, kang J Ho, Park C and Mcmullen A M 2012 Radiation Shielding Materials Containing Hydrogen, Boron, and Nitrogen: Systematic Computational and Experimental Study – Phase I NIAC Final Report

[3] Iguchi D, Ohashi S, Abarro G J E, Yin X, Winroth S, Scott C, Gleydura M, Jin L, Kanagasegar N, Lo C, Arza C R, Froimowicz P and Ishida H 2018 Developmant of Hydrogen-Rich Benzoxazine Resins with low Polymerization Temperature for Space Radiation Shielding *ACS Omega* **3**pp 11569-11581

[4] Badavi F F, Blatting S R, Atwell W, Nealy J E and Norman R B 2011 A detreministic electron, photon, prpton and heavy ion transport suite for the study of the jovian moon Europa *Nucl. Instrum. Methods Phys. B* **269** pp 232-238

[5] Singleterry R C 2013 Radiation enegineering analysis of shielding materials to assess their ability to protect astronauts in deep space from enegetic particle radiation *Acta Astronautica***91** pp 49-54

[6] Atwell W 2012 ExtraVehicular Activity (EVA) Operation in the Geostationary Environemnt *42nd, Int., Conf.,on Environmental Systems* 15-19 July 2012 San Diego Califomia

[7] Mansur L K, Frame B J, Gallego N C, Guetersloh S B, Johnson J O, Klett J W and Townsend L W 2005 Assessment of Shielding Materials Performance for Deep Space Missions *Mater. Res. Soc. Symp. Proc.* 851

[8] Singleterry R C, Blattnig S R, Clowdsley M S, Qualls G D, Sandridge C A, Simonsen L C Norbury J W Slaba T C, Walker S A, Badavi F F, Spangler J L, Aumann A R, Zapp E N, Rutledge R D, Lee K T and Norman R B 2011 OLTARIS: on-line Tool for the Assessment of radiation in space *Acta Astronaut.* **68** pp 1086-1097

[9] Wilson J W, Slaba T C, Badavi F F, Reddell B D and Bahadori A A 2014 Advance in NASA radiation transport research: 3DHZETRAN *Lifr Sci. Space Res.* **2** pp 6-22

[10] Wilson J W, Slaba T C, Badavi F F, Reddell B D and Bahadori A A 2015a 3DHZETRN: Shielding ICRU Spherical Plantom *Lifr Sci. Space Res.* **4** pp 46-61

[11] Wilson J W, Slaba T C, Badavi F F, Reddell B D and Bahadori A A 2015bSolar Proton Transport within an ICRU Sphere Surrounded by a Complex Shield: Combinatorial Geometry NASA TP 2015-218980

[12] Slaba T C, Wilson J W, Badavi F F, Reddell B D and Bahadori A A 2015b Solar Proton Transport within an ICRU Sphere Surrounded by a Complex Shield: Ray-Trace Geometry NASA TP 2015-218994

[13] Norman R B, Slaba T C and Blattnig S R 2013 An Extension of HZETRN for cosmic ray initiated electromagnetic cascades *Adv. Space Res.* **51** pp 2251-2260

[14] Slaba T C 2013 Faster Heavy Ion Transport for HZETRN NASA Technical Paper 2013-217803

[15] Slaba T C Blattnig S R and Tweed J 2012 Reduced Discretization Error in HZETRN *Journal of Computational Physics* **234** pp 217-229

[16] O'Neill P M, Golge S and Slaba T C 2015 Badhwar – O'Neill 2014 Galactic Cosmic Ray Flux Model Description NASA Technical Paper 2015-218569

[17] Li X, Warden D and Bayazitoglu Y 2018 Analysis to Evaluate Multilayer Shielding of Galactic Cosmic Rays *Journal of Thermodynamics and Heat Tranfer* **32(2)**pp 525-531

[18] Aghara S K, Sriprisan S I, Singleterry R C and Sato T 2015 Shielding evaluation for solar particle events using MCNPX, PHITS and OLTARIS codes *Lifr Sci. Space Res.* **4** pp 79-91

[19] ICRP, Internalnational Commission on Raiological Protection 1991 The Recommendations of the International Commission on Radiological Protection. Annals of the ICRP 21 1-3

IOP Conf. Series: Materials Science and Engineering 1221 (2022) 012004 doi:10.1088/1757-899X/1221/1/012004

Synthesis and Investigation of Ferroelectric Curie Transition in $BaTiO_3$

Ajay Kumar, Sudip Naskar and Dipankar Mandal*
Quantum Materials and Devices Unit, Institute of Nano Science and Technology,
Knowledge City, Sector 81, Mohali 140306, India
Email: 1234ajaysingh01@gmail.com

Abstract: Ferroelectric phase comprising $BaTiO_3$ nanoparticles and nanorods have been prepared by high temperature sintering. The XRD spectra of sintered powder was used to confirm the ferroelectric phase crystallisation. The non-ferroelectric cubic phase to ferroelectric tetragonal phase transition of $BaTiO_3$ was studied by temperature dependent dielectric study. The Curie temperature of 105 °C in sintered nanoparticles was observed from dielectric study of tetragonal phase $BaTiO_3$

1. Introduction:

Ferroelectric materials are point of interest due to piezoelectric and pyroelectric properties of these materials. Barium titanate ($BaTiO_3$) is well known ferroelectric material with very high piezoelectric and pyroelectric coefficient.[1]$BaTiO_3$ mainly exists cubic, orthorhombic, rhombohedral and tetragonal phases.[2,3] Out of these Tetragonal is most ferroelectric and cubic is non-ferroelectric phase of $BaTiO_3$.[4] There are different approaches such as Hydrothermal, solid-state processing and vapour deposition technique used to prepare ferroelectric phase $BaTiO_3$.[5–7] The shape and size of material can be controlled in hydrothermal process but it is not possible to achieve required phase in case of $BaTiO_3$ .There are various literature showing synthesis process for different phase of $BaTiO_3$, still it is a tedious task to prepare ferroelectric phase of $BaTiO_3$. Here, we have synthesised ferroelectric phase $BaTiO_3$ nanoparticles and nanorods by two step processes and did temperature and frequency dependent dielectric study to investigate the relaxation mechanism and ferroelectric to paraelectric phase transition (Curie temperature) in $BaTiO_3$.

2. Experimental Section:

2.1. Raw material:

Barium chloride ($BaCl_2.2H_20$), titanium dioxide (TiO_2), barium acetate ($Ba(OOCCH_3)_2$) and titanium isopropoxide ($Ti(OCH(CH_3)_2)_4$) were supplied by Sigma Aldrich, NaOH was supplied by Merk and Titanium isopropoxide was supplied by TCI chemicals.

2.2. $BaTiO_3$ nanoparticles preparation:

Initially cubic phase $BaTiO_3$ were synthesised by hydrothermal process, where $BaCl_2.2H_2O$, TiO_2 and NaOH were used as precursor material. As prepared cubic phase $BaTiO_3$ nanoparticles were further sintered at 1200 °C for 4 h with ramp rate of 4 °C/min for ferroelectric phase crystallization.

2.3. $BaTiO_3$ nanorods preparation:

$BaTiO_3$nanorods were prepared by high temperature calcination of electrospinning fibres. To prepare the electrospinning solution, 1M barium acetate was added to 3ml acetic acid and kept on hot plate at 40 °C to prepare uniform solution. Titanium isopropoxide was added dropwise to control the Ba/Ti ion ratio. Separate solution of 2g PVP in 3ml of ethanol was prepared and both solutions were mixed together and kept on hot plate 40 °C for 2h. The solution was loaded in syringe and electrospinning process was carried out to by applying 9.2 kV voltage between tip and collector. Flowchart for the synthesis of nanorod and nanoparticles is shown in figure 1. The electrospinning fiber was collected on drum collector and dried overnight to remove the solvent. As prepared fiber was further sintered at different temperature for crystallization of $BaTiO_3$.

2.4. Characterisation:

The crystallographic study of $BaTiO_3$ powder and fiber was done by X-ray diffractometer (D8 Bruker Advance) with CuK_{α}, X-ray radiation (0.154178 nm) in the 2θ range of 20 – 80°. Keysight 49910E impedance analyser was used to record the temperature and frequency dependent dielectric properties.

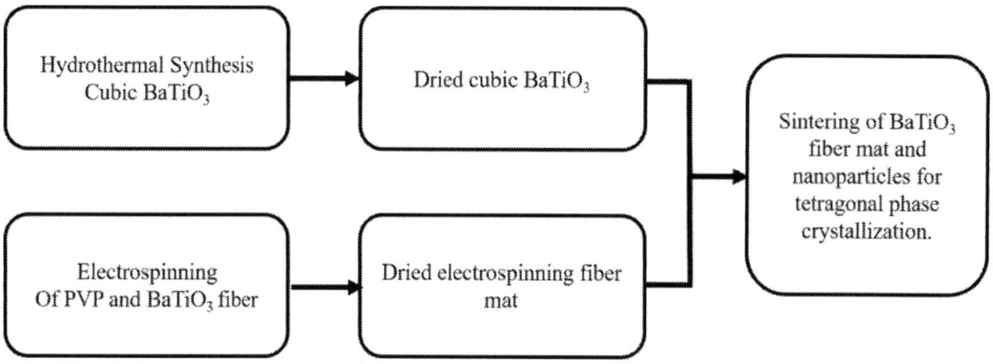

Figure 1. Flowchart of $BaTiO_3$ nanoparticles and nanorod synthesis.

3. Result and discussion:

The crystallographic study of $BaTiO_3$ nanoparticles and nanorod was done by X-ray diffraction technique. The XRD pattern in figure 2a, indicates that $BaTiO_3$ prepared by hydrothermal process crystallize in cubic phase due to appearance of (202) plane.[8] While after calcination the appearance of (200) and (002) plane indicates the tetragonal phase crystallization in $BaTiO_3$.[9] XRD pattern of as prepared electrospinning fiber in figure 2b don't show any $BaTiO_3$ peak, while with increase in calcination temperature above 500 °C crystallisation

Process of $BaTiO_3$ take place. The decrease in intensity of planes in tetragonal phase is due to change in particle and grain size of $BaTiO_3$.

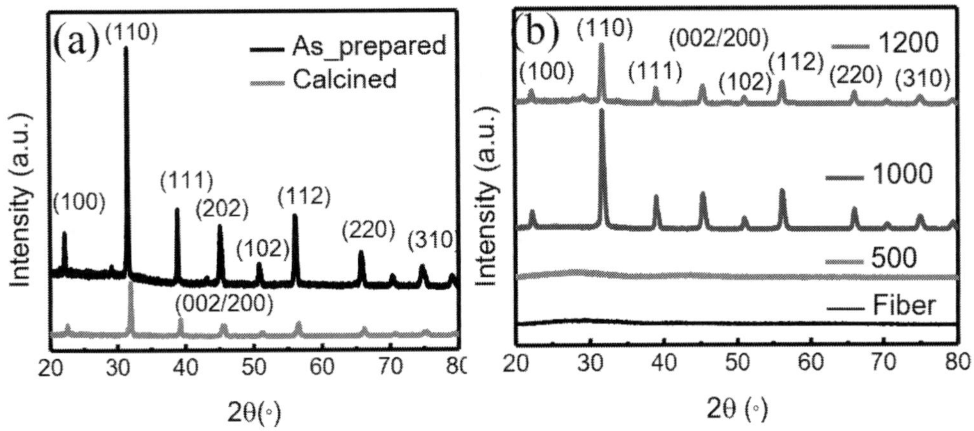

Figure 2.XRD pattern of (a) $BaTiO_3$ nanoparticles prepared by hydrothermal process and sintered at 1200 °C for tetragonal phase crystallisation (b) electrospinning fiber of $BaTiO_3$ and PVP fibers sintered at 500, 1000 and 1200 °C for 4h.

The temperature dependent dielectric study confirms the ferroelectric phase crystallisation in $BaTiO_3$. Prepared tetragonal phase particles were used to prepare pellet for temperature dependent dielectric study. Frequency dependence of dielectric constant is shown in figure 3a and it can be observed that lower frequency region shown higher value of dielectric constant and decreases with increase in frequency. This increase in dielectric constant in lower frequency region is due to interfacial polarisation created in pellet.[10]Figure 3b indicates that ferroelectric to paraelectric phase transition is observed at 105 °C in sintered pellet while no such transition is observed in powder prepared by hydrothermal process. Duong et.al. reported that Curie temperature of $BaTiO_3$ is 120 °C and varies with change in grain size. Hence, the observed change in Curie transition due change in grain size of $BaTiO_3$ powder and observation of curie transition in sintered powder confirms the ferroelectric phase crystallisation.[11,12]

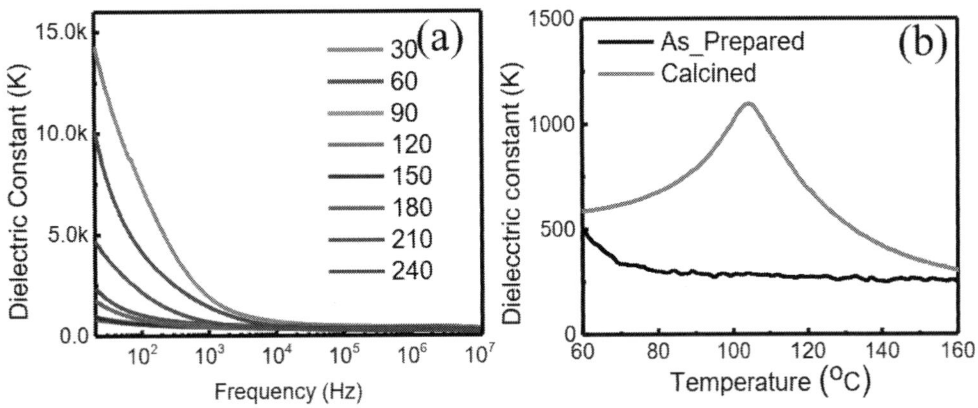

Figure 3. Variation of dielectric constant(a) withfrequency at temperature varying from 30 to 240 °C in ferroelectric phase $BaTiO_3$and (b)withtemperature in both non ferroelectric and ferroelectric $BaTiO_3$nanoparticles at 100 Hz.

Conclusion:

In this study, we have prepared tetragonal phase $BaTiO_3$ nanoparticles and nanorods to investigate the temperature dependent dielectric properties.The tetragonal phase crystallisation of $BaTiO_3$at high temperature sintering was confirmed by XRD and temperature dependent dielectric study.

References:

[1] Hoshina T, Hatta S, Takeda H and Tsurumi T 2018 Grain size effect on piezoelectric properties of $BaTiO_3$ ceramics *Jpn. J. Appl. Phys.***57**0902.

[2] Jiang B, Iocozzia J, Zhao L, Zhang H, Harn Y W, Chen Y and Lin Z 2019 Barium titanate at the nanoscale: Controlled synthesis and dielectric and ferroelectric properties *Chem. Soc. Rev.* **48** 1194–228.

[3] Kwei G H, Lawson A C, Billinge S J L and Cheong S W 1993 Structures of the ferroelectric phases of barium titanate *J. Phys. Chem.***97** 2368–77.

[4] Zhang Q, Cagin T and Goddard W A 2006 The ferroelectric and cubic phases in BaTiO3 ferroelectrics are also antiferroelectric *Proc. Natl. Acad. Sci. U.S.A.***103** 14695–700.

[5] Chen C, Wei Y, Jiao X and Chen D 2008 Hydrothermal synthesis of $BaTiO_3$: Crystal phase and the Ba2+ ions leaching behavior in aqueous medium *Mater. Chem. Phys.* **110** 186–91.

[6] Simon-Seveyrat L, Hajjaji A, Emziane Y, Guiffard B and Guyomar D 2007 Re-investigation of synthesis of $BaTiO_3$ by conventional solid-state reaction and oxalate coprecipitation route for piezoelectric applications *Ceram. Int.* **33** 35–40.

[7] Wang T H, Hsu P C, Korytov M, Genoe J and Merckling C 2020 Polarization control of epitaxial barium titanate ($BaTiO_3$) grown by pulsed-laser deposition on a MBE-SrTiO3/Si(001) pseudo-substrate *J. Appl. Phys.***128** 104104.

[8] Selvaraj M, Venkatachalapathy V, Mayandi J, Karazhanov S and Pearce J M 2015 Preparation of meta-stable phases of barium titanate by Sol-hydrothermal method *AIP Adv.***5** 117119.

[9] Zhu D, Liu H, Luo H, Sun S and Chen J 2020 Tetragonal phase and enhanced depolarization temperature in Ba-rich (Bi,Na)TiO_3–$BaTiO_3$ lead-free piezoelectrics *Ceram. Int.***46** 3708–14.

[10] Vayunandana Reddy Y K and Mergel D 2007 Frequency and temperature-dependent dielectric properties of $BaTiO_3$ thin film capacitors studied by complex impedance spectroscopy *Phys. B Condens. Matter***391** 212–21.

[11] Duong N X, Bae J S, Jeon J, Lim S Y, Oh S H, Ullah A, Sheeraz M, San Choi J, Ko J H, Yang S M, Kim K H, Kim I W, Ahn C W and Kim T H 2019 Polymorphic phase transition in $BaTiO_3$ by Ni doping *Ceram. Int.***45** 16305–10.

[12] Zhao Z, Buscaglia V, Viviani M, Buscaglia M T, Mitoseriu L, Testino A, Nygren M, Johnsson M and Nanni P 2004 Grain-size effects on the ferroelectric behavior of dense nanocrystalline $BaTiO_3$ ceramics *Phys. Rev. B Condens. Matter***70.**

IOP Conf. Series: Materials Science and Engineering 1221 (2022) 012005 doi:10.1088/1757-899X/1221/1/012005

Real-time detection of 1-butanol VOCin an indigenously developed Gas chamber

Pratima Bhat[1,2] and Naveen Kumar S.K[1]

[1]Department of Electronics, Mangalore University, Mangalagangotri, Mangalore-574199

[2]Department of Electronics, JyotiNivas College Autonomous, Bangalore-576090

Email.id: pratimabhat1@rediffmail.com

Abstract: In this paper, we discuss the sensing studies of 1-butanol volatile organic compound (VOC) in an indigenously developed gas chamber. Designing of the portable gas chamber is discussed with the integration of thin-film ZnO sensor on an Inter-digitated electrodefabricated on a flexible Teflon substrate. Exposure of 1-butanol VOC inside the chamber causes a change in the resistance of the ZnO thin film. This change in the resistance is converted to voltage with a Wheatstone bridge network and an instrumentation amplifier circuit designed using precision components.The sensitivity of the sensor for 50 ppm and 100 ppm of 1-butanol is tested at room temperature (27°C) and the results are compared with our early studies using Keithley instruments.

1. Introduction

There is an escalating need to develop gas sensors to monitor all aspects of our environmentwhich has a concern with pollution, health and safety. Toxic gases like H_2S (Hydrogen sulfides), oxides of Nitrogen (NO, NO_2, N_2O), and VOCs like methanol, benzene, toluene, xylene, propanol, butanol are hazardous substances that are released intothe air as waste from chemical industries and by automobile engines. They aredangerous to human health in the short term and the long term can lead toglobal warming [1, 2]. In the health care sector, gas/VOC sensors can sense the biomarker for the early diagnosis of several diseases[3].Semiconductor gas sensors are energy control type of sensor, as the physical or electrical properties of a sensing material change upon exposure to gas molecules. This change is transmitted to external device as electric energy with the help of electrode connected to it [4]Thesensingof gas involvesreceptor, interface and transducer phenomenon. When the sensing film is exposed to air, it interacts with oxygen molecules in the air and acts as a receptor. The oxygen molecules gently diffuse and take electrons from the grain boundary of the film and are converted into oxygen anions. This interface mechanism leads to the loss of electrons causes a depletion region or space charge region across the grain boundary of the sensing layer which leads to a change in the resistance at grain boundaries [5]. This resistance further changes upon exposure to target gas, depending upon the type of semiconductor material and type of target gas used. Most of the chemical or physical processes occur at the sensor's surface and that is influenced by the design aspects of the testing chamber. Studying reliable and stable results from a sensor is not possible before optimizing all design aspects of the testing chamber [6, 7]. There are very few testing centers available for the researchers, andhave restrictions for sensing different kinds of gas or VOC due to contamination or corrosive problems. Most of these chambers are bulky because ofthe complex vacuum process. Testing requires manual effort and consumes valuable time which serves as the motivation here to develop a portable and economical gas sensing chamber and measurement system.

IOP Conf. Series: Materials Science and Engineering 1221 (2022) 012005 doi:10.1088/1757-899X/1221/1/012005

This paper briefly describes the gas chamber design and gas measurement system that includes the signal acquisition circuit, signal processing circuit and amplifier circuit, with the help of which sensor's output is processed and displayed through the LED.

2. Materials and Methods

2.1 Design and development of Gas chamber

The proposed test chamber providesa study of static measurement of the gas/VOC. Figure 1(a) shows the CAD design of the chamber and figure 1(b) is the developed gas chamber. Table1 lists the functionality and parameters of the indigenous gas chamber.

Table 1: ***Components of the Gas chamber –Functionality and Parameters.***

Sl.no	Components	Functionality	Parameters
1.	Chamber	It's thermally conducting and electrically insulator, not reactive with any gas, non-corrosive.	21 cm –diameter, 5ltr – volume capacity Make- Indallium
2.	Specimen holder	Holds the sensor	Polished Steel test bench, Dimensions-70X 50 X 20 mm
3.	Specimen heater Pad	Attached to specimen holder and provides temperature to the specimen film	Tubular pencil heater - 2 Nos. Diameter -8mm Length-60 mm Capacity -70 watt
4.	Solvent holder	Holds the solvent of the target VOC	Polished curved crucible, Size-50X50X10 mm.
5.	Solvent heater (evaporator) Pad	Attached to the solvent holder, to evaporate the target solvent	Tubular pencil heater-2 Nos. Diameter -8mm Length-60 mm Capacity -70 watt
6.	Thermostat_2	Measures temperature of specimen heater. This temperature sets the sensor's operating temperature.	6 mm size (Elcon Make) 30°C to 300 °C
7.	Thermostat_1	This temperature evaporates the solvent into its vapor state.	6 mm size (Elcon Make) 30°C to 300 °C
8.	Control Panel	Housing the thermostats and ON/OFF power switch	MS sheet fabricated box
9.	Rotary switches	Sets the temperature of thermostat_1 and thermostat_2 and controls theHeater	Range - 0°C to 300 °C Range - 0°C to 300 °C
10.	DC Voltage testing terminals (Electrodes)	Placed on the IDE of the sensor, to measure the variations in the resistance of sensing material.	Adjustable rod of 8mm diameter insulated outside connected to the outer Brass bolt assembly
11.	PU tube_1	Inlets the gas from the cylinder act as a flow controller.	6 mm diameter with anopen and closing valve
12.	PU tube_2	Outlets the gas from the chamber	6 mm diameter with an open and closing valve

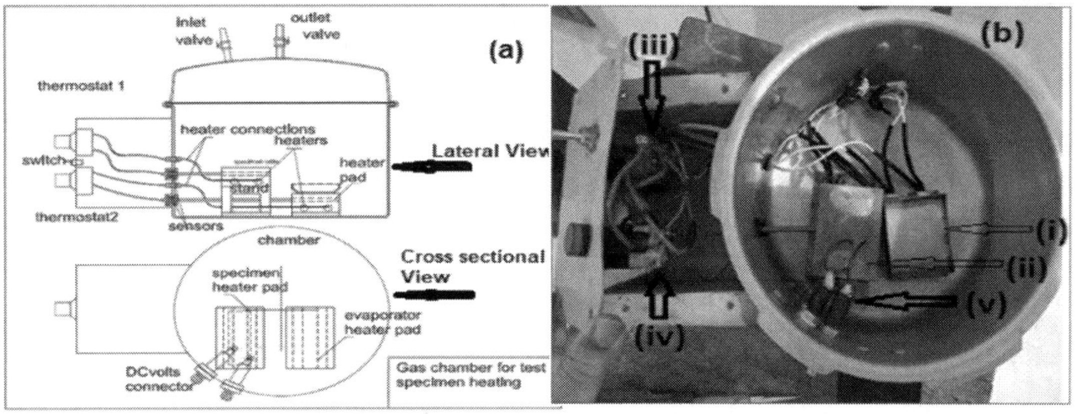

Figure.1(a) CAD design of proposed gas chamber (b) Developed Gas chamber-(i) solvent holderwith heater (evaporator), (ii) specimen holder with heater; (iii) Switch for solvent heater thermostat_1), (iv) Switch for substrate heater (thermostat_2), (v) Electrode connection

2.2 *Working of the Gas chamber*

For gas testing, the ZnO sensor is placed on the specimen holder and the temperature ofthe specimenheater is at an off state to maintain room temperature. To study the sensing property of the sensor towards different concentration levels in ppm (parts per million) of target VOC (1butanol) inside the chamber, the volume needed to be injected into the solvent holder using a pipette is calculated using Equation 1, and is tabulated in Table 2.

$$Vx\ (ml) = \frac{CppmXV_G XM\ .WtXP}{\rho XRXT} \qquad (1)$$

Where C ppm – Concentration of the target VOC/gas ppm (mg/m^3), Vx – volume of the solvent injected ,VG –volume of the gas chamber (0.005m^3), M. wt. – Molecular weight of the solvent g/mol, ρ –density g/m^3, R-gas constant =8.31, P= 1 Pascal (atmospheric pressure) and T= Room temperature (273 K).

Table 2: Volume (μl) of 1-butanol needed to produce 1-butanol VOC ppm (Equation 1)

Volume of solvent (Vx) (μl))	Concentration of VOC in C ppm (mg/m3)
20.16	100
10.08	50

After loading the solvent, the temperature of the solvent evaporator is set to 120°C which is the evaporating temperature of 1-butanol. The variations in the resistance of ZnO thin film are further studied through signal processing circuit and instrumentation amplifier circuit discussed in the next section. After testing of VOC, the lid of the gas chamber is opened for the sensors to recover in air.

2.3 *Implementation of signal acquisition circuit*

The interaction of the gas/VOC on the surface of the material is a characteristic of a semiconducting metal oxide sensor. When the surface of ZnO comes in contact with the target VOC, the resistance of the surface film varies depending upon the type of gas-oxidizing or reducing type. 1-butanol is a reducing type of gas and ZnO being n-type of semiconductor, when comes in contact reduces the resistance of the ZnO surface. This variation in the resistance decides the concentration of 1-butanol

IOP Conf. Series: Materials Science and Engineering 1221 (2022) 012005 doi:10.1088/1757-899X/1221/1/012005

VOC in the chamber. This change in the resistance of the ZnO sensor has to be measured with a suitable circuit. For this data acquisition(resistance) from the ZnO nanosensor, a wheat stone bridge network connected to an instrumentation amplifier is constructed as shown in figure 2a. ZnO sensor is connected in parallel to one of the arms of the bridge as shown in figure 2b. The network can pick small variations in the sensor's resistance when the sensor comes in contact with the target gas. This is further amplified by the amplifier.The bridge and the amplifier is driven by a 9volt battery. The output voltage at the amplifier is adjustedby a potentiometer (Rpot)in the circuit as given in Equation 2.

$$\text{Vout} = \frac{R3}{R2}\frac{(2R1+Rpot)}{Rpot}(V_A - V_B) \quad (2)$$

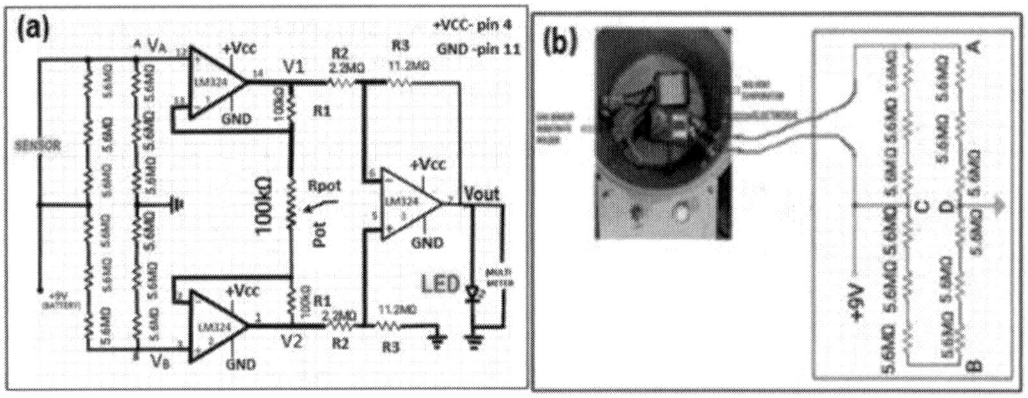

Figure 2: (a) Wheatstone bridge with an Instrumentation amplifier and a LED at the output (b) ZnO sensor connection replacing shunt resistor (Rs) through electrical feeds from the chamber.

The designed circuit is initially tested with a shunt resistor (Rs in M Ω range), connected parallel to one of the arms of the bridge which is later replaced by the sensor. The output voltage w.r.t the change in the shunt resistance is noted for the designed circuit and is plotted, as shown in figure 3.

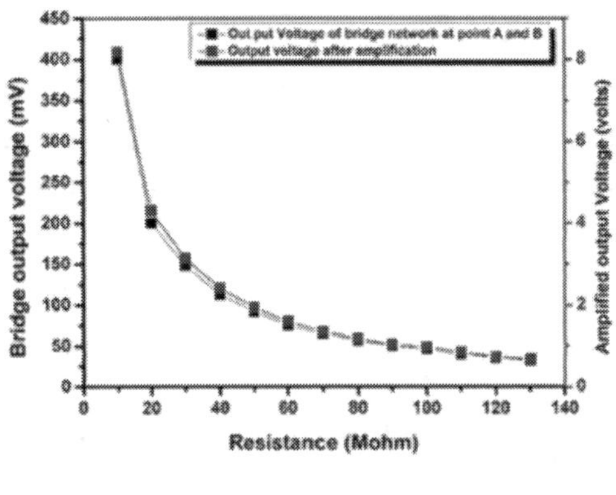

Figure 3: Variation of the output voltage of bridge network and amplifier circuit w.r.t change in the shunt resistance (Rs in MΩ)

3. Results and discussion

Synthesis, deposition, fabrication of ZnO sensor and the characterization study of ZnO as thin-film sensing layer is discussed in detail in our previous paper [8, 9].

IOP Conf. Series: Materials Science and Engineering 1221 (2022) 012005 doi:10.1088/1757-899X/1221/1/012005

3.1 1-butanol sensing in the test chamber

The experimental setup is shown in figure 4b, wherein ZnO is placed on the specimen holder with an electrode on the contact pads of IDE structure to establish electrical connection.The sensitivity study of the ZnO sensor for 1-butanol VOC at 100 ppm and 50 ppm level is performed in a closed chamber by observing the variation in the output voltage of the instrumentation circuit connected to the sensor. This voltage is measured using a digital multimeterat the output of the circuit.
The transient voltage response of ZnO sensor for 100 ppm and 50 ppm is shown in figure 4a.

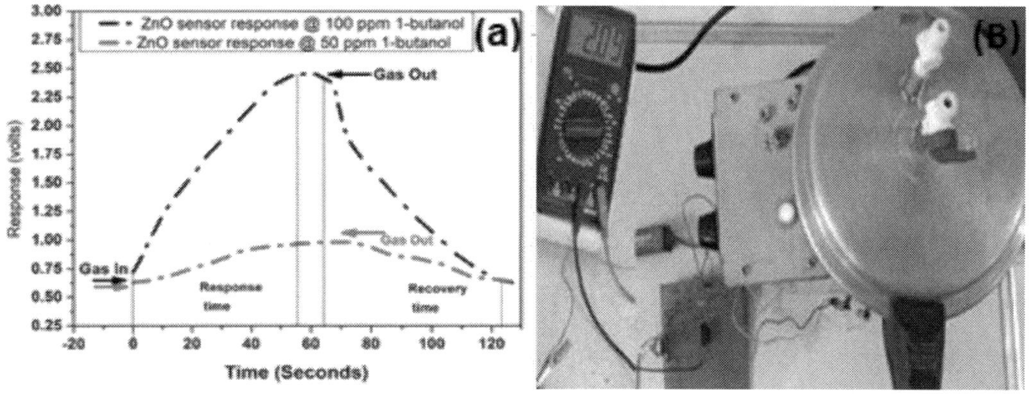

Figure 4:(a) Transient response of output voltage for the 1butanol sensing (b) Experimental setup of the chamber with a sensor placed inside the chamber with instrumentation amplifier circuit

The output voltage is around 0.69 volts in ambient air and the voltage gradually increases and stabilizes to 2.32 volts on exposure to 100 ppm of 1-butanol inside the chamber. It stabilizes there and when 1-butanol is released out, the voltage decreases back to 0.67 V.
Similarly, it was observed that on exposure to 50 ppm of 1-butanol, the voltage rises to 0.98V from 0.67V and stabilizes there and gradually decreases to 0.66 V when the gas is released out by removing the lid.
Using reverse engineering concept, the resistance corresponding to these voltagesare observed and noted from the graph of figure 3. From figure 3,thecorresponding resistance value of ZnO film in the air (Ra) and resistance on exposure to 1-butanol gas (Rg) is noted. The corresponding values are Ra= 125 MΩ and Rg = 38 MΩ approximately for 100 ppm and Ra = 135 MΩ and Rg = 98 MΩ for 50 ppm of 1-butanol.
The sensitivity of the sensor is defined as the ratio of sensor's film in air (Ra) to the resistance of the film in presence of the 1-butanol (Rg) to (Ra) [10] (see Equation 2),

$$S = \frac{Ra}{Rg} \text{When Ra >Rg} \qquad (2)$$

Sensitivity towards 50 ppm is observed as S= Ra/Rg = 135/98 = 1.3 and the sensitivity for 100 ppm is observed as S=Ra/Rg = 125/38 = 3.28. (See Table 3).

Table 3:*Comparison of results obtained by the custom-designed system with Keithleysystem*

Results	Present work through custom-designed chamber		Previous work through Keithley Instrument[12]	
Parameters	50 ppm	100 ppm	50 ppm	100 ppm
Sensitivity	1.3	3.28	1.28	3.27
Response time(Sec)	62	55	60	50
Recovery time (Sec)	70	60	74	55

The response time (in seconds) (Rs) is defined as the time required for the sensor to reach 90% of saturation response value and the recovery time is defined as the time taken for the sensor to reach back 10% of reference value from a minimum value. Response and recovery time are calculated in Table 3.

These parameters are compared with those obtained in our previous work [12] and it has closely matched. Thus a portable, low-cost, simple operationbased instrumentation system can be realized. In the future different VOCs like acetone, ethanol, methanol etc., can be tested using a ZnO sensor or other sensorsin static mode of measurements.

Conclusion

A cost-effective, portable gas chamber is designed with theassociated data acquisition circuit containing a precision wheat stone bridge network and an instrumentation amplifier circuit. ZnO thin film sensor is connected in the circuit and is testedfor 1-butanol VOC by optimizing the parameters like specimen temperature(RT) and solvent temperature (120°C). 1-butanol VOC was sensed at different concentration levels-50 ppm and 100 ppm inside the chamber. Sensing parameters like sensitivity, response time and recovery time ofthe ZnO sensor werestudied and calculated. The sensitivity was found to be 1.3 and 3.28 for 50 ppm and 100 ppm of 1-butanolrespectively, which matched with the previous work.

References

[1] Nazemi H, Joseph A, Park J and Emadi A. 2019 Advanced Micro and Nano-Gas Sensor Technology: A Review. *Sensors* **19** (6) 1285.

[2] Haocheng Ji, Wen Zeng and Yanqiong Li 2019 Gas sensing mechanisms of metal oxide semiconductors: a focus review. *Nanoscale* **47** (11) 22664-22684.

[3] Das S and Pal M 2020 Review—Non-Invasive Monitoring of Human Health by Exhaled Breath Analysis: A Comprehensive Review *J. Electrochem. Soc* **167**, 037562.

[4] Wei Guan, Na Tang, Kuang He, Xiaoying Hu, Mingshan Li and Kaiming Li 2020. Gas-Sensing Performances of Metal Oxide Nanostructures for Detecting Dissolved Gases: A Mini-Review. *J Front. Chem.*

[5] Zaki M, Hashim U, Md. Arshad M.K, Fathil M.F.M, Ruslinda A.R, Ayub R.M, Subash C.B, Voon C.H., Foo K.L, R. Adzhri, Azman A.H. 2015 Real-Time Detection by Properties of Tin Dioxide for Formaldehyde Gas Sensor. *RSM Proc.*

[6] Afzal M, Naik P. S, Nadaf L. I, 2015. Cost-Effective Experimental Setup for Gas Sensing Applications. *Journal of Applied Chemistry*(IOSR-JAC) **8** (12) 37-44.

[7] Jayasree T, Multan S, 2014 Study of Gas Sensors for the Detection of Volatile Organic Compounds in Breath. *Applied Mechanics and Materials* **573**, 785-790.

[8] Bhat P, Pradhan U and Kumar S.K, 2018 Synthesis and Characterization of ZnO Nano Discs Using Wet Chemical Method for Sensing Applications. *Materials Today: Proceedings* **5**, 10763–10770.

[9] Bhat P, Kumar S.K, Nagaraju P, 2020 Fabrication of ultrasensitive hexagonal disc structured ZnO thin-film sensor to trace nitric oxide. *Journal of Asian Ceramic Societies.*

[10] Anafi Nuraini and Ilwhan Oh, 2020 Volatile organic compound gas sensors based on methylammonium lead iodide perovskite operating at room temperature. *RSC Adv.*, **10,** 12982-12987.

[11] Bhat P, Kumar S.K, Nagaraju P, 2019 Effect of annealing temperature on microstructure, optical, morphology and gas sensing characterization of nanodisc ZnO thin films. *AIP Conference Proceedings* **2244**, 110008.

[12] Bhat P, Kumar S.K, Nagaraju P, 2019 Synthesis and characterization of ZnO-MWCNT nanocomposites for 1-butanol sensing application at room temperature. *Physica B: Condensed Matter* **570**, 139–147.

IOP Conf. Series: Materials Science and Engineering 1221 (2022) 012006 doi:10.1088/1757-899X/1221/1/012006

Effect of the hole concentration on the polaronic mobility for the strongly interacting electrons

Debika Debnath* **and Ashok Chatterjee**
School of Physics, University of Hyderabad, Hyderabad, Telangana, India, 500046
*debika.physics@gmail.com

Abstract. The effective hopping of the polaron has been studied for different hole concentrations. Holstein-Hubbard model is used for the half-filled and less than half-filled cases. A series of canonical transformations are imposed on the generic Hamiltonian and considering the strongly interacting electrons, the transformed Hamiltonian has been reconstructed to an effective t-J model. Using the mean-filled Hartree-Fock approximation the electronic Hamiltonian has been solved with the Zuberev's Green function technique. The effective polaronic transport is found to be strongly dependent on the electron and hole concentration of the system.

1. Introduction

The polaron physics [1, 2] has potential application in studying the electronic and transport properties of various materials, such as high temperature superconductor (HTSC), colossal magneto-resistance materials, cuprates etc. The electron-phonon coupling coefficient has a pivotal role to the polaronic mobility. For the weak electron-phonon interaction coefficient, the polaron can move through the lattice. For the stronger electron-phonon interaction strength, the mobile polaron may get trapped in its own potential and become localized. The transition from this mobile large polaron to the localized small polaron is known as the self-trapping transition. It has been also observed that the electron-hole concentration and the electron-phonon interaction strength both have a competitive effect towards the polaronicmobility.

Sankar and Chatterjee (SC) [3, 4] have studied the self-trapping transition for the strongly correlated electronic system using a variational technique followed by the zero phonon average to the Hamiltonian. In our present work, we have modified the SC's work and the effects of the electron density towards the self-trapping transition have been studied. We have considered the two-dimensional Holstein-Hubbard model followed by a series of canonical transformations and a general many phonon averaging to eliminate the phonon degrees of freedom from the system. The electronic Hamiltonian is then transformed to the *t-J* model and it is solved using the mean-filled Hartree-Fock approximation. Gutzwiller approximation and Zuberev's Green function method have been the tools to solve the system. Next we have finally calculated the ground state energy of the system using numerical variational method and the effective hopping parameter for the polaron have been studied with respect to the electron-phonon interaction coefficient and electron-hole concentration in the system.

2. The model

The Hamiltonian for the system is described as:

$$H = -t \sum_{<ij>\sigma} c_{i\sigma}^{\dagger} c_{j\sigma} + U \sum_{i} n_{i\uparrow} n_{i\downarrow} + \omega_0 \sum_{i} b_i^{\dagger} b_i + V \sum_{<ij>\sigma\sigma'} n_{i\sigma} n_{j\sigma'} + g_1 \sum_{i\sigma} n_{i\sigma} \left(b_i^{\dagger} + b_i\right)$$
$$+ g_2 \sum_{<ij>\sigma} n_{i\sigma} \left(b_j^{\dagger} + b_j\right) \quad (1)$$

where the parameter t refers to the nearest-neighbour hopping integral and U is the onsite Coulomb correlation energy. $c_{i\sigma}^{\dagger}$ $(c_{i\sigma})$is the creation (annihilation) operator for an electron with spin σ at site i and $b_i^{\dagger}$ (b_i) is the creation (annihilation) operator for a phonon at the site i with frequency ω_0. The number of an electron at site i with spin σ is $n_{i\sigma} = c_{i\sigma}^{\dagger} c_{i\sigma}$. V measures the strength of the NN electron correlation. The on-site e-p interaction coefficient is g_1 and NN e-p interaction constant is g_2.

2.1. Formulation

A series of canonical transformations are performed on the Hamiltonian H, using the generators,

$$(i) G_1 = \frac{g_1'}{\omega_0} \sum_{i\sigma} n_{i\sigma} \left(b_i^{\dagger} - b_i\right) + \frac{g_2'}{\omega_0} \sum_{<ij>\sigma} n_{i\sigma} \left(b_{i+\delta}^{\dagger} - b_{i+\delta}\right),$$

$$(ii) G_2 = \alpha_s \sum_{i} \left(b_i b_i - b_i^{\dagger} b_i^{\dagger}\right),$$

$$(iii) G_3 = \alpha_d \sum_{i\sigma} n_{i\sigma} \left(b_i b_i - b_i^{\dagger} b_i^{\dagger}\right),$$

$$(iv)\ G_4 = \frac{1}{2} \sum_{i \neq j} \beta \left(b_i b_j - b_i^{\dagger} b_j^{\dagger}\right).$$

To eliminate the phonon degrees of freedom from the system, the transformed Hamiltonian is averaged over the all phonon state,

$$|\Phi_{ph}\rangle = \sum_{n=0}^{M} c_n\ |\varphi_n(x)\rangle.$$

The electronic Hamiltonian is then transformed to an effective *t-J* model and using the Gutzwiller approximation [5, 6], where we write $(1 - n_{i\bar{\sigma}}) c_{i\sigma}^{\dagger} c_{j\sigma} (1 - n_{j\bar{\sigma}}) \cong \varphi_t c_{i\sigma}^{\dagger} c_{j\sigma}$, where $\varphi_t = 2x_h/(1 + x_h)$. Here the hole concentration in the system is denoted by,$x_h = (1 - n)$, and the electronic concentration is denoted by$n$. Finally the Hartree-Fock approximation is used to solve the Hamiltonian. The average occupation number per site (n), the Hartree correction to the kinetic energy term (p) are calculated using the Zuberev's Green function technique [7]. In our work we have considered the system at $T \to 0$ with square density of states for the polarons.

Therefore the ground state (GS) energy per site is obtained as,

$$\varepsilon_0 = n\varepsilon_e - \left(\varphi_t t_{eff} + p\tilde{J}\right) zp + Nz \left[\frac{1}{4}(\tilde{J} - 4V_1^e) n^2 + \tilde{J} p^2\right] + K. \quad (2)$$

Here $\tilde{J} = \varphi_J J = 4J/(1 + x_h{}^2)$. The GS energy is calculated by numerically minimising the equation (2) with respect to all the variationalparameters g_1', g_2', α_s, α_d, β, and c_n's. The parameters ε_e, t_{eff}, $\tilde{J}$, $V_1^e$and K are complicated equations (not written here).

3. Results and discussion

The polaronic mobility is plotted in Figure 1, with respect to the electronic concentration for different e-p interaction strength. It is found that the effective mobility started to decrease from the bare Hubbard parameter t, as the electronic concentration and the electron-phonon interaction coefficient increases. After a certain value of n, the polaron becomes trapped in its own potential, resulting the self-trapping transition and t_{eff} becomes zero. Here our result is compared with the Sankar& Chatterjee results [3] and it is found that the effective polaronic mobility is higher at the present case and stronger electron-phonon interaction strength is required to trap the polaron.

The effect of the e-p interaction on the polaronic mobility has been studied in Figure 2. As the on-site electron-phonon interaction coefficient increases, the effective polaronic mobility decreases.For weak electron-phonon interaction strength it is found the effective hopping parameter is higher when the hole concentration is more. The higher hole concentration implies lower electronic density, that helps the polaron to be more mobile. But as the on-site e-p interaction increases further, the polaron mobility seizes faster for higher hole concentration.

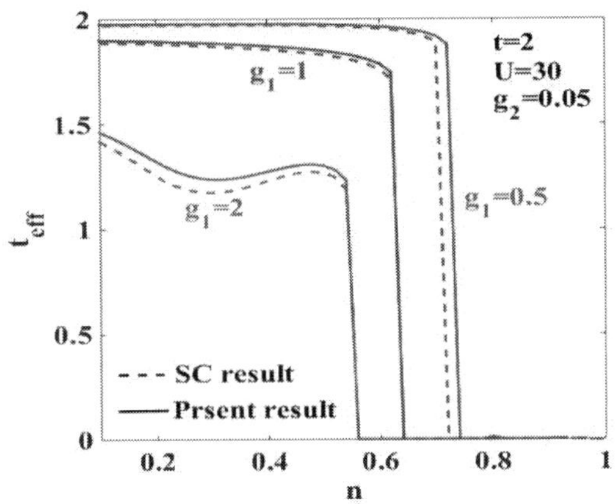

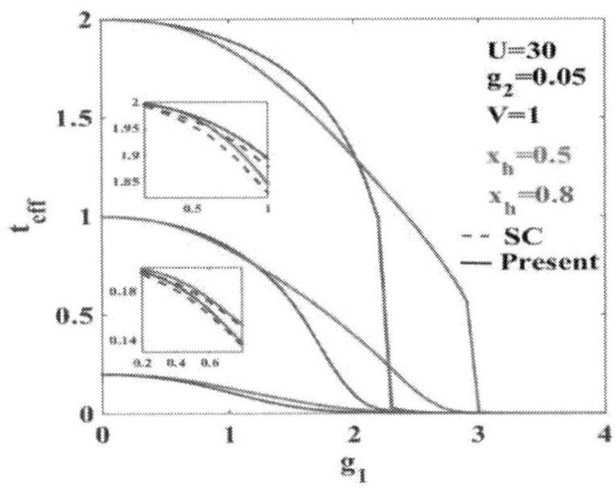

Figure 1. Effective hopping parameter(t_{eff}) **Figure 2.** Effective hopping parameter (t_{eff}) with respect to the electron concentration (n).with respect to the e-p interaction strength(g_1).

4. Conclusion

The change of the effective hopping parameter with respect to the electronic concentration shows the present result denotes higher effective hopping than the SC result which carries important information of the polaron mobility, band width and effective mass. Therefore this result has pivotal role in studying the transport properties of the high temperature superconductors and cupratemeterials.

Acknowledgement

DD acknowledges the financial support from Department of Science and Technology, India (DST/INSPIRE Fellowship/2017/IF170064).

5. References

[1] Mitra T K and Chatterjee A Mukhopadhyay S 1987 *Phys. Rep.* **153** 91
[2] Devreese J T and Alexandrov A S 2009 *Rep. Prog. Phys.* **72** 066501
[3] Sankar I V and Chatterjee A 2014 *Eur. Phys. J. B* **87** 154.
[4] Sankar I V and Chatterjee A 2013 *AIP Conference Proceedings* **1512** 1136
[5] Gutzwiller M C 1963 *Phys. Rev. Lett.***10**, 159
[6] Gutzwiller M C 1965 *Phys. Rev.* A., **137**, 1726
[7] Zubarev D N and Usp. Fiz.Nauk. 1960 *Phys.* ***71**, 71*

Titanium Nitride based multi-functional thin films

M.Ghanashyam Krishna[1*] and K.A. Padmanabhan[2]

[1]CASEST, School of Physics, University of Hyderabad, Hyderabad, Telangana, India
[2]Materials Science and Engineering Program, Department of Mechanical Engineering, Anna University, Chennai, India

***Corresponding Author:** mgksp@uohyd.ac.in

Abstract The aim of the present work is to demonstrate that the applications of TiN thin films can be expanded well beyond the traditional domains. For example, although TiN is known to exist in several sub-stoichiometric forms their properties have not been exploited fully. Using a patented sputtering process technology, we show that the Ti_2N phase is semiconducting, possesses a band gap of the order of 3.5eV and a hardness in the region of 5-7 GPa. Thisimplies that Ti_2N thin films can be candidates for light emitting diode applications. In some cases, reflectivity <2% in the visible region was also achieved which is useful for optical shielding applications. The nitrogen stoichiometry and thickness of TiN thin films can be tuned to achieve colours from blue to brown to golden yellow for decorative coating applications. TiN thin films can be a cost-effective replacement for Au since they also display a surface plasmon resonance at the same wavelength. The conductivity of TiN is sufficiently high to replace gold as an electrode material in electronic devices such as diodes. The properties of TiN-Si_3N_4 and TiN-Polyaniline composite thin films are also reported. The current work, thus, demonstrates the multi-functionality of TiN as an optical, opto-electronic and electronic material.

1. Introduction

Titanium Nitride (TiN) is one of the most popular transition metal nitrides that has attracted attention due to its high conductivity and hardness. Therefore, much of the literature focuses on these two properties [1-5]. However, TiN thin films have recently been shown to exhibit many properties that make them suitable for other applications as well. Surface plasmon resonance (SPR) in TiN thin films at a wavelength comparable to that of gold nano- particles is a good example [6-8]. Another application is in the microelectronics industry as an ohmic contact and a diffusion barrier for metals in semiconductor devices [9,10]. In addition, TiN exists in several sub-stoichiometric forms which makes it possible to manipulate properties by controlling the nitrogen stoichiometry. This leads to variations in conductivity, color, optical behavior and hardness that are dependent solely on nitrogen concentration. Ti_2N is one such sub-stoichiometric form that has proved challenging to stabilize in bulk or thin film form. The current authors have developed a patented technology that enables the stabilization of Ti_2N in thin film form [11-13]. The compound is shown to have semiconducting nature with a band gap that compares well with that of GaN and displays reasonable hardness as well as scratch resistance.

In this study, TiN thin films are grown by RF and DC magnetron sputter deposition from a Ti target in Ar+N_2 or pure N_2 atmosphere. By controlling the process, amorphous, nanocrystalline or crystalline TiN_x ($0.5 < x \leq 1$) films could be obtained. It is also established that TiN exhibits a nitrogen-stoichiometry-dependent reflectivity minimum centred around 530 nm. Calculations of optical

IOP Conf. Series: Materials Science and Engineering 1221 (2022) 012007 doi:10.1088/1757-899X/1221/1/012007

constants indicate that this feature corresponds to SPR and is observed in both undoped and Nb-doped TiN thin films [14, 15]. The electrical, mechanical and optical behaviour of TiN-Si_3N_4 nanocomposite thin films grown on Phynox (Co-Cr-Ni) alloy and IN718 substrates display excellent conductivity, hardness and high optical reflectivity [16]. The properties of TiN-polymer composite thin films produced by a simple chemical synthesis method is also discussed. Thus, this work expands the range of applications of TiN thin films into the electrical, optical, opto-electronic and sensing domains by controlling non-stoichiometry and doping.

2. Results and Discussion

2.1. Non-stoichiometric TiN films

The nitrogen stoichiometry of the TiN films is controlled by varying the nitrogen (N_2) pressure during RF sputtering. The porosity is controlled to achieve almost 0% level by changing the pressure from 12 to 25 mTorr apart from substrate temperature and thickness. The hardness of these films was between 5-7 GPa [11-13]. The stoichiometry was estimated from the position of the reflectance minimum relating to the charge transfer between the Ti-2p and N-2s which, for stoichiometric TiN occurs at 2.33 eV or approximately 530 nm [17]. The change in the position of this feature is attributed to nitrogen non-stoichiometry. For example, at a power density of 3 $W.cm^{-2}$, the peak position was observed at 3.1, 2.85, 2.7 and 2.65 eV when the films were deposited at 12.15, 20 and 25 mTorr respectively. These values signify deviation from the 2.33 eV value indicating non-stoichiometric TiN. The films were confirmed to have the Ti_2N stoichiometry on annealing at 873K for one minute [11]. Optical transmission of the order of 60-80% in the visible region and a band gap in the range of 3.6 -4.5 eV, which is similar to that of GaN [11,14] were achieved. The band gap region of these films in the as-deposited state, is shown in figure 1.

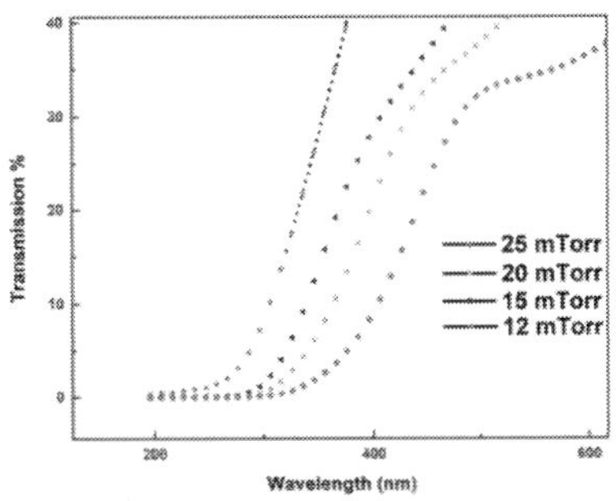

Figure 1: Measured optical transmission of TiN_x films deposited at different N_2 pressures showing the large band gap

Notwithstanding the impressive optical properties of the Ti_2N phase, the challenge is that this phase naturally makes a transition to the stoichiometric form at thicknesses > 150 nm and when the films are deposited at high substrate temperatures. Thus, if these processing challenges are overcome, as done here, Ti_2N can rival GaN as a cost-effective alternative for opto-electronic applications such as light emitting diodes. Under favourable conditions, the reflectivity over the wavelength range of 400-550 nm was < 2% [17].

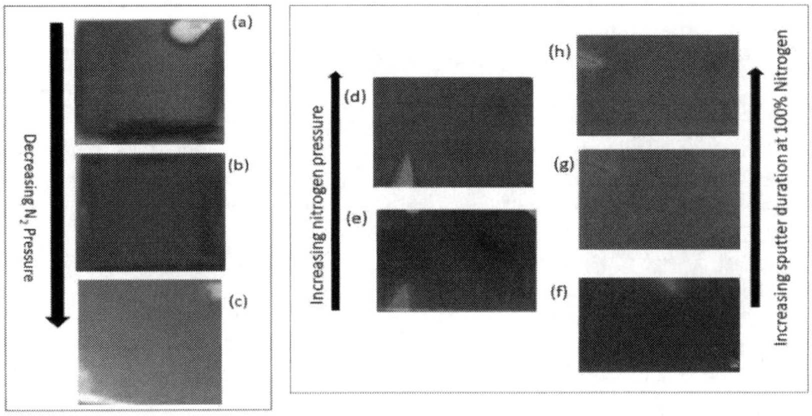

Figure 2: Different colors obtained in TiN_x thin films by (a), (b), (c) decreasing N_2 pressure during sputtering (films are deposited on stainless steel substrates). (d), (e) increasing N_2 pressure and (f), (g), (h) increasing duration of deposition at 100% nitrogen in the process gas (films in (d)-(h) are deposited on brass substrates, hence the difference in colors)

It is known that stoichiometric TiN is golden yellow in color. However, not much is known about the color of nitrogen deficient TiN. In our study we could show that the colors of the films on stainless steel substrate can be controlled from black, brown, blue to golden yellow as the nitrogen pressure is changed during film deposition, as shown in figure 2a-c. The colors of the TiN films deposited on brass substrates are controlled by changing the percentage of nitrogen in the sputter gas mixture of argon and nitrogen as displayed in figure 2d-e. The colour can also be controlled by increasing the duration of deposition, at a fixed value of 100% N_2 in the sputter gas (figure 2f-h). In addition, since the TiN_x films are scratch-resistant with medium hardness, these films are expected to be very useful for decorative coatings. Stainless steel and brass are commonly used in the manufacture of jewelry products and these films can provide value addition both in terms of attractive colors and durability.

2.2 Stoichiometric TiN films.

Stoichiometric TiN films display an absorption peak at 2.33 eV nm, which is attributed to the charge transfer between the Ti-2p and N-2s states. In nanostructured thin films, the origin of this absorption peak is considered to be due to surface plasmons with a resonance frequency at this energy [6-8]. Indeed, several studies, including those reported by the present authors, show that the real part of the dielectric function for TiN is negative at optical frequencies[6-8,14,15] which is a pre-requisite to be classified as SPR.SPR is generally observed in noble metals such as Au and Ag, but its observation in TiN thin films combined with oxidation resistance and bio-compatibility signifies that they can be considered as cost-effective alternatives for such applications. The specular reflectance curves for TiN thin films deposited at 100 W RF power for different durations, on glass substrates are shown in fig.3. It is evident from these figures that there is a strong reflectance minimum at 530 nm (2.3 eV) for all the films corresponding to the SPR peak. The maximum hardness achieved in the stoichiometric TiN films was 25 GPa [18].

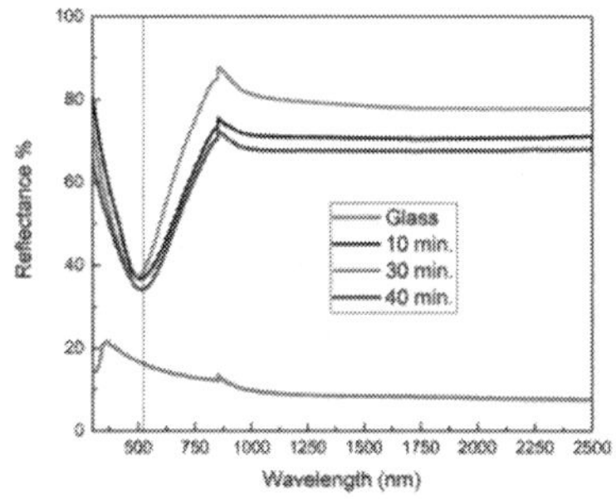

Figure 3: Optical reflectance spectrum showing the surface plasmon resonance peak corresponding to TiN at 530 nm. There is no deviation in the peak position with increase in duration of deposition indicating that all the films are stoichiometric.

The application of high electrical conductivity of TiN thin films is discussed now. As stated earlier, TiN is inherently as electrically conductive as Au. This fact has been exploited recently to replace Au in many electronic devices with TiN as the electrical contact. The ability to use TiN thin films as electrical contacts and the processes by which they are produced being CMOS technology compatible make them ideal cost-effective alternatives to Au. In our work we have fabricated Schottky diodes using SiC as the semiconductor material. Two diode structures were investigated for comparison: (1) Au-SiC-Ag, and (2) TiN-SiC-Ag, wherein Au or TiN is the bottom electrode and Ag of 500μm diameter deposited through a shadow mask by thermal evaporation is the top electrode. The temperature dependence of the device characteristics is investigated over the range 300-423K. The typical I-V characteristics are presented in figure 4a-b.

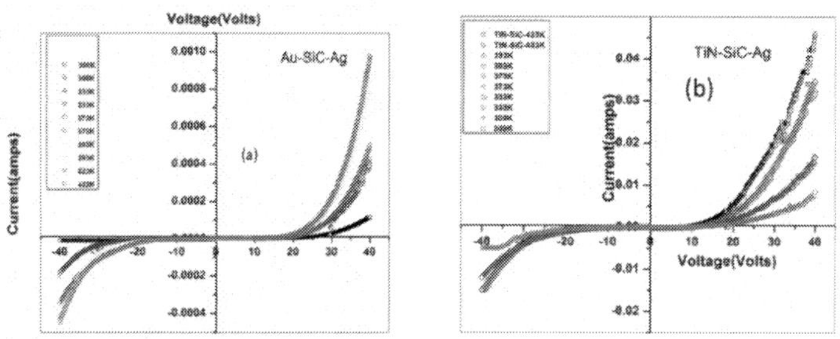

Figure 4: I-V characteristics of Schottky diodes fabricated using SiC thin film as the semiconductor, and (a) Au, and (b) TiN as the bottom electrode with Ag being the top electrode in both cases. Clearly the performance with TiN as the bottom electrode matches that of the device with Au as the bottom electrode

Based on the current generated it can be inferred that the TiN based diodes are superior in performance to that of the Au based diode. Thus, even on this count TiN can replace Au as the material of choice.

2.3 TiN based composite thin films

Another approach that we have adopted to exploit the multi-functionality of TiN is to synthesize composite thin films. In this work TiN-Si_3N_4 thin films in the ratio of 1:1 were deposited on borosilicate glass (BSG), quartz, single crystal Si, phynox alloy and Inconel718 alloy substrates [16]. The TiN is nanocrystalline while Si_3N_4 is amorphous. Typical specular reflectance curves of the TiN-Si_3N_4 films on the BSG, quartz and Si substrates shown in figure 5 indicates that the SPR peak of pristine TiN that occurs at 530 nm is red-shifted by 30-70 nm in the presence of Si_3N_4. Thus, the SPR can be tuned by synthesizing composite thin films.

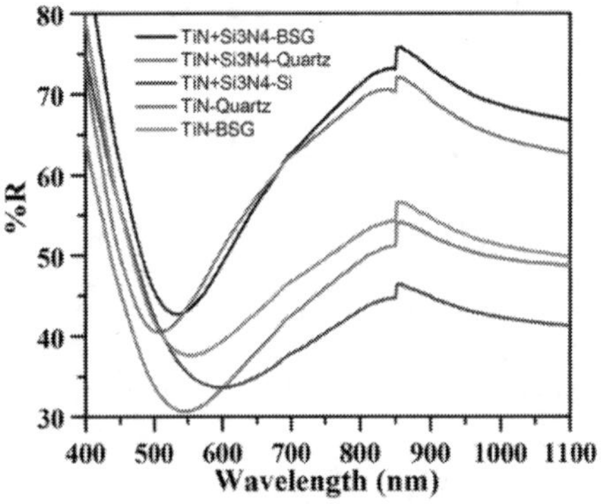

Figure 5: Optical reflectance spectra of TiN-Si_3N_4 composite thin films show the presence of SPR resonance peaks which are red-shifted compared with those of pristine TiN thin films. Another noteworthy point is the high reflectance in the infrared region of the spectrum.

The typical hardness of the nanocomposite films is of the order of 8 GPa on these substrates. The same nanocomposite films when deposited on phynox alloy substrates exhibited a maximum hardness of 11 GPa and 13 GPa on IN718 substrates [16].

The techniques for processing the films discussed in the foregoing sections are based on physical vapour deposition, which is very expensive. Hence, simpler cost-effective chemical routes have also been explored for the synthesis of TiN based nanocomposite films. One example is TiN-polyaniline(PANI) films that were synthesized by mixing commercial TiN powders with PANI solution. The resulting solution is then applied on a substrate by dipping to achieve thin films of thickness from 100 to 1000 nm. The x-ray diffraction pattern in figure 6a clearly shows that the TiN is nanocrystalline in the composite. Since PANI is a conducting polymer with a lower conductivity than

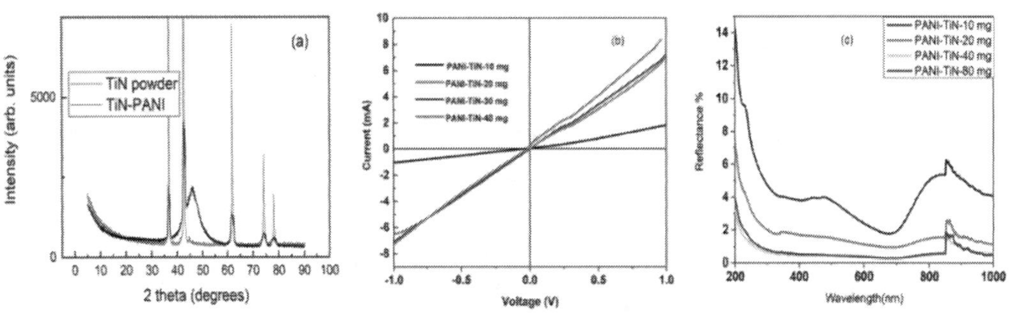

Figure 6: (a) X-ray diffraction pattern of TiN powder and TiN-PANI composite thin film, (b) electrical conductivity, and (c) optical reflectance spectra of the composite thin films.

TiN, the composite conductivity can be tuned by varying the loading of PANI as shown in figure 6b. The resistance of the composites varies from 0.5 kΩ to 1 kΩ at 0.25 V bias and 1kΩ to approximately 8kΩ at the highest bias of 1V showing that these materials can be used for electronic applications. The specular reflectance spectra displayed in figure 6c indicate that the reflectance is less than 10% in the region from 300-1000 nm. This is in contrast to >80% reflectance observed for pristine TiN and 40-70% observed for TiN-Si_3N_4 composite thin films at wavelengths > 750 nm. Thus, the optical response of the TiN thin films can be tuned by fabricating composites.

3. Summary:

The present work demonstrates the versatility of TiN as a multi-functional material beyond the traditional domain of mechanical properties. It is shown that the optical behaviour can be tuned by changing the stoichiometry leading to films with a band gap in the UV region. Nitrogen stoichiometry and process parameter-controlled variation of colours of TiN is also demonstrated. It is also shown that TiN can replace Au as an electrode in electronic devices. Tunability of these properties when TiN based composite thin films are fabricated is an additional feature of the current work.

Acknowledgment:

The authors acknowledge the contributions of M.S.R.N. Kiran, K.Vasu, V. Shankernath, K.L.Naidu, P.C.Akshara and K. Aravind in carrying out the experiments and providing data for this work. Facilities provided by the School of Physics and Centre for Nanotechnology University of Hyderabad under the DST-PURSE, UGC-NRC/DRS/CAS programmes are also acknowledged.

References:

[1] VeprekS, ReiprichS, 1995 *Thin Solid Films* **64** 268
[2] GaoL, GstottnerJ *et al.*, 2004 *Microelectron. Eng.* **76** 76
[3] TsetserisL, LogothetidisS, PantelidesST, 2009 *Appl. Phys. Lett.* **94** 161903
[4] Rosmaninho RRF, Rizzo G *et al.*, 2007 *Chem. Eng. Sci.* **62** 3821
[5] GreeneJE, SundgrenJ-E, HultmanL *et al.*, 1995 *Appl. Phys.Lett.* **67** 2928
[6] Steinmüller-NethlD, KovacsR, *et al.*, 1994 *Thin Solid Films*, **237** 277
[7] NaikG V, SchroederJL, NiX, KildishevAV *et al.*, 2012 Opt. Mater. Exp., **2** 478
[8] PatsalasP, KalfagiannisN, KassavetisS, 2015 *Materials*, **8** 3128
[9] DiddenA, BattjesH, MachunzeR, 2011 *J. Appl. Phys.*, **110** 033717
[10] LiL, NakamuraR, WangQ *et al.*, 2014 *Nanoscale Res Lett.*, **9** 590
[11] KiranMSRN, KrishnaM G, PadmanabhanKA, 2008 *Appl. Surf. Sci.*, **255** 1934
[12] VasuK, KrishnaM G, PadmanabhanKA, 2011 *Appl. Surf. Sci.* **257** 3069
[13] PadmanabhanKA, KrishnaM G, KiranMSRN, 2014 *US Patent No. 8,888,965 B2*
[14] VasuK, GopikrishnanGM, KrishnaM G, PadmanabhanKA, 2012 *Appl Phys A* **108** 993
[15] ShankernathV, NaiduKL, KrishnaM G, PadmanabhanKA, 2017 *Mater. Res. Bull.* **85** 121
[16] ShankernathV, 2017 *Ph.D. thesis* University of Hyderabad
[17] KiranMSRN, KrishnaM G, PadmanabhanKA, 2008 *Int. J. Nanomanuf.*, **2** 420
[18] KiranMSRN, KrishnaM G, PadmanabhanKA, 2011 *Solid State commun.* **151** 561

Optoelectronic, microstructural, ecofriendly and photo-catalytic evaluation of Aspartic acid cross-linked Poly (vinyl alcohol) /Copper Oxidenanotubescomposite films

Yamanappagouda Amaregouda[1],Tilak Gasti[2], and Kantharaju Kamanna[1]*
[1]Department of Chemistry, Rani Channamma University, Belagavi – 591 156, India
[2]Department of Chemistry, Karnatak University, Dharwad – 580003, India
*Corresponding author: E-mail: kk@rcub.ac.in

Abstract

The present work describes design and preparation of aspartic acid (Asp) cross-linked Poly (vinyl alcohol) (PVA) /Copper oxide nanotubes (CuO NTs) composite filmsvia solution casting method.The CuO NTswere prepared by novel biogenic method using *RobiniaPseudoacacia*Leaf Extract.Theprepared CuO NTs and PVA/CuO/Asp nanocomposite filmsarecharacterized by various techniques.The adsorption and photo-catalytic efficiencies of PVA/CuO/Asp nanocomposite films were compared with pristine CuO NTs. The adsorption capacity of PVA/CuO/Asp nanocomposite films showed higher due to presence of active sites, high surface activity and semiconducting nature of CuO NTs and Asp embedded in PVA matrix. The comparative photocatalytic activities of the CuO NTsand PVA/CuO NTs/Asp nanocomposite filmsare examinedagainst methylene blue (MB) as a substrate degradation, and film PVA/CuO NTs/Asp showed highest (up to 77%) degradation compared to other films. The improved photocatalytic behavior of PVA nanocomposite film attributed to its narrower band gapreduces the recombination rate of exited electron-hole, enhanced UV-Vis absorption intensity, extinction coefficient and refractive index, improved utilization of sunlight and enhanced adsorption of dye due to functional groups, biodegradable, surface wettability, hydrophilic surface and self–cleaning nature.

Keywords: Polyvinyl alcohol, CuO NTs, Aspartic acid, photocatalysis, optical property.

1. Introduction

From the last two decades, Polymeric nanocomposites are attracting more study attention as a result of advancements in electrical, thermal, optical, and mechanical properties, as well as their tremendous potential for highly useful materials [1]. To create a polymer–inorganic nanocomposite, inorganic metal oxide nanoparticles were additionally added to the polymer matrix. Sensors, medical devices, medication delivery, coatings, adhesives, optical integrated circuits, cars, microelectronic packaging, injection molded products, packaging materials, aerospace, and other uses for nanocomposites are all promising [2]. Polymers have a number of advantages, including low cost, ease of production, high strength, flexibility, and outstanding mechanical qualities, as well as being environmentally friendly. Nanocomposites based on semiconductor nanoparticles and a polymer matrix are promising materials for use in optoelectronics and photocatalysis, as well as the development of environmentally friendly products [3]. The present work reported, PVA/CuO NTs/Asp nanocomposites are better materials for the pollutant MB dye removal because of its high; photocatalytic activity, optical transparency, refractive index, optical absorption, photovoltaic performance of low-bandgap, adsorption, hydrophilic surface and self-cleaning. Polyvinyl alcohol (PVA) is hydrophilic synthetic polymer which has now received a lot of attention from researchers due to its excellent film-forming properties, nontoxicity, biodegradability and optical transparency [4]. Adsorption of dyes using PVA based polymer composite adsorbents are well studied in the literature. PVA has been widely investigated as a host matrix for various types of dopants such as nanofillers and bioorganic molecules to modulate the effective photocatalytic properties [5]. PVA along with semiconducting material composite films are demonstrated efficient dye degradation. Researchers reported that incorporation of the nanofiller into PVA matrix and biopolymer blended with PVA showed efficient removal of MB dye. Researchers also have reported that nanofiller and biomolecule doped PVA composites show excellent optical properties.

Low dimensional CuO nanomaterials, particularly 1D CuO nanomaterials have received great attention because of their huge surface area, narrow bandgap with p-type semiconductor, high refractive index, enhanced optical property, ability to spread uniformly in polymers and potential

application in converting solar energy through photocatalysis for the degradation of organic pollutants in water [6]. In particular CuO NTs belong to one dimensional nanomaterial, it also act as a electrode performed as both a P-type semiconductor for generating the cathodic photo current signal and a biomimetic catalyst to synergize the biological catalyst XOD to constitute the dual catalysis system. In another paper reported CuO NTs, having better bonding strength and, hence, a longer lifetime. Researcher investigated many applications of CuO NTs and its doped films used for photocatalysis, optoelectronic properties. The strong hydrogen bonding between CuO and PVA reported [7]. The physical properties and application of CuO NPs with PVA composite films for optical properties is well documented [8].
Amino acid is a fundamental component of protein and has a series of potential attributes for effective dye removal, due to the following properties such as presence of amine (–NH2) and carboxyl (–COOH) functional groups, surface adsorption, complexation (coordination), ion exchange, pI value, hydrophobicity, hydrophilicity and exhibits dielectric and high refractive index [9]. Among the twenty amino acids Asp is unique properties, due to the presence of side chain containing additional carboxyl group resulted negative charge. Researcher demonstrated the strong formation of complex between CuO NPs and Asp is documented. Literature reported semiconducting properties of Asp is alone 5.02 eV and also exhibit very good hydrophilic property which helps potential removal dye is established [10]. To reach our purpose, we first synthesized CuO NTs by using biogenic method, which is then loaded onto PVA polymer matrix using aspartic acid as a cross-linking agent. The structural, morphological and optical properties of the prepared films are investigated. In addition, the adsorption and photocatalytic activities of the prepared transparent nanocomposite films have been subjected for the degradation of methylene blue (MB) dye.

2. Results and Discussion

2.1 UV-Visible spectroscopy

The absorption spectra are used to monitor the formation ofCuO NTs. The prepared films and the absorption spectra of the plant extract, copper nitrate and CuO NTs are presented in Figure1a and Figure 1b.As seen in Figure 1b, plant extract exhibited absorption band at λ_{max}= 298 nm was due to presence of phytochemicals whileabsorption band at 293 nm for the copper nitrate whereas the increased absorption band at λ_{max} = 295 nm was corresponded to the CuO NTs the increase in the absorption as well as change in the λ_{max}compared to copper nitrate solution was due to the surface plasmon resonance of CuO NTs [11]. FTIR, XRD, SEM EDX analysis given in the ESI.

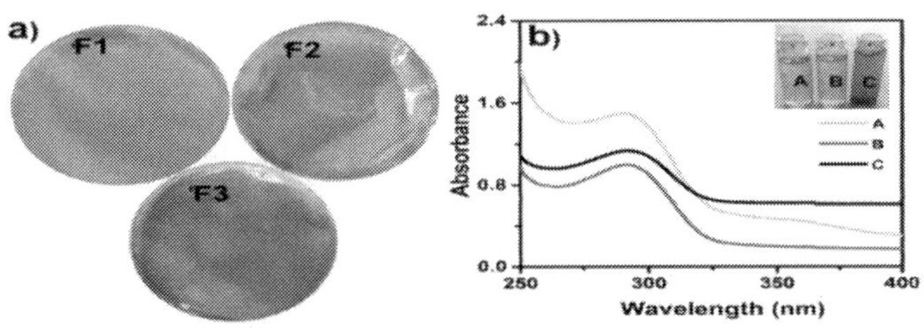

Figure 1a) Image of prepared PVA and its composite films, b) UV-Vis spectra of: A) plant extract, B) Cu $(NO_3)_2.3H_2O$ and C) CuO NTs

2.2 Water contact angle (WCA)

The surface wettability of the prepared composite films was analyzed by static contact angle measurement and summarized in Figure2Sd (ESI).The WCA of F1 film was found to be 69.5±4.1°, which indicates that, the film has hydrophilic surface. The film surface that has contact angle < 90° and > 90° are said to be hydrophilic and hydrophobic surfaces respectively. Higher the contact angle lower will be wettability and vise-versa [12].The WCA of F2& F3 films were found to be 50.1±6.3° and 39.8±2.9° respectively; it shows that, the incorporation of CuO NTs and Asp onto a PVA enhances the hydrophilicity, surface wettability, surface energy and self-cleaning nature of the film surface.

2.3 Soil degradation

The soil degradation of the packing material is the most important property for the eco-friendly application concern [4]. The soil degradation of the prepared films in 10 days experimental data was shown in Fig. 2.The degradation rate of the film F1 shows 25%. This degradation rate was increased after incorporation of the CuO NTs and Asp onto PVA. The percentage degradation of the prepared films in soil found to be 65.5% and 68%, for F2 and F3 respectively was shown in Fig. 1S (Supporting information) The compositefilm F2 and F3 shows higher degradation (>50%) than the film F1.

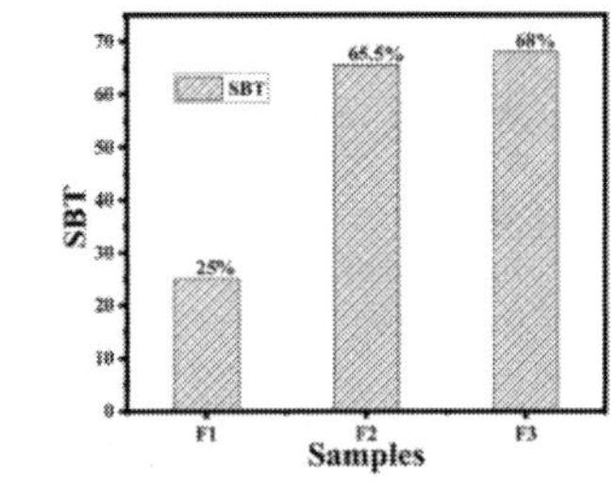

Figure 2: Soil degradation of prepared films F1-F3

3. Optical properties

3.1 Optical absorption and transmission

Figure 3a shows the UV–Vis absorption spectra of the PVA nanocomposite films. An absorption band of pure PVA is observed at 260 nm, and it is assigned to π-π^* electronic transition which may due to the acetate group in the PVA back bone or the semicrystalline nature of PVA. F2 film shows the large and broad absorbance peak at 264 nm due very good interaction between the –OH group of PVA and CuO NTs. F3 film shows the large and broad two absorbance bands in range of 300 nm and 460 nm ($n \rightarrow \pi^*$ electronic transition indicates the presence of carbonyl group) due to the complex formation between CuO NTs and Asp, and it was confirmed by the formation of blue color of the film (Figure 1a). The absorbance of the polymer prepared films increases with the addition of CuO NTs and Asp to the PVA, the absorption edge shifts towards higher wavelengths (red shift) [8]. The redshift of absorption edge of the composite films indicated hydrogen bonding between molecules of CuO NTs, PVA (–OH groups), Asp (–NH_2, –COOH groups), red shift is also for the increase in the crystallinity films with addition of CuO NTs and Asp to the PVA.Figure 3b shows the transmission spectra of PVA nanocomposite films. It is clear that, the transmission slight changes with the addition of CuO NTs and Asp to the PVA. This indicates very good transmission spectra in the UV region (200–250 nm) related to the PVA bandgap [12]. The practically good transmittance of the polymer films in the visible region (400–800 nm) is due to the normal dispersion of the light.

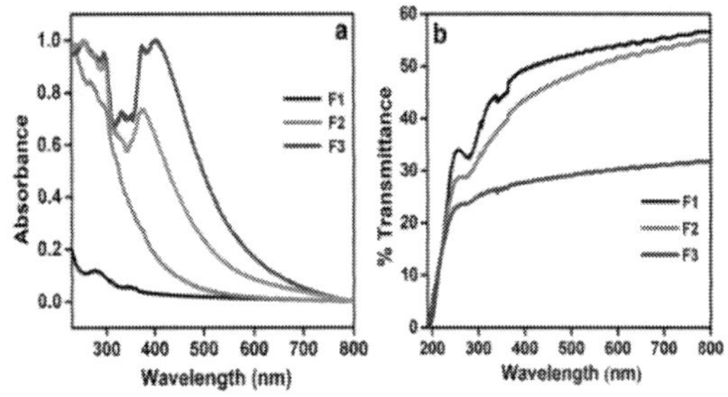

Figure 3: UV-Vis absorption spectra (a) and UV-vis transmittance spectra (b) of prepared nanocomposite films

3.2 Optical energy bandgap

The optical absorption coefficient was investigated *via* photon energy to identify the band structure and the type of electron transition take place for the polymer nanocomposites using Tauc's power law (eq.3):

$$\alpha h v = B(hv - Eg)^n \quad (1)$$

where α is the absorption coefficient (α = 2.303×A/d), A is the absorbance, d is the polymer film thickness, hυ is the incident photon energy, B is constant, Eg is the optical bandgap of the material, type of transitions allowed direct, allowed indirect, forbidden direct and forbidden indirect transitions and their exponent n can be equal to 1/2, 2, 3/2 or 3 respectively [8].The direct optical band gap obtained by $(\alpha h\upsilon)^2$ v/s (hυ) plot illustrated in Figure 4a, the optical band gap for pure F1 is equal to 5.6 eV, which decreases to 3.0 eV for F2 and 2.3 eV for F3. The indirect optical band gap obtained by $(\alpha h\upsilon)^{1/2}$ v/s (hυ) plot is illustrated in Figure 4b, it can be seen from the figure7b that, the optical band gap for pure F1 is equal to 5.3 eV, which decreases to 2.7 eV for F2 and 2.1 eV for F3. In both direct and indirect band gaps, the decrease in the optical band gap from F1–F3 is an indicator of the change in the structure of the polymer matrix and is associated with the creation of localized state in the bandgap. This decreased bandgap is very much necessary to enhance the photocatalytic activity of the film.

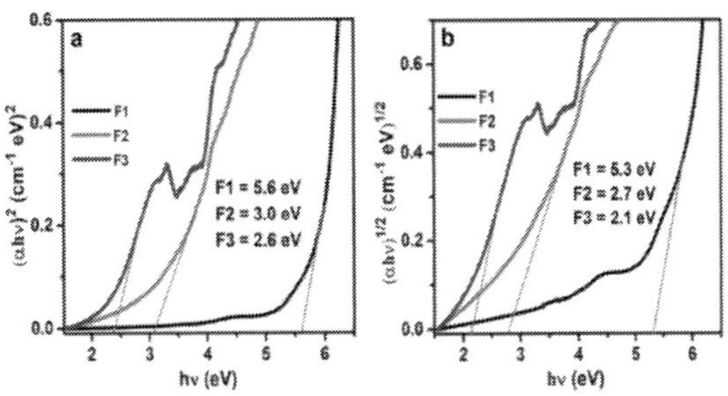

Figure 4: The plots of direct (a) and indirect (b) optical band gaps polymer nanocomposite films

3.3 Extinction coefficient and refractive index

The most important parameters of optical application are the refractive index (n) and the extinction coefficient (k). These optical parameters were determined from the reflectance and absorption coefficient (α) data according to the following equations (4), (5) and (6):

$$k = \frac{\alpha\lambda}{4\pi} \quad (2)$$

$$n = \frac{1+\sqrt{R}}{1-\sqrt{R}} \quad (3)$$

$$R = 1 - (\sqrt{T * \exp(A)} \quad (4)$$

Where R is the reflectance calculated from the transmission (T) and absorption (A) data. Soliman, *et al.*, reported the extinction coefficient decreases with increase of wavelength in the UV region, the incident photon has high energy in the UV region and this energy is enough to excite electrons to overcome the band gap to transfer electron from the valence band to the conduction band. While, in the visible region (400–800 nm), the k value increases with the increase in the wavelength of the incident light, this can be because of the photon energy is unable to excite electron from one state to another state, and this energy can be lost due to scattering or reflection by polymer films. But in our present work (Figure5a), the extinction coefficient decreases continued in the visible region also (extinction coefficient of film F2 decreases up to 430 nm, film F3 decreases up to 600nm of the visible region), due to this, incident photon have enough energy in the visible region and this energy is enough to excite electrons to overcome the band gap (to transfer electrons from the valence band to the conduction band), due to this reason, the prepared films showed effective photocatalytic degradation of MB in the visible light.

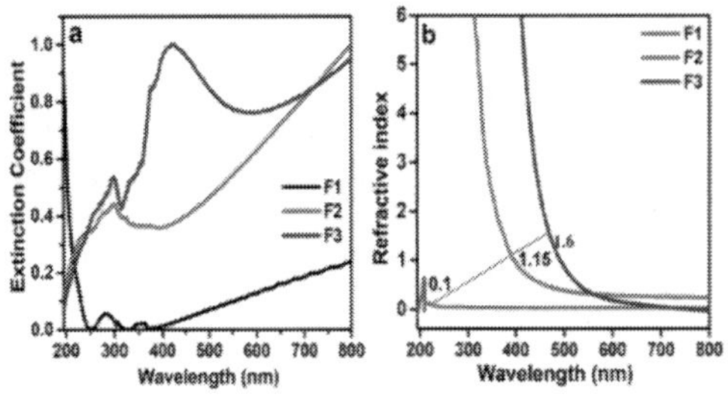

Figure 5: *The plots of (a)* Extinction coefficient*(b)* refractive index *polymer nanocomposite films*

Figure5b shows the dependences of the refractive index on the wavelength for PVA polymer nanocomposite films. It is evident that n values increase (F1=0.1, F2=1.15 and F3=1.6) as the addition of CuO NTs and Asp. This is associated with the high density with which CuO NTs and ASP are packed in the polymer matrix. The increase in refractive index (n) is assigning the material optical systems and electronic features [13].

3.4 Sorption and photocatalytic degradation of MB dye

The prepared composite films and CuO NTs were evaluated for their photocatalytic performance of the dye MB degradation by exposure to sunlight shown in the Figures 6(b–f). There was no MB degradation observed in the absence of the catalysts, which indicated the stability of the MB under self-photolysis [14].The prepared film F3 have a negatively charged surface with carboxylic and amine groups, as well as a pH of 7–9 (based on individual substrate pH), which enhances the adsorption of cation species due to electrostatic attraction between the composite film negative charge surface and the positive charge on the MB dye. The net sunlight irradiation time was exposed up to 70 min, and the absorption intensity measured at 664 nm, and the absorption intensity was monitored for every 10 min interval. The results revealed that, the bare CuO NTs and F1 film shows low photocatalytic activity against degradation of the MB for 70 min and 50 min was 52% and 20% respectively. Meanwhile F2 and F3nanocomposite films gave much higher degradation of 74.99% and 77.15% respectively compared to PVA and CuO NTs alone (Figure8a). The experimental data revealed that, the photocatalytic activity of F1 boosts with the addition of CuO NTs and Asp. The film F3 composite degraded dye 77.15% in 50 min and F2 took 74.99% in 70 min dye degradation. This is because of strong interaction and complex formation between the CuO NTs and Asp in film F3 (confirmed by the color change to blue in the film and FT–IR spectra), which enhanced photocatalytic activity.

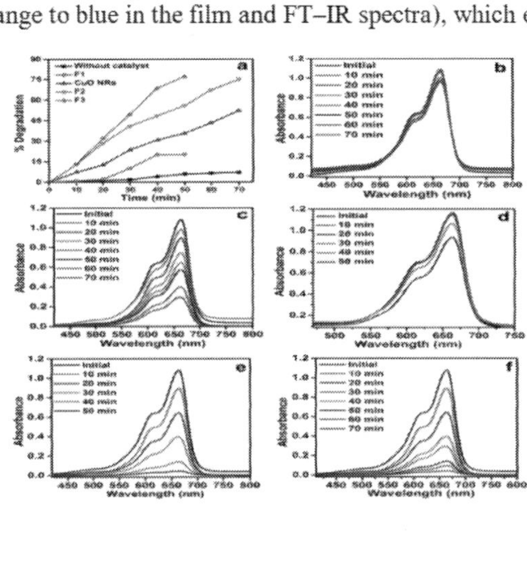

Figure 6(a) Degradation percentage of Methylene blue, Changes in the UV-Vis absorption spectra of MB aqueous solution in the presence of (b) Without catalyst, (c) CuO NTs, (d) PVA, (e) PVA–CuO NTs, and (f) PVA–CuO NTs-Asp composite films

3.5 Photocatalytic mechanism

The superior photocatalyticactivity of the PVA/CuO NTs/Asp composite filmunder Sun light may be ascribed due to its by: (i) strong interaction between the CuO NTs and Asp, (ii) reducing the band gap bycreation of localized state in the bandgap, (iii) enhance the efficiency of photo generated electron–hole pair separation, (iv) very good adsorption ability, (v) decreased extinction coefficient in the visible region due to this, incident photon have enough energy in the visible region and this energy is enough to excite electrons to overcome the band gap, (vi) optically transparent.

$$CuO\ NTs + h\nu \rightarrow h^{+} + e^{-}\ (CuO\ NTs\ pristine)$$
$$CuO\ NTs - Asp + h\nu \rightarrow h^{+} + e^{-}(Composites)$$

The possible mechanisms can be put forward for the degradation of MB by the prepared composites and CuO NTs under visible light irradiation (Figure7) which may proceed through dye sensitization and CuO NTs–Asp sensitization. During sensitization, MB must be adsorbed onto the surface of the composite film. The adsorbed MB and CuO NTs–Asp will absorb visible light and get excited from its ground state to its excited state and migrated to the conduction band (CB) generating holes on the VB (Figure. 14). The generated holes in the VB react directly with the organic dye (MB) or with the surrounding water molecules to generate hydroxyl radicals (•OH). The electrons collected on the CB react with the surrounding water and oxygen molecules to generate hydroxyl (•OH) and superoxide radicals (•O_2^-). In such a way, produced radicals and holes are able to contribute essentially to the degradation of the color MB and other natural toxins, within the extreme response of the photodegradation of the MB, the final products of the response are CO_2 and H_2O.

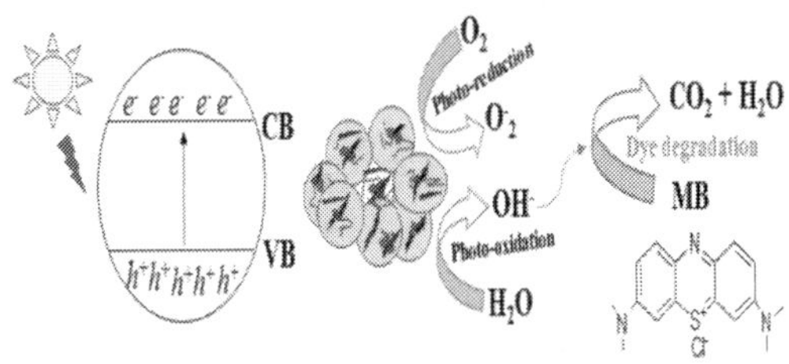

Figure 7: Plausible photo-degradation mechanism of MB

4. Conclusion

In summary, biosynthesized CuO NTs successively incorporated into PVA matrix using aspartic acid is cross-linking agent and the properties of prepared nanocomposite films were investigated by different characterization techniques. The use of Asp allows the CuO NTs to form complex layer on the surface of PVA film is confirmed by the FT-IR and UV-vis absorption and the blue color appearance of the prepared film. The FESEM images confirmed the morphology of CuO NTs and PVA/CuO NTs/Asp nanocomposite film. The optical properties of PVA/CuO NTs/Asp nanocomposite film such as absorption, transmittance, direct band gap, indirect band gap, extinction coefficient, refractive index exhibited good agreement with the reduction in band gap. The adsorption behavior of PVA/CuO NTs/Asp nanocomposite film was studied by Langmuir, Freundlich adsorption isotherm, and pseudo second order kinetics. The photo-catalytic activities of the CuO NTs, PVA and PVA nanocomposite films are examined for the methylene blue (MB) degradation, and film PVA/CuO NTs/Asp showed highest (up to 77%) degradation compared to other films in 50 min time duration. These results suggest that the prepared PVA/CuO NTs/Asp nanocomposite films could find use in photo-catalytic removal of toxic organic contaminants from wastewater due to its simplicity, self-cleaning nature and low cost.

Acknowledgement

Authors are thankful to UGC for the award of Major Research Project and VGST, Govt. of Karnataka for SMYSR award and K-FIST-Level-II for financial support to KK.

Reference

[1] El-Shamy, A.G., Zayied, H.S.S., 2020. New polyvinyl alcohol/carbon quantum dots (PVA/CQDs) nanocomposite films: Structural, optical and catalysis properties. Synthetic Metals **259,** 116218.
[2] Selvi, J., Mahalakshmi, S., Parthasarathy, V., 2017. Synthesis, Structural, Optical, Electrical and Thermal Studies of Poly(vinyl alcohol)/CdO Nanocomposite Films. Journal of Inorganic and Organometallic Polymers and Materials **27**, 1918–1926.
[3] Bdewi, S.F., Abdullah, O.G., Aziz, B.K., Mutar, A.A.R., 2016. Synthesis, Structural and Optical Characterization of MgONanocrystalline Embedded in PVA Matrix. Journal of Inorganic and Organometallic Polymers and Materials **26,** 326–334.
[4] Gasti T, Hiremani VD, Kesti SS, et al (2021) Physicochemical and Antibacterial Evaluation of Poly (Vinyl Alcohol)/Guar Gum/Silver Nanocomposite Films for Food Packaging Applications. J Polym Environ.
[5] Zayed, M.F., Eisa, W.H., Anis, B., 2016. Removal of methylene blue using Phoenix dactylifera/PVA composite; an eco-friendly adsorbent. Desalination and Water Treatment **57**, 18861–18867.
[6] Zhu, D., Wang, L., Yu, W., Xie, H., 2018. Intriguingly high thermal conductivity increment for CuO nanowires contained nanofluids with low viscosity. Scientific Reports **8,** 1–12
[7] Dey, K.K., Kumar, P., Yadav, R.R., Dhar, A., Srivastava, A.K., 2014. CuO nanoellipsoids for superior physicochemical response of biodegradable PVA. RSC Advances **4**, 10123–10132.
[8] Aslam, M., Kalyar, M.A., Raza, Z.A., 2020. Fabrication of nano-CuO-loaded PVA composite films with enhanced optomechanical properties. Polymer Bulletin.
[9] Tulip, P.R., Clark, S.J., 2006. Lattice dynamical and dielectric properties of L-amino acids. Physical Review B - Condensed Matter and Materials Physics **74,** 15–19.
[10] Zanatta, G., Gottfried, C., Silva, A.M., Caetano, E.W.S., Sales, F.A.M., Freire, V.N., 2014. L-asparagine crystals with wide gap semiconductor features: Optical absorption measurements and density functional theory computations. Journal of Chemical Physics **140.**
[11] Felix, S., Chakkravarthy, R.B.P., Grace, A.N., 2015. Microwave assisted synthesis of copper oxide and its application in electrochemical sensing. IOP Conference Series: Materials Science and Engineering **73**, 1–6.
12] Cholant, C.M., Rodrigues, M.P., Peres, L.L., Balboni, R.D.C., Krüger, L.U., Placido, D.N., Flores, W.H., Gündel, A., Pawlicka, A., Avellaneda, C.O., 2020. Study of the conductivity of solid polymeric electrolyte based on PVA/GA blend with addition of acetic acid. Journal of Solid State Electrochemistry **24**, 1867–1875.
13] Aslam, M., Kalyar, M.A., Raza, Z.A., 2018. Investigation of Zinc Oxide-Loaded Poly(Vinyl Alcohol) Nanocomposite Films in Tailoring Their Structural, Optical and Mechanical Properties. Journal of Electronic Materials **47,** 3912–3926.
14] Rahman Khan, M.M., Pal, S., Hoque, M.M., Alam, M.R., Younus, M., Kobayashi, H., 2019. Simple Fabrication of PVA-ZnS Composite Films with Superior Photocatalytic Performance: Enhanced Luminescence Property, Morphology, and Thermal Stability. ACS Omega **4**, 6144–6153.

Magnetic and magneto caloric properties of melt-spun rare earth intermetallic compound $DyAl_2$

Mitali Madhusmita Prusty[1], J. Arout Chelvane[2], A. V. Morozkin[3] and R. Nirmala[1]*

[1]Department of Physics, Indian Institute of Technology Madras, Chennai 600 036, India
[2]Defence Metallurgical Research Laboratory, Hyderabad 500 058, India
[3]Department of Chemistry, Moscow Lomonosov State University, Moscow, 119992, Russia
E-mail: nirmala@physics.iitm.ac.in

Abstract. Laves phase rare earth intermetallic compound $DyAl_2$ has been prepared by melt-spinning under argon atmosphere. The melt-spun $DyAl_2$ sample is crystalline (cubic structure, space group *Fd-3m*) and is nanostructured. The crystallite size of melt-spun $DyAl_2$ calculated from the room temperature powder X-ray diffraction data is about 23 nm. Transmission electron microscopy image reveals particles of average size, 14 nm. The melt-spun $DyAl_2$ undergoes a paramagnetic to ferromagnetic transition at ~29 K (T_C). This value is about 32 K lower than the ferromagnetic transition temperature of $DyAl_2$ sample prepared by conventional arc-melting technique. Magnetization of the melt-spun $DyAl_2$ does not saturate at 2 K in field of 70 kOe. Magnetic entropy change near T_C has been calculated using the field dependent magnetization data. The maximum value of isothermal magnetic entropy change (ΔS_m) of melt-spun $DyAl_2$ is ~ -10.5 $Jkg^{-1}K^{-1}$ at 30 K for a field change of 70 kOe. The ΔS_m value is as large as -24 $Jkg^{-1}K^{-1}$ at 63 K for the arc-melted $DyAl_2$ compound for the same field change. The formation of nanograins upon melt-spinning has led to the reduction of T_C as well as the magnetocaloric effect around T_C.

1. Introduction

Magnetocaloric effect (MCE) is related to the heating or cooling of ferromagnetic materials upon magnetization or demagnetization under suitable thermodynamic conditions [1, 2, 3]. Magnetocaloric effect has also been an intense research topic of the last few decades aiming at practical applications such as magnetic cooling devices and heat pumps. In this regard, rare earth intermetallic compounds, transition metal alloys and ceramic oxides are considered as prominent candidate materials. Among these, rare earth based compounds are of interest because of the large intrinsic magnetic moment and molar magnetic entropy associated with the tripositive rare earth ions. Rare earth intermetallics and alloys are usually prepared by electric arc-melting of constituent elements. Alternatively, the arc-melted compounds could be remelted and the molten liquid could be rolled over a cold, spinning copper wheel into thin ribbons. This is called melt-spinning and is considered useful to stabilize metastable phases. Since the rate of quenching is very large, melt-spinning technique also leads to formation of materials with micro- and nano-granularity. Melt-spinning method has been used to synthesize a few giant magnetocaloric materials in highly crystalline form that show improved magnetic and magnetocaloric properties [4, 5, 6]. Laves phase rare earth intermetallic compounds RAl_2 (R = heavy rare earth) order ferromagnetically at low temperatures and exhibit large magnetocaloric effect [7, 8]. The magnetic transition temperature spans from ~10 K for $ErAl_2$ to ~165 K for $GdAl_2$. Recent studies on melt-spun $GdAl_2$ compound revealed nanogranularity but the large magnetocaloric effect property was conserved [9]. In the present work, the rare earth dialuminide $DyAl_2$ has been prepared by melt-spun technique and its magnetocaloric properties are studied. Small

$DyAl_2$ particles of ~14 nm are obtained and ferromagnetic transition temperature is reduced by about fifty percent upon melt-spinning. However, substantial isothermal magnetic entropy change values are obtained near the magnetic transition temperature in melt-spun $DyAl_2$.

2. Experimental details

Polycrystalline $DyAl_2$ compound has been prepared by electric arc-melting under argon atmosphere starting from stoichiometric amounts of constituent elements [Dy (3N pure), Al (4N pure), Goodfellow, UK] under Ar atmosphere. The sample was melted a few times to ensure homogeneity. The mass loss after melting was less than 0.5%. The arc-melted ingot was used as a starting material for melt-spinning under argon atmosphere. A cold copper wheel rotating with speed of ~17 m/s was used to spin fine ribbon samples. The melt-spun samples were studied in as-prepared condition, without any annealing. Arc-melted and melt-spun $DyAl_2$ have been characterized by powder X-ray diffraction (XRD) (Rigaku, CuK_α radiation, λ=1.5406 Å). Powder XRD data were analysed using Rietan program within isotropic approximation [10]. Transmission electron microscope (Technai G2 T20) was used to study the melt-spun sample. DC magnetization has been studied as a function of temperature and applied magnetic field using commercial magnetometers (MPMS 3, Quantum Design). Isothermal magnetization data obtained near the magnetic transition have been used to compute the magnetocaloric effect.

3. Results and Discussion

Powder XRD data of arc-melted and melt-spun $DyAl_2$ (R=Dy, Er) compounds confirm that these samples crystallize in $MgCu_2$-type cubic structure (Space group *Fd-3m*). The lattice parameters are listed in table 1. The melt-spun treatment leads to a slight shrinking of unit cell with a small decrease in lattice parameter (a) and unit cell volume (V).The broadening of XRD peaks in the data of melt-spun sample suggests small crystallite sizes. The crystallite size of the melt-spun $DyAl_2$ compound is estimated using Scherrer formula and the average value is ~23 nm. Transmission electron microscopy image also revealed small particles of average size ~14 nm [Fig. 1].

Table 1

Lattice parameter (a), unit cell volume (V) and reliability factor (R_F) of arc-melted and melt-spun $DyAl_2$ compounds ($MgCu_2$-type, space group *Fd-3m*, N 227, *cF*24). #XRD contains a few unindexed reflections of low intensity.

Compound	***a*** (Å)	V(Å^3)	R_F (%)
Arc-melted $DyAl_2$	7.8449(4)	482.79	7.8
Melt-spun $DyAl_2$ #	7.8234(12)	478.84	12.2

Magnetization vs temperature has been measured in the temperature range of 300 K to 5 K [Fig. 2]. A ferromagnetic order is observed at temperatures 61 K and 29 K (T_C) respectively in arc-melted and melt-spun $DyAl_2$. Arc-melted $DyAl_2$ is known to order ferromagnetically at 61 K and is a second order transition [7, 11]. The hysteresis observed between the zero-field-cooled and field-cooled magnetization data measured in 100 Oe field below T_C could be attributed to domain wall pinning effects in the ferromagnetic state. This irreversibility vanishes in the magnetization data obtained in 5 kOe field. The reduced magnetic transition temperature of melt-spun $DyAl_2$ could be attributed to the nanogranularity of the sample.

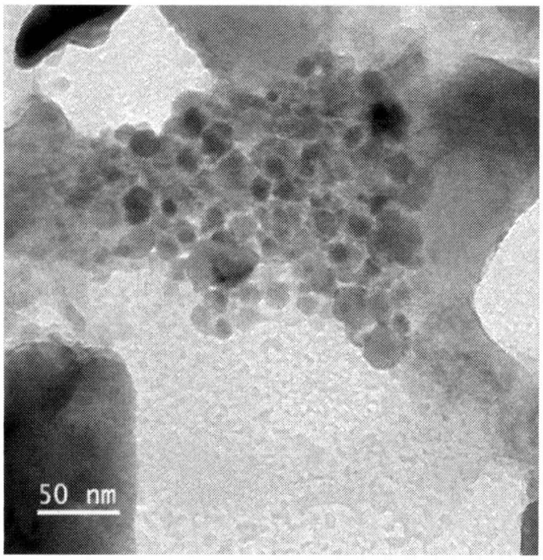

Figure 1 Transmission electron microscopy image melt-spun $DyAl_2$ indicating nanosized particles.

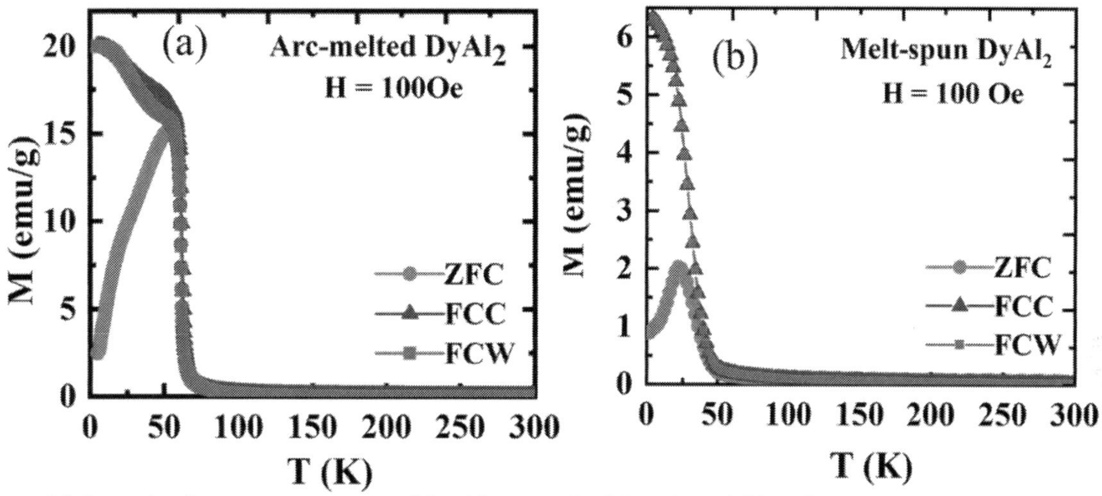

Figure 2 Magnetization vs temperature of the (a) arc-melted $DyAl_2$ and (b) melt-spun $DyAl_2$ in 100 Oe field measured during zero-field-cooled (ZFC), field-cooled cooling (FCC) and field-cooled-warming (FCW) conditions.

The paramagnetic susceptibility is fitted to the Curie-Weiss law. From the fit, the paramagnetic Curie temperature (θ_p) is obtained as +77 K and +12 K respectively for the arc-melted and melt-spun $DyAl_2$ samples. The corresponding effective paramagnetic moment values are 11 μ_B/f.u. and 10 μ_B/f.u. respectively. The magnetization vs field measured at 5 K depict the soft ferromagnetic nature of arc-melted $DyAl_2$ [Fig. 3]. The hysteresis is negligible also for the melt-spun sample. However, the magnetization does not saturate even in fields of 70 kOe. Particle size induced disorder could play a role.

IOP Conf. Series: Materials Science and Engineering 1221 (2022) 012009 doi:10.1088/1757-899X/1221/1/012009

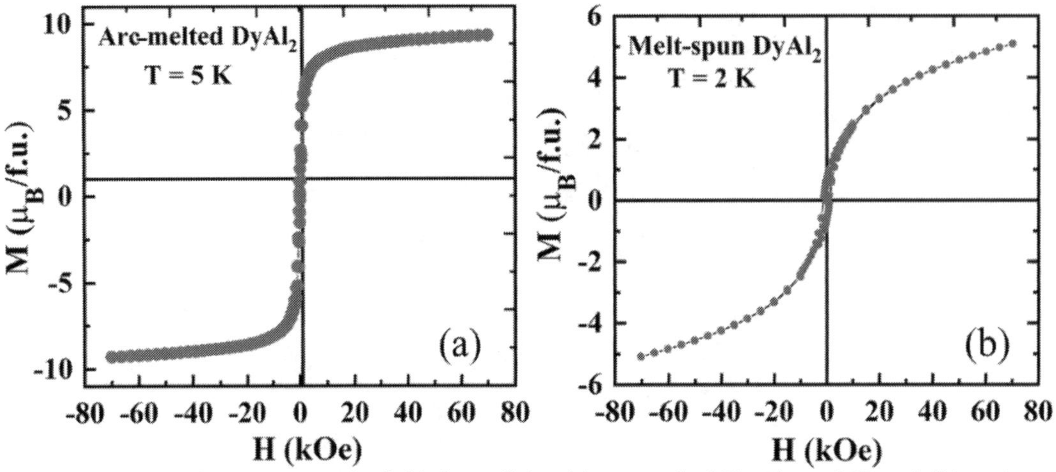

Figure 3 Magnetization vs magnetic field data of the (a) arc-melted $DyAl_2$ at 5 K and (b) melt-spun $DyAl_2$ at 2 K.

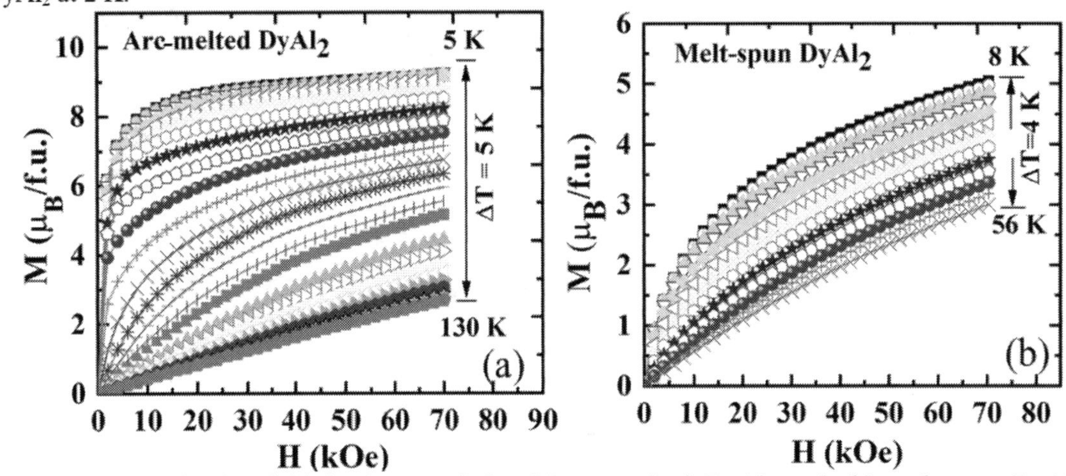

Figure 4 Magnetization-field isotherms of the (a) arc-melted $DyAl_2$ and (b) melt-spun $DyAl_2$, measured around T_C in zero-field cooled condition.

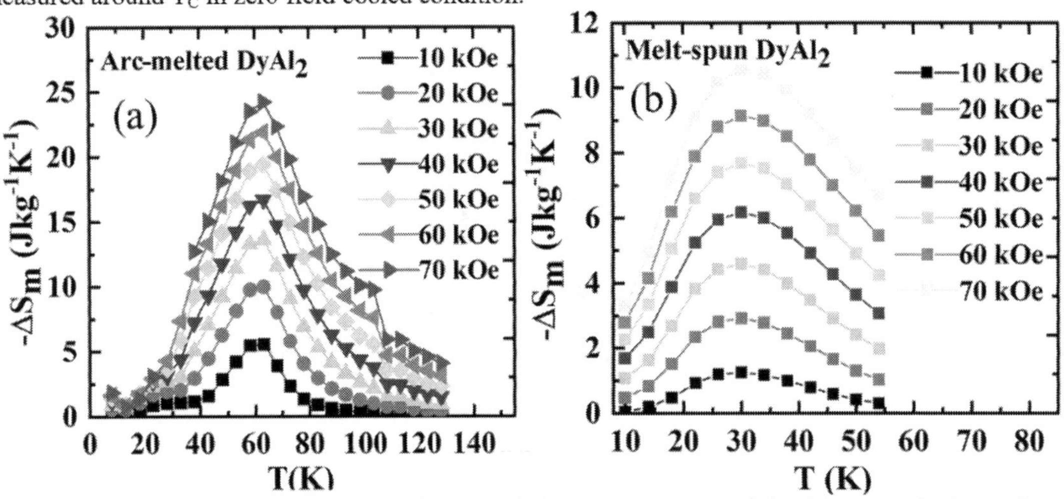

Figure 5 Isothermal magnetic entropy change (ΔS_m) vs temperature of the (a) arc-melted $DyAl_2$ and (b) melt-spun $DyAl_2$ for typical magnetic field changes.

Using the magnetization vs field data obtained around T_C, isothermal magnetic entropy change (ΔS_m) has been calculated. [Fig. 4]. Starting from thermodynamic Maxwell relation the following expression is arrived at.

$$\Delta S_m = \mu_0 \int_{H_i}^{H_f} \left(\frac{\partial M}{\partial T}\right)_H dH$$

Here, H_i and H_f are the initial and final value of applied magnetic fields and μ_0 is the permeability of free space [12]. The maximum value of ΔS_m for 70 kOe field change at 30 K is about -10.5 $Jkg^{-1}K^{-1}$ for melt-spun $DyAl_2$ [Fig. 5]. This is much smaller than that in the arc-melted $DyAl_2$.The ΔS_m value is -24 $Jkg^{-1}K^{-1}$ at 63 K for arc-melted $DyAl_2$ for 70 kOe field change.

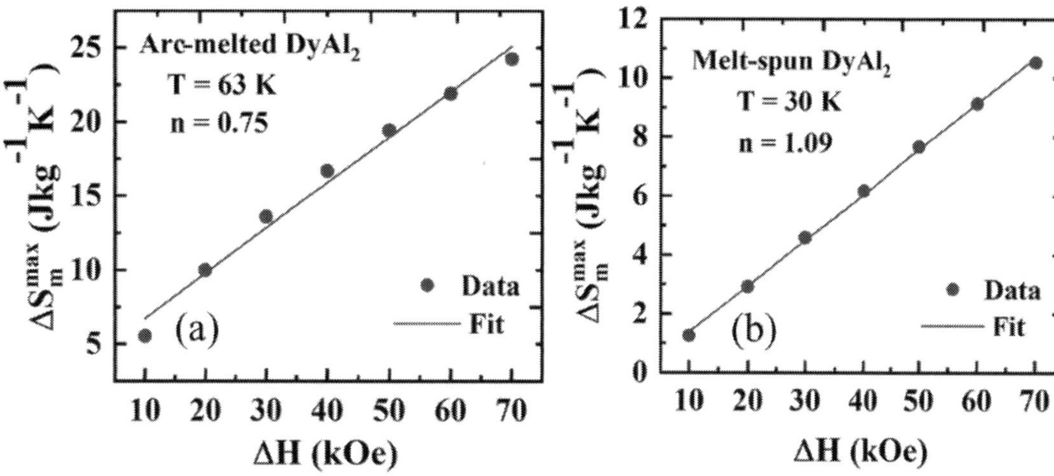

Figure 6.Field dependence of maximum value of isothermal magnetic entropy change (ΔS_m^{max}) of the (a) arc-melted $DyAl_2$ and (b) melt-spun $DyAl_2$.The power law fit ($\Delta S_m^{max} \propto \Delta H^n$) is shown.

The magnetic field dependence of magnetocaloric effect is studied by fitting the maximum isothermal magnetic entropy change values for a given field change (ΔH) to a power law i.e. $\Delta S_m^{max} \propto \Delta H^n$ [Fig. 6]. The value of 'n' is found to be 1.09 for the melt-spun $DyAl_2$ sample at 30 K whereas a value of n = 0.75 is found for arc-melted $DyAl_2$. A value of n = 2/3 is expected for a mean-field ferromagnet [13]. The substantial deviation of the exponent from the mean-field value in the melt-spun $DyAl_2$ could be related to the spin disorder in the nanosized sample.

Earlier study on melt-spun $GdAl_2$ and ball-milled $GdAl_2$ show a disorder broadened magnetic entropy change spread over a wide temperature range around T_C [9, 14]. Nanostructured $GdAl_2$ prepared by long-duration mechanical milling is atomically disordered and exhibits spin glass-like behaviour [15, 16]. Likewise, $DyAl_2$ nanoparticles of average size 50 nm (coated with alumina) prepared by arc-discharge method show decreased magnetic entropy change values compared to that of bulk [17]. Present work reveals that melt-spinning with moderate wheel speeds results in crystalline Laves phase intermetallics RAl_2 (R = Gd and Dy) of nanosized particles retaining the low temperature ferromagnetism.

4. Conclusions

Crystalline, nanosized $DyAl_2$ (cubic, *Fd-3m*) compound has been prepared by melt-spinning method. The ferromagnetic ordering temperature is ~29 K and is about half of the arc-melted $DyAl_2$. Using magnetization vs field data, isothermal magnetic entropy change has been estimated. The maximum value of ΔS_m of melt-spun $DyAl_2$ is about -10.5 $Jkg^{-1}K^{-1}$at 30 K for 70 kOe field change. This value is much smaller than in arc-melted $DyAl_2$ and that could be attributed to size induced disorder.

5. Acknowledgement

Authors thank Prof. A. K. Nigam, TIFR for helping us to record a couple of low temperature magnetization vs field data.

References

[1] Gschneidner Jr K. A, and Pecharsky V. K., Annu. 2000, Rev. Mater. Sci. **30** (387
[2] Lyubina J., 2017, J. Phys. D: Appl. Phys. **50** 053002.
[3] Franco V., Blázquez J.S., Ipus J.J., Law J.Y., Moreno-Ramírez L.M. and A. Conde, Prog 2018 Mater.Sci. **93** 112.
[4] Yan A., Müller K. □ H, Gutfleisch O., 2005, J. Appl. Phys. **9** **7**036102
[5] Lyubina J., Schäfer R., Martin N., Schultz L. and Gutfleisch O., 2010, Adv. Mater. **22** 3735
[6] Bez H. N., Pathak Arjun K., Biswas A., Zarkevich N., Balema V., Mudryk Y., Johnson D. D. and Pecharsky V. K., 2019, Acta Materialia, **173** 225
[7] Purwins H. and Leson A., 1990, Adv. Phys. **39** 309
[8] Tishin A.M. and Spichkin Y.I., 2003, The Magnetocaloric Effect and its Applications, Institute of Physics publishing, Bristol,
[9] MitaliMadhusmitaPrusty, AroutChelvane J., and Nirmala R., 2020, Mater. Res. Express **7** 064001
[10]. Izumi F., and Young R.A., 1993, Oxford University Press, Oxford 13
[11] Lima A. L., Gschneidner K. A., Jr.,. Pecharsky V. K, and Pecharsky A. O., 2003, Phys. Rev. B **68** 134409
[12] Pecharsky V. K., Gschneidner K. A., and Magn Jr., J, 1999, Magn.Mater.**200** (44
[13] Franco V., Blázquez J. S., and Condea A., 2006, Appl. Phys. Lett.**89** 222512
[14] Paula V. G. de, Silva L. M. da, dos Santos A. O., Lang R., Otubo L., Coelho A. A., and Cardoso L. P., 2016, Phys. Rev. B **93** 094427
[15] Zhou G. F. and Bakker H., 1994, Phys. Rev. Lett. **73** 344
[16] Zhou G. F. and Bakker H., 1995, Phys. Rev. B **52** 9437
[17] Zhang W.S., Bruck E., Zhang Z.D., Tegus O., Li W.F., Si P.Z., Geng D.Y., Klaasse J.C.P. and Buschow K.H.J., 2006, Alloys J.. Compd. **143** 29

IOP Conf. Series: Materials Science and Engineering 1221 (2022) 012010 doi:10.1088/1757-899X/1221/1/012010

The green and chemical synthesis, and characterization of ZnO nanoparticles by aqueous leaf extract of aloe vera

K M Kavya Prakash*,T Vinodkumar and V Prakash

Department of PG Studies & Research in Physics, Payyanur College, Edat,Kannur, Kerala-670327, India

*Email: kavyaprakashkm2018@gmail.com

Abstract. The exact physical and chemical properties of nanoparticles depend on the way in which they are synthesized. The common techniques are hypothermal methods, electrochemical deposition, sol-gel method, chemical vapour deposition, co-precipitation method etc. In the present study, zinc oxide nanoparticles have been synthesized by botanical extract and chemical reducing agents with major objective to compare the nature and average particle size of the nanoparticles synthesized by these two methods. An attempt is also made to understand the influence of molarity on the chemically synthesized NPs by varying the molar concentration of precursor. The structure and morphology of synthesized NPs have been studied using X-ray diffraction (XRD) technique. For chemically synthesized NPs the major peak was observed at 36.147 for both the precursor at 0.2M and 0.3M. The average crystalline size was found to be 13.09 nm and 11.09 nm for precursor at 0.2M and 0.3M respectively. The major peak for green method was observed at 36.147 with an average crystalline size of 9.2 nm. The study clearly indicates that, green method is particularly suitable for the synthesis of smaller sized NPs.In chemical method, the molarity of the precursor influences the particle size, as the molarity decreases particle size also decreases.

1. Introduction

Nanotechnology deals with the study of particles having dimension from 1 nm to 1000 nm. Physical, chemical and biological methods are the commonly used methods for synthesizing nanoparticles [1]. The physical methods are sometimes inappropriate because they require large amount of energy, time and space. The chemical methods are considered as an unfriendly approach as they use toxic and harmful substances [2]. The green methods are generally cheap and safe for the synthesis of nanoparticles. Biodegradable raw materials such as plant parts, microbes, fruit peels and vegetables are used for green synthesis [3,4].The NPs so synthesized find their applications in various fields such as electrochemistry, catalysis, sensors, biomedicines, food packaging, health care, cosmetics, optical devices etc [5].Preparations of inorganic nanocrystalline metal oxides have wide interest due to the extremely large surface area and wide biological applications.

Zinc oxide NPs are type of metal oxide nanoparticles having band gap 3.3 eV and high binding energy (60MeV), and considered to be a very hardest material [2]. The zinc oxide nanoparticles are non-toxic, nearly insoluble in water, show high degree of cancer cell selectivity and appear as white powder [6]. Due to their excellent biocompatibility, synthesis of ZnO NPs became focused area of many research works nowadays [7]. The efficiency and applications of ZnO NPs depend on size, shape, surface state, crystal structure and the way of synthesis [7]. The ZnO NPs are commonly used in sunscreens as it reflects the ultraviolet light and also being used to kill harmful microorganisms in packaging. The ZnO NPs have good photocatalytic properties in the UV-Visible region [8]. Many enzymes and ointments also use zinc NPs as an essential ingredient.

There are several methods used for the preparation of zinc oxide nanoparticles viz. sol-gel technique, chemical precipitation method, thermal method and green route method. In the present study, the biosynthesis of ZnO NPs from leaf extract of aloe brabadens is miller commonly known as aloe vera belongs to asphodelaceae (liliaceae) family and mostly found in subcontinent of India, dry regions of Africa, Asia, Europe and America was preferred. The gel part of aloe vera leafs are widely used in the preparation of traditional medicines and showed antifungal and skin healing properties [9]. The leaf extracts of aloe vera is rich in vitamins, enzymes, minerals and aminoacids [10]. The ZnO NPs were also synthesized by chemical precipitation method for comparison and to understand the efficacy.

2. Materials and methods

Aloe vera leaves were collected locally from the study area and commercially available zinc acetate was used as a precursor. The characterization was done using XRD technique and average particle size calculation by debye-scherrer formula.

The aloe vera is a widely used medicinal plant commonly seen in hot and semi-tropical climates.It is a juicy plant of the genus aloeand originated from Arabian Peninsula. It contains healthy compounds which are used for various medicinal purposes. It is also known as a multifunctional plant as all parts of the plant can be used for various applications.

2.1. Synthesis

*2.1.1. Preparation of aloe vera leaf extract.*Fresh leaves of aloe vera were collected and washed thoroughly with running water and then air dried. The leaf extract was prepared by the gel part of aloe vera. About50 gm of aloe vera was boiled in 50 ml of deionized water for half an hour. The resulting extract was taken for further analysis.

2.1.2. Biosynthesis of ZnO NPs. In biosynthesis, 0.2 M zinc acetate (precursor) was prepared in 50 ml of deionized water. Aloe vera leaf extract (10 ml)was added drop wise under constant stirring using a magnetic stirrer for 10 minutes. In order to adjust the pH of the prepared solution, NaOH (2M) was added drop wise while stirring. The obtained white crystalline precipitate was washed well with distilled water, filtered and dried in an oven at 60°C. The process ultimately ends with the formation of a white smooth powder.

2.1.3. Chemical synthesis of ZnO NPs. In chemical method, commercially available zinc acetate with different molarity (0.3M and 0.2M) was used as precursor and sodium hydroxide as reducing agent.

2.2. Sample preparation

The direct precipitation method was adopted for the chemical production of zinc oxide nanoparticles. Solution of zinc acetate (0.2 M) and sodium hydroxide (0.4 M) were prepared with 100 ml of deionized water. The NaOH solution was slowly added into freshly prepared zinc acetatesolution at room temperature under vigorous stirring. The stirring results a white suspension which indicates the formation of zinc oxide nanoparticles. The white suspension was collected using centrifugation (20 minutes) and washed three times with distilled water, then dried in oven at 150°C for 2 hours. The same process was repeated by changing the molarity of precursor to 0.3 M.

2.3. Zinc oxide nanoparticles (ZnO NPs)
Zinc oxide is a white solid which is insoluble in water. It is considered to be a hazard chemical which is dangerous to the environment also [6]. The inhalation of zinc oxide nanoparticles results in fever, nausea, muscle pain and vomiting. Table 1gives the general properties of zinc oxide nanoparticles.

Table 1. Properties of ZnO[3].

Item	Property
Chemical structure	ZnO
Molecular weight	81.389 $gmol^{-1}$
Appearance	White solid
Odour	Odourless
Crystal structure	Wurtizite
Band gap energy	3.37 eV
Density	5.606 gcm^{-3}

3. Structural and optical characterization
The X-ray diffraction is used to determine crystallographic structure and morphology of nanoparticles. The X-ray powder diffraction method is also used for identification of phase and can provide information about cell dimensions. Average particle size can be calculated using debyesherrer's formula.

$$D = \frac{K\lambda}{\beta\cos\theta} \quad (1)$$

Where, D is average particle size, λ is wave length of the radiation, β is full width at half maximum (FWHM) of the reflection peak and K is Scherrer constant.

The scherrer constant K used in the formula changes with the shape of the particle and commonly used value is 0.9. The most common wave length of X-ray is 1.54Å. The XRD operates at 40kV voltage and 30mA current.

4. Results and discussion
The XRD pattern of zinc oxide nanoparticles synthesized using both chemical and green methods are given below. Figure 1 below shows the XRD pattern of zinc oxide nanoparticle prepared by chemical precipitation method at 0.3 M of zinc acetate as a precursor. The major peaks observed at 31.77, 34.13, 36.08, 47.61 and 56.68. The diffraction pattern and interplanar spacing closely matched with the JCPDS value of ZnO NPs in the standard diffraction pattern. The average particle size calculatedwas found to be 13.09 nm with 1.11405 FWHM.

IOP Conf. Series: Materials Science and Engineering 1221 (2022) 012010 doi:10.1088/1757-899X/1221/1/012010

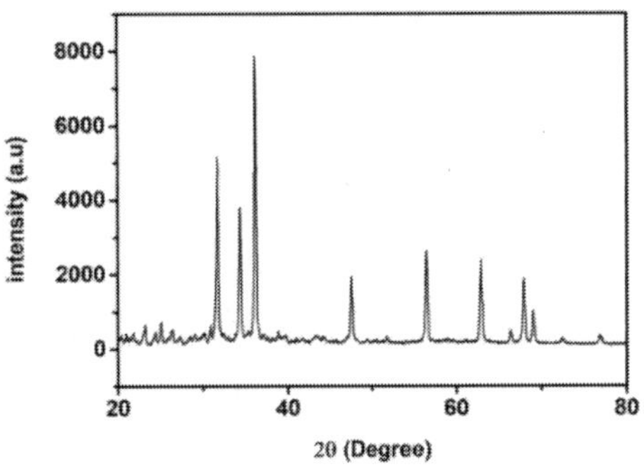

Figure 1. The XRD pattern of ZnO at 0.3 M zinc acetate.

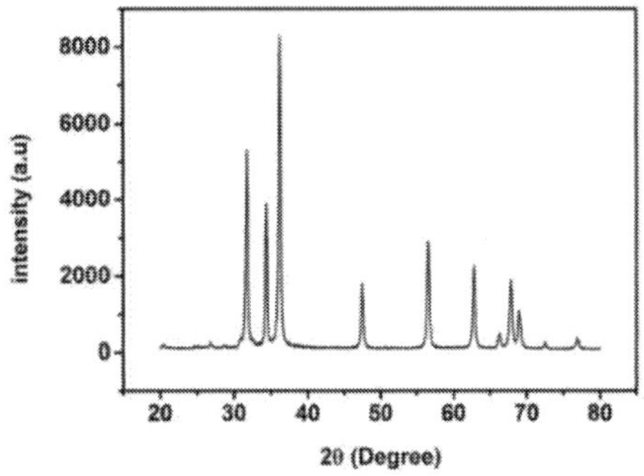

Figure 2. The XRD pattern of ZnO at 0.2 M zinc acetate.

Figure 2 above shows the XRD pattern of zinc oxide nanoparticleprepared by chemical precipitation method at 0.2 M of zinc acetateas a precursor. The major peaks observed at 31.87, 34.3, 36.04, 47.48 and 56.41. The diffraction pattern data obtained were compared with original JCPDS data and found good matching. This confirms the synthesized nanoparticlesare ZnO. The average particle size obtained was 11.95 nm.

Figure 3 below show the XRD pattern of zinc oxide nanoparticles synthesized using green method. The major peaks observed at 31.63, 34.27, 36.14, 47.48 and 56.53. Confirmation was done using JCPDS value of ZnO NPs. The average particle size was found to be 9.2 nm.

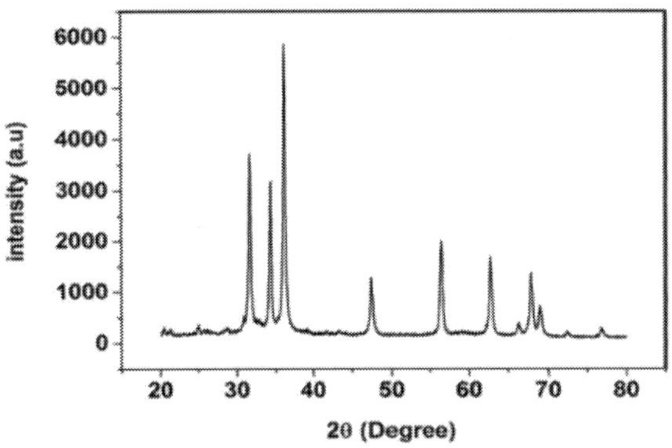

Figure 3. The XRD pattern of ZnO by green method.

4.1. Factors influence the synthesis

There are many factors that influence the synthesis, characterization and application of nanoparticles viz. pH of the solution, temperature of the reaction medium, time, pressure, concentration of the extract solution etc. The green synthesis of nanoparticles uses nontoxic and ecofriendly raw materials and hence got wide acceptance among many other methods. The size of the nanoparticles synthesized depends on the solution pH, however pH has less effect in physical and chemical synthesis. The green synthesis requiresa temperature less than 100°Conly. The exact nature of the synthesized nanoparticles depends on the temperature of the reaction medium.

5. Conclusion

In the present work, it is introduced a cost-effective and ecofriendly method for the synthesize of zinc oxide nanoparticles from aloe vera plant extract. The green synthesis method was effectively implemented to synthesis ZnO nanoparticles. The study clearly indicates that, green method is particularly suitable for the synthesis of smaller sized NPs. In chemical method, the molarity of the precursor influences the particle size, as the molarity decreases particle size also decreases. The particles synthesized can be used as photocatalysts and the biologically synthesized nanoparticles are expected to be used in bio medical fields.

Acknowledgements
The first author is grateful to the Head of the Department of Physics, PRNSS College for providing the lab facility.

REFERENCES

[1] Jeeva Lakshmi Sharath R, ChandraprabhaMN, NeelufarE, AbhishikaHazra and MalyasreePatra 2012 *J Biochem Tech.* **3(5)** S151-s154
[2] Ayeshamariam A, Kashif M, Vidhya V S, Sankaracharyulu M G, Swaminathan V, Bououdina M and Jayachandran M 2014 *Journal of optoelectronics and Biomedical Materials.***6** 85-99
[3] PrathibhaVasudevan and Arun Prasad S 2020 *J. Cluster Science.***8**
[4] Saurabh Sharma, Kuldeepkumar, Naveen Thakur, Chauhan S and Chauhan M S 2020 *IndianAcademy of Sciences.***43** 20
[5] Shakeel Ahmed, Annu, Saif Ali Chaudhary and SaiqaIkram 2017 *J. photochemistry & photobiology, B:Biology.* **166** 272-284
[6] MohdFadhlan Shah Hermandy, MohdZakiMohdYusoff, Mohammad SyarifuddinYahya and RabiulAwal M 2020 *Inernational Journal of Emerging Trends in Engineering Research.* **8(10)** 6896-*6902*
[7] Dheivamalar S and Bansurabanu K 2021 *Indian J. Pure and applied phys.***59** 11-22
[8] Partiwi Y, Yulizar Y and Apriandanu D O B 2020 *IOP conf. Series :Materials Science and Engineering.* **763** 012022
[9] Anbukkarasi V, Srinivasan R and Elangovan N 2015 *International J. pharmaceutical Science Reviewand Research.* **33** 110-115
[10] Noorjahan C M, Jasmine shahina S K, Deepika T and Summers Rafiq 2015 *International J.Scientific Engineering and Technology Research.* **04** 5751-53

IOP Conf. Series: Materials Science and Engineering 1221 (2022) 012011 doi:10.1088/1757-899X/1221/1/012011

Structural and Mechanical characteristic study of HPMC polymer composite films

Sunil Kumar [1,2], Sandeep Dongre[1], Raghu S[3], T. Demappa[4] and J Sannappa[1,*]

[1]Department of P G Studies and Research in Physics, Jnana Sahyadri, Kuvempu University, Shimoga, India
[2]Department of Physics, S.S Arts College and T.P Science Institute, Sankeshwar, Belagavi, India
[3]Department of Physics, KLE Society's, Basavaprabhu Kore Arts, Science and Commerce College, Chikodi, India
[4]Department of Polymer Science, Sir M V, P G Centre, University of Mysore, Mandya, India
Email : sannappaj2012@gmail.com

Abstract. Hydroxypropyl methylcellulose (HPMC) were extensively known material in packaging, agriculture, medicine, electrolytes in solid state batteries and other areas due to the exceptional biocompatibility and easy production, however shows lower mechanical behaviour. The current investigation was embraced to delineate the effect of potassium iodide (KI) on structural and mechanical properties of biopolymer electrolyte films grounded on Hydroxypropyl methylcellulose (HPMC). The liquidation of the salt into the polymer crowd is done by solution cast technique and the X – Ray diffraction (XRD) analyses were used to fix the structural properties of pristine HPMC and composite HPMC films. The crystallinity (X_C) was calculated, the pristine HPMC films shows high level of crystallinity. However improved amorphous spaces of HPMC polymer crowd were exposed in HPMC composite films. The addition of KI had a considerable impact on break elongation. In HPMC, KI showed good intercalation, and large concentrations of KI resulted in crater-like pits on the film surfaces. Reductions in the tensile strength and elastic modulus along with an increase in elongation were noticed in film integrated with KI.

1. Introduction

Biodegradable polymers have attracted a lot of attention in recent years because of their clear environmental benefits over nondegradable or slowly degradable conventional petrochemical-based polymeric products. [1,2]. Flexible electronics on polymer substrates are now being developed for a variety of uses, including electronic textiles and paper-like displays. Due to the flexible material employed as a substrate, such electrical devices are generally subjected to significant external pressures and substantial strains. To obtain a functional and dependable electrical equipment, understanding of the mechanical properties of materials is required. As a result, some research has been done into the mechanism of biopolymer reinforcement. [3,4], particularly in mechanical applications [5,6]. Cellulosic polymer has attracted a great deal of attention due to its low cost and its status as the most available resource in nature. HPMC is the most often utilised hydrophilic, biodegradable polymer for film formation, owing to its ability to operate as a pH-independent and viscous gelling agent..Cellulosic polymers, particularly hydroxypropyl methylcellulose (HPMC) are a generally applicable biopolymer matrix for drug administration, coating and food packaging [7-9]. HPMC is approved by the FDA for use as a film former, emulsifier, stabilizer, protective colloid, suspending agent, or thickener in the food industry [10-12].

However, few basic investigations on the HPMC matrix exist, such as the impact of its molecular weight on mechanical properties. Several papers on bio-based research subjects have recently been developed, and they show tremendous promise for future bio-based coatings applications[13]. Potassium iodide is an inorganic compound with the formula KI, is an ionic compound that contains 23.55 percent potassium and 76.45 percent iodine [14]. KI is a well-known material for its wide range of applications. Potassium Iodide (KI) is one of permissible inorganic salts used in medication, especially in treatment of thyroid

IOP Conf. Series: Materials Science and Engineering 1221 (2022) 012011 doi:10.1088/1757-899X/1221/1/012011

gland [15]. KI is approved by the FDA for use as a food additive [16,17]. Potassium iodide (KI) is used in Commercial and industrial water-cooling tower systems, emergency drinking water purification/disinfecting, fresh food sanitization, food-contact surface sanitization, hospital surface disinfection [18]. KI is also used as polymer composites for battery and electrochemical cell applications. Therefore, the study was based on potassium salts because of their high efficiency compared to other salts.

2. Experimental Techniques

The mechanical properties of the HPMC and the doped films were determined using The mechanical performance of complex film was gauged by the Universal Testing Machine (UTM) (Zwick Roell Z020, Germany). The trial was made with a load cell of 5kN. The thicknesses of the films were measured using a digital screw gauge and it was in the range 300 μm. The X-ray diffractometer (Rigaku Miniflex-II) through Ni filtered CuKα radiation (=1.540 Å) was used to obtain the diffraction patterns at Bragg's angle (2Θ) in the range of 10°-80° with a scanning speed rate 5°/ min and step size 0.02°. The surface morphological of prepared polymer complex films were observed by using Carl Zeiss field emission scanning electron microscopy (FE-SEM), with a resolution image 2.7 nm at 1 Kv and 1.5 nm at 15 Kv.

2.1Materials and sample preparation methods

The salt form of potassium iodide (KI) was used in this study and powder Hydroxypropyl methylcellulose (HPMC-E15LV) polymer was used in our preparation. All of the chemicals were bought from Sigma Aldrich and are of analytical quality. A solution casting approach using double distilled water as a solvent was used to prime polymer composite films [8]. In this sample preparation, three grams (3gm) of HPMC mixed with 100 mL double distilled water (3 wt%) and stirred for 30 minutes at room temperature (30 °C) with a magnetic stirrer. Pure HPMC with a suitable weight percentage ratio (1 wt% and 1.6 wt%) of KI salt was prepared by adding KI salt into each prepared 3wt percent (100mL) of HPMC polymeric solution under constant mixing for 10 to 12 hours. The obtained viscous solutions were poured into flattened glass plates/petri dishes (10cmX15cm) and kept evaporation for 5 days in a vacuum chamber. Then after evaporation of solvent water, polymer films are collected at the base of glass plates and kept in a high-discharge desiccator to avoid dust and moisture uptake. Finally, composite films were prepared, and the thickness of these films was determined using a Screw gauge by checking the thickness at five different parts of each film sample and found to be around 300 μm.

3. Results and Discussions

3.1. Structural Study

XRD is a method for determining whether a substance is crystalline or amorphous. Figure 1 shows how XRD measurement techniques were used to determine the structural integrity of the prepared composite film samples. In figure 1 (T), The pure KI salt peaks were clearly visible between 22 and 78 (PCPDF file no. 780750), indicating the crystalline nature of salt. [17,18]. Following that, these peaks vanish from the composite films (T1, T2), indicating that the dopant salt has completely dissolved in the host matrix. The broad diffraction peak for pure film (T) appeared at around 21.240 (T0), suggesting the semi-crystalline and more amorphous existence of HPMC [8-10].

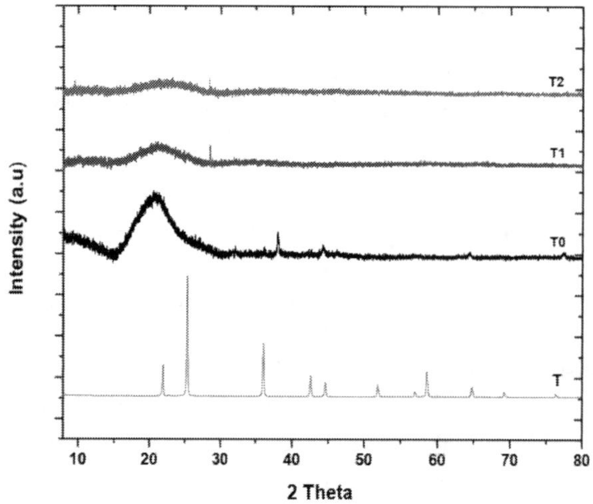

Figure 1. XRD of HPMC/KI composites

3.2. Morphology Investigation (SEM)

SEM is generally used to explore the compatibility of polymer composite materials by identifying interfaces and phase separation. [11]. The compatibility of the polymer matrix and the inorganic dopants has a significant effect on mechanical properties. The surface morphology of the pure (T0) and doped films (T1, T2) with same magnifications were explored by scanning electron microscopy (SEM). From these SEM micrographs of pure HPMC film samples, where one can see a plane morphology, The consistency of the bonds in the film matrix has a major effect on the structure of HPMC film [21]. Complexed polymer composite film samples figure 2 (T1, T2) shows different degree of smoothness, which is indication of segregation of the dopant in that substrate polymer matrix but surface morphologies are small change [8,9], the smooth appearance of the surface is generally associated with a reduction in crystallinity. This is due to salt dispersion and dissociation, which can cause topological disruption in the electrolyte, resulting in a more amorphous phase and makes the electrolyte more flexible. [7-9].

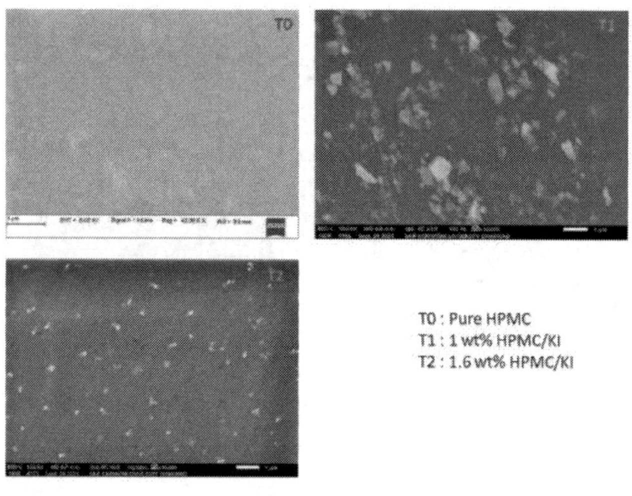

Figure 2.SEM of HPMC/KI composites

3.3. Mechanical properties

The figure 3 showed is the stress-strain curves of virgin (T0) and doped films with different concentrations of salt (T1 and T2). Tensile tests were carried out to define the mechanical behavior of the polymer matrix, namely the Young's Modulus (YM), Tensile strength (TS), and Elongation at break (EB) (tabulated in Table 1) [28,29]. The highest tensile strength value of 37 MPa obtained for pure film matrix, and decline in the tensile strength was resulted for complex films. The tensile modulus data correlates the decrease in intensity observed from structural studies. Crystallinity does have an effect on mechanical properties because chains aligned side-by-side in the crystalline region are closely linked to each other. They become more cripple and stiffer as compared to amorphous materials [30,31]. The material's stiffness is measured by its elastic modulus, it was clear observation from the figure 3 that modulus value was varied with the salt percentage. This can be connected to the partly exfoliated and partly intercalated morphology of the complex films observed in SEM.The elongation at break is a relative difference in the material's length before fracture. It is a ductility index [32]. Elongation at break value of 5.1 % was observed for pure film and its value raises to peak value of 18.1% for T2 which supports the observed structural and morphological data.

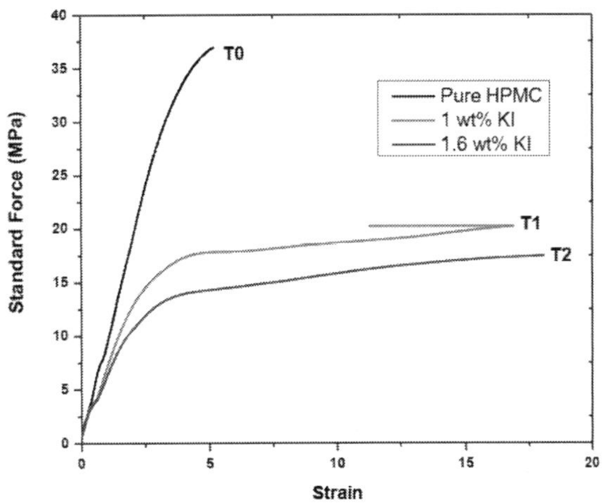

Figure 3. Stress versus strain curve for HPMC/KI composites

Table 1. Mechanical parameters of HPMC/KI composites.

Sample Code	FMAX (N)	Young's Modulus (MPa)	Tensile Strength (MPa)	Elongation at break (%)	Stress at break
T0	42.42	715	37	5.1	37
T1	53.93	869	20.2	16.9	20.2
T2	41.56	782	17.5	18.1	17.4

4. Conclusion

In the current workHPMC and KI salt composite film of different weight percent have been successfully made by a simple protocol using solution casting method. The host polymer was semicrystalline in nature and the Crystallinity was strongly dependent on the chemical composition of the films, which critically influences the mechanical and other properties of the resulting films. XRD spectra shows the structural behavior of films samples, the relative intensity of these peaks decreased with the increase in the KI salt concentration, which is indicating the decrease of crystalline nature. SEM images indicates the surface morphological behavior of the film samples which is in good agreement with the structural studies. According to the tensile test, the variations in the Young's modulus, tensile strength and elongation at break was observed. The highest doping film sample showed substantial variance, high amorphosity, and a solid value of elongation at break. The findings are consistent with structural, morphological studies. We can deduce that, films made from HPMC polymers can be used in a number of industries (packaging and coverage), but this is contingent on knowing the physico-chemical behaviour of the materials in question.

5. References

[1] Chiellini E and Solaro R 1996 *Advanced Materials*, **8** 305–313,.
[2] AlbertssonA C and Karlsson S *1995 Acta Polymerica***46** 114–123
[3] FahsA,BroglyM, Bistac S, SchmittM **2010***Carbohydr. Polym.* ***80*** 105–114.
[4] Ghosh T K,Gope S, MondalD,BhowmikB, Mollick M M R, MaityD, RoyI, SarkarG, SadhukhanS, RanaDJ**2014** *Biol. Macromol.* ***66***338–345.
[5] ShiSC, JiangSZ **2020** *Surf. Coat. Technol.* ***389***125655.
[6] ShiSC, TsaiXN, PekSS**2020** *Surf. Coat. Technol.* ***389*** 125617
[7] Burdock G A, 2007 *Food Chem. Toxicol.* **45**2341
[8] Liesegang T J, 1993 *Int. Ophthalmol.-Clin.* **33** 127
[9] Olin B R, 2001 *Facts and Comparisons Inc.* St. Louis 698-707
[10] EU. 2004 Regulation (EC) No. 1935/2004 European Parliament and the Council of 27 October 2004 on materials and articles intended to come into contact with food repealing.
[11] FDA. 2001. FDA/CFSAN/OPA: Agency response letter: GRAS notice no. GRN 000065.
[12] WangBL, Mai YW, ZhangXH **2004***Acta Mater.* **52**4961–4972.
[13] VaeziK, AsadpourG, SharifiH **2020** *Int. J. Biol. Macromol.* ***146*** 374–386.
[14] Potassium Iodide in the Merck Index an Encyclopedia of Chemicals, Drugs, and Biologicals, 14th Ed., p. 1317 (Monograph No. 7643). Merck & Co., Inc., Whitehouse Station, NJ, 2006.
[15] BarkerWH and Wood Jr., W.B. 1940 *J. Am. Med Assoc.***114(12)** 1029-1038.
[16] FDA, 21 CFR§ 172.375 Potassium iodide.
[17] FDA, 21 CFR§178.1010 Sanitizing solutions
[18] EPA, Iodine and Iodophor Complexes: RED Document. Reregistration Case No.: 3080. November 2006. Docket ID: EPA-HQ-OPP-2006-0599-0006
[19] Sandhya Rani N, Sannappa J, Demappa TMahadevaiah2014 *OSR Journal of Applied Physics (IOSR-JAP)* e-ISSN: 2278-4861. **6(3)** 30-41.
[20] Krishna Jyothi N, VenkataRatnam K K, NarayanaMurthy P, VijayaKumara K, 2016, *Mater. Today***3** 21-30
[21] Krishna Jyothi N, VenkataRatnam K K, Narayana Murthy P, Vijaya Kumara K 2016 *Bull. Mater. Sci.*Indian Academy of Sciences**39(4)** 1047–1055
[22] Sandhya Rani N, Sannappa J, Demappa T Mahadevaiah, 2014*Ionics* **20**201-207
[23] Effect of strong electrolyte containing gelling aids on the solgel transition temperature of Hypromellose 2910, Elanz Sadeghi.2019
[25] SjostromE 1981 Wood Chemistry: Fundamentals and applications, *Academic Press, New York,* 169-189.
[26] Rabinovitch EB2003 *J. Vinyl. Addit. Technol.***9(2)** 61-64.
[27] Zhang S, Lee JY, Hong L 2004 *J Power Sources* 115:288
[28] ReddyN, YangY 2005. *Biotechnology***23(1)** 22–27.
[29] KanteshBalani, VivekVerma, Arvind Agarwal, Roger Narayan. Biosurfaces: A Materials Science and Engineering Perspective, First Edition. Edited by © 2015 The American Ceramic Society. Published 2015 by John Wiley & Sons, Inc

[30] DjafariPetroudy S R 2017*Advanced High Strength Natural Fibre Composites in Construction* 59-83.

IOP Conf. Series: Materials Science and Engineering 1221 (2022) 012012 doi:10.1088/1757-899X/1221/1/012012

Photophysical Properties of Rhodamine Dyes in Aqueous Organic Solvents: Evaluation of Ground and Excited State Dipole moments

Vikram Pujari[1], Mallikarjun K. Patil[1], T.S. Tilakraj[1], Vighneshwar S. Bhat[1] and Sanjeev R. Inamdar[1*]

[1]Laser Spectroscopy Programme, Department of Physics, Karnatak University, Dharwad, India-580003.

[*]Corresponding author: *him_lax3@yahoo.com

Abstract: In this present work, we have determined the ground and excited state dipole moments of two Rhodamine dyes with similar molecular skeleton and having different functional groups, namely Rhodamine 560 chloride (R560) and Rhodamine 610 Chloride (R610). Absorption and emission spectra of R560 and R610 were recorded in binary mixtures of DMSO-Water and 2 Propanol-Water at varying compositions. The two probes exhibit hypochromic / bathochromic shift in absorption and fluorescence spectra for different compositions of aqueous organic solvents. This shift arises mainly because of polarity of the solvent. By employing solvatochromic shift method we have determined the ground and excited state dipole moments. The experimentally obtained results suggest that the excited state dipole moments of two probes are much higher than their ground state dipole moments.

Keywords: Rhodamine, Solvatochromic shift, Dipole moment, Binary mixture, Solvent polarity.

1. Introduction

The rhodamines are class of dyes containing Xanthane core. The rhodamine dyes are used for Biomarking, as water tracing agent, in fluorescence microscopy etc.[1,2]. Rhodamine derivatives having different functional groups may change their electric dipole moment [3].The study on photophysical properties of various Rhodamine derivatives were made in series of alcohols, alkanes and general solvents[4]. In present work we have chosen two Rhodamine dyes with similar molecular skeleton and having different functional groups, namely Rhodamine 560 chloride (R560) and Rhodamine 610 Chloride (R610). Absorption and emission spectra of R560 and R610 were recorded in binary mixtures of DMSO-Water and 2 Propanol-Water at varying compositions. Moreover, the obtained spectral data is used for evaluating the ground and excited state dipole moments of the probes by employing solvatochromic shift method.

2. Materials and methods

The laser dyes R560 and R610 were procured from Exciton Inc., (USA).The solvents used in present work are dimethyl sulphoxide (DMSO) and 2Propanol. The dyes and solvents were used as received.

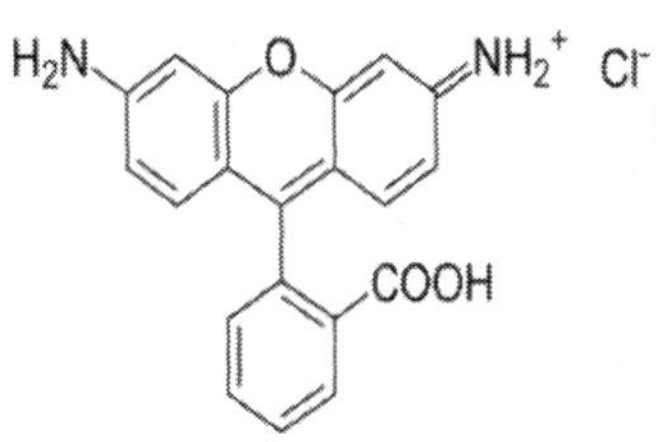

Figure 1. Molecular structures of R560

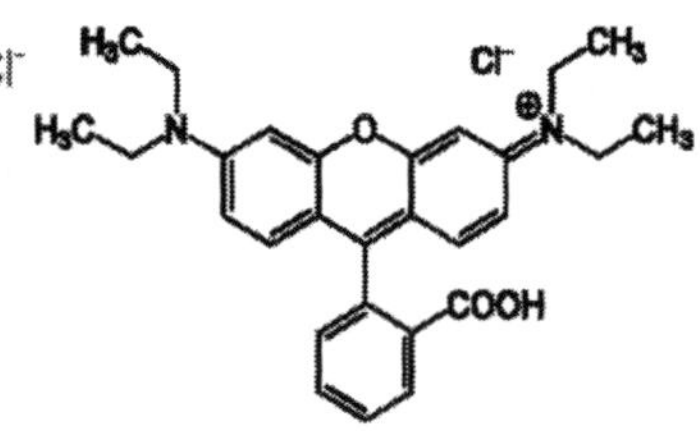

Figure 2. Molecular structure of R610

The absorption spectra were recorded using dual-beam UV-Visible spectrophotometer (JASCO, model V-670) and fluorescence emission spectra were recorded using Spectrofluorometer (JY Horiba, model Fluoromax4). The values of dielectric constants (ϵ) and refractive indices of solvent mixtures were taken from the literature [5-7]. All the measurements were taken at room temperature (298K). The concentration of solute was kept low to reduce self-quenching.

3. Theory

The interaction of solute molecules with surrounding solvent environment strongly depends upon the refractive index (n), the bulk dielectric constant (ϵ) of solvent molecule.

By making use of quantum-mechanical perturbation theory of second-order, subsequent equations have been obtained [8-10] for the maximum value of absorption (ν_a) and emission (ν_f) in terms of wave numbers in different composition of solvent mixtures of varying permittivity (ϵ) and R.I. (n).

$$\nu_a-\nu_f = m_1 f(\epsilon,n) + \text{const} \quad (1)$$

$$\nu_a+\nu_f = -m_2\,\phi(\epsilon,n) + \text{const} \quad (2)$$

Where,

$$f(\varepsilon,n)=\frac{2n^2+1}{n^2+2}\left[\frac{\varepsilon+1}{\varepsilon+2}-\frac{n^2-1}{n^2+2}\right] \quad (3)$$

Is the solvent polarity function and

$$\phi(\varepsilon,n) = f(\varepsilon,n) + 2g(n) \quad (4)$$

With

$$g(n)=\frac{3}{2}\left[\frac{n^4-1}{(n^2+2)^2}\right] \quad (5)$$

$$m_1=\frac{2(\mu_e-\mu_g)^2}{hca_0^3} \quad (6)$$

And

$$m_2=\frac{2(\mu_e^2-\mu_g^2)}{hca_0^3} \quad (7)$$

In equations 1 and 2, parameters m_1 and m_2 are the slopes of linear plots of f(ϵ,n) versusν_a-ν_f and ϕ(ϵ,n) versus ν_a+$\nu_f$respectively. $\mu_g$and $\mu_e$are ground and excited state dipole moment of the molecule, h is Planck's constant, c is the velocity of light in vacuum and a_0 is the cavity radiusof the solute molecule. The molecular volumes of dyes were calculated by using atomic increment method [11]. From calculated molecular volumes of dyes a_0 can be calculated by using the relation [12]:

$$a_0 = \left(\frac{3M}{4\pi\delta N}\right)^{1/3} \quad (8)$$

Where M is the molecular weight of the solute, δ is the density of the solute molecule and N is Avogadro'snumber.

Formulations of Kwaski [10] assumes that symmetry of probe is unaltered upon electronic transition and the ground and excited state dipole moments are parallel, then based on Eqns. 6 and 7, following equations are obtained.

$$\mu_g = \frac{(m_2-m_1)}{2}\left[\frac{hca_0^3}{2m_1}\right]^{\frac{1}{2}} \quad (9)$$

$$\mu_e = \frac{(m_2+m_1)}{2}\left[\frac{hca_0^3}{2m_1}\right]^{\frac{1}{2}} \quad (10)$$

4. Results and Discussions

The observed absorption spectra and emission spectra of R560 and R610 in binary mixtures of DMSO-Water are shown in Figure 3 and 4 and in binary mixtures of 2Propanol-Water are shown in Figure 5 and 6,respectively. The spectral data obtained of R560 and R610 in binary mixtures of DMSO-Water are tabulated in Table 1 and 2 and in binary mixtures of 2Propanol-Water were tabulated in Table 3 and 4 along with the refractive index (n), dielectric constant (ϵ) of solvent mixtures, solvent polarity function f(ϵ,n) and ϕ(ϵ,n), respectively.

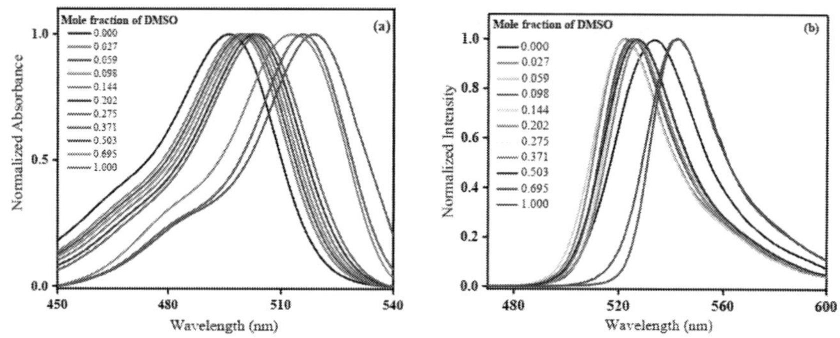

Figure 3. Normalized (a) Absorption spectra and (b) Fluorescence spectra of R560 in different mole fractions of DMSO in Water.

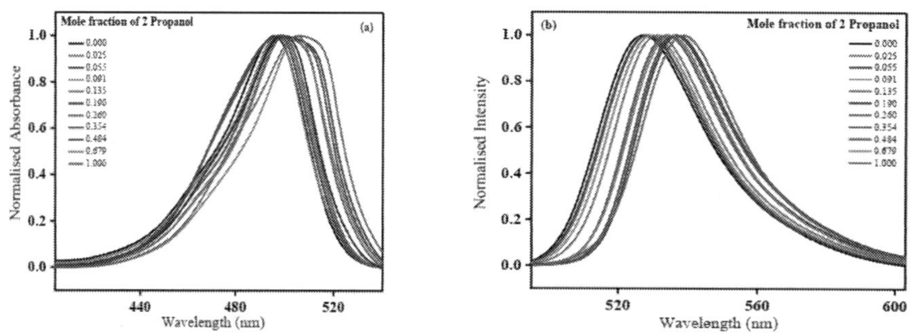

Figure 4. Normalized (a) Absorption spectra and (b) Fluorescence spectra of R560 in different mole fractions of 2 Propanol in Water.

For the probes R560 and R610 the variation of (ν_a-ν_f) with solvent polarity function f(ϵ,n) and variation of (ν_a+ν_f) with ϕ(ϵ,n) in DMSO-Water mixtures are shown in Figure 7 and in 2 Propanol-Water mixtures are shown in Figure 8 respectively. The observed data were fit to a straight line by using linear progression method, the slopes were taken as m_1 for the graph (ν_a-ν_f) v/s f(ϵ,n) and m_2 for (ν_a+ν_f) v/s ϕ(ϵ,n). Cavity radius (a_0), correlation coefficients (r), and determined values of ground (μ_g) and excite state (μ_e) dipole moments for R560 and R610 in DMSO- Water mixtures are tabulated in Table 5.1 and in 2Propanol-Water mixtures are tabulated in Table 5.2 respectively.

The emission wavelength of the molecule is longer than the absorption wavelength as there is more dynamical processes occur like vibrational relaxation, internal conversion, formation of hydrogen bonding etc.[15]. The two probes R560 and R610 are having -C=O, -N-H_2 and -O-H functional group which can form hydrogen bonding with solvent molecules [14].In DMSO-Water mixtures both the probes R560 and R610 can have tendency to form hydrogen bonding with solvent molecules [13]. Hypsochromic and bathochromic shifts were observed in emission spectra of both the probes in DMSO-Water mixtures. It may be due to n-π* or π- π* transition [16]. In 2Propanol-Water mixtures R560 and R610 probes behave differently as solvents are polar protic. Here also we observed hypsochromic and bathochromic shifts in emission spectra.

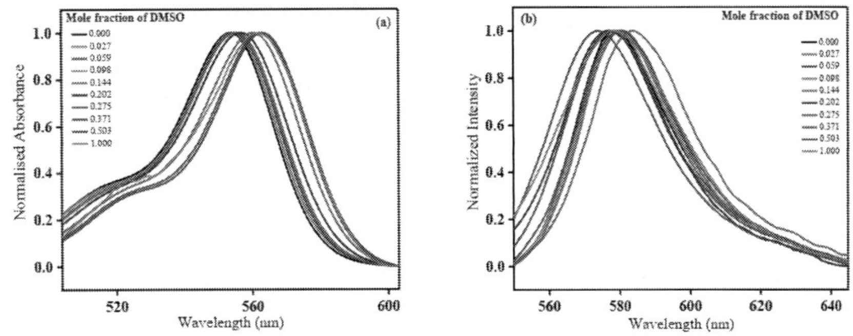

Figure 5. Normalized (a) Absorption spectra and (b) Fluorescence spectra of 610 in different mole fractions of DMSO in Water.

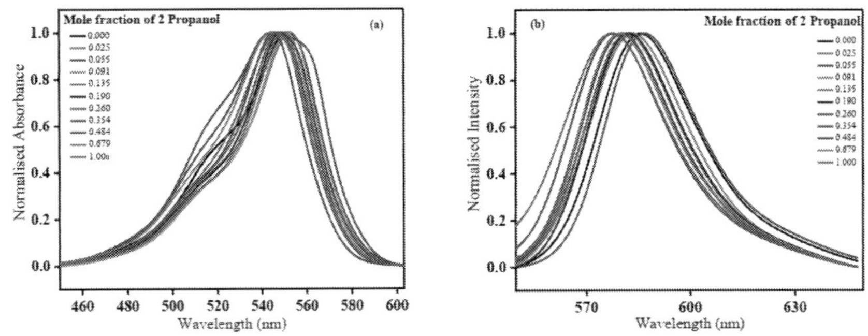

Figure 6. Normalized (a) Absorption spectra and (b) Fluorescence spectra of R610 in different mole fractions of 2 Propanol in Water.

Table 1. Experimental data along with calculated values of f(ϵ,n) and ϕ(ϵ,n) for R560 in aq. DMSO

Mol.fract. of DMSO	ν_a (cm^{-1})	ν_f (cm^{-1})	ν_a-ν_f (cm^{-1})	ν_a+ν_f (cm^{-1})	n	ϵ	f(ϵ,n)	ϕ(ϵ,n)
0.0000	20161.29	19157.08	1004.20	39318.38	1.33	77.56	0.9129	1.3626
0.0596	20040.08	19047.61	992.46	39087.70	1.34	77.02	0.9101	1.3732
0.0980	20000.00	19011.40	988.59	39011.41	1.34	76.03	0.9085	1.3770
0.1446	19960.07	19011.40	948.67	38971.49	1.35	74.49	0.9057	1.3837
0.2023	19920.31	18975.33	944.98	38895.65	1.35	73.34	0.9029	1.3917
0.2756	19880.71	18975.33	905.38	38856.05	1.37	71.47	0.8987	1.4023
0.3717	19841.26	18975.33	865.93	38816.60	1.38	68.73	0.8921	1.4184
0.5036	19531.25	18450.18	1081.06	37981.43	1.40	64.70	0.8836	1.4352
0.6953	19379.84	18726.59	653.25	38106.44	1.43	58.46	0.8690	1.4603
1.0000	19305.01	18587.36	717.65	37892.38	1.47	48.11	0.8446	1.4843

Table 2. Experimental data along with calculated values of f(ϵ,n) and ϕ(ϵ,n) for R610 in aq. DMSO

Mol.fract. of DMSO	ν_a (cm^{-1})	ν_f (cm^{-1})	ν_a-ν_f (cm^{-1})	ν_a+ν_f (cm^{-1})	n	ϵ	f(ϵ,n)	ϕ(ϵ,n)
0.0000	18050.54	17241.37	809.16	35291.92	1.33	77.56	0.9129	1.3626
0.0596	18018.01	17301.03	716.98	35319.06	1.34	77.02	0.9101	1.3732
0.0980	18018.01	17241.37	776.63	35259.40	1.34	76.03	0.9085	1.3770
0.1446	18018.01	17241.37	776.63	35259.40	1.35	74.49	0.9057	1.3837
0.2023	17953.32	17331.02	622.29	35284.34	1.35	73.34	0.9029	1.3917
0.2756	17857.14	17211.70	645.43	35068.85	1.37	71.47	0.8987	1.4023
1.0000	17730.49	17301.03	429.45	35031.53	1.47	48.11	0.8446	1.4843

Table 3. Experimental data along with calculated values of f(ϵ,n) and ϕ(ϵ,n) for R560 in aq.2 Propanol

Mol. fract. of 2Propanol	ν_a(cm^{-1})	ν_f(cm^{-1})	ν_a-ν_f (cm^{-1})	ν_a+ν_f (cm^{-1})	n	ϵ	f(ϵ,n)	ϕ(ϵ,n)
0.0000	20202.02	18968.13	1233.88	39170.15	1.33	77.29	0.9113	1.3690
0.0254	20161.29	18925.05	1236.23	39086.35	1.34	68.76	0.9037	1.3723
0.0555	20120.72	18932.22	1188.50	39052.95	1.35	60.81	0.8945	1.3766
0.0916	20080.32	18882.17	1198.14	38962.50	1.35	55.90	0.8888	1.3749
0.1356	20040.08	18832.39	1207.68	38872.47	1.36	50.40	0.8803	1.3757
0.1904	20040.08	18733.60	1306.47	38773.69	1.36	42.94	0.8678	1.3673
0.2608	20120.72	18733.60	1387.11	38854.33	1.37	36.07	0.8519	1.3567
0.3544	20080.32	18656.71	1423.60	38737.04	1.37	30.67	0.8347	1.3449
0.4848	20080.32	18628.91	1451.40	38709.23	1.37	25.47	0.8124	1.3267
0.6792	19801.98	18698.57	1103.40	38500.56	1.37	20.37	0.7814	1.2970
1.0000	19762.84	18573.55	1189.29	38336.40	1.38	16.64	0.7481	1.2650

Table 4. Experimental data along with calculated values of f(ϵ,n) and ϕ(ϵ,n) for R610 in aq.2 Propanol

Mol. fract. Of 2Propanol	ν_a(cm^{-1})	ν_f(cm^{-1})	ν_a-ν_f (cm^{-1})	ν_a+ν_f (cm^{-1})	n	ϵ	f(ϵ,n)	ϕ(ϵ,n)
0.0000	18148.82	17070.67	1078.14	35219.49	1.33	77.29	0.9113	1.3690
0.0254	18083.18	17135.02	948.15	35218.21	1.34	68.76	0.9037	1.3723
0.0916	18148.82	17176.22	972.59	35325.05	1.35	55.90	0.8888	1.3749
0.1356	18181.81	17182.13	999.68	35363.95	1.36	50.40	0.8803	1.3757
0.2608	18248.17	17229.49	1018.6	35477.67	1.37	36.07	0.8519	1.3567
0.3544	18348.62	17229.49	1119.12	35578.12	1.37	30.67	0.8347	1.3449
0.4848	18348.62	17307.02	1041.59	35655.65	1.37	25.47	0.8124	1.3267
0.6792	18348.62	17247.32	1101.29	35595.95	1.37	20.37	0.7814	1.2970

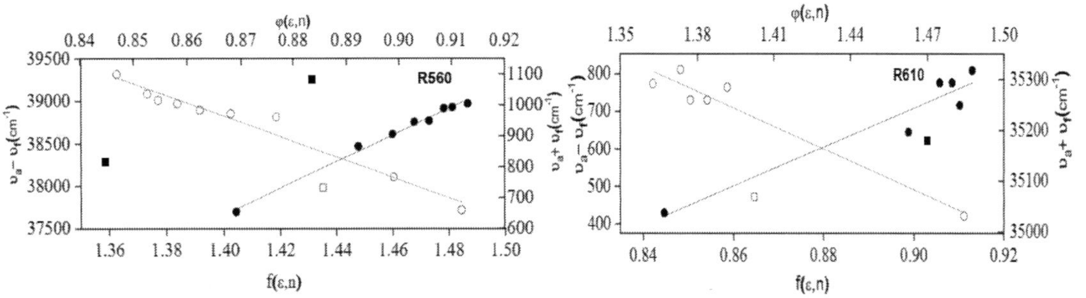

Figure 7. Plot of (ν_a-ν_f) vs. f(ε,n) (•) and (ν_a+ν_f) vs. ϕ(ε,n) (o) of R560 and R610 for different mole fractions of DMSO in Water. (■) and (□) represent data was excluded from calculations.

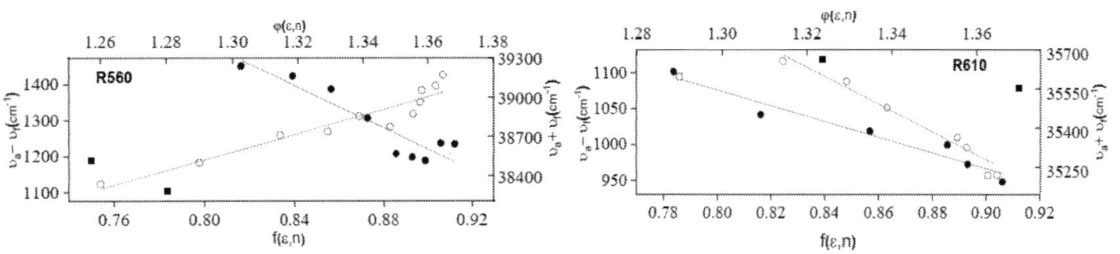

Figure 8. Plot of (ν_a-ν_f) vs. f(ε,n) (•) and (ν_a+ν_f) vs. ϕ(ε,n) (o) of R560 and R610 for different mole fractions of 2Propanol in Water. (■) and (□) represent data was excluded from calculations.

Table 5. Cavity radius (a_0), correlation coefficient (r) and determined values of ground (μ_g) and excite state (μ_e) dipole moments for R560 and R610 in different mole fractions of DMSO in water.

Solute molecule	a_0 (Å)	Correlation Coefficients (r)		μ_g(D)	μ_e(D)
R560	4.95	0.99[a]	0.98[b]	2.43	12.3
R610	5.63	0.95[a]	0.97[b]	2.69	6.91

[a]Correlation coefficient of the graph ν_a-ν_f v/s f(ε,n).
[b]Correlation coefficient of the graph ν_a+ν_f v/s ϕ(ε,n).

Table 6. Cavity radius (a_0), correlation coefficient (r) and determined values of ground (μ_g) and excite state (μ_e) dipole moments for R560 and R610 in different mole fractions of 2 Propanol in water.

Solute molecule	a_0 (Å)	Correlation Coefficients (r)		μ_g(D)	μ_e(D)
R560	4.95	0.92[a]	0.94[b]	4.17	10.1
R610	5.63	0.97[a]	0.98[b]	14.7	19.1

[a]Correlation coefficient of the graph ν_a-ν_f v/s f(ε,n)
[b]Correlation coefficient of the graph ν_a+ν_f v/sϕ(ε,n).

5. Conclusions:

The Photophysical properties of Rhodamine dyes were studied in aqueous DMSO and aqueous 2 Propanol by employing solvatochromic shift method. It is observed that the excited state dipole moments of both the probes in both the mixtures are higher than their ground state dipole moments. Also the ground state dipole moments of both R560 and R610 are higher in aqueous 2Propanol mixtures as compared to aqueous DMSO mixtures because polarity of the molecules in DMSO-Water is lesser than in 2Propanol-Water mixtures. Both the probes have higher tendency of forming hydrogen bond with the solvent molecules in case of 2Propanol-Water mixtures as both solvents contain -O-H group. So experimentally obtained ground state dipole moments of both the probes are slightly higher in 2Propanol-Water mixtures.

Acknowledgement:
The authors would like to thank the University Science Instruments Centre (USIC), Karnatak University, Dharwad for providing the UV-Visible absorption spectrophotometer.

References:

[1]LV Johnson, ML Walsh, and LB Chen. 1980 *Proc. of the National Aca.ofSci.,***77(2)**, *990–994.*
[2] V Jeannot, JM Salmon, M Deumié, and P Viallet. 1997 *The J. of Histochem. &Cytochem.,***45(3)**, 403–412,
[3] MA Rauf, JP Graham, SB Bukallah, MAS Al-Saedi. 2009*Spectrochim. Acta Part A* 72 133.
[4] MS Zakerhamidi, M Moghadam , A Ghanadzadeh , S Hosseini. 2012 *J. of Lumi.***132,** 931–937
[5] BR Gayathri, JR Mannekutla, SRInamdar.2008 J. *of Mol. Structure* **889** 383–393
[6] PSKadolkar, SA Patil, MNWari. SR Inamdar 2020*J. of Mol. Liqu.***312**, 113452
[7]P Navarro,MLarriba, S García, J García, F Rodríguez. 2002*J. Chem. Eng. Data* 2012, **57**, 1165−1173
[8] AKawski, Z Naturforsch. **57a**255–262.
[9] AKawski, P Bojarski, B Kuklin. 2008*ski, Lett.***463** 410–412,
[10]AKawski, BKukli´nski, P Bojarski, H Diehla, ZNaturforsch. 2000**55a**817–822.
[11] JTEdward. *Journal of Chemical Education*. April 1970,**47**, 4,
[12] PSuppan. 1983, *Chem. Phys. Lett.***94** 272.
[13] D Martin, HHanthal, Dimethyl Sulfoxide.1975, *Wiley*, New York,
[14] K Aggarwal, JM Khurana. 2014,*J Photochem.Photobiol. A: Chem.***276** 71–82,
[15] CReichradt. 2004, **third ed.,***Wiley VCH, Verlag,* GmbH and Co.
[16] SAPatil, PS Kadolkar, MN Wari, SRInamdar. 2021*Materials Today Comm.***29,** 102733

IOP Conf. Series: Materials Science and Engineering 1221 (2022) 012013 doi:10.1088/1757-899X/1221/1/012013

Phase transition and ferroelectric properties study of lead-free $0.95BaTiO_3$-$0.05CaSnO_3$ ceramics synthesized by high energy ball milling assisted solid state reaction route

B Kar[1], S Swain[1], P Kumar[1*] and S Panigrahi[1]

[1] Department of Physics and Astronomy, National Institute of Technology, Rourkela, Odisha, India, 769008,
E-mail: pawankumar@nitrkl.ac.in

Abstract. Lead-free $0.95BaTiO_3$-$0.05CaSnO_3$ ceramics were synthesized by high energy ball milling (HEBM) assisted solid state reaction route. Calcination temperature and time for phase formation was optimized and found to be relatively less than other $BaTiO_3$-doped system. X-ray diffraction (XRD) and Raman study were performed for phase examination of the synthesized material. Rietveld analysis confirmed double-phase co-existence of this 95/05 system at room temperature. Sintering temperature and time for better densification were optimized to be 1250°C for 3hr. From dielectric study it was found that ferroelectric-paraelectric phase transition temperature($T_{c,\ observed} \approx$ 55°C) was reduced considerably in compared to the pure $BaTiO_3$ system ($T_c \approx$ 125°C). No relaxorbehaviour was found from dielectric study. Polarization vs. electric field hysteresis loop study showed the development of saturated hysteresis loops but the remnant polarization was found decreased compared to pure $BaTiO_3$ system.

1. Introduction

$Pb(Zr,Ti)O_3$-based materialshave been the leading candidates for ferroelectric applications due to excellent ferroelectric properties and good temperature stability [1]. However, the high content of toxic lead in PZT-based ceramics has raised environmental and health concerns and research trend in lead-free piezoelectric materials has attained considerable attention [2].$BaTiO_3$, $(Bi,Na)TiO_3$, $(K,Na)NbO_3$-based ceramics are found to be good substitutes for PZT-based ceramics. Among them, $BaTiO_3$-based lead-free perovskite materials like $(Ba,Ca)(Zr,Ti)O_3$, $(Ba,Ca)(Sn,Ti)O_3$, and $(Ba,Ca)(Hf,Ti)O_3$are being mostly studieddue to their excellent piezoelectric properties. The ultrahigh piezoelectric property in these systems is due to the modification of the phase boundary by the substitution of Ca^{2+} and Sn^{4+}/ Zr^{4+}/Hf^{4+}ions at Ba^{2+} and Ti^{4+} sites respectively of pure $BaTiO_3$ system. It is well studied that pure $BaTiO_3$ exhibits three phase transitions : rhombohedral to orthorhombic phase transition ($T_{RO} \approx$ -90°C), orthorhombic to tetragonal phase transition ($T_{OT} \approx$ 5°C) and tetragonal to cubic phase transition ($T_C \approx$ 125°C) [3]. The whole effect of substitution of Ca^{2+} at Ba^{2+} site and Sn^{4+}/ Zr^{4+}/Hf^{4+} at Ti^{4+} site shift the T_{RO} and T_{OT} towards higher temperature region and T_C towards lower temperature region simultaneously [4]. By varying the doping percentage, a multiphase co-existencecan be attainedand this enhances the piezoelectric property of the system.In 2009, Liu and

Ren reported d_{33} value as high as 620pC/N by co-existing rhombohedral-pseudo cubic-tetragonal (R-PC-T) phases [4]. So, the piezoelectricproperties of these BT-based lead-free perovskite materials can be enhanced by forming multiphase co-existence.

In recent years lead-free bulk (1-x)$BaTiO_3$-x$CaSnO_3$ [abbreviated as $BTCS_x$] ceramics hasattained considerable attention due to good piezoelectric properties comparable to BZT-BCT system. However, the ferroelectric study of$BTCS_x$ shows decrease in the remnant polarization and coercive field with increasing Ca^{2+} and Sn^{4+} content [5]. As stated earlier the T_C for $BTCS_x$ system shifts towards lower temperature region with increasing Sn^{4+} content and consequently for a higher percentage of Sn doping can make this system paraelectric in nature at room temperature that hinders practical application. For bulk$BTCS_{x=0.05}$system prepared by conventional ball milling, reported values of T_C are around 75°C [5].In this study the effect of high energy ball milling on the structural and electrical properties of $BTCS_{x=0.05}$ has been investigated.

2. Experimental procedures

Lead-free0.95$BaTiO_3$-0.05$CaSnO_3$ceramics were synthesized by high energy ball milling (HEBM) assisted solid state reaction route using starting chemicals as $BaCO_3$(99%), $CaCO_3$(99%), TiO_2(99%), SnO_2(99.5%). The measured chemicals in the HEBM chamber were dry milled with zirconia balls for 5 hrs.at 300 rpm speed with 15 minutes milling time and consecutively 20 minutes break. The formation of pure phase of milled sample was found at calcination temperature of 1100°C for 3 hrs.which is relatively less than other BT-based system prepared by solid state synthesis method. PVA binder was mixed with calcined powder and after being dried the mixture was grinded to get fine powder for pelletization. For the evaporation of binder, the pellets were pre-sintered at 600°C for 7 hrs. The pellets were finally sintered at 1250°C for 3 hrs. X-ray diffraction (XRD)was performed of the calcined sample at a scanning rate of 4°/minandReitveld refinement was donefor the detailed phase analysis. Raman spectra was taken at a laser wavelength of 532nm. The average diameterand thickness of the sintered pelletswere measured to be8.5mmand1mm. Microstructure analysis of the $BTCS_{x=0.05}$ ceramic was checked from scanning electron microscope(SEM) micrograph. Variation of dielectric constant with temperature was recorded in a LCR meter at five different frequencies. PE loop was measured by using Radiant precision premier II under a 20kV bias field at 10Hz frequency.

3. Results and discussions

3.1. Structural study

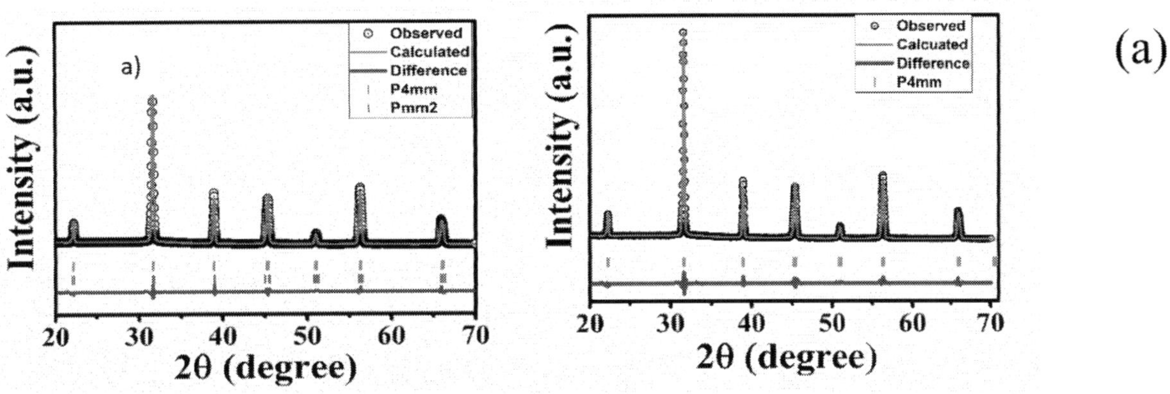

Figure 1. Reitveld refinement of $BTCS_{x=0.05}$a) tetragonal phase b) tetragonal and orthorhombic phase 3 hrs.

Figure 1 shows the XRD pattern of the $0.95BaTiO_3$-$0.05CaSnO_3$ceramic powder calcined at 1100°C for

a pure perovskite structure was observed without any impurity phase.Reitveld refinement confirmed the co-existence of tetragonal (P4mm) and orthorhombic (Pmm2) phase at room temperature (27°C). The details of refined parameters are listed in the table 1.Where pure $BaTiO_3$ exhibits only tetragonal phase at room temperature [6], the small amount (5%) of Ba^{2+} and Ti^{4+} site replacement by

Table 1. Refined parameters of tetragonal phase and (tetragonal+orthorhombic) phase

	Space Group	a (Å)	b (Å)	c (Å)	α=β=γ	χ^2	R_p	R_{wp}	Content (%)
a)	P4mm	3.99857	3.99857	4.01421	90	5.32	4.93	6.53	100
b)	P4mm+Pmm2	3.99962	3.99962	4.01185	90	3.51	4.05	5.30	88.32
		3.98747	3.98517	4.02811					11.68

$Ca^{2+}$and Sn^{4+}might shift the lower transition temperature ($T_{O\text{-}T}$) towards room temperature region [4].Although the percentage of orthorhombic phase is only 11.68% the chi square value decreased from 5.32 to 3.51 and also the values of other goodness of fit parameters (R_p, R_{wp})diminished implying better fit and confirmed a co-existence of dual phase of tetragonal and orthorhombic. The lattice parameters of tetragonal phase of $BTCS_{x=0.05}$ have found decreased from the pure tetragonal phase of bulk $BaTiO_3$ (a=b=3.9999Å, c=4.017Å), where an increase of the c parameter has been observed for the orthorhombic phase thereby reducing a and b parameter. This can be understood as the Ca^{2+} ionic radius (0.99Å) is less than Ba^{2+} ionic radius (1.33Å) but the ionic radius of Ti^{4+} (0.68Å)and Sn^{4+} (0.71Å) are almost comparable.

3.2. Raman study

Room temperature Raman spectrum of $BTCS_{x=0.05}$ ceramics is shown in figure 2. The three distinct sharp peaks at 301.53 cm^{-1}, 513.83 cm^{-1} and 717.34 cm^{-1} corresponds to the vibrational modes of $E(TO_2)$, $A_1(TO_3)$ and $A_1(LO_3)$ due to the presence of tetragonal phase. However, the peaks at 167.5 cm^{-1}(A_1TO_1)and 264.2 cm^{-1}(A_1TO_2) due to the tetragonal phase are diffused and almost flattened whichconfirms a finite presence of orthorhombic phase associated with tetragonal symmetry [5].

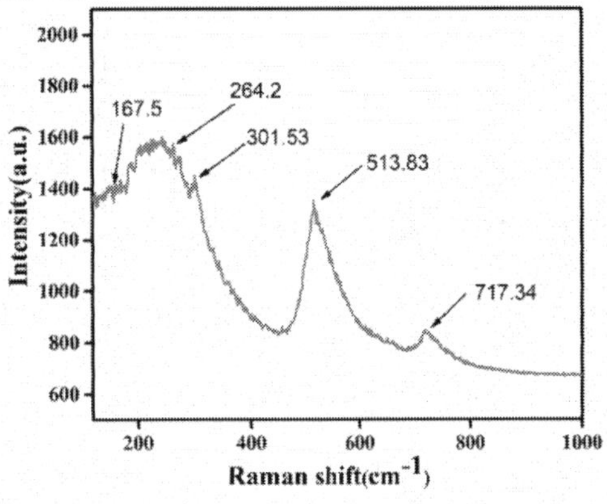

Figure 2. Room temperature Raman spectrum ofBTCS$_{x=0.05}$ ceramics **Figure 3.** SEMmicrograph ofBTCS$_{x=0.05}$ pellet

3.3. SEM study

Figure 3 shows the scanning electron microscope image of BTCS$_{x=0.05}$ pellet sintered at 1250 °C for 3h.

It can be seen that although grain formation has occurred the sizes of grains are not equal. Grain sizes vary from 0.594μm to 5.83μm. The average grain size was found to be 2.31 μm which is smaller in compared to other BTCS$_x$ [5] and BT-based systems prepared by conventional ball milling or solid state method rather than HEBM [7,8]. The grains are closely packed with low porosity which suggests well densification of the sintered pellets.

4. Electrical properties study

4.1. Dielectric study

Variation of dielectric constant (ε_r) with temperature (T) at five different frequencies (100Hz, 1kHz, 10kHz, 100kHz, 1MHz) of BTCS$_{x=0.05}$ ceramics prepared by HEBM method is shown in figure 4. Ferroelectric to paraelectric transition temperature (T_c) was found to be 55°C for this system. A decrease in the Curie temperature has been observed in compared to pure BT ($T_{c,\ BT}$≈125°C)[3]and BTCS$_{x=0.05}$ ($T_{c,BTCSx=0.05}$≈75°C)system prepared by conventional ball milling method [5]. Sn^{4+} substitution on Ti^{4+}site and the presence of smaller grains might cause the shifting of T_c to such a lower temperature region.

Value of maximum dielectric constant (ε_r) is 7886.6 at the Curie temperature for 100Hz frequency.Such a high value of ε_r near room temperature signifies the advantage of BTCS$_{x=0.05}$ ceramics prepared by HEBM method. From the figure 4 it can be noticed the gradual decrease in ε_r with increasing frequency which is due to the filtration of polarization with increasing frequency [9]. At 1 MHz frequency there is an abrupt decrease in ε_r due to the decrease in polarization value. However, with a doping of 5% Ca^{2+} and Sn^{4+} norelaxorbehaviour has been observed.

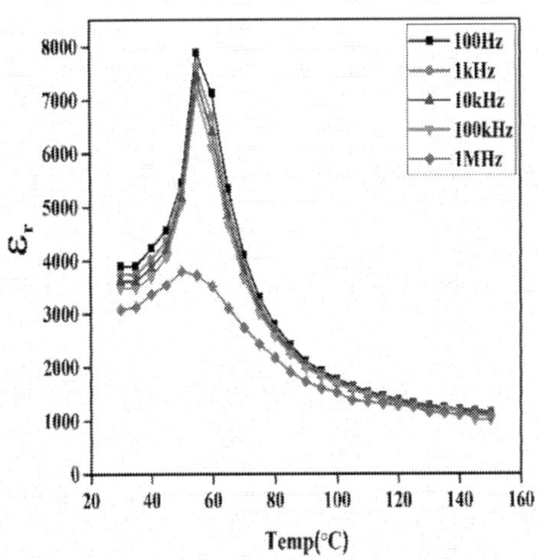

Figure 4. Variation of $\varepsilon_r$with temperature at different frequency

5. PE (hysteresis) loop study

Variation of room temperature polarization with electric field measured at 10Hz frequency of $BTCS_{x=0.05}$ ceramicsis shown in figure 5.Nearly saturated slender hysteresis loop has been observed at a bias field of 20kV indicating ferroelectric phase of the material at room temperature. The values of saturation polarization (P_s), remnant polarization (P_r) and coercive field (E_c) are 12.394 μC/cm^2, 2.1376 μC/cm^{2}and 1.667kV/cm respectively. These values are smaller in comparison to $BTCS_{x=0.05}$ system prepared by conventional ball milling [5]. An asymmetry in the PE loop is easily observed which is due to the domain pinning effect caused by defects [10].

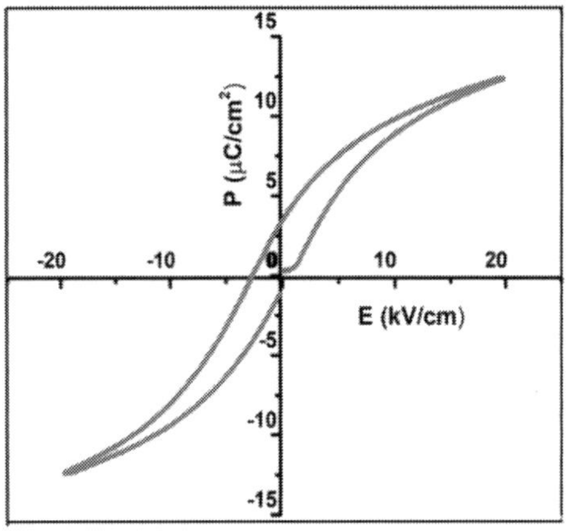

Figure 5. PE hysteresis loop

Conclusion

Low calcination and sintering temperature of $BTCS_{x=0.05}$ ceramics compared to other solid state method emphasizes the usefulness of high energy ball milling (HEBM) method. Above room temperature T_c (≈55°C), high value of ε_r(≈7886.6), P_s (≈12.394 μC/cm^2) and P_r(≈2.1376 μC/cm^2) make $BTCS_{x=0.05}$ ceramics an useful room temperature applicable ferroelectric material. Further, by changing Sn^{4+} and Ca^{2+} content multiphase co-existence can be achieved and the overall performance of ferroelectricity and piezoelectricity can be improved.

References

[1] Saito Y, Takao H, Tani T, Nonoyama T, Takatori K, Homma T, Nagaya T and Nakamura M 2004 *Nature***432** 84
[2] Zhang S J, Xia R and Shrout T R 2006 *J. Appl. Phys.***100** 104108
[3] Tian Y, Wei L, Chao X, Liu, Z and Yang Z 2013*J Am Ceram Soc.* **96** 496-502
[4] Liu W F and Ren, X B 2009 *Phys. Rev. Lett.***103** 257602
[5] Liu G, Li Y, Gao J, Li D, Yu L, Dong J, Zhang Y, Yan Y, Fan B, Liu X and Jin L 2020 *J. Alloys Compd.***826**15416
[6] Bechmann R 1956 *AcoustSoc Am***28** 347-350

[7] Siddiqui A M, Hussain F, Hanif M S, Mohamad A A and Tufail M 2018 *Int. J. Electroactive Mater.* **6** 42-47
[8] Shao S, Zhang J, Zhang Z 2008J *Phys D: Appl Phys* **41** 125408
[9] Sahu R and Kumar P 2020 *Phase Transitions* **93** 91–99
[10] Swain S, Kumar P and Sonia 2021 *Ceram. Int.* **47** 26511–26518

The Structural and Electrical Studies on Cu-Co Ferrites

Ch Venkateshwarlu[1], M Ramesh[1], G Vinod[2], Y Suresh Reddy[2], K Rajashekhar[2], B Naresh[3], P Ramesh[3], U Dasharatha[3], B Venkatesh[3] and J Laxman Naik[*]

[1]University College of Science, Saifabad, Osmania University, Hyderabad, T.S, India

[2,3]Research scholars, Department of Physics, University College of Science, Osmania University, T.S, India

[*]University College of Science, Osmania University, Hyderabad, T.S, India

[*]Corresponding author Email address: laxmannaikj@gmail.com

Abstract: The system Cobalt substituted copperferritesampleswere synthesized by a conventional double sintering ceramic method.A series of doped Copper ferrites withcomposithion $Cu_{1-x}Co_xFe_2O_4$ (x = 0.2, 0.4, and 0.6) are synthesized. The X-raydiffraction pattern at room temperature show that the sample exists in single phase with a spinel structure.Copper-Cobalt ferrites seriesaretakenand their electrical conductivityproperty are studied with a variation of composition and temperature. Log (σT) vs $10^3/T$ is almost linear with a variation close to the temperature of Curie.In general, the paramagnetic region activation energy is greater than that of the ferromagnetic region's.

***Keywords*:** Cu-Co ferrites, Electrical conductivity, Seebeckcoefficient, Energy of activation, and Temperature of curie.

1. Introduction

Copper ferrites have been used as high-frequency devices such as radiofrequency coils, transformer cores, antennas, and read-write heads in high-speed optical tapes [1, 2] nd so on,because of high electrical resistivity, high permeability, negligible eddy present losses [3,4], magneto-resistive and magneto optical characteristics, it is very helpful for wide range of energy applications [5–7].The Co^{+2} ion have good electrical conductivity and paramagnetic in nature, so it may help the materials to increase the conductivity. In the present work $Cu_{1-x}Co_xFe_2O_4$ (x = 0.2, 0.4, and 0.6) are synthesized by a conventional double sintering ceramic methodAs a result, we conducted a thorough investigation of the electrical conductivity of Cu-Co ferrites as a function of temperature. The resultsare presented in this communication.

2. Preparation methods

Cu-Co ferrites with the formula $Cu_{1-x}Co_xFe_2O_4$, where x=0.2, 0.4, and 0.6 were made using a traditional double sintering ceramic process with CuO, $Co_3O_4$and Fe_2O_3 99% pure AR grade were used. The chemicals are precisely measured to the tenth of a milligram and thoroughly mixed in acetone medium using steel balls in a ball mill for 48 h to achieve a fine particle size.The mixture is then placed in a China dish and carefully placed in an oven at 100°C for around 8 hours to remove the traces of acetone.Later the finest powder has been achieved, pellets are madein a cylindrical disc shape using hydraulic pressure at a rate of 10 tons/sq. inch. The pellets are pre-sintered in the air for 4 hours at around 800°C. The pellets are grinded again for 48 h and pellets are maid for final sintering at 1000°C in air. Following the same procedure, poly-crystalline ferrites with the compositions are prepared.

IOP Conf. Series: Materials Science and Engineering 1221 (2022) 012014 doi:10.1088/1757-899X/1221/1/012014

3. Results and discussion

3.1. X-Ray Diffraction

An X-ray diffractro-meter (Philips PW1140) with Cu k_{α}radiation(1.54Å) was used to conduct X-ray diffraction studies on the given samples.The provided relations measure the planes (hkl) of spacing 'd' and the lattice constant 'a'.

$$n\lambda = 2d_{hkl}Sin\theta$$

$$d_{hkl} = n\lambda/2Sin\theta,$$

$$a = d/(h^2+k^2+l^2)^{1/2}$$

When 'n' is a reflecting order, ' λ' is the 2adioed wavelength, ' θ ' is the x-ray beam's glancing angle, (h, k, l) is a miller indices, and 'a' is a lattice constant.The variation in lattice parameter with 'Co' content inthe Cu-Co is presented in Fig.1

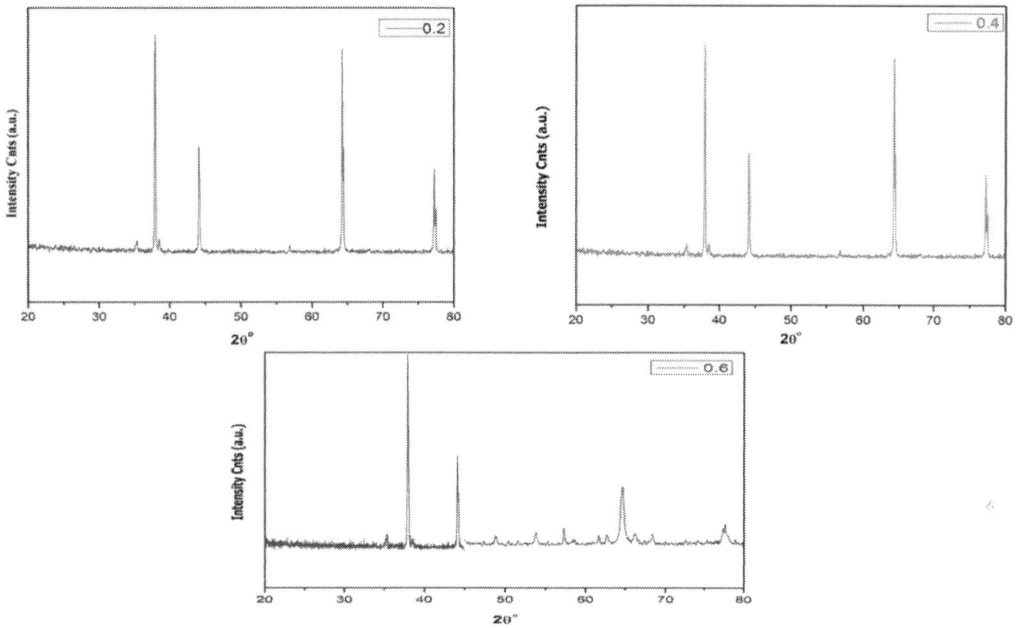

Fig.1 The variation in lattice parameter with 'Co' content in the Cu-Co

The XRD pattern shows a single phase with a spinel structure [8].Bragg's angle (2θ), in the range of 10°to 80°, was used to record the XRD figures.Table.1 shows that with rising dopant Co in Cu, the grid continues to increase.With regard to the ionic radius, this concentration can be predicted for the Co^{2+}(0.82 Å) and Cu^{2+}(0.70Å). The Co^{2+} ions are larger than that of Cu^{2+} ions.The lattice parameter increases from 8.335Åto 8.361Åis shown in Fig.2.A similar variation in the Ni-Gd ferrite composition of the lattice parameter was founded Hemeda*et al*[9].

Table 1: Lattice parameters of $Cu_{1-X}Co_xFe_2O_4$ Ferrites.

Composition (x)	0.2	0.4	0.6
Lattice parameter (A^0)	8.335	8.346	8.361

3.2. X-Ray Density, of $Cu_{1-X}\,Co_x\,Fe_2o_4$ Ferrites

The formula was used to calculate X-ray densities from the values of the Lattice parameters[10].

$$\rho = \frac{8M}{Na^3}$$

where ρ - X-ray density, 8 – Number of molecules in a unit cell of spinel lattice, M–Molecular weight of the ferrites samples,N – Avogadro's number (6.023×10^{23}) anda – Lattice parameters of the ferrite samples.

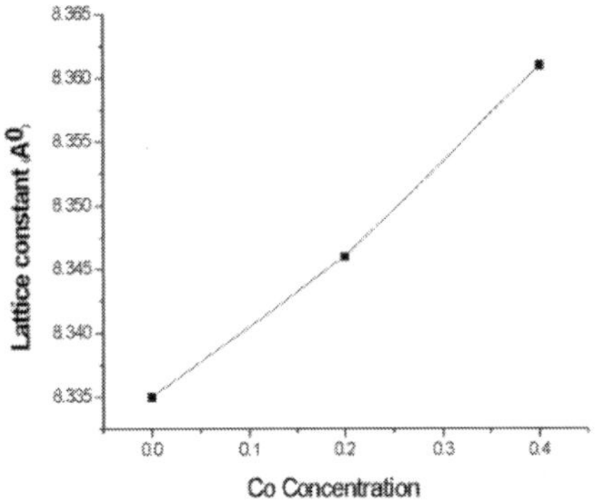

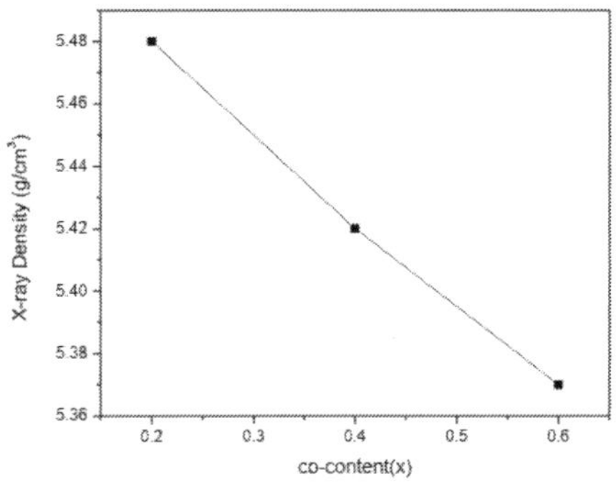

Fig 2. The lattice parameter v/s concentration. **Fig 3**. X-Ray density v/s co-content.

As shown in Table.2, the X-Ray density of Cu-Co mixed ferrites decreases as the 'Co' content increases. Where the X-Ray density values decrease from 5.48 to 5.37 gm/cm^3 as shown in Fig.3.

Table 2: X-Ray Density of $Cu_{1-X}\,Co_x\,Fe_2O_4$ Ferrites

Composition (x)	0.2	0.4	0.6
X-RayDensity (grm/cm^3)	5.48	5.42	5.37

3.3.SEM and EDAX

From the SEM image, the grains are homogeneous and agglomerated as shown in Fig. 4. The EDS diagram for Cu-Co shown in figure5. Cu, Co, Fe and Oelementshave been detected in the EDSdiagram for Cu-Co-Fe_2O_4.Thus the proving that dopant Co^{2+} replaced Cu^{2+} in the lattice as shown in the figure 4. The EDS pattern obtained for all samples gives element weight percentage and atomic percentage as shown below.

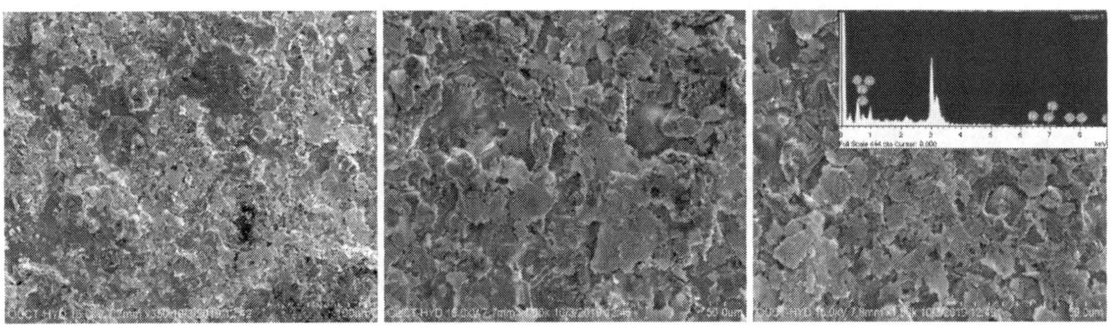

Fig.4. SEM pattern of $Cu_{1-x}Co_xFe_2O_4$ (inset EDS pattern).

IOP Conf. Series: Materials Science and Engineering 1221 (2022) 012014 doi:10.1088/1757-899X/1221/1/012014

3.4.Electrical Conductivity

The two probe methods were used to test electrical conductivity[11]. Using a Keithley electrometer model HR6517A, from room temperature to well below the Curie temperature. We were using the relation

$$\sigma = \frac{It}{VA}$$

The Cu-Co ferrites under this study had their electrical conductivity (σ)measured.Where 'I' denotes the current flowing through the sample, 'V' denotes the voltage applied to the sample, 't' denotes the thicknessofthe sample, and 'A' denotes the sample area, and Table 3 shows electrical conductivity data on mixed Cu-Co ferrites at 343Ktemperature.

Table 3. Electrical conductivity data on mixed Cu-Co ferrites at 343K temperature

S.No.	Ferrite composition	Electrical conductivity(σ) in $\Omega^{-1}cm^{-1}$	Seebeckcoefficient(S)in μV/K
1.	Cu0.8Co0.2 Fe2O4	9.637×10^{-10}	-192
2.	Cu0.6Co0.4 Fe2O4	8.801×10^{-10}	-166
3.	Cu0.4Co0.6 Fe2O4	7.940×10^{-10}	-138

3.5.Seebeck coefficients

The Seebeck coefficients(S) were determined using a differential method from room temperature to well beyond the Curie temperature[12,13]. The sample was mounted on two well-split copper blocks with silver paint. The differences in temperature between the two ends of the sample were held at 10K.The Keithly 181 nano voltmeter was used for voltage measurements.The following equation was used to calculate the thermoelectric power, also known as the Seebeck coefficient (S).

$$S = \frac{\Delta E}{\Delta T}$$

ΔE is the thermal electromotive force(e.m.f) provided by temperature difference ΔT across the sample. The experiments for mixed Cu-Co ferrites, including ferrite compositions, in this analysis, are presented in Table 3. Electrical conductivity values decrease with increasing cobalt contentRezlescu et al. [14] discovered that as the zinc content in Li-Zn ferrites increased, the resistivity increased, which is consistent with this observation. Table 3 also contains the Seebeck coefficients of mixed Cu-Co ferrites at room temperature calculated from the thermal-e.m.f. values.The Seebeck coefficient has a negative sign, suggesting that conduction is the dominant mechanismin these ferrites is of n-type semiconductors. The hopping of electrons [15] from Fe^{2+} to Fe^{3+} is the dominant conduction mechanism in all cobalt substituted copper ferrites, based on this observation.

$$Fe^{3+} + e^{-} \Leftrightarrow Fe^{2+}$$
$$Fe^{2+} \rightarrow Fe^{3+} + e^{-}$$

Electrical conductivity of Cu-Co ferrites of various compositions at temperatures ranging from 343K to far above the temperature of the Curie. The plots showed in Figure 6 are log (σT) versus temperature (10^3/T).

The value of log (σT) rises linearly with increasing temperature up to a point, T_c (K), where the slope changes, as shown in the figures. Curie temperatures for Cu-Co ferrite samples under study were determined using the Gravity process [16]. Table 4 shows the transition temperatures T_σ(K) and temperature of cuire T_c(K) calculated using the gravity process. Table 4 shows that for the ferrites tested, the T_σ(K) magnetic transition temperature has resembled with T_c(K) transition temperature. The value of T_c(K) increases as the cobalt content increases, as can be seen in the graph. The increase in Curie temperature with increasing cobalt content could be explained by the number of magnetic ions present in the two sub-lattices and their mutual interactions.

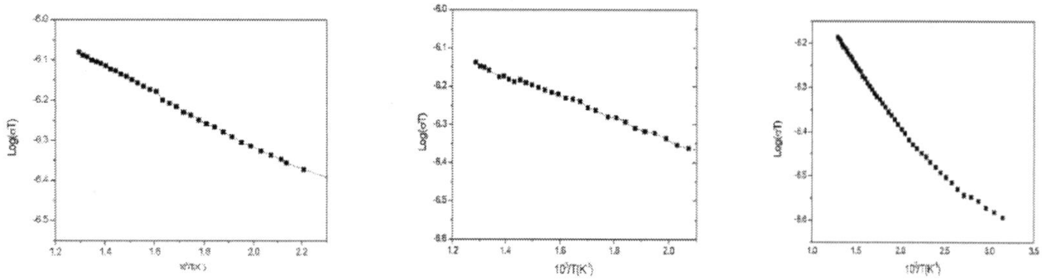

Fig.5. The log (σT) v/s Temperature for $Cu_{0.8}Co_{0.2}\,Fe_2O_4$, $Cu_{0.6}Co_{0.4}\,Fe_2O_4$ and $Cu_{0.4}Co_{0.6}\,Fe_2O_4$ ferrites

Table 4. Transition temperatures and Cu-Co ferrite activation energy.

S.No.	Ferrite composition	T_c (K)	T_σ(K)	Activation Energy		ΔE = E2-E1	Resistivity (ρ) Ohm.Cm (x10^{10})
				Ferri-magnetic region (E_1) eV	Para-magnetic Region (E_2) eV		
1	$Cu_{0.8}Co_{0.2}\,Fe_2O_4$	738	736	0.357	0.678	0.321	0.103
2	$Cu_{0.6}Co_{0.4}\,Fe_2O_4$	750	745	0.453	0.751	0.298	0.113
3	$Cu_{0.4}Co_{0.6}\,Fe_2O_4$	758	760	0.568	0.834	0.266	0.125

Irkin and Turov's theory [17] describes the existence, nearby the Curie Point, of kinks or transitions.In theory, the Curie pass induces a shift in the gradient of the straightline[18], and The exchange relationship between the external and internal electrons, which changes on the Curie point, determines the magnitude of the effect.The observation of the experiment Curie point transition in mixed Cu-Co ferrites agrees with Irkin and Turov'stheory [17-20].Activation energies determined from log (σT) versus (10^3/T) in the ferrimagnetic and paramagnetic areas.As the cobalt content increases, the activation energy values begin to decrease. The table shows that the activation energy is higher in the paramagnetic zone than in the ferromagnetic area.As a consequence, in accordance with Irkin and Turov's theory.

When it comes to Li-Ti [21] and Cu-Cd ferrites, similar findings have been observed[22]. The kink would be greater in situations where the outer and inner electrons have a stronger exchange interaction[23-25]. Depending on the structural peculiarities of the ferrites, as well as the importance of electrical resistance, This kink's scale may be smaller or bigger[26].Previous studies have shown that bigger kinks are typical of low-specific resistance ferrites as the activation energy in such materials is similar to the magnitude of the energy of spontaneous magnetization produced by magnetic transition[27].The extent of the kink is shown in column 6 of Table 4 according to the difference in the active energy (ΔE)of ferromagnetic (E_1) and paramagnetic(E_2) regions. The seventh column of the table gives the Cu-Co ferrite resistivity values of different compositions under analysis[28].When these two columns are compared, it is clear that ΔEdecreases as resistivity increases.The change in the path is usually due to the change in the conductivity mechanism.Electron hopping between the Fe^{2+} and Fe^{3+} ions results in lower conduction (under Curie)while hopping from the Polar Hopping at higher temperatures (>higher curie temperature)[29-31]. The calculated paramagnetic activation energy is greater than that of the hopping polaron conduction.

Acknowledgments

The authors gratefully thankto CSIR for providing the funding to do the research (file No: 09/132(0879)/2018-EMR-I).

IOP Conf. Series: Materials Science and Engineering 1221 (2022) 012014 doi:10.1088/1757-899X/1221/1/012014

References

[1] KulikouskiJLexiecoskiA and MagnJ1980 Mater. 19 -117
[2] RichesE E in and Cook J G 1972Ferrites: A review of Mate.&appls., Miles and Bones, Londan, ,p.17.
[3] VermaA GoelT C MendirattaR Gand AlamM I1999Mater. Sci. Eng. 60 156–162
[4]ThangjamBand SoibamINanomaterJ20172–8
[5] MsomiJ MoyoTand MagnetJ 2009 MagMater. 321 1246–1250
[6] KaiserM and Ata-AllahS S 2009 Mater. Res. Bull. 44 1249–1255
[7] BhandareM RJamadarH V PathanA T ChouguleB K and ShaikhA M2011 J.AlloysCompd. 509 L113–18.
[8] AhmedM A El M K ANimr and Tawfik 1991 Phys. Stat. Solidi A 123 501.
[9] HemedaO MSaidM Z and Barakat 2009J.Magn Materi,224 132.
[10] SmithJ and VijoH P J 1959Ferrites, Philips Tech. Library Clever - Home Press Ltd,The Netherlands .
[11] Murthy V R K and SobhanadriJ 1976 *Phys.Stat.Solidi A* 38 647.
[12] Simsa Z and Czech1996*J.Phys*. B16 919.
[13] ReddyV D MaliM A and ReddyP V1991*Mter.Sci.Eng*. B 8 295.
[14] RezlescuNCondurachDPetairuP and LucaE1974*J.Am.Ceram.Soc*. 57 40.
[15] VanUitertL G 1955*J.Chem.Phys*. 23 1883.
[16] RavinderD Thesis, Osmania University 1988.
[17] IrkinY P and TurovE A1957 *Sov.Phys. JETP* 33 673.
[18] SmithJ and WijnH P J1959*Ferrites,* Philips Technical Library, London p.234.
[19] LanjeN Y and KulkarniD K 2001*J.Magn.Magn.Mater*. 234 114.
[20] VijayaK Kumar and RavinderD2002*Materi.Letts*. 52 166.
[21] ManjulaR and MurthyV R K1986*J.Appl.Phys*. 59 2929.
[22] RavinderD2000*Materi. Letts*. 43 129.
[23] KringerM I 1979*Phys. Status Solidi* B 79 9.
[24] KlingsM I1975*J.Phys.C*8 3595.
[25] MottN F andGureyR W1948 Electronic Process in Ionic Crystals, Oxford University Press, Oxford,
[26] Regaty Anitha Reddy and Kattepogu Rama Rao 2019 Rare Metals.
[27] MicheleBastianello and Silvia Gross2019Royal Society of Chemistry Adv.9 33282
[28] Saadia Rasheed and Rafaqat Ali Khan 2019 et.Al, JMMM, Volume Pages 4711 236-241
[29]Zhihui Ye Zaixing Deng and Lin Zhang 2020 IOP Publishing, Mater. Res. Express 7 035007
[30] Zena Mohammed and Ali Abbas 2018 I JAER ISSN 0973-4562 Volume 13, Number 19 pp. 14231-14235
[31]Alireza Samavati andMustafaM. K Materials Express, American Scientific Publishers

Dielectric And Relaxation Studies in Multi Oxides Doped Borotellurite Glasses

Amarkumar Malge[1], T.Sankarappa[1]*, G.B.Devidas[2], J.S.Ashwajeet[3], Ashwini Devidas[1] and [1]Mohansingh Heerasingh[1]

[1]Department of Physics, Gulbarga University, Kalaburagi 585106, India
[2]Department of Physics, Kuvempu University, Shankaraghatta, India
[3]Department of Physics, Davanagere University, Davanagere, India
*Corresponding author:sankarappa@rediffmail.com

Abstract

Melt quenching technique has been adopted to synthesize a set of borotellurite glasses consisting of ZnO, Li_2O, WO_3 and Dy_2O_3, with a varied content of WO_3. After confirming their amorphous nature by XRD, they were investigated for dielectric properties over a wide range of freqency and temperature. Observed variations in dielectric parameters with frequency, temperarure and composition suggests the presence of mobility of charge carriers, increase of loss due to ionic migration and involvement of dc conductivity. Single phase and conducting nature of the samples have been obtained by sketching cole-cole plots. DC conductivity estimated from impedance spectra indicated semiconducting nature and its activation energy increased with increase of WO_3 concentration. Dielectric relaxation time and its thermal activation energy were determined independently from modulus and impedance and found that the former decreased with increase of temperature and the later increased with WO_3 content. Master curves drawn for electric moduli revealed that relaxation mechanism in these glasses is temperature independent.

Key words: Glasses, dielectric relaxation, relaxation time, electric moduli, conductivity.

1.INTRODUCTION

Boric oxide has low melting temperature, low thermal expansion and high chemical durability. It is a good host for rare earth elements and have applications in phosphors, solar energy converters and moisture resistance, etc [1,2]. At the same time, TeO_2 is the most stable oxides of heavy metals which when mixed with borate oxide, stability and conductivity of the glasses get enhanced. Borate and tellurite oxides together assure high glass-forming ability, high mechanical strength and high resistance to corrosion [3]. ZnO acts as both conditional glass former and at lower mole percentage it acts as network modifier. In addition to reducing the viscosity of the melt it reinforces network integrity by controlling crystallization [4]. Li_2O increases electrical conductivity and improves mechanical properties by converting BO_3 units into BO_4 units. The Li_2O doped glasses are used in high energy density solid-state batteries, super ionic conductors, and solar cell applications [5].

Rare earth ion doped materials are know to have high co-ordination number and high polarizability. Presence of rare earth ions in tellurite glasses reduced their dielectric loss and increased dielectric constant [6]. Glasses of high dielectric constant are useful for developing high energy capacitors. Dy^{3+} is an interesting ion as there will be non-spherical distribution of charges in dysprosium, Dy^{3+} ions and that in turn can increase dielectric constant of the glasses. Tungstic oxide (WO_3) is a conducting semi-transition-metal oxide that exists in W^{6+}, W^{5+} and W^{4+} states. It facilitates multi component doping and contributes to polaron conduction [7-9]. WO_3 helps to improve chemical stability and enhance glass transition temperature which allows use of glasses under high optical intensities without any thermal damage.

In view of the utility of various oxides as described above and that there are no many reports in the literature on the dielectric properties and relaxation studies of WO_3 doped multi component borotellurite glasses, a set of borotellurite glasses [B_2O_3–TeO_2 –ZnO–Li_2CO_3 – Dy_2O –WO_3] have been thoroughly investigated for dielectric properties and relaxation mechanism and, the results obtained are discussed in the present article.

2. EXPERIMENTAL

By following melt quenching and using analytical reagent grade chemicals H_3BO_3, Li_2CO_3, H_2WO_4, ZnO, Dy_2O_3 and TeO_2, the glasses of composion, $(B_2O_3)_{0.2}$–$(TeO_2)_{0.3-x}$–$(ZnO)_{0.28}$–$(Li_2CO_3)_{0.2}$ – $(Dy_2O_3)_{0.02}$–$(WO_3)_x$ ($0.02 \leq x \leq 0.12$ mol%) were synthesized and labeled as W1, W2, W3, W4, W5 and W6 respectively. Chemicals in fixed quantities were ground and melted in alumina crucibles at 1273K. Melt was quenched between two stainless steel plates and then annealed at 473K for 24 hours. Powdered samples have been subjected to XRD studies in a Rigaku make X-ray diffractometer [10]. Samples of known dimensions were readied for dielectric measurements by polishing and silver pasting their top and bottom surfaces. The capacitance, C, and dissipation factor, tanδ were measured in the temperature range from 300K to 620K and frequency range from 50Hz to 15MHz, in an Impedance analyzer (Wayne Kerr 6500B).

The dielectric constant, $\varepsilon' = (\frac{Cd}{\varepsilon_0 A})$ and dielectric loss, $\varepsilon'' = \varepsilon' tan\delta$ were determined. Real and imaginary parts of electric moduli $M' = \varepsilon'/(\varepsilon'^2 + \varepsilon''^2)$ and $M'' = \varepsilon''/(\varepsilon'^2 + \varepsilon''^2)$) and, impedance $Z' = M''/(\omega\varepsilon_o)$ and $Z'' = M'/(\omega\varepsilon_o)$ [11] were also estimated. Here, ε_o is the permittivity of free space, ω the frequency, d the thickness and A the cross-sectional area of the sample.

3. RESULTS AND DISCUSSION

3.1 Dielectric properties:

In the measured range of temperature and frequency, ε' and ε'' for the present glasses are found to be in the range from 5.37 x10^2 to 3.97 x10^3 and from 2.89 to 1.993x10^4 respectively. Fig.1 demonstrates typical plots of frequency dependence of ε' and ε" for W1 glass at different temperatures. It can be noted that both ε' and ε" decreases with increasing frequency and increases with increasing temperature. The remaining glasses W2, W3, W4, W5, and W6 showed similar trends of variations in ε' and ε". These observations agree with reported papers [12-14].

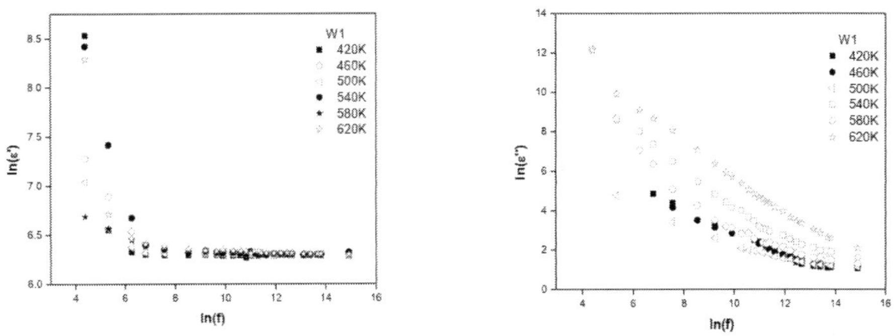

Figure1. Plots of **(a)** ln(ε') versus ln(f) and **(b)** ln(ε") versus ln(f), for W1 glass at different temperatures.

The behavior of ε' observed in Fig.1 (a) at a lower frequency can be ascribed to contributions such as space charge polarizations, motion of conducting charges etc. Polarizability is largely driven by ionic and orientation sources, which decrease with frequency increase and eventually become almost constant because of the inertia of ions and molecules [6,14,15]. Consequently, the modulus representation has been employed to analyze data as ε' at higher frequencies appears to be temperature independent.

In Fig.1 (b), the high values of ε" at lower frequency can be characterized by the contribution of both ion migration and electron polarization [15]. In the present glasses, polaronic motion between different valence sites of tungsten and Li^+ ions may be the main reason for dielectric loss. The variations in ε' and ε" with temperature are linked to bond energies. The behavior of both ε' and ε" at a lower temperature can be assignable to contributions of electronic and ionic components plus negligible orientation components. The temperature rise decreases bond energy and that in turn facilitates orientation and space-charge polarizations [11,15].

3.2 Modulus (M):

Information about conduction mechanism and relaxation behavior can be obtained by splitting the electric modulus, M into real (M') and imaginary (M") parts. Fig.2 shows the frequency dependence of (a) M' and (b) M" for W1 glass at different temperatures. In Fig.2(a), we see that M' increase with increase in frequency and temperature and reaches a constant value. At lower frequency, M' is small and tends to be zero, which implies the absence of electrode polarization effect and the observed dispersion of M' with frequency is mainly due to conductivity relaxation [16]. That is, the ions are free to move in the region between the plateaus of low and high frequency [11].

Fig.2 (b) shows M" variation with ln(f) for selected temperatures for W1 glass. Assymetric relaxation peaks with M''_{max} at characteristic frequencies, f_c can be observed. Peak is shifting towards higher frequency with increase in temperature. The shift in peak towards higher frequency indicates mobility of charge carriers. And the shift of peak with temperature signals a thermally activated process (dc conductivity) in which hopping mechanism of charge carriers dominates [17,18]. On the lower frequency side of the M'_{max}, the ions will drift for long distance and on the higher frequency side of the M'_{max} region they will be confined to the potential well and free to move within the well only [19]. Similar moduli spectra were obtained for remaining glasses of the present set.

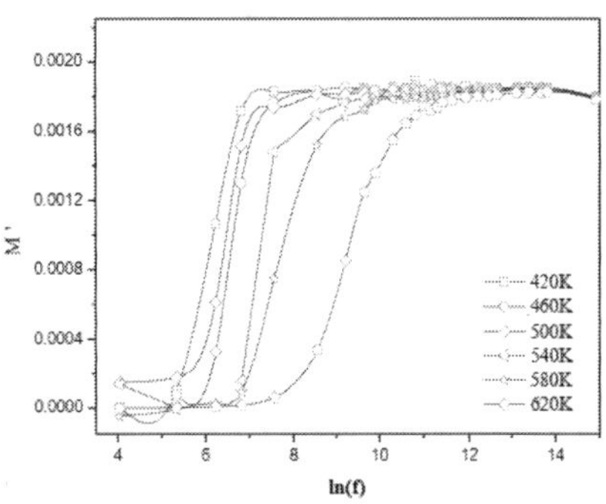

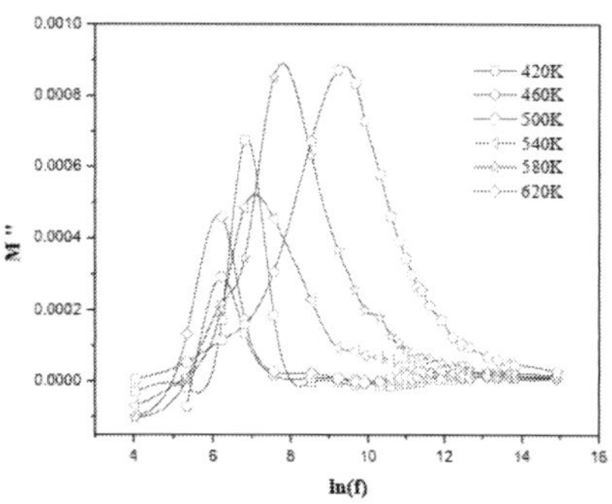

Figure 2. Frequency dependence of (a) M' and (b) M" for W1 glass at different temperatures.

3.3 Impedance (Z):

The complex impedance of W glasses has been analyzed by sketching the plots of Z' versus Z". Fig.3 displays cole-cole plot for W1 glass at selected temperatures. A well shaped semicircle can be found at all the temperatures of interest. The remaining glasses of the present series showed similar nature of variations in Z' and Z" suggesting the single phase nature of the samples [11]. It is also clear that intercept of semicircle shifts towards lower Z' with increasing temperature, which imply that bulk resistance of the sample is decreasing with temperature. The observed decrease in the radius of the semi-circles with increase of temperature reveals an increase of loss due to ionic migration with temperature [20,21]. By measuring intercepts of semicircles (R_{dc}) onto Z' axis, DC conductivity has been estimated as $\sigma_{dc} = \frac{t}{A}\left(\frac{1}{R_{dc}}\right)$ where, t is the thickness and A cross-sectional area of glass [22,23]. Variation of $\ln(\sigma_{dc})$ with reciprocal of temperature, T for W glasses is shown in Fig.4. By fitting linear lines as per the Arrhenius expression, activation energy, E_{dc} was determined. Obtained E_{dc} values are mentioned in figure. These E_{dc} are of the range as those reported for WO_3 doped alkali borate glasses in reference [12].

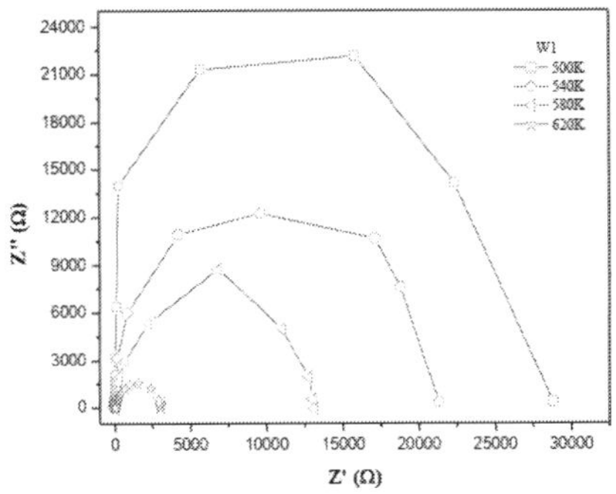

Figure 3. Cole-Cole plot for W1 glass at four different temperatures.

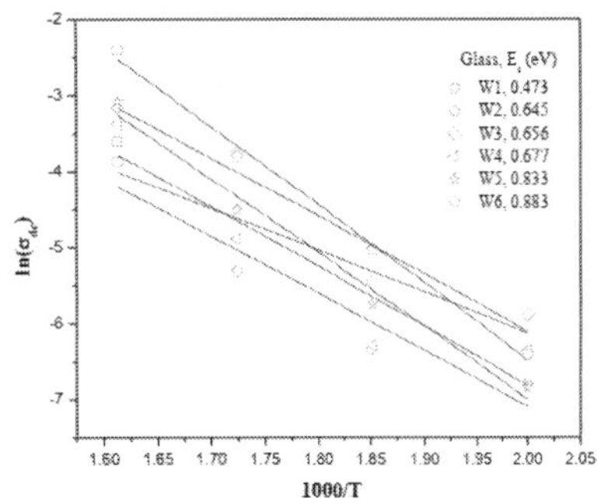

Figure 4. Arrhenius plots of dc for W-glasses. Activation energy, E_{dc} is mentioned along with sample labels.

3.4 Relaxation mechanism

Dielectric relaxation can understood by taking hopping of charge carriers into account. Transition metal oxide doped glasses contain small polarons with the same mobility as ions. Present glasses have both polarons and ions as charge carriers.

Dielectric relaxation time can be determined using the condition, $\omega_m\tau_m = 1$ by substituting ω_m values extracted from frequency versus M" or Z" plots. The activation energy, E_τ for relaxation is determined on the assumption that it is a thermally activated process and is expressed as [13],

$$\tau = \tau_o \exp\left(E_\tau / k_B T\right) \qquad (1)$$

The estimated values of τ from both modulus and impedance are presented in Table-1 and it can be observed that τ decreases with temperature. This agrees with Nd_2O_3 doped alkali borate glasses [19]. According to Eq.(1), the plots of $\ln(\tau_{M''})$ and $\ln(\tau_{Z''})$ against $(T)^{-1}$ are drawn and shown in Fig.5 and Fig.6 respectively. The E_τ obtained from M" and Z" are found to increase with increase in WO_3 content and are close to each other (Table 1). Activation energy for dc conduction and relaxation mechanism are found to be of the same size. Therefore, it can reasoned that the charge carriers faced the same potential barrier both in conduction and relaxation process [19].

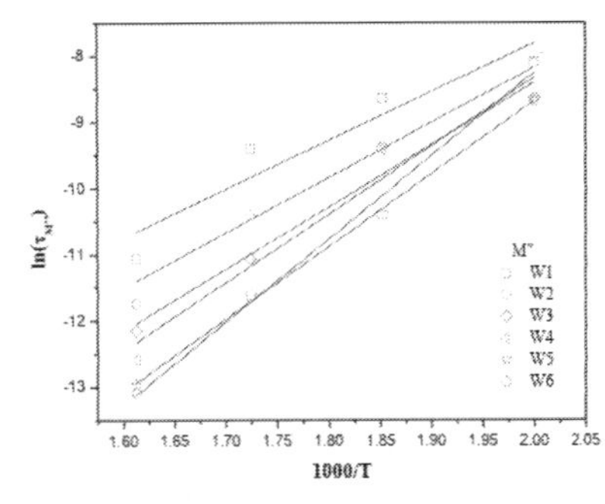

Figure 5. Plots of $\ln(\tau_{M''})$ versus (1/T) for all W-glasses.

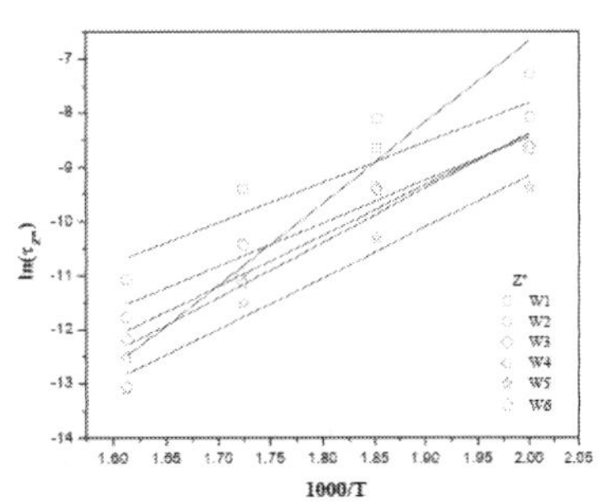

Figure 6. Plots of $\ln(\tau_{Z''})$ versus (1/T) for all W-glasses.

Table 1. Dielectric relaxation time, τ and activation energy, Eτ obtained from moduli, M" and impedance, Z" respectively.

Glass	$\tau_{M''}$ (ms)				$\tau_{Z''}$ (ms)				Activation energy (eV)	
	500K	540K	580K	620K	500K	540K	580K	620K	$E\tau_{M''}$	$E\tau_{Z''}$
W1	0.307	0.176	0.082	0.016	0.307	0.174	0.082	0.016	0.634	0.637
W2	0.171	0.176	0.030	0.008	0.171	0.085	0.030	0.008	0.716	0.683
W3	0.176	0.084	0.016	0.005	0.180	0.082	0.018	0.005	0.808	0.807
W4	0.176	0.082	0.016	0.003	0.171	0.082	0.015	0.004	0.888	0.872
W5	0.171	0.031	0.009	0.002	0.084	0.033	0.010	0.002	0.945	0.815
W6	0.299	0.030	0.009	0.002	8.047	2.369	0.155	0.001	1.082	1.297

From Fig.3, f_{max} corresponding to M''_{max} for different temperatures have been noted and $\omega_{max} = 2\pi f_{max}$ were estimated. Assuming relaxation process to be thermally activated one Arrhenius expression for this can be written as [11,24],

$$\omega_{max} = \omega_o \exp\left(-\frac{E_{\omega_{max}}}{k_B T}\right) \quad (2)$$

As per this expression, $\ln(\omega_{max})$ versus (1/T) for W glasses are plotted in Fig.7. Linear fits to the data yielded activation energy, $E\omega_{max}$ values as 0.626 eV, 0.702 eV, 0.797 eV , 0.871 eV , 0.937 eV and 1.011 eV for W1, W2, W3, W4, W5 and W6 glasses respectively. It may be observed that $E\omega_{max}$ increases with increase of WO_3 content similar dc activation energy.

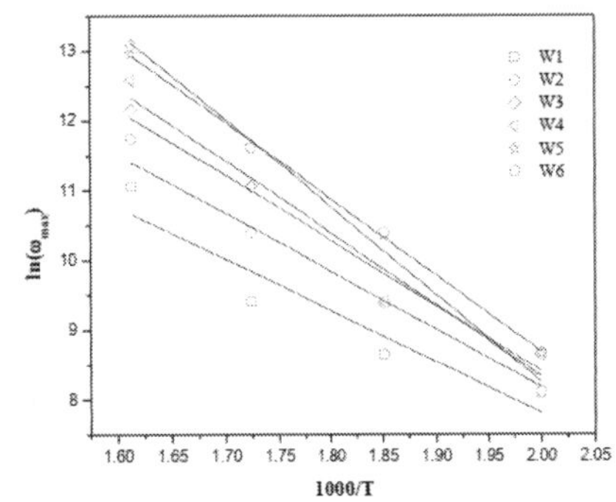

Figure 7. Plots of $\ln(\omega_{max})$ versus (1/T) for all D-glasses.

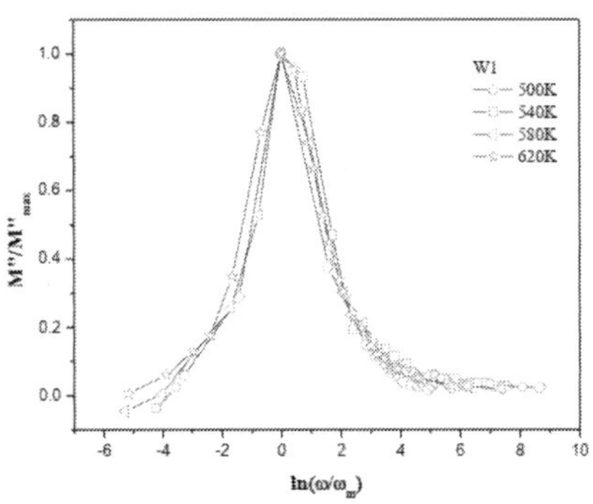

Figure 8. Normalized plots of electric moduli, M" for W1 glasses at different temperatures.

Ion transport in oxide glasses is best understood by scaling analysis. Fig.8 shows the reduced plot of imaginary component of dielectric modulus (M''/M''_{max}) versus reduced frequency (ω/ω_{max}) for W1 glass at different temperatures. It is evident from the modulus spectra that the relaxation processes that occur in W1 glass are temperature independent. Master curves drawn for other five glasses of present series appeared to be same as Fig.8 and revealed the same information. These results are very much in agreement with those quoted in references [12,13,16].

4. CONCLUSIONS

i) ZnO, Li_2O, WO_3 and Dy_2O_3 doped borotellurite glasses were probed for dielectric properties and the observed changes in these properties at lower frequency were ascribed to migration of Li^+ ions in the glass.

ii) Electric moduli and impedance have been analyzed to fine detail. The cole-cole plots revealed the conducting and single phase nature of the prepared material.

iii) The overall dielectric behavior of samples suggests the presence of charge carrier's mobility, increase of ionic migration loss and involvement of DC conductivity.

iv) DC conductivity has been estimated and variation with temperature reveals the semiconducting nature. The activation energy for all glass samples was determined and noticed to be increasing with the increase of WO_3 content.

v) Relaxation time, τ and thermal activation energy, $E\tau$ for relaxation time has been extracted from both M" and Z". τ decreased with the rise of temperature whereas $E\tau$ increased with the increase of WO_3 content.

vi) Temperature-independent relaxation process is revealed by master curves for present glasses.

ACKNOWLEDGEMENT

Authors acknowledge financial assistance received from the DST (SERB), Government of India, New Delhi in the form of a major research project sanctioned to Prof. T. Sankarappa, Principal Investigator.

References

[1] Shuhaimi SA, Hisam R, Yahya A K 2020 Solid State Sciences.
[2] Al-B. Mohammed F A, Lakshminarayana G, Baki S O, Halimah M K, Kityk I V, Mahdi M A, 2017 Optical Materials 73 686-694.
[3] Amarkumar M, Sankarappa T, Ashwajeet J S, Sujatha T, 2019 J. Phys.: Conf. Ser. 1172 (012012.
[4] Amarkumar Malge, Sankarappa T, Sujathab T, Ashwajeet J S, Devidas G B 2019 Mater. Today: Proc. 18 5340–5344.
[5] Vijayalakshmi L, Naveen Kumar K, Bhaskar Kumar G, Pyung Hwang 2017 J Non-Cryste Solids, 475 28-37.
[6] Azlan M N, Shafinas S Z, Halimah M K, Suriani A B, 2019 J Mater Sci: Mater Electr 30 18015–18024.
[7] Ersundu A E, Karaduman G, Celikbilek M, Solak N, Aydin S, 2010 J.Alloys Compd, 508, 266-271
[8] Ersundu A E, Celikbilek M, Aydin S, 2012.J. Non-cryst solids, 358, 641-647.
[9] Tandon R P, Hotchandani S, 2001 Physica status solidi (a)185, 2, 453–460
[10] Amarkumar Malge, Sankarappa T, Sujathab T, Devidas G B, Abdul Azeem P, 2020 Optical Materials 109 110282.
[11] Amarkumar Malge, Sankarappa T, Sujathab T, Devidas G B, 2020 J. Phys.: Conf. Ser. 1451 012014.
[12] Sheoran A, Sanghi S, Rani S, Agarwal A, Seth V P, 2009 J. Alloys Comp 475 804–809.
[13] Ali A A, Shaaban M H, 2010 Solid State Sciences 12 2148-2154.
[14] Ashwajeet J S, Sankarappa T, 2017 Ionics 23, 3 627.
[15] TAHA T A, AZAB A A, J. electr. mater. 45 5170-5177.
[16] Dhankhar S, Kundu R S, Dult M, Murugavel S, Punia,N Kishore R, Indian J Phys, DOI 10.1007/s12648-016-0850-9
[17] Ragab M. Mahani, Samir Y. Marzouk, 2013 J of Alloys Comp 579 () 394–400.
[18] Monisha M, Nimitha, Prabhu S., Ashwitha Nancy D'Souza, Suresh Bharadwaj, R.J. Chowdary, M.I. Sayyed, Abdullah Alhuthali M S, Al-Hadeethi Y, J. 2021) J Non-Cryst Solids 560 120733.
[19] Ramteke D D, Swart H C, Gedam R S, 2017 Journal of Rare Earths, 35, 480.
[20] Suresh S, Prasad M, Chandra V, Mouli, 2010 J Non-Cryst Solids 356 1599–1603
[21] Barde R V, Waghuley S A, Ceram Inter, 39(6), 6303–6311.
[22] Singh D P, Shahi Kamal K, Kar K, 2016 Solid State Ionics 287 89–96.
[23] Veeranna Gowda V C, Anavekar R V, 2007 J Mater Sci 42:3816–3824
[24] Hutchins M G, Abu-Alkhair O, El-Nahass M M, Abdel-Hady K, 2007 J Non-Cryst Solids 353 (4137–4142

IOP Conf. Series: Materials Science and Engineering 1221 (2022) 012016 doi:10.1088/1757-899X/1221/1/012016

Thermochemical Characterization of Cycas Circinalis Seed Shell to Evaluate their Potential as Biofuel Source

Vanisree G S and Aparna K

Department of Chemical Engineering National Institute of Technology Calicut, Kerala,India

Corresponding Email:vanisree_p190120ch@nitc.ac.in

Abstract:Increased environmental pollution anxieties and quest for high energy demand from renewable energy sources forced to use lignocellulosic biomass to produce Bio-fuels. Cycas circinalis seed shell is a plant-derived bio-waste readily available in the southern coastal part of India. This research aims to explore the thermochemical characteristic of Cycas circinalis seed shells. Substantial volume reduction of feedstock and low processing time makes the thermal conversion route more attractive than any other method of waste to energy conversion. Biomass degradation behavior is studied by Thermogravimetric and derivative thermogravimetric analysis (TGA and DTG); this depicts biomass degradation in three primary stages, each corresponding to the decomposition of hemicellulose, cellulose, and lignin components. Functional groups were determined by Fourier transform infrared (FTIR) spectroscopy. The obtained calorific value of biomass was found to be 16.17 MJ/Kg. Proximity analysis results showed the percentage of moisture content, volatile matter, ash content, and fixed carbon as7.95%,77.35%,2.05%,12.65%respectively. The ultimate analysis determines the composition of significant elements present in Cycas ciricinails. Carbon, Hydrogen, and Oxygen are found to be 45.54%, 6.5 %, and 47.16 %, respectively. Sulfur and nitrogen component values were observed to be lower, which is also a promising result in view of environmental concerns.

1. Introduction:

The fundamental target of Paris climate agreement 2015,is to bound the rise of global warming by 2050 at a maximum of 2^0C and reducing the visible climate change. It can be possible achieved by defossiling transport revolution[1]. An alternate to fossil fuel is required as it provides environmental deterioration and depletion of fossil hydrocarbon. Bio-fuel gained attention due to its advantages: biodegradability, low carbon dioxide emission, non-toxicity, sulfur-free, unburned hydrocarbon, and particulate materials[2].The first energy source used by humankind to meet their daily needs is biomass. Biomass is a carbonaceous material vital as a renewable energy source mainly composed of cellulose, hemicellulose, lignin, and a small fraction of inorganic matters. Its composition and components differ for each plant material [3,4]. Out of the total global biomass, around 55%utilize as feed and food products, 27% as bioenergy, 8% as biomaterial, and 10% for other purposes. A significant portion of the world's energy demands is satisfied by biomass, primarily in rural areas where people use it in traditional ways to cook and other purposes. Energy generation from biomassexpects to be tripled within the first quarter of this century[5,6].The thermo-chemical process can convert lignocellulosic biomass into energy and chemical compound in two ways[7]. The first process is gasification, which involves keeping the biomass at a high temperature with minimal oxygen to produce syngas, which is a mixture of CO, H_2, CO_2, and CH_4[8]. Pyrolysis is the second route of biofuel conversion. Without oxygen, an intermediate temperature of 300-600^{0}C is applied to the biomass to convert it into liquid pyrolytic oil, solid char, and light gases akin to syngas[9].

Recognizable proof and characterization of biofuel property is the first and significant stagefor biomass usage. The above study is the one essential code, that identifies the qualities, potential, quality, and environmental viability of any biomass source[10–12]. Thermochemical characterization of biomass is a common and practical technique for determining its suitability as a fuel for energy production. Numerous biomass characterizations are reported in the literature, including wood biomass, agricultural waste biomass, sewage sludge, fruit tree residue biomass, domestic waste biomass, microalgal biomass, and others [4,6,13–18].

Cycas circinalis (Queen Sago) is a common ornamental gymnosperm found in tropic and subtropic areas that belongs to the cycadacea family[19]. The flour obtained from*Cycas circinalis* seed is used for food[20].Sago waste has different applications reported by researchers. Sago pith waste ash is used for making geopolymer mixture concrete[21], *Cycas circinalis* seed shell is used as a reinforcing filler in rubber composite[22], and other applications include heavy metal adsorbent, activated carbon,nanomaterials,composites[23].Figure1shows the *Cycas circinalis* seed shell used for this study.

Figure1. *Cycas circinalis* seed shell

2. Materials and methods:

Cycas circinalis seed shell were collected from plants located in *cheliya region*, which is a village in Kozhikode district of Kerala state. These were washed a number of times with water to remove any impurities present. After that, it is air-dried for 14 days to remove the moisture content. The dried shell is crushed and ground to make it powder and sieve to get 85-micron size. The powder is stored in an air-tight container to avoid biomass moisture close contact.

2.1. Proximate Analysis

Proximate analysis is done using appropriate ASTM procedure to determine the moisture content, ash content, volatile matter, fixed carbon present in the *Cycas circinalis* seed shell

2.1. Moisture content

Standard ASTM D3173M(ASTM International 2013) procedure was used to determine the moisture content. About 1gm of sample is weighed in a silica crucible which is kept in a muffle furnace without keeping the lid at a temperature 105^{0}C for 1 hr. After that sample is weighed and the amount of moisture content is calculated by the equation (1).

$$\text{Moisture (M),\%}=\frac{W1-W2}{W3}\times 100 \quad (1)$$

Where W_1 is the weight of sample and Crucible before heating(g),W_2 is the weight of sample and Crucible after heating(g),W_3 is the weight of taken(g).

2.1.2. Volatile matter

Volatile matter in the sample was found using the procedure outlined standard ASTMD3175 (ASTM International 2007).1g of sample weighed in a silica crucible is placed in a preheated muffle furnace at 950^{0}C with its lid closed for 7 minutes. The samples were allowed to cool with the lids on after being removed from the surface. The sample plus crucible is then weighed to calculate the volatile matter present in the sample using equation(2).

$$\text{Weight loss\% (L)} =\frac{W4-W5}{W3}\times 100 \quad (2)$$

Where W_4 is the weight of Crucible with lid and sample before heating (g), W_5 is the weight of Crucible and lid with content after heating (g).

$$\text{Volatile matter (V), \%} = \text{L-M} \quad (3)$$

2.3. Ash content
Standard ASTMD3174 technique is used to determine the ash content of the biomass sample. About 1 g of sample is accurately weighed in a silica crucible and is placed inside a muffle furnace without lid then heated progressively to 500°C at 1hr. Then heating is continued to 950°C. After 1hr the temperature is kept constant for 2hr. Then the sample is taken out from the furnace and allowed to cool with its lids closed. Ash content is calculated using equation (4).

$$\text{Ash(A), \%}=\frac{W6-W7}{W3}\times 100. \quad (4)$$

Where W_6 is the weight of crucible with lid and sample before heating (g), W_7 is the weight of Crucible with lid (g).

2.4.Fixed carbon
The amount of fixed carbon present in the sample is calculated using equation (5)[18].

Fixed carbon F,% =100-(M+A+V) (5)

3.Ultimate Analysis:
Elementar Vario EL-III was used to estimate the amount of carbon (C),nitrogen (N),hydrogen (H), and sulphur (S) present in the sample.The oxygen content calculated using equation(6).

Oxygen content,% =100-(carbon+nitrogen+hydrogen+sulphur) (6)

3.1. Calorific Value Analyis:
Athena Bomb Calorimeter –ATB-01 is used to find the higher heating value of the sample based on ASTM2015 procedure.

3.2.Fourier transform infrared analysis(FTIR):
The functional group present in the sample is analyzedby Jasco 4700 Fourier transform infrared spectroscopy. The obtained spectra were using scan rate of 8 with a step size of 1 cm^{-1} within the wave number range of 400 to 4000 cm^{-1}.

3.3.ThermogravimetricAnalysis(TGA and DTG):
The change in feedstocks mass over different temperature range is a critical characteristic that must be thoroughly investigated in connection with any thermochemical conversion process. Hitachi TGAThermogravimetric analyzer was used to investigate thermochemical degradation behavior under pyrolysis and combustion working circumstances.

4.Result and Discussion:
4.1.Proximate analysis:
Table 1 enlisted the proximate analysis of *Cycas circinalis* seed shell.Moisture content in a fuel absorbs some of the heat released during the combustion process; as a result, the amount of heat available from the fuel has lowered. The higher the water content, the lower the heating value, and hence the worse the fuel efficiency [24].The result showed that*Cycas circinalis* seed shell belongs to low moisture content biomass with7.95% moisture content.

Volatile matter is condensable and noncondensibleduring heating. It influences the igniting flame stability and the amount of unburned carbon in ash[18]. Volatile matter present in the *Cycas circinalis* seed shell calculated as 77.35%seems high but comparable with other biomass reported in literature [25].

Low ash content of 2.05% was found in the *Cycas circinalis* seed shell.A high ash content may reduce the heat and mass transfer and produce particle aggregation in large-scale production[26]. Low ash in the feedstock indicate greater ease of processing during pyrolysis of biomass hence *Cycas circinalis* seed shell is a potential waste energy option.

Fixed carbon is the solid carbon that remains in the char after the moisture and volatile materials have been removed. It contains elemental carbon in the biomass as well as carbonaceous residue after heating. It was found that12.65 % fixed carbon in *Cycas circinalis* seed shell.

Table 1. Proximate analysis of *Cycas circinalis* seed shell

	%
Moisture content	7.95

Volatile matter	77.35
Ash	2.05
Fixed carbon	12.65

4.2.Ultimate Analysis:

Chemical constituents present in the *Cycas circinalis* seed shell found by ultimate analysis and tabulated in Table2. This analysis helps in calculating the theoretical amount of air necessary for complete combustion. Fuel gas produced per kilogram of fuel burned, the amount of water vapor in the flue gas, the amount of S in the flue gas, and the calorific value of the fuel are all affected by the C and H content of the fuel. Biomass with higher C content and a lower O content tends to have a higher heating value. The H content discovered to be 6.5 wt%.

N and S content in biomass is essential in terms of environmental emissions since they result in the formation of NOX and SOX.The absence of N and S in the seed shell of the *Cycas circinalis* makes it suitable for bioenergy production[6].

Table 2. Ultimate analysis of *Cycas circinalis* seed shell

Element	%
Oxygen	47.16
Carbon	45
Hydrogen	6.5
Nitrogen	0
Sulphur	0

4.3.Calorific value:

The calorific value measures the amount of heat that can be produced from unit weight of biomass. It plays a remarkable role in any biomass feedstock for energy production. The calorific value of *Cycas circinalis* seed shell was calculated as 16.17 MJ/kg. Calorific value is comparable with prevailing raw material for biofuel generation[17]

4.4.Fourier transform infrared analysis(FTIR):

Figure 2 illustrates the FTIR spectra of the *Cycas circinalis* seed shell. The functional groups present in the biomass was analyzed.The peaks present in spectra of 3300cm^{-1} revealed the OH stretching vibration and hydrogen bond of hydroxyl groups[27]. It was reported in literature, peaks at 2917cm^{-1}reprents CH group starching of aliphatic group present in the cellulose,hemicellulose,ligin components[28].C=C aromatic ring vibrationl starching found at apeaks of 1732cm^{-1}and 1603cm^{-1}[29]. C-O carboxylgroup vibrational stretch is reported at 1360cm^{-1}.Sharp peak observedat 1016cm^{-1} isattributed to C-O stechingof primary alcohol[22].

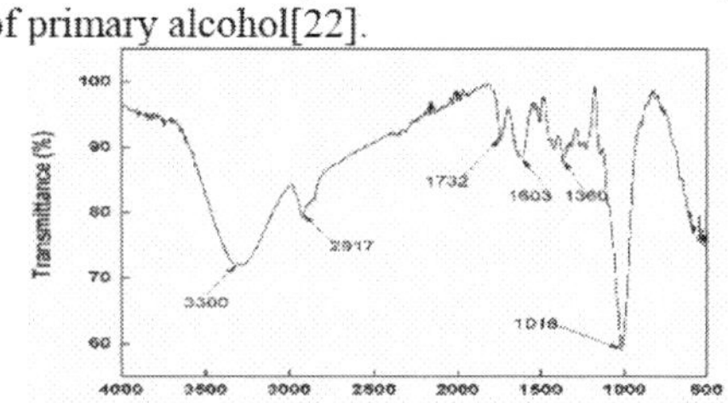

Figure2. FTIR spectra of *Cycas circinalis* seed shell

4.5.Thermogravimetric Analysis(TGA):

This analysis is done to study the thermal degradation behavior of biomass. Mass loss of the sample is measured and ploted versus temperatureThe decomposition of the *Cycas circinalis* seed shell was observed to takeplace in 3 different phases, as shown in figure 3(a).The first phase indicates moisture loss, whereas the others correspond to the decomposition of hemicellulose, cellulose, and lignin components. At a temperature of 185^{0}C, the first stage of decomposition was seen. A total weight loss of 5.87% was noted.

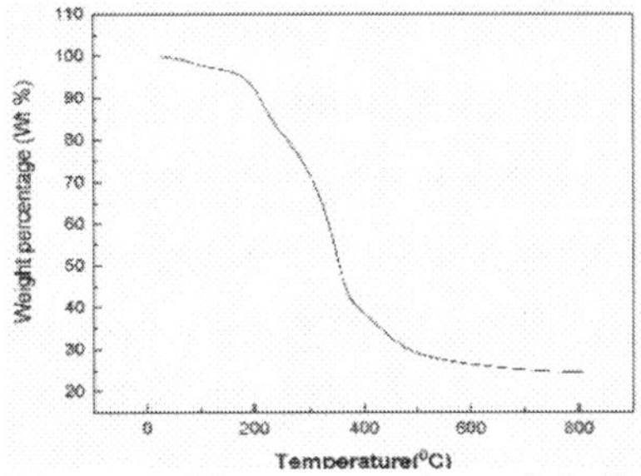

Figure 3. TGA curve of*Cycas circinalis* seed shell.

The decomposition of hemicellulose and cellulose components. The weight loss recorded during the second phase was53.5%. The peak conversion temperature is recorded during the second stage.The result obtained in the stage agrees with the decomposition pattern of different biomass studied in other literature[30].The lignin decomposition is very complex, continuous, and slow. Third phase decomposition was observed starting at 475^{0}C and continuing up to 800^{0}C. The slow and continued degradation of lignin component along with charring process of the non volatile components.The mass loss for the third stage was found to 15.32%.The decompositionindicates a good Stability of the biomass to be used as a feedstock to produce biofuel via thermochemical conversion process.

5. Conclusion:

Sustainability to access biomass material as solid biofuel is studied by its thermo chemical characterization.In this paper, we did an extensive study on the characterization of *Cycas circinalis* seed shell.The obtained calorific value of biomass was found to be 16.17 MJ/kg. Proximity analysis results showed the percentage of moisture content, volatile matter, ash content, and fixed carbon as7.95%,77.35%,2.05%, 12.65% respectively. The ultimate analysis shows Carbon, Hydrogen, and Oxygen are found to be 45.54%, 6.5 %, and 47.16 %, respectively. The absence of Sulfur and Nitrogen is also a promising result in environmental perspective. Thus it can be concluded that the *Cycas circinalis* seed shell is a environmentaly benign source for biofuel. Further studies can be employed using *Cycas circinalis* seed shell as feedstock for biofuel production.

References:

[1] Unglert M, Bockey D, Bofinger C, Buchholz B, Fisch G, Luther R, Mueller M, Schaper K, Schmitt J and Schroeder O 2020 Action areas and the need for research in biofuels *Fuel***268** 117227

[2] Laskar I B, Deshmukhya T, Bhanja P, Paul B, Gupta R and Chatterjee S 2020 Transesterification of soybean oil at room temperature using biowaste as catalyst; an experimental investigation on the effect of co-solvent on biodiesel yield *Renewable Energy***162** 98–111

[3] Ahorsu R, Medina F and Constantí M 2018 Significance and challenges of biomass as a suitable feedstock for bioenergy and biochemical production: A review *Energies***11** 3366

[4] Demirbas A 2008 Biofuels sources, biofuel policy, biofuel economy and global biofuel projections *Energy conversion and management***49** 2106–16

[5] Popp J, Kovács S, Oláh J, Divéki Z and Balázs E 2021 Bioeconomy: Biomass and biomass-based energy supply and demand *New Biotechnology***60** 76–84

[6] Ahmed A, Hidayat S, Abu Bakar M S, Azad A K, Sukri R S and Phusunti N 2021 Thermochemical characterisation of Acacia auriculiformis tree parts via proximate, ultimate,TGA,DTG, calorific value and FTIR spectroscopy analyses to evaluate their potential as a biofuel resource *Biofuels***12** 9–20

[7] Panwar N, Kothari R and Tyagi V 2012 Thermo chemical conversion of biomass–Eco friendly energy routes *Renewable and Sustainable Energy Reviews***16** 1801–16

[8] Ren J, Liu Y-L, Zhao X-Y and Cao J-P 2020 Biomass thermochemical conversion: A review on tar elimination from biomass catalytic gasification *Journal of the Energy Institute***93** 1083–98

[9] Chowdhury Z, Kaushik P, Wageeh A, Suresh S, Syed T, Ganiyu A, Emy M, Rahman F and Rafie B 2017 Pyrolysis: A sustainable way to generate energy from waste *IntechOpen Limited, London*

[10] Vassilev S V, Vassileva C G and Vassilev V S 2015 Advantages and disadvantages of composition and properties of biomass in comparison with coal: An overview *Fuel***158** 330–50

[11] Akhtar N, Goyal D and Goyal A 2016 Physico-chemical characteristics of leaf litter biomass to delineate the chemistries involved in biofuel production *Journal of the Taiwan Institute of Chemical Engineers***62** 239–46
[12] Raveendran K, Ganesh A and Khilar K C 1996 Pyrolysis characteristics of biomass and biomass components *Fuel***75** 987–98
[13] Cheng F, Dehghanizadeh M, Audu M A, Jarvis J M, Holguin F O and Brewer C E 2020 Characterization and evaluation of guayule processing residues as potential feedstock for biofuel and chemical production *Industrial Crops and Products***150** 112311
[14] Arif M, Li Y, El-Dalatony M M, Zhang C, Li X and Salama E-S 2021 A complete characterization of microalgal biomass through FTIR/TGA/CHNS analysis: An approach for biofuel generation and nutrients removal *Renewable Energy***163** 1973–82
[15] de Oliveira Maiaa B G, Souzab O, Marangonib C, Hotzaa D, de Oliveiraa A P N and Sellinb N 2014 Production and characterization of fuel briquettes from banana leaves waste *CHEMICAL ENGINEERING***37**
[16] Hallac B B, Sannigrahi P, Pu Y, Ray M, Murphy R J and Ragauskas A J 2009 Biomass characterization of Buddleja davidii: a potential feedstock for biofuel production *Journal of Agricultural and Food Chemistry***57** 1275–81
[17] Hossain N 2020 Characterization of novel moss biomass, Bryum dichotomum Hedw. as solid fuel feedstock *BioEnergy Research***13** 50–60
[18] Raveendravarrier B, Palatel A and Chandrasekharan M 2020 Physico-chemical Characterization of sewage sludge for thermochemical conversion processes *Energy Sources, Part A: Recovery, Utilization, and Environmental Effects* 1–25
[19] Hussain M M, Chakraborty P and Bhattacharya K 2012 Pollen grains of queen sago (Cycas circinalis L.), a source of aeroallergen from West Bengal, India: an immunochemical approach *Aerobiologia***28** 39–47
[20] Noora B, Sudheesh C, Sangeetha N and Sunooj K V 2019 Effect of isolation methods on the crystalline, pasting, thermal properties and antioxidant activity of starch from queen sago (Cycas circinalis) seed *Journal of Food Measurement and Characterization***13** 2147–56
[21] Johari I B, Md Said M A, Snin M A B M, Ayob N F A B, Jamaluddin N S B and Rashid M R B M 2021 Effect of Treated Sago Pith Waste Ash and Silica Fume to the Mechanical Properties of Fly Ash-Based Geopolymer Brick Key Engineering Materials vol 879 (Trans Tech Publ) pp 100–14
[22] Jinitha T, Sreejith M, Balan A K, Shaniba V, Subair N and Purushothaman E 2018 Chemical modification of sago seed shell powder: surface characterization and thermal studies *Materials Today: Proceedings***5** 16293–9
[23] Amin N, Sabli N, Izhar S and Yoshida H 2019 Sago wastes and its applications *Pertanika J. Sci. & Technol***27** 1841–62
[24] Patel B and Gami B 2012 Biomass characterization and its use as solid fuel for combustion *Iranica Journal of Energy & Environment***3** 123–8
[25] Odetoye T, Onifade K R, AbuBakar M and Titiloye J O 2013 Thermochemical characterisation of Parinari polyandra Benth fruit shell *Industrial Crops and Products***44** 62–6
[26] Tahir M H, Zhao Z, Ren J, Rasool T and Naqvi S R 2019 Thermo-kinetics and gaseous product analysis of banana peel pyrolysis for its bioenergy potential *Biomass and Bioenergy***122** 193–201
[27] Naduparambath S and Purushothaman E 2016 Sago seed shell: determination of the composition and isolation of microcrystalline cellulose (MCC) *Cellulose***23** 1803–12
[28] Yusuf A A and Inambao F L 2020 Characterization of Ugandan biomass wastes as the potential candidates towards bioenergy production *Renewable and Sustainable Energy Reviews***117** 109477
[29] Ahmad R K, Sulaiman S A, Yusup S, Dol S S, Inayat M and Umar H A 2021 Exploring the potential of coconut shell biomass for charcoal production *Ain Shams Engineering Journal*
[30] Menon S D, Sampath K and Kaarthik S S 2021 Feasibility studies of coconut shells biomass for downdraft gasification *Materials Today: Proceedings***44** 3133–7

Investigations on dislocation density and strain of polyaniline withWO_3nanocomposites

B T Vijaykumar[1], B Bharati [2], K Priyanka [3], Ramabai[4], BManjunatha[5] and Basavaraja Sannakki[1]

[1-4]Department of Post-Graduate Studies and Research in Physics Gulbarga University, Kalaburagi-585106, India
[5]Dept. of Physics, PG Centre, Yeragera, Raichur-584133, India
E-mail: sannakki.phy@gmail.com, vijjutalvar456@gmail.com

Abstract. Polyaniline is equipped by in-situ chemical oxidative polymerization method using ammonium persulphate as an oxidizing agent. The prepared Polyaniline is doped with the nano particles of WO_3 at dissimilar weight percentage. The nano-compsites are characterized using XRD and SEM. The crystallite sizes are obtainvia Debye-Scherer relation from the XRD spectra for nano-composites of different weight percentages. The crystallite sizes of nano-particles decreased with increase in weight percentage of WO_3 with the PANI. The dislocation density due to the defects in the crystals of nano-composites is obtained. The dislocation density increased as wt % of WO_3 in PANI increased. Further, the strain produce at the grain boundary is obtained by FWHM of the XRD spectra. The strain of nano-composites at the grain boundary increased as dislocation density increased. Strain at the grain boundary is also increased as weight percent of WO_3 increased in the nano-composites of PANI. Further, the surface morphology of the nano-composites of PANI/WO_3 at different weight percentages were observed using SEM. The particles are appeared granular in shape and size. The agglomeration in the sizes of the nano-composites was observed as weight percent of WO_3 with PANI increased.

1. Introduction

The tendency of having promising applications from the well-known conducting polymers of polyaniline, polypyrrole, polythiophene etc, is having importance in advanced technological fields since from few decagon. PANI is used in various fields, such as light emitting diodes, electronics,electrochemistry, optoelectronics, and electromagnetic interference shielding andelectro chromic displays. additionally its preparation is easy, lowcost, environmental stability is good,lowdensity, enhanced electronic properties which gives possible applications such assensors, rechargeable batteries,electro catalysis, and bio sensors, drug delivery etc.[1-5].Conductive polymeric materials was conjugate organically chain compound that shows large electrically conducting like to metal since here huge carrier concentration of extensive π-electron,well-known as polaron, which permit the charging mobilities along the strength of character of the polymeric chains. Polyaniline (PANI) survivesin different forms,that is, it can be different in the chemical and physical properties [6-8].The PANI/WO_3 nano-composite has large properties like nontoxicaswellaspossessinghighlyhardenings, superiorpuritiesandhighmelting point.The Tungsten oxide maybean add-on advantage in these fields of polymer; in exacting, it may givesuperiorapplicationsindopingtoPANI.Theproperty of the Tungsten oxides like high hardening and high melting points can be usedasrefractory'sagent.Nanostructuremetaloxidesarepromisingmaterial for blendswithpolymer nanocomposites,due to their mechanical,electrical,thermal, optical andmultifunctional properties.The polymer nanocomposites withTungsten oxides have shown good physical properties [9].The result of manipulative characteristic principles such as Dislocation density and Lattice strain of WO_3 composite of different weight % are important factors in determining physical properties to make sure the success of formation of these nanocomposites[10]. X-ray diffraction is one of the ultimateefficient instruments for estimation of avg. crystal size, dislocationdensity and lattice strain. These estimated values are based on the FWHM (full width at half

IOP Conf. Series: Materials Science and Engineering 1221 (2022) 012017 doi:10.1088/1757-899X/1221/1/012017

maximum) data [11-12]. In this paper, the analysis and determination of micro structural properties, the size of crystals, dislocation density and lattice strain of PANI/WO_3 of different weight percentages are studied.

2. Experimental

2.1 Materials and methods

The chemicals such as aniline, Ammonium persulphate ($(NH_4)_2S_2O_8$), Hydrochloric acid (HCl) and Tungsten Oxide (WO_3) were procured from Sigma Aldrich, Mumbai. All the chemicals wornwere in analytical reagent (AR) grade for the synthesis of polyaniline and its nanocomposites.

2.2 Synthesis of Polyaniline

In-situ polymerization chemical oxidative method is used to prepare the polyaniline and its WO_3nanocomposites. Polyaniline was synthesized by mixing0.2 moles of Aniline with 100 ml of distilled water and 0.2 moles of HCl is added to the solution to form aniline hydrochloridein the 1:1 ratio. The solution is kept for stirring on magnetic stirrer to get clear homogeneous solution. The oxidizing agent of 0.4 moles of Ammonium persulphate [$(NH_4)_2S_2O_8$] solution is added drop wise continuously to the aniline hydrochloride solution by keeping the solution in ice bath at 5 °C. The color of the mixer solution changesfrom brownish to the dark green. The solution is stirred continuously for 4-5 hours for proper mixing after that the solution is kept 24 hours for further polymerization. The solution was filtered by use of the whatman filter paper, so theresulting precipitate on the filter paper be washed with distilled water and acetone repeatedly until to get clear solution and to remove the residual impurities. The precipitate was firstly dried in the air and the reserved in the hot air oven at 60 °Ckeptfor a period of one day for drying. The final product of polyaniline substance was obtained.

2.3 Synthesis of Polyaniline/WO_3 Nanocomposites

The similar procedure is used for groundingof the polyaniline nano composites by adding 5 wt % of WO_3 nanoparticles to the solution of aniline hydrochloride and then APS is added drop wise by keeping the solution into the ice bath at 5°C. The color of the mixture solution changes from brownish to the dark green. A solution is stirred continuously for 4-5 hours for proper mixing after that the solution is kept 24 hours for further polymerization. The solution was filtered using Whatman's filter paper, the resulting precipitate on the filter paper was washed with distilled water and acetone repeatedly until to get clear solution and also to remove the residual impurities. The precipitate was firstly dried in air and the kept in hot air oven at 60 °C for a period of one day for drying. The final product of polyaniline/WO_3 nanocomposite was obtained. Similarly the other polyaniline nanocomposites of 10 and 15 weight percent of WO_3 were prepared. In this way 3 different Polyaniline/WO_3 (5, 10, and 15) in polyaniline have been synthesized. All polymer nanocomposites are separately crushed into fine powder in an agate mortar.

3. Result and Discussions

3.1 XRD Analysis

XRD pattern of Polymer of PANI and its nanocompositessamples were obtained using an advance powder X-ray diffractometer [Rigaku,Model: Ultima IV] with monochromatic source of CuKα radiation (λ=1.54056 A^O) with scan rate of 4° per minute to study the crystalline/amorphous nature of the samples. The XRD spectrum of the polyaniline is recorded with scan rate of 4° per minute and the spectra is given in figure1. From figure 1 observed that, the X-ray diffraction patter of pure polyaniline is an amorphous in nature. The wide peak was detect at an angle of 2θ = 25-27°, which is the main characteristics peak of polyaniline and the broadness of the peak clearly shows the sample is of amorphous in nature [13].The XRD spectra of polyaniline with WO_3 nanocomposites at 5, 10 and 15 different weight percent is recorded with scan rate of 4° per minute and the spectra is given in Figure

2. From figure-2, 2θ worth with (hkl) planes of PANI/WO_3 at 5 wt % are 24.14 (200), 33.14 (022), 35.62 (212), 40.83 (222), and 49.46° (400), were observed and confirmed by JCPDS No. 01-071-0131.

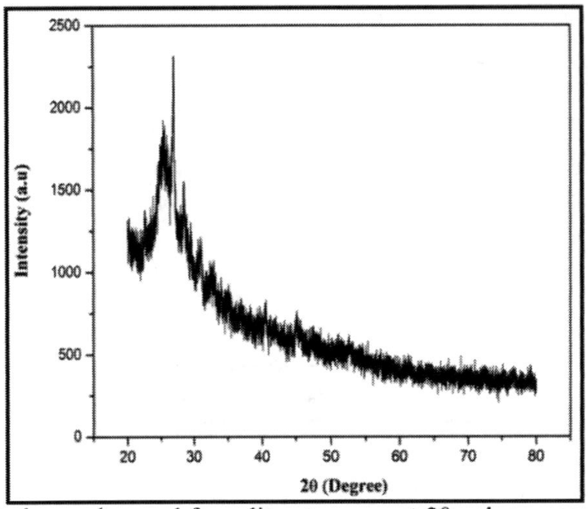

These peaks with hkl planes observed from literature are at 2θ values were 23.12 (002), 23.67 ((020), 24.35 (200), (28.91 (112), 33.27 (022), 34.17 (220), 49.92 (040) and 55.98 (400), which are attributed to monoclinic and orthorhombic phase of $WO_3$and the potent intensity of the diffraction peaks indicates the highly crystalline structure.

Figure1.X-ray diffractionspectrum of PANI

Hence it is observed that the indexed (hkl) values are comparable with present observed values and hence the structure of the PANI/WO_3 nanocompsite is confirmed monoclinic and orthorhombic phase. The figure2clearly indicatescrystalline natureof the PANI nanocomposites and the same were observed[14].From figure2a the XRD pattern of PANI/WO_3nanocomposite at 5 wt% shows, peaks are take place at an angles of 24.14^0, 33.14^0, 35.62^0, 40.83^0, and 49.46^0, are the characteristic intense peaks of PANI/WO_3. The sharp intensive peaks of XRD, clearly understand a crystallinity nature of nanocomposites. The values of the(hkl) planes are (200), (022), (212), (222), and (400). The avg. crystallite size (D)is calculated using Debye-Scherer'sformula known in equation (1). The crystallite sizes of 14.81nm, 18.24nm,23.52nm, 29.65nm, and 26.89nm were observed for PANI/WO_3 at 5 wt% and the mean crystallite size was found to be 22.62nm.

$$D = \frac{K\lambda}{\beta cos\theta} \qquad (1)$$

where, the constant K is grasp to be 0.94, λ is the wavelength of X-ray used, which is CuKα radiation (λ=0.154184nm), θ is the Bragg angle in radians, and β is full width at half maximum (FWHM) of the diffraction peaks.

Fromfigure 2(b) the peaks of XRD spectra occurred at angles of 2θ are 24.11^0, 33.15^0, 35.63^0, 40.86^0, and $49.49^0$with the (hkl) values for the planes are (200), (022), (212), (222), and (400) respectively. The average crystallite size of 22.27nm is obtained for PANI/WO_3 at 10 wt%.The same natures of peaks were observed for PANI/WO_3 at 15 wt% and are shown in figure 2(c). In this case the XRD intense peaks with (hkl) values occurred at an angles of 24.14^0(200), 33.15^0(022), 35.64^0 (212), 40.88^0(222), and 49.48^0(400). The average crystallite size of 23.30nm is obtained.

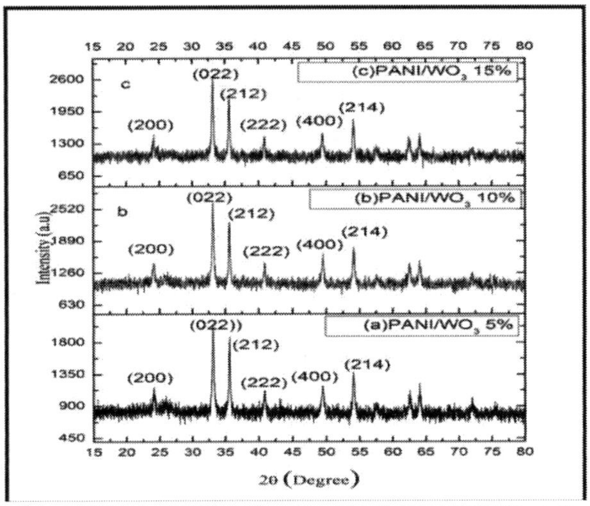

Figure 2.X-ray diffraction spectra (a)PANI/WO_3at 5 wt%, (b)PANI/WO_3at 10 wt%, (C)PANI/WO_3 at 15 wt%.

3.2 Dislocation density (ρ):

The values of mean particle size obtain from equation (1), can be used to obtain the values of dislocation density of the PANI/WO_3 at different wt%. The dislocations density value (ρ)denotes the number of dislocation lines per unit area of crystal defects possessed by a crystal, which is the size of crystal defects[15].In other words, the degree of crystalline profile will be illustratedby the dislocations density value of the nanoparticles.The dislocation density (ρ)can be calculated using equation (2)

$$\delta = \frac{1}{D^2} \qquad (2)$$

The dislocation density (ρ)of nanocomposites prevail in the present study is quite small, which indicates that the PANI/WO_3 have been a high degree of crystallinity.The dislocation density of PANI/WO_3 nanocomposite of 5 wt % is 18.4 x 10^{14}lines/m^2, where as the dislocation density of the PANI/WO_3 nanocomposite of 10 wt% is 19.5 x 10^{14}lines/$m^2$and the dislocation density of the PANI/WO_3 nanocomposite of 15 wt % is 20.1 x 10^{14}lines/m^2. It is observed that, as weight percentage of WO_3 with the PANI increased the dislocation density also increased in the nanocomposites of PANI/WO_3. The dislocation density occurs due to defects in the crystal of nanocomposites of different weight percent [16].The avg. grain size obtained with the help of SEM morphology and the different wt% of polymer nanocomposites of dislocation density (ρ)is known in table 1.

Table 1. The average grain size and dislocation density of PANI/WO_3 nanocomposites.

Sample at different Wt%	Average grain size From SEM (μm)	Dislocation density $\rho x10^{14}$(lines/m^2)
PANI/WO_3 5%	0.59	18.4
PANI/WO_3 10%	0.11	19.5
PANI/WO_3 15%	0.06	20.1

3.3 Lattice Strain of the crystallite

The lattice strain, ε of the PANI/WO_3 composites at different weight percentage of WO_3 is determined from the equation (3) using the experimental values of FWHM(β) and the corresponding peak angle θ;

$$\varepsilon = \frac{\beta cos\theta}{4} \qquad (3)$$

Table 2. The values of Lattice strain at grain boundary for PANI/WO_3 nanocomposites.

Wt% of material	Lattice strain x 10^{-3}(ε)
PANI/WO_3 5%	1.5
PANI/WO_3 10%	1.7
PANI/WO_3 15%	1.8

The lattice strain in nanocomposites of PANI/WO_3at 5 wt percentage is obtained as $1.5 x 10^{-3}$, in case of PANI/WO_3at 10 wt percentages,the strain is$1.7x10^{-3}$and that of PANI/WO_3at 15 wt percentage the strain is $1.8x10^{-3}$. It is found that the strain at the grain boundary is increased with increase in weight percent of WO_3 in the nano-composites of PANI.

3.4 Morphology Analysis

The micrographs of SEM image for pure polyaniline isgiven in figure 3(a), and the agglomerated granular in shape are observed. The avg. grain size was found to be 0.27μm.The grains are well interconnected with oneanother, which shows that they have enough biding energy to combine with neighbouring grains ofthe polyaniline molecules.The surface morphologicalstudies of polyaniline indicate that the orientation of polyaniline chains, it can be commemorating from the SEM micrographs [17].

Figure 3 (a-d). Shows the SEM image of PANI and PANI/WO_3nanocomposites of 5, 10 and 15 different weight percentages

The SEM image at 5, 10 15 wt % of WO_3 in PANInanocompositesareshown in figure3(b-d). The grains are highly agglomerated, irregular in shape but they are well interconnected with each other's. The average grain size was found to be 0.59μm, for 10 wt%, 0.11μm and for that of 15wt%, the grain size was 0.06 μm.By comparing the figure3(a-d), it can be concluded that the gradual increase in granular size and change in morphology, which enables the transportation of charged particles through the carbon back bone of the polymer chains [18].

4. Conclusions

The PANI and PANI/WO_3 composites were prepared by chemical oxidative route. The XRD analysis confirmed the formation of PANI and its nanocomposites indicate that PANI/ WO_3 composite have crystalline in nature. The crystallite sizes of nano-particles decreased with increase in weight percentage of WO_3 with the PANI. The dislocation density due to defects in the crystals of nano-composites was obtained. The dislocation density increased as weight percent of WO_3 in PANI increased. Further, the lattice strain produced at grain boundariesof the nano-composites increased as dislocation density increased. The SEM images helped to draw the inference that, doping of WO_3 had an effect on PANI morphology. In case of PANI nanocomposites as WO_3wt%, increased the agglomeration of granular size also increased.

Acknowledgements
The authors are thankful to the Dept of Physics, GUK, for providing University fellowship to carry out the research work.

References

[1] Chandrakanthi R L N and Careem M A 2002 *Thin Solid Films***514**17
[2] Somani P R, Marimuthu R, Mulik S R, Sanikar S R and Amalnerkar D 1999 *Synth. Met.* **45**106
[3] He Y 2005 *J. Mater. Chem. Phys.***268** 134
[4] Geng L, Zhao Y, Huang X, Wang S and Zhang S and Wu J 2007 *Sens.Actuators***34** 568
[5] Yu S, Xi M, Jin X, Han K, Wang Z and Zhu H 2010 *J. Catal. Commun.* **57** 1125
[6] MacDiarmid A G, Rev. 2001 *Mod. Phys.***73**, 701
[7] Stejskal J, Kratochvil P, Jenkins A D1996 *Polymer***37**, pp- 367
[8] Trivedi D C In Handbook of Organic Conductive Molecules and Polymers, H.S. Nalwa(Ed.)**2**,(1997) *Wiley*,Chichester,pp.505–572
[9] Green,A.G.,&Woodhead,A.E.(1912).CXVII-aniline-blackandallied compounds. Part II*Journal of the Chemical Society, Transactions,* ***101***,1117–1123
[10] Gubicza J, Szepvolgyi J, Mohai I, Zsoldos L, Ungar T 1999 *Materials Science and Engineering*A**280** (2000) 263-269
[11] Bindu P and Thomas S 2014 *J. Theor. Appl. Phys.***8** 123-34
[12] Venkkateswarlu K, Sandhyarani M, Nellaippan T A, Rameshbabu N, 2014 *Procedia MaterialsScience***5** 212-221
[13] Elmetwally M, Abdelrazek, Amar M Abdelghany, Shalabya I Badr, Mohamed A Morsi 2018*Journal of Materials Research and Technology* Volume **7** Issue 4 419-431
[14] Kumar R and Yadav B C 2016 *J. Inorganomet Polym***26**,1421-1427
[15] Khan Z R, Zulfequar M and Khan M S 2010 *Mater. Sci. Eng. B* 174 145-9
[16] Sutapa I W, Abdul Wahid Wahab, Taba P and Nafie N L 2018 *J. Phys.: Conf. Ser.***979** 012021
[17] Abell L, Pomfret S J,Adams P N and Monkman (1997) *Synth. Metals***849**127
[18] Ramesh Patil, Aashis. S. Roy, Koppalkar R, Anilkumar, Srikant Ekhelikar, 2011 "Electricalconductivity of polyaniline/NiZno3 composites: *A Solid State Electrolyte", Ferroelectric,***Vol. 423**, pp.77-85

IOP Conf. Series: Materials Science and Engineering 1221 (2022) 012018 doi:10.1088/1757-899X/1221/1/012018

Study of stagnation layer of laterally colliding laser produced aluminium plasmas

Shilpa S. [1] and Pramod Gopinath [1,2]

[1]International School of Photonics and [2]Inter University Centre for Nanomaterials and Devices, Cochin University of Science and Technology, Kochi 682022, India

E-mail: pramod@cusat.ac.in.

Abstract. When two dense laser produced plasmas collide, a layer of plasma stagnates at the collision front exhibiting special properties, is called stagnation layer. The characteristics of stagnation layer formed due to the collision of two laser produced aluminium plasmas were studied using spectroscopic and the fast imaging techniques. Time gated Intensified Charge Coupled Device (ICCD) imaging was used to study the temporal evolution of stagnation region and time resolved spectroscopy was used to obtain information about the distribution of neutral as well as the ionic species in collision process of laterally colliding aluminum plasmas. Electron density and temperature of the stagnation layer were also calculated from the emission spectra. The overall experiment results clearly depict the peculiar properties of stagnation layer which is more advantageous than the conventional laser produced plasma.

1. Introduction

Laser produced plasma has been an attractive research area for the past few decades due to its enormous applications. Plasma produced by the collision of two laser produced plasmas exhibit some spectacular characteristics. The notable significance is its better control and tuning of plasma parameters compared to that of single plasma. When two plasmas collide, there exist two possibilities where the plasma can either interpenetrate each other or can stagnate at the collision front. In the case of hard stagnation, rapid accumulation of plasma material at the collision front leads to the formation of a dense layer of material between the two plasmas called stagnation layer[1]. The study of properties of stagnation layer has attracted the curiosity of researchers for the past four decades because of its broad range of applications in various fields of science such as inertial confinement fusion[2], laser induced breakdown spectroscopy[3], pulsed laser deposition[4], EUV lithography[5], understanding the formation of supernovae and its remnants[6] etc.

When two streaming plasmas intersect, it may either be collisionless type or collision dominated type which is generally determined by the collisionality parameter of plasmas depending on the density, temperature, ionic charge and relative velocities. Rambo et al introduced the concept of collisionality parameter which determines whether interpenetration or stagnation is dominant in the colliding laser produced plasmas[7]. The collisionality parameter is given by

$$\xi = \frac{L}{\lambda_{ii}} \qquad (1)$$

where L is the separation between the two seed plasmas (plasma dimension) and λ_{ii} is the ion-ion mean free path and is given by[8]

$$\lambda_{ii}(1 \rightarrow 2) = \frac{m_i^2 v_{12}^2}{4\pi e^4 Z^4 n_i \ln \Lambda_{12}} \qquad (2)$$

IOP Conf. Series: Materials Science and Engineering 1221 (2022) 012018 doi:10.1088/1757-899X/1221/1/012018

where m_i is the ion mass, v_{12} is the relative collision velocity, e is the charge of the electron, n_i is the average plasma ion density, Z is the average ionization state of the plasma and $\ln \Lambda_{12}$ is the so-called Coulomb logarithm for collisions between the individual plasma plumes. In standard definition, when $\xi > 1$ it is termed as highly collisional stagnation of plasmas and if $\xi < 1$ it represents interpenetration of plasmas[9].

The interactions between laser produced plasmas were first reported by Rumbsy et al in 1970s[10]. Later there have been numerous studies done on the interaction between plasmas produced in solid targets, thin films, metal foils etc. of different target geometries (lateral collision, orthogonal collision etc.) and also different laser pulse durations (femtosecond to nanosecond) [11–15]. Stagnation layer and its properties attract the curiosity of many researchers and it still requires further investigation to understand the complex physical processes underlying in it.

2. Experimental setup

Plasmas are generated by 1064nm laser radiation from a Q switched Nd:YAG laser with a pulse width of 10ns, 130mJ and a repetition rate of 10 Hz , allowed to pass through a focusing lens of 60cm on an aluminium target of 99.99% purity. The target kept in a vacuum chamber is rotated by means of an electric motor in order to avoid surface etching. The seed plasmas are produced by splitting the laser pulse into two by using a wedge prism of wedge angles $1^0 3^1 0^{11}$.
The distance between the two foci is related to the lens and wedge prism by the relation,
$d = \gamma f (n-1)$
where n is the refractive index of the wedge prism (here fused silica with n=1.496), f is the focal length of the lens and γ is the acute angle of the wedge in radians [16]. Here seed plasmas are produced by a separation of 4mm. The pressure inside the chamber is kept at $3x10^{-5}$mbar. An intensified charge coupled device (ICCD) camera (Andor New iStar ICCD) with a time resolution of 1 ns is used to record the plasma plume images which is triggered externally using the Q- switch SYNC output of laser. The time delay between the gate opening of ICCD camera and the Q-Switch output is monitored using a photodiode and an oscilloscope. The plasma plume images are recorded using intensified charge coupled device (ICCD) at various time delays with a gate width of 20ns. Optical emission spectroscopy is used to study the characteristic emissions in the stagnation region. The focal length of spectrometer is 500mm and a wavelength resolution of 0.01nm.

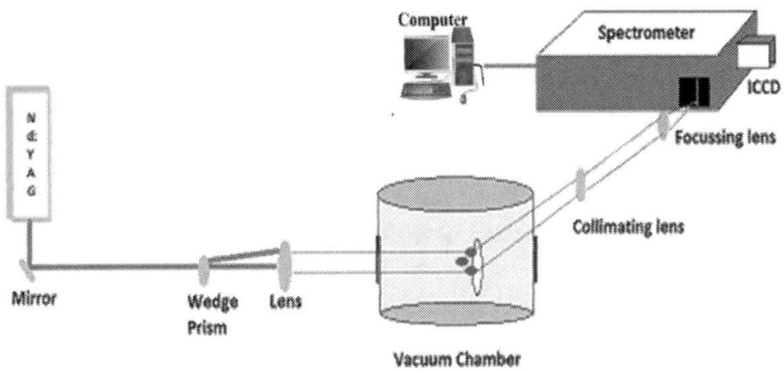

Figure 1. Experiment setup

IOP Conf. Series: Materials Science and Engineering 1221 (2022) 012018 doi:10.1088/1757-899X/1221/1/012018

3. Results and discussion

Fast imaging and Optical emission spectroscopy are the techniques used to study the characteristics of stagnation layer produced by the collision of two laser produced aluminum plasmas. Spatial separation between two plasmas is fixed as 4mm. Figure 2 shows the two dimensional ICCD camera images of plume interaction at different time delays. At initial time (~100ns) the two seed plasmas expand independently and as time elapses (~125ns) an interaction region started to appear at the middle of two plasmas. At this time delay both seed plasmas and interaction region are visible and as time elapses the two seed plasmas undergo complete collision to form a strong luminous interaction region called stagnation layer. The intensity and dimension of stagnation layer increases with the time delays.

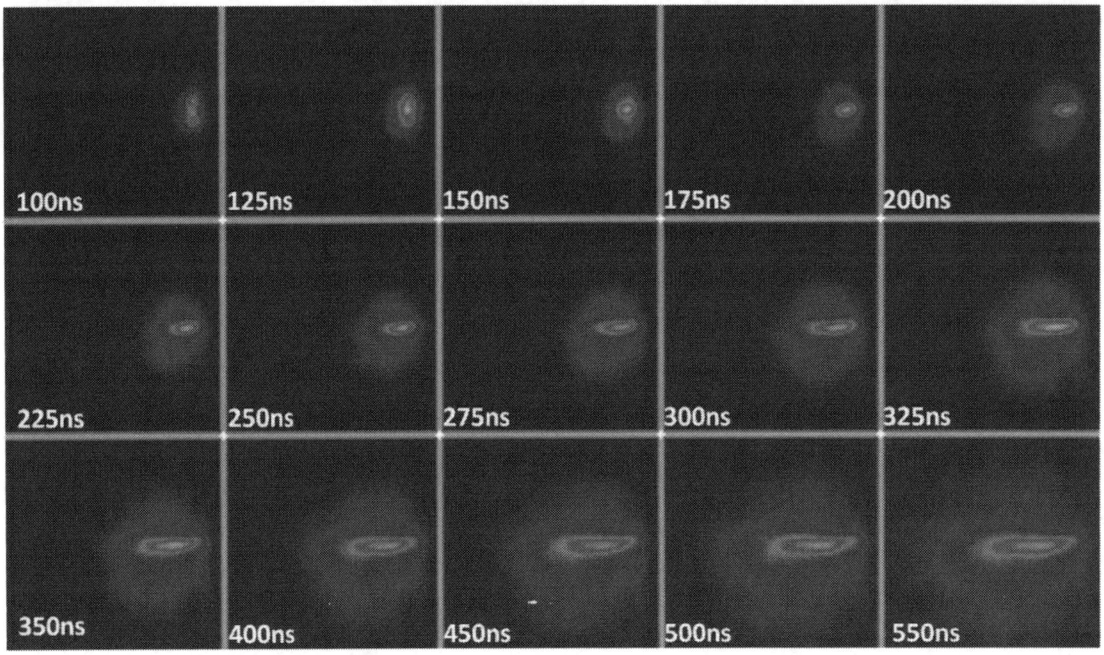

Figure 2. Sequence of ICCD images of laser produced colliding plasma plumes

A long luminous stagnation layer is formed by the interaction of aluminium plasma plumes with a separation of 4mm. Figure 3(a) is the plume position versus time graph of the stagnation layer. The linear dependence of axial dimension with time suggests the free expansion of stagnation region. The plume front positions are fitted by a line and its slope gives the velocity. The estimated axial velocity of stagnation region is 4.8×10^6cm/s which is greater than the lateral velocity 1.33×10^6cm/s.

The aspect ratio (length/width) of stagnation layer at different time delays is shown in figure 3(b). The initial increase in value is due to the greater axial expansion of stagnation region and for t > 400ns the aspect ratio exhibits less variations. It means as time elapses, the stagnation region laterally expanded in proportion to that of greater axial expansion. Larger value of axial dimension indicates the linear expansion of stagnation layer.

IOP Conf. Series: Materials Science and Engineering 1221 (2022) 012018 doi:10.1088/1757-899X/1221/1/012018

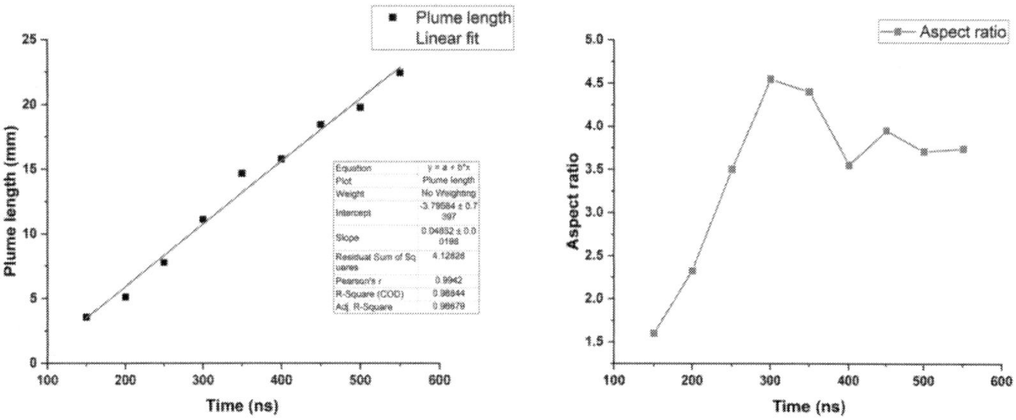

(a) (b)

Figure 3. (a) R-t plot and (b) Aspect ratio of stagnation region

In order to get more insight into the spectacular properties of stagnation region, optical emission spectroscopy has been employed to record the line emission of neutral as well as the ionic species in collision process. Emission spectrum of Aluminum consists of prominent transitions from ionic and neutral species including Al II 358.7nm (4f $^3F_3$-3d $^3D_3$), Al II 466.3nm ($3s^2$2p-3s4p), Al III 360.19nm ($2p^6 3d^3 D_{3/2}$-$2p^6 4p^2 p_{3/2}$), Al III 361.23nm ($2p^6 3d^3 D_{3/2}$-$2p^6 4p^2 P_{1/2}$), Al I 394.4nm ($3s^2 4s^2 S_{1/2}$-$3s^2 3p^2 P_{1/2}$), Al I 396.15nm ($3s^2 4s^2 S_{1/2}$-$3s^2 3p^2 2p_{3/2}$).Temporal evolution of these lines are recorded.

The emission intensity of ionic lines in the stagnation region initially builds up and later decreases where as the neutral line intensity grows up with time. Initial increase of ionic lines intensities are due to the increased ionization and the later increase of neutral lines intensities are due to the recombination processes takes place in stagnation layer[15].

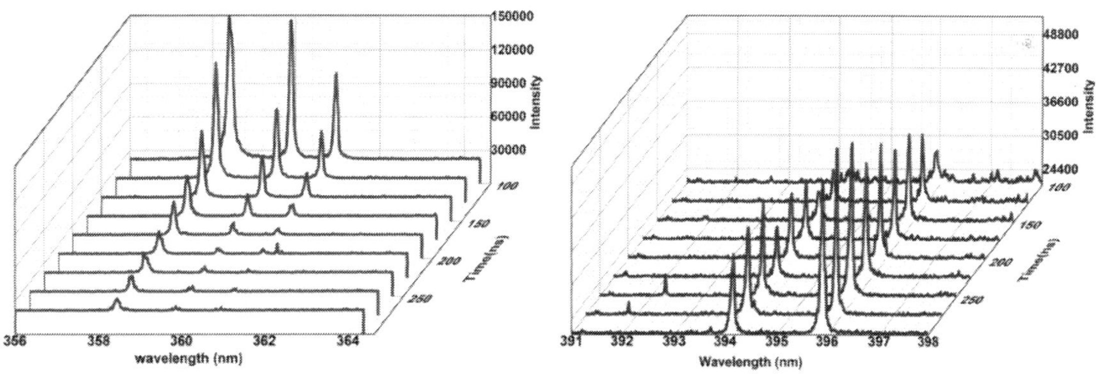

Figure 4. Spectrum of interaction zone at 4mm distance from the target.

The plasma parameters like electron density and electron temperature are estimated to study its evolution in stagnation region. The full width at half maxima (FWHM) of Stark broadened line is related to the electron density by a simplified expression:

$$\Delta\lambda_{(1/2)} = 2W\left(\frac{n_e}{10^{16}}\right) \qquad (3)$$

where n_e is the electron density, $\Delta\lambda_{(1/2)}$ is the FWHM and W is the electron impact parameter[17,18]. The electron density is estimated using stark broadened line of Al I transition (394nm). The observed emission line profile is fitted with Lorentzian function to obtain the line width.

Temporal evolution of electron density at a distance 4mm from the target surface is shown in figure 5(a). The ejected species from the seed plasmas interact and stagnate at the collision front results an increase in electron density.

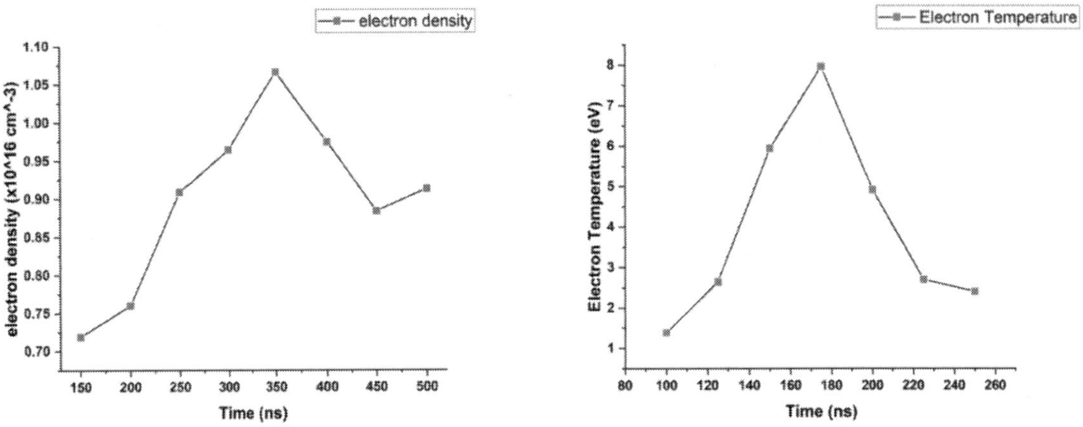

(a) (b)

Figure 5. Temporal variation of (a) electron density and (b) electron temperature of stagnation layer.

The electron temperature of plasma in the local thermodynamic equilibrium (LTE) approximation is obtained by substituting the emission intensities of Al II (358.7nm and 466.3nm) into the Boltzmann equation (4),

$$\frac{I_{ij}}{I_{kl}} = \frac{\nu_{ij} A_{ij} g_i}{\nu_{kl} A_{kl}\, g_k} \exp\left(\frac{-(E_i - E_k)}{K_B T_e}\right) \qquad (4)$$

where I is the emission intensity due to the transition between two energy levels, ν is the frequency of the spectral lines, A is Einstein's coefficient, g is the statistical weight of the energy level, E is the energy, K_B is the Boltzmann constant, T_e is the electron temperature, and the subscripts i, j, k, and l indicate the energy levels. Figure 5(b) shows the temporal evolution of electron temperature in the stagnation region. The temperature initially increases up to 175ns and then decreases. It is due to the increased collision in the interaction zone resulting in heating of plasma.

4. Conclusion

The characteristics of stagnation layer of laterally colliding laser produced aluminium plasmas have investigated. By using fast imaging technique it has been observed that an intense stagnation region is developed at the interaction zone. Optical emission spectroscopy was employed to record the ionic and neutral species of Al. Intensity of ionic lines are initially higher due to the increased ionization and the increase in neutral emissions at later times are due to the recombination processes in the plasma. Electron density and temperature of the stagnation layer were estimated from the emission spectra. Long luminous stagnation layer at the interaction zone is highly advantageous in various fields of laser plasma applications.

Acknowledgement

The authors acknowledge the financial support from DST- FIST, Gov. of India. SS acknowledges KSCSTE, Govt. of Kerala, for KSCSTE Research Fellowship.

References

[1] Hough P, McLoughin C, Kelly T J, Hayden P, Harilal S S, Mosnier J P and Costello J T 2009 Electron and ion stagnation at the collision front between two laser produced plasmas *J. Phys. D. Appl. Phys.* **42** 055211

[2] Nishimura K, Shibata R, Yabuuchi T, Sunahara A and Tanaka K A 2014 Laser scattered images observed from carbon plasma stagnation and following molecular formation *Appl. Phys. Lett.* **104** 1–4

[3] Tognoni E 2005 Characterization of a collinear double pulse laser-induced plasma at several ambient gas pressures by spectrally- and time-resolved *Appl. Phys. B Lasers Opt.* **80** 559–68

[4] Camps E, Escobar-Alarcón L, Haro-Poniatowski E and Fernández-Guasti M 2002 Spectroscopic studies of two perpendicularly interacting carbon plasmas generated by laser ablation *Appl. Surf. Sci.* **197–198** 239–45

[5] Oliver J and Sizyuk T 2019 Stagnation layer development from two colliding Sn plasmas for an efficient EUV source *Phys. Plasmas* **26** 1–11

[6] Velas K M and Milroy R D 2014 Probe measurements of the three-dimensional magnetic field structure in a rotating magnetic field sustained field-reversed configuration *Phys. Plasmas* **21** 1–7

[7] Rambo P W and Denavit J 1994 Interpenetration and ion separation in colliding plasmas *Phys. Plasmas* **1** 4050–60

[8] Chenais-Popovics C, Renaudin P, Rancu O, Gilleron F, Gauthier J C, Larroche O, Peyrusse O, Dirksmöller M, Sondhauss P, Missalla T, Uschmann I, Förster E, Renner O and Krousky E 1997 Kinetic to thermal energy transfer and interpenetration in the collision of laser-produced plasmas *Phys. Plasmas* **4** 190–208

[9] Mondal A, Kumar B, Singh R K, Joshi H C and Kumar A 2019 Spectroscopic investigation of stagnation region in laterally colliding plasmas : Dependence of ablating target material and plasma plume separation *Phys. Plasmas* **26** 1–9

[10] Rumsby P T, Paul J W M and Masoud M M 1974 Interaction between two colliding plasmas *plasma Phys.* **16** 969–75

[11] Kumar B, Singh R K, Sengupta S, Kaw P K and Kumar A 2014 Propagation dynamics of laterally colliding plasma plumes in laser-blow-off of thin film *Phys. Plasmas* **21** 1–7

[12] Dardis J and Costello J T 2010 Stagnation layers at the collision front between two laser-induced plasmas : A study using time-resolved imaging and spectroscopy *Spectrochim. Acta Part B* **65** 627–35

[13] Fallon C, Hayden P, Walsh N, Kennedy E T and Costello J T 2015 Target geometrical effects on the stagnation layer formed by colliding a pair of laser produced copper plasmas *Phys. Plasmas* **22** 093506

[14] Fallon C, Hayden P, Walsh N, Kennedy E T and Costello J T 2014 The Effect of Wedge Angle on the Evolution of a Stagnation Layer in a Colliding Plasma Experiment *J. Phys. Conf. Ser.* **548** 012036

[15] Mondal A, Kumar B, Singh R K, Joshi H C and Kumar A 2019 Spectroscopic investigation of stagnation region in laterally colliding plasmas: Dependence of ablating target material and plasma plume separation *Phys. Plasmas* **26** 022102

[16] Fallon C, Hayden P, Walsh N, Kennedy E T and Costello J T 2015 Target geometrical effects on the stagnation layer formed by colliding a pair of laser produced copper plasmas *Phys. Plasmas* **22** 093506

[17] Griem H R 1962 Stark broadening of isolated spectral lines from heavy elements in a plasma *Phys. Rev.* **128** 515–23

[18] Harilal S S, Bindhu C V, Issac R C, Nampoori V P N and Vallabhan C P G 1997 Electron density and temperature measurements in a laser produced carbon plasma *J. Appl. Phys.* **82** 2140–6

Pharmacological screening of heartwood of Tectona grandis Linn.for antibacterial activity by Agar-Well Diffusion method

R Singh

Chemistry Department, M.L.B. Govt College, Nokha (Bikaner), Rajasthan
Email: chemdrranveersingh2015@gmail.com

Abstract. Medicinal plants has always been a subject of interest for human being from ancient times due to health benefits. Researchers found that, people in different regions of the world used the same or similar plants for health benefits. Medicinal plant research includes many fields like drug discovery, ethnobotany, traditional and indigenous medicines. During corona pandemic we realised that there are no effective therapies available for new and remerging infectious diseases. Public health badly affected due to drug resistant microorganism and appearance of new disease causing microbes. Due to increasing side effects of synthetic drugs and uncontrolled infectious diseases researchers have taken interest in natural sources of medicines and have screened a large number of plants for new bioactive compounds. Natural products have been a well-known source of various molecules in the process of drug discovery and many bioactive natural product derivatives. Stereo chemical and structural properties of natural products are characteristics which are responsible for exploring novel molecular diversity. One of the successful strategies for the investigation of new medicinal agents from plants includes screening of plant extract, fractions followed by the purification of the constituents. Several plants and their products have been evaluated for their antimicrobial and antioxidant activity. In present study, the ethanolic extract of the heartwood of Tectona grandis Linn. was screened for antibacterial activity.

Keywords: Tectona grandis Linn.Antibacterial activity, agar-well diffusion method, gentamycin.

1. Introduction

In Ayurveda and homeopathic medicinal system, natural products of flora have been used as a source of medicines, some of which are used traditionally for health benefits in their locality [1-2]. Medicinal plants also have a wide potential for being used as antioxidants [3], anti-diabetic [4], anticaries agents [5] and Indian flora also have the wide potential for being used against corona for prophylaxis or as adjuvant [6]. Natural products always have interest in screening for drug discovery because of ease of availability, lower cost, and lesser side effect [7]. Scientist, researchers, pharmacists, and academicians have a global challenge for the search of new and more effective natural antibiotic agents since many infectious agents becoming resistant to synthetic drugs and also create side effects. For this medicinal herbs, shrubs and plants have been growing more interested in the new era because of their biological activities like antimicrobial [8], antimutagenicity [9], antioxidant [10-11], and applicability to treat several diseases [12-13]. Plants also play a very important role in modern medicinal sciences, in the introduction of new therapeutic agents [14] and anti-HIV agents [15]. Tectona grandis Linn. Belongs to the family Verbenaceae, commonly known as "sagwan" or "teak"..Tectona grandis Linn. is native to India, Myanmar and South-East Asian countries and is probably the most widely cultivated high value heartwood (HVH) in the world[16,17]. This plant is medicinally important and many reports claim to cure several diseases according to Indian traditional systems of medicines. Joshi, Singh and Pardasaniphytochemically investigate its root and isolated several chemical components [18]. Its wood is acrid, laxative, sedative and beneficial in piles, leucoderma and dysentery. It has anthelmintic, analgesic, anti-diabetic, anti-inflammatory and expectorant properties [19-21]. Its stem bark also reported for analgesic and anti-inflammatory activity [24]. The world's major medical threats are infectious diseases and the situation has further been complicated with time as we feel during the corona pandemic.

This study was there-fore designed to investigate the antibacterial activity of the heartwood extracts of Tectona grandis Linn. Against some infectious bacteria like Gram +ve : Bacillus subtilis, Staphylococcus aureus and Gram –ve : Escherichia coli, Enterobacter aerogenes, Pseudomonas aeruginosa and Raultellaplanticola by agar-well diffusion method [25,26] using gentomycin as standard.

2. Experimental

2.1 Preparation of test extracts
For antibacterial activity, powdered heartwood of Tectona grandis Linn. were extracted with ethanol. The ethanolic extract was concentrated in vacuo, fractionated with benzene, pet. Ether, ethyl acetate and the residue was re-extracted (2 x 8 hr) for complete exhaustion. Then the extracts and the fractions were pooled individually and dried in vacuo. All the extracts were stored at 4°C in a refrigerator until screened for a particular activity. However, their final concentration was prepared in the respective solvents, before use.

2.2 Source of test organisms
Pure cultures of test bacteria, namely Bacilllus subtilis, Enterobacter aerogenes, Escherichia coli, Pseudomonas aeruginosa, Raoultellaplanticola and Staphylococcus aureus, were obtained from S.M.S. lab. Jaipur (Rajasthan). These cultures were grown and maintained on Nutrient Broth Medium (NBM) at 27°C for 48 hr.

2.3 Cultures of test microbes
Stock cultures were maintained at 4°C on slops of nutrient agar. Active cultures for experiments were prepared by transferring an inoculating loop of cultures from the stock cultures to test tubes of NB Medium, which were incubated without agitation for 24 hr at 37°C.

2.4 Antibacterial assay
For antibacterial assay Agar- well diffusion method was adopted, because of its reproductivity and precision. This method also used previously to study antibacterial activity of Acacia raddiana against same selected test bacterias [26]. By pouring 20 ml of molten media into sterile petri plates, the plates were prepared. The plates were allowed to solidify for 8 min. and then, 30 μl suspension was spread uniformly with the help of a sterile glass spreader. After that, the plates were allowed to dry for 5 min. By using a sterile stainless steel borer, the wells of 6mm diameter were punched in the plates. The test sample was allowed to diffuse for 30 min after loading the test extract and control in the wells. The plates were kept for incubation at 37°C for 24 hr. At the end of incubation, inhibition zones formed around the well, which were measured in millimetres with a transparent scale. The experiments were performed in triplicate and the mean value of the diameter of inhibition zones with ± standard deviation were calculated.

IOP Conf. Series: Materials Science and Engineering 1221 (2022) 012019 doi:10.1088/1757-899X/1221/1/012019

3. Result and discussion

In the present study, six different microbial species were used to screen the possible antibacterial activity of plant extract. *E. coli*, which causes frequent urinary tract infections, is the most common organism, *B. subtilis, E. aerogenes, E. coli, P. aeruginosa, R. planticola*and *S. aureus* were chosen as the test organisms because of their high resistance to the antibacterial agents. Thus, any plant which would exhibit pronounced activity against these organism might yield an important antibiotic. The results of antibacterial activities of the selected plant extracts have been presented in Table 1..

Table 1. Antibacterial activity of the heartwood of *Tectona grandis Linn.*

S. No.	Plant species	Type of extract/fraction	Dose (mg/disc)	Test Microbes											
				B. subtilis		*S. aureus*		*E. coli*		*P. aeroginosa*		*E. aerogenes*		*R. planticola*	
				IZ	AI	IZ	AI	IZ	AI	IZ	AI	IZ	AI	IZ	AI
1.	*Tectona grandis* (heartwood)	EtOH	4	8.51	0.38	8.48	0.40	8.26	0.43	7.96	0.40	6.00	0.40	6.90	0.34
		Pet. Ether	4	7.22	0.32	9.23	0.48	9.05	0.40	6.78	0.34	6.34	0.45	8.17	0.40
		C_6H_6	4	9.13	0.41	10.94	0.51	9.72	0.50	8.05	0.41	7.00	0.50	7.76	0.38
		EtOAc	4	11.05	0.50	12.98	0.62	12.30	0.60	10.93	0.56	7.42	0.53	10.08	0.50

IZ = Inhibition zone (in mm) including the diameter of disc (6 mm);
AI = Activity index = Inhibition zone of sample/Inhibition zone of standard;
Standard: Gentamycin;
(-) = No activity.

All the test extracts from the selected plant species were found to be moderate and most active against mostly test bacterias. The ethanolic extract of heartwood of *Tectona grandis Linn.* Exhibited moderate activity against selected bacteria (Table 1). The various fractions of the heartwood of *Tectona grandis Linn.*Exhibited significant activity against E. *aerogenes, E. coli and S.aureus*(Table 1). The ethyl acetate fractions of the heartwood of*Tectona grandis Linn.*Exhibited significant activity against all test becterias(Table 1). All the fractions exhibiting more effectiveness against *S.aureus*but ethyl acetate fraction show best efficiency against*S.aureus.*

4. Conclusion

In this study, antibacterial screening of the heartwood of *Tectona grandis Linn.* were done by the agar-well diffusion method. The result showed that the extract/fractions of the heartwood of *Tectona* *grandis Linn.*Have effectiveness against all test bacterias but ethyl acetate fraction finds better efficiency than others against *S. aureus*. This concluded that the ethyl acetate fraction of the heartwood has a wide antibacterial potential against *S.aereus*. Thus, an important antibiotic against *S.aureus* might be yield from the ethyl acetate fraction. Further investigations are necessary to evaluate the potential against other infectious diseases, especially for antiviral diseases.

References

[1] S.K. Jain, "Indian system of medicine and Homoeopathy",Medicinal Plant of India (vol. I&II).
[2] Mbuni Y M, Wang S, Mwangi B N, Mbari N J, Musili P M, Walter N O, Hu G, Zhou Y and Wang Q 2020 Plants (Basel) Medicinal plants and their traditional uses in Local 9(3) 331.
[3] Naik G H, Priyadarsini K I, Satav J G, Banavalikar M M, Sohani D P, Biyani M K, Mohan H 2003 Phytochemistry, Comparative antioxidant activity of individual herbal components used in Ayurvedic medicine. 63 97.
[4] Grover J K, Vats V, Rathi S and Dawar R 2001 J Ethnopharmacol. Traditional Indian anti-diabetic plants attenuate progression of renal damage in streptozotocin induced diabetic mice,76(3)233-8.
[5] Jagtap A G, Karkera S G R 1999 J Ethnopharmacol, Potential of the aquous extract of Terminalia cheoula as an anticaries agent, 68 299-306.
[6] Ahmad S, Zahiruddin S, Parveen B, Basist P, Parveen A, Gaurav, Parveen R and Ahmad M 2021 Frontiers in Pharmacology Indian medicinal plants and formulations and their potential against Covid-19-preclinical and clinical research. 11 578970.
[7] Foye W O 1981 Principals of Medicinal Chemistry, Screening of natural products for drug discovery" 697.
[8] Ji W, Asada Y, Yoshikawa T 1998 Planta Med. Antimicrobial flavonoids from Glycerrhiza glabra Hairy root cultures. 64(8) 746-47.
[9] Kaur S, Grover I S, Singh M, 1998 Mutat. Res. Antimutagenicity of hydrolyzable tannins from Terminalia chebula in Salmonella typhimurium. 419 169–179.
[10] Konovalova G G, Tikhaze A K, Lankin V Z 2000 Bull. Exp. Biol. Med. Antioxidant activity of parapharmaceutics containing natural inhibitors of free radical processes., 130(7) 658-60.
[11] Scartezzini P, Speroni E 2000 J. Ethnopharmacol. Review on some plants of Indian traditional medicine with antioxidant activity 71 23-43.
[12] Polva G M, Foo L Y 1994 Phytochemistry. Inhibition of eukaryote signal-regulated protein kinese by plant derived catechin-related compounds 6 1399-1405.
[13] Sies H 1993 Eur. J. Biochem. Strategies of antioxidant defense 215 213-19.
[14] Dev S 1989 Heigher plants as a source of drugs. In: Plant and Society (Macmillan publishers Ltd., London) pp 267-292.
[15] Tewtrakul S, Subhadhirasakul S, Kummee S 2006 J. Sci. Technol, Anti-HIV-1integrase activity of medicinal plants used as self medication by AIDS patients Songklanakarin 28(4) 785-790.
[16] Keiding H, Wellendorf H and Lauridsen E B 1986 Evalulation of an international series of teak provenance triale (Danida Forest seed centres Publication, Humlebeak, Denmark).
[17] Kjaer E D, Lauridsen E B and Wellendorf H 1995 Second series of an international series of teak provenance triale (Danida Forest seed centres Publication, Humlebeak, Denmark).
[18] Joshi K C, Singh P and Pardasani R T 1977 Planta Medica ,Chemical components of roots of Tectona grandis and Gmelina arborea., 32(1) 71-75.
[19] Nayeem N, Karvekar M D 2010 Internet J Pharmacol, Analgesic and anti-inflammatory activity of the methanolic extract of frontal leaves of Tectona grandis.. 8 ISSN 1531-2976.
[20] Ghaisas M, Navghare K, Takawale A, Zope V, Tanwar M and Deshpande A 2009 J.Ethanepharmacol, Tectona grandis on dexamethasone –induced insulin resistance in mice. 122(2) 304-07.
[21] Diallo A, Gbeassor M, Vovor A, Eklu G K and Aklikokou K 2008 Fitotherapy. Effect of

Tectona grandis leaves on phenylhydrazine-induced anemia in rats. 79(5) 332-36.
[22] Ragasa C Y, Lapina M C, Lee J J, Mandia E H and Rideout J A 2008 Nat. Prod. Res. Secondary metabolites from Tectona philippinensis., 22(9), 820-24.
[23] Krishna M S and Nair J A 2010 Int. J. Pharma. Sci. Drug. Res.,Antibacterial, Cytotoxic and antioxidant potentials from different extracts of leaves, bark and wood of Tectona grandis. 2(2), 155-58.
[24] Mohammad A 2011 Malays J. Pharm. Sci., Malays. In Vivo analgesic and anti-inflammatory effect of Tectona grandis linn. stem bark extract. 9(1), 1-11.
[25] Boyanova L, Gergova G, Nikolov R, Derejian S, Lazarova E, Katsarov N, Z Mitov I and Krastev K 2005 J. Med. Microbiol., Activity of Bulgarian propolis against 94 Helicobacter pylori strains in vitro by agar-well diffusion, agar dilution and disc diffusion methods. 54, 481-83.
[26] Singh R 2021 STD, Chemical examination and antibacterial activity of heartwood of Acacia raddiana;.10(7),86-94.

IOP Conf. Series: Materials Science and Engineering 1221 (2022) 012020 doi:10.1088/1757-899X/1221/1/012020

Continuum removal and self-absorption correction incalibration-free laser-induced breakdownspectroscopy for efficient elemental analysis

Lekha Mary John[1], **BalakrishnaPrabhu B. N.**[1,2] **and K. K. Anoop**1,*
[1] Department of Physics, Cochin University of Science and Technology, Cochin-682022, India
[2] Centre for Integrated Studies, Cochin University of Science and Technology, Cochin-682022
E-mail: anoopkk@cusat.ac.in

Abstract. Even though laser-induced breakdown spectroscopy (LIBS) has emerged as a powerful analytical technique, the broad continuum emission and self-absorption effects in laser-produced plasmas (LPP) limit the accuracy of the LIBS technique in multi-elemental compositional analysis. In this work, we developed an algorithm to detect and remove the broad continuum emission, which usually originates from free-free and free-bound transitions. To eliminate the continuum, the segment-wise background correction method (using identified continuum parts of varied range) was used. The spectral interference of lines is more likely to be found in LIBS spectra, especially with low-resolution spectrometers. A Lorentzian curve fitting method was used to resolve closely spaced emission lines. The `internal reference self-absorption correction (IRSAC)' method was introduced to correct the reabsorption effects in LPPs. When these methods are applied to LIBS data of bronze alloy, more accurate quantitative findings are obtained, with a major component accuracy error of less than 10% when compared to its reference abundance.

1. Introduction

Laser-induced breakdown spectroscopy (LIBS) is a rapid spectrochemical technique based on atomic emission spectroscopy[1]. In LIBS technique, a pulsed laser is focused on the surface of the sample, create a transient microplasma of the sample. From the emission spectra of the plasma, we can determine the elemental composition of the sample as well as the plasma parameters. It has shown great potential in many fields due to its unique features, such as in situ, real-time analysis of multi-elemental samples without any sample pre-treatment. 'Nowadays, LIBS has been widely used for the analysis of alloys, especially for alloy classification, production control and quality assurance'[2]. In industry, Bronze alloys are used for the construction of massive instruments as well as the thin-layer coating over steel wire for protection of surface and improvement of adhesion to resin[3]. Exact quantification of bronze is needed in most cases since the material properties of bronze is mainly depend on the amount of copper in the alloy.Thisstudy used specific spectral pre-processing techniques for the abundance estimation of binary bronze alloy sample while using a low-resolution portable spectrometer. In the initial stage of laser-produced plasma, intense continuum emission overwhelms the isolated characteristic lines [4]. The attenuation speed of continuum background is faster than that of characteristic emission lines; therefore, we can eliminate it by using a time-gated detection system for an extent. While using the LIBS system without intensified CCD, the underlying continuum persists in most of the situations. The spectral profile of emission lines is distorted and shows spectral interference between nearby emission lines when using a low-resolution spectrometer, which needs to be taken into account. If an experimental emission line can be fitted by a Lorentzian profile, these problems can be eliminated up to an extent [5]. Essentially, a plasma is optically thin when the emitted radiation gets away from the plasma without notable re-absorption. When the plasma is optically thick, the emitted radiation is significantly re-absorbed by the cold atoms along the optical path length. It causes the self-absorption of spectral lines, resulted in the reduction of spectral line intensity and increased line width. Therefore, for an accurate LIBS measurement, the elimination of the self-absorption effect is crucial. We have developed an algorithm for segment-wise background correction, using identified continuum sections of varying ranges. Developed an algorithm based on the `internal reference self-absorption correction (IRSAC)' method via regressive intensity adjustment to eliminate the self-absorption effect [6].

2. Experimental Details

The optical emission spectrum from the laser-produced plasma (LPP) of bronze alloy was recorded using the laboratory LIBS equipment, which is depicted schematically in Figure 1. A frequency-doubled Q-switched Nd: YAG laser (λ = 532 nm, τ= 10 ns, E = 40 mJ) was used to ablate the target at rough vacuum (1 Torr) condition. The applied fluence$F \approx 75\ J/cm^2$ was sufficient to produce intense light emission from plasma plume. The characteristic emission lines from LPP are then collected using a lens system and directed to a portable fiber optics spectrometer (Ocean HDX-XR, resolution: 0.6 nm). The target used in the experiment was standard Copper Tin alloy (Bronze) with a known concentration (70:30) by wt.%. The alloy target was positioned on a rotating stage to avoid local drilling, and to get uniform plasma on each ablation.

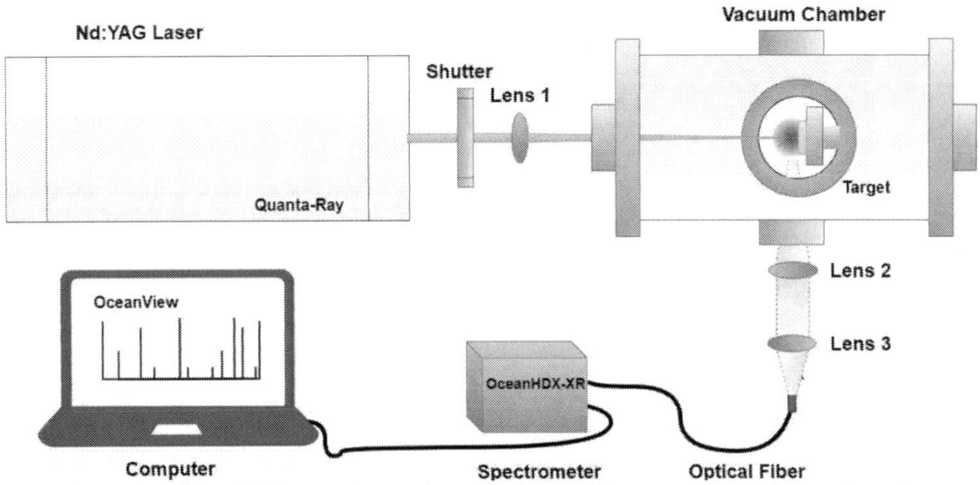

Figure 1: Schematic of the LIBS experimental setup. A portable spectrometer is employed to record the optical emission spectra

3. Result and Discussion

3.1. Continuum Removal of LIBS Spectrum

The ideal spectra for LIBS analysis consist of characteristic emission lines give both qualitative and quantitative information of the spectral sample. During the initial stages of laser-induced plasma formation, free-free and free-bound transitions adds a continuum to the spectra that changes the intensity values and hence affects the precision of quantitative analysis of sample. The quantification of elements is done using the intensity values of characteristic lines of each element. The addition of a continuum causes the intensity values to shift and hence causes inaccurate results. Therefore, it is crucial to eliminate the continuum from the spectrum before analysis [4].

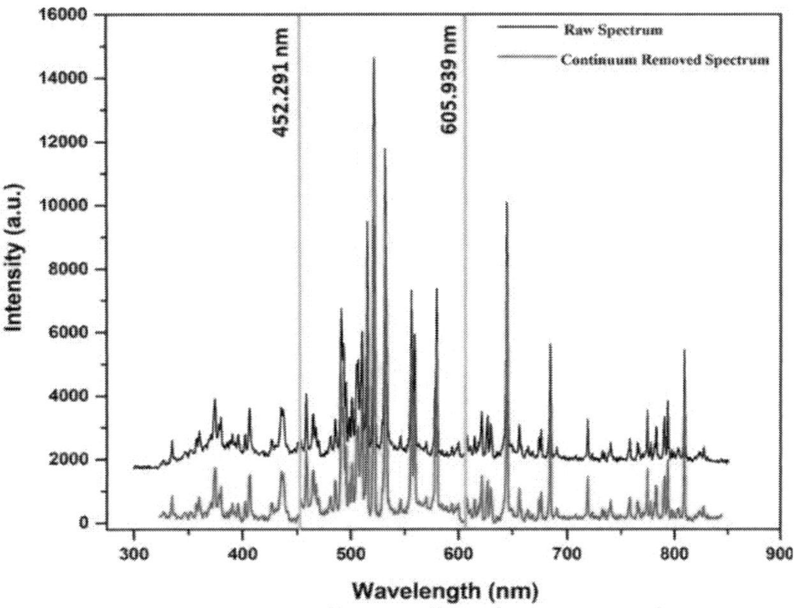

Figure 2: LIBS spectrum of bronze alloy taken at 1 Torr air pressure

The spectrum in consideration consists of 3 major segments, which can be characterized as continuums [323.008 nm to 452.291 nm; 452.291 nm to 605.939 nm; 605.939 nm to 844.633 nm]. The method of continuum correction identifies the continuum of the spectra by fitting a curve enveloping the continuum. In order to do this, we first identify all the valleys of the spectra, which can be used to create an envelope of the continuum. However, as the spectrum is irregular, some of the valleys lie much deviate from the envelope of the continuum. In order to correct this, we ignore all points that are much deviated by considering the valleys of the selected points, i.e. second-order valleys. These points are then fitted with a 4thorder polynomial to fit curves to the continuum and then subtract from the actual spectra [4]. The process of segmentation of the spectra is vital as applying the method to the whole spectra does not create a proper envelope for the continuum and hence does not effectively remove the continuum. Once the continuum is removed and each segment is baseline corrected, we stitch back each segment to obtain the total continuum removed spectra. This continuum corrected spectra can now be used for quantification of the sample.

3.2. Lorentzian Fitting of Emission Line

LIBS spectrum is taken in the rough vacuum condition and the identified elements from the continuum corrected spectrum are Copper (Cu I: 402.262, 406.264, 427.510, 458.697, 465.112, 510.554, 515.323, 521.820) and Tin (Sn II: 357.532, 556.191, 558.881, 579.886, 607.763, 645.354, 676.081, 684.418, 719.077, 774.142). The emission line profile depends on the dominant broadening mechanism. In LIBS, Stark broadening is the major broadening mechanism compared to Doppler broadening and natural broadening, leading to the symmetric Lorenz profile to the emission lines [7]. While using a low-resolution spectrometer, the spectral profile is distorted and shows spectral interference between emission lines, which needs to be considered. If an experimental emission line can be fitted by a Lorentzian profile, these problems can be eliminated up to an extent [5]. Figure 3a show the typical Lorentzian fitted profile of experimental spectral lines of Sn II at 556.191 nm and 558.881 nm. The Lorentzian fitted emission lines are used for finding plasma parameters and the elemental composition of bronze alloy. The electron number density $n_e$is determined from the Stark broadening of the emission line at 521.820 nm (Cu I), and full width at half maximum (FWHM) is taken from the Lorentzian fitted emission line shown in figure 3b [7]. The value of n_e in the bronze plasma is obtained to be 2.4 x 10^{16}cm^{-3}.

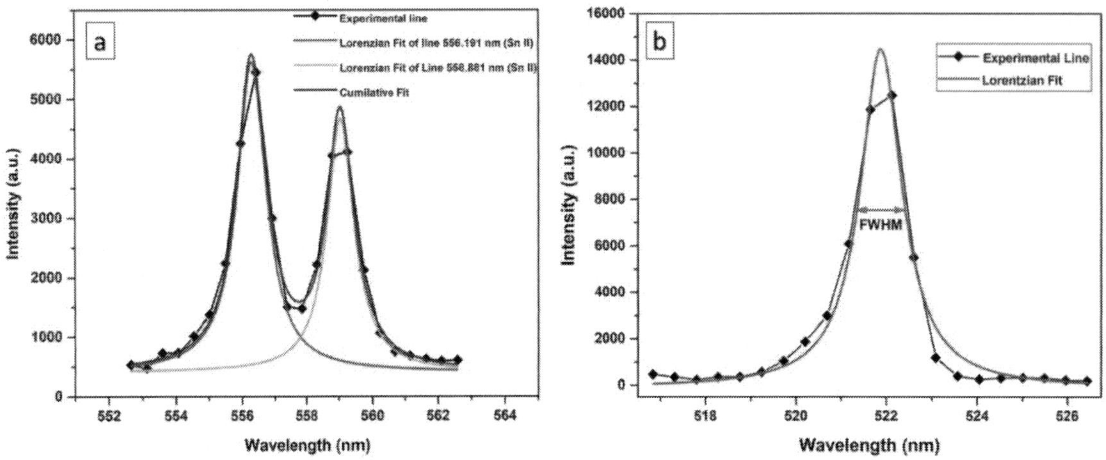

Figure 3:a. Lorentzian fitted spectral lines at 556.191 nm (Sn II) and 558.881 nm (Sn II), b. Lorentzian fitted Cu I line at 521.8 nm

3.3. Self-Absorption Correction

In LIBS, the elemental compositional analysis and plasma characterization fundamentally depend on the optically thin emission lines. Hence, an increase in linewidth and reduction in line intensity occurred due to SA effect of optically thick plasma. For eliminating SA effects in spectral-line intensities, many research has carried out in the last few decades, including SA correction based on the principle of `curve of growth (COG)' of the spectral lines proposed by Bulajic et al. [8], `internal reference for self-absorption correction (IRSAC)' method [6], `internal reference– external standard with the iteration correction (IRESIC) method [9], `blackbody radiation referenced self-absorption correction (BRR-SAC)' [10], etc.

We have employed the IRSAC method to estimate and correct the self-absorption effect [6]. The internal reference line for each species was selected to evaluate the self-absorption coefficient of other spectral lines. The selected internal reference line having higher excitation energy or low transition probability is slightly affected by the self-absorption effect, and thus the self-absorption of the reference line can be ignored. Using the self-absorption coefficient of each line, spectral line intensity can be iteratively corrected[2]. This self-absorption coefficient (f) of each line of the same species can be evaluated using the expression:

$$f = \frac{I_{\lambda}^{ki}}{I_{\lambda R}^{mn}} \frac{A_{mn}\, g_m}{A_{ki}\, g_k} e^{-((E_m - E_k/k_B T)} \tag{1}$$

Here, I^{ki}, A_{ki}, E_k, g_k are the integrated intensity, transition probability, upper state energy, and statistical weight of the line (to be corrected for SA) and I^{mn}, A_{mn}, E_m, g_m are the respective parameters of the internal reference line. Plasma temperature is required for calculating the self-absorption coefficient, which is calculated from the Boltzmann plot built using uncorrected emission line intensities. Once we get the self-absorption coefficient (f), we can evaluate the corrected line intensity by dividing measured line intensity by f. We can again calculate the self-absorption coefficient by using the new temperature determined from the Boltzmann plot built by corrected intensities, and this process can be repeated till the data points on the Boltzmann plot align linearly (or correlation coefficients converged). Figure 4 shows the Boltzmann plot of Copper and Tin lines before and after SA correction of spectral line intensities, which shows a good correlation between the data points and fitted lines. The plasma temperature, determined from the slope of the Boltzmann plot built from intensity corrected spectral lines, is 11590 ±20 K (≈1 eV). The final corrected intensity is then used for calibration-free LIBS (CF-LIBS) calculation.

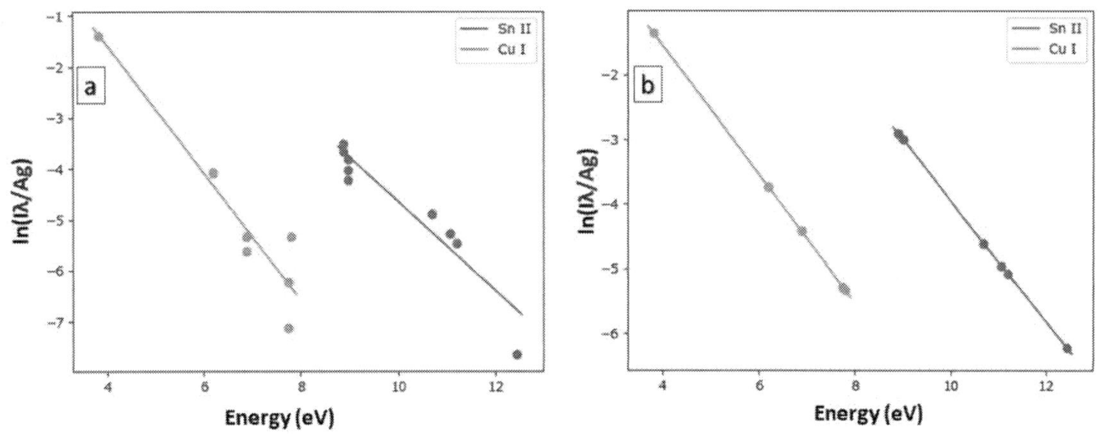

Figure 4: Boltzmann plot before and after SA correction using IRSAC method

3.4. Quantitative Analysis

LIBS is an analytical technique to determine the precise concentration of a sample with high accuracy. For reliable quantitative analysis, plasma should satisfy Local Thermodynamic Equilibrium (LTE), and the McWhirter criterion is the necessary condition for plasma to be in LTE [11]. Calibration-free LIBS (CF-LIBS) is introduced to estimate the concentration of constituent elements without plotting the calibration curve [12–15]. We quantify the alloy sample by CF-LIBS proposed by Ciucci et al. [12]. The integral intensity of emission line is given by,

$$I_{\lambda}^{ki} = FC_sA_{ki}\frac{g_k e^{-(E_k/k_BT)}}{U_s(T)} \tag{2}$$

Here, g_k, C_s, E_k, $U_s(T)$, A_{ki} and T are the statistical weight, concentration of the species, upper state energy of the line, partition function, transition probability and plasma temperature respectively. When we take logarithms on both sides and rearranging, a familiar form of Boltzmann plot equation is obtained,

$$y = \frac{\ln I_{\lambda}^{ki}}{g_k A_{ki}},\ x = E_k,\ a = -\frac{1}{k_BT},\ b^s = \frac{\ln(FC_s)}{U_s(T)} \tag{3}$$

The y-intercept b^s is directly related to the logarithm of the species concentration. Using the above expression, we calculated the concentration Cu I and Sn II species using self-absorption corrected emission lines. The emission spectra of the bronze alloy also contain singly ionized lines of Cu and neutral lines of Sn. To estimate the concentration of Cu II and Sn I, we use the Saha-Boltzmann equation while the electron number density ne and plasma temperature T are known,

$$\frac{n_e N_s(z+1)}{N_s(z)} = 6.04 \times 10^{21}\, T_{eV}^{3/2}\frac{U_s(z+1)}{U_s(z)}\exp\left(-\frac{\chi_z^s}{T_{eV}}\right) \tag{4}$$

Where N_s and $N_s(z+1)$ are the population of the ground state of the neutral atomic species and that of the singly ionized species respectively and $\chi_z^s$is the ionization energy of the species. From this expression, we have determined the weight percentage of Cu and Sn in the bronze alloy sample. The ratio of Cu and Sn concentrations in the bronze alloy obtained from the CF-LIBS analysis is 65 ± 5: 35 ± 5 and that of the certified ratio is 70: 30. We can further improve the accuracy of results by using time-gated spectrometers.

4. Conclusion

We used a nanosecond laser for LIBS to investigate different spectral pre-processing techniques to get trustable compositional analysis of multi-elemental alloy sample while using a low-resolution portable spectrometer. For a more accurate analysis, the underlying continuum of the spectrum is removed, and the selected emission lines have been Lorentzian fitted. IRSACmethod is used for determining the self-absorption in spectral-lines. We calculated the electron number density and plasma temperature using the Stark broadening and Boltzmann plot method, respectively. Furthermore, we estimated the abundance of Copper and Tin in Bronze alloy using LIBS. The accuracy error of the major component (Copper) is 7%, and we can reduce it further by using gated spectrometers. Our investigations reveal that LIBS is a versatile tool for elemental identification and alloy classification in industry and material science.

Acknowledgments
K. K. Anoop acknowledges the Science and Engineering Research Board (SERB) for the financial assistance through a TARE Grant (No. TAR/2018/000340).

References

[1] Cremers D A and Radziemski L J 2013 *Handbook of Laser-Induced Breakdown Spectroscopy* (John Wiley & Sons)
[2] Shakeel H, Haq S U, Aisha G and Nadeem A 2017 *Phys. Plasmas***24** 063516
[3] Ermalitskaia K F, Voropay Y S and Zajogin A P 2010 *J. Appl. Spectrosc.***77** 153–9
[4] Liu J, Zhang R, Li X, Chen J, Liu J, Qiu J, Gao X, Cui J and Heshig B 2018 *Opt. Express***26** 16171–86
[5] Chu Y, Zhang Z, He Q, Chen F, Sheng Z, Zhang D, Jin H, Jiang F and Guo L 2020 *J. Adv. Res.***24** 353–61
[6] Sun L and Yu H 2009 *Talanta***79** 388–95
[7] Sarkar A and Singh M 2017 *Plasma Sci. Technol.***19** 025403
[8] Bulajic D, Corsi M, Cristoforetti G, Legnaioli S, Palleschi V, Salvetti A and Tognoni E 2002 *Acta Part B At. Spectrosc.***57** 339–53
[9] Dong J, Liang L, Wei J, Tang H, Zhang T, Yang X, Wang K and Li H 2015 *J. Anal. At. Spectrom.***30** 1336–44
[10] Li T, Hou Z, Fu Y, Yu J, Gu W and Wang Z 2019 *Anal. Chim. Acta***1058** 39–47
[11] Luo W F, Zhao X X, Sun Q B, Gao C X, Tang J, Wang H J and Zhao W 2010 *Pramana***74** 945–59
[12] Ciucci A, Corsi M, Palleschi V, Rastelli S, Salvetti A and Tognoni E 1999 *Appl. Spectrosc.***53** 960–4
[13] Gomba J M, D'Angelo C, Bertuccelli D and Bertuccelli G 2001 *At. Spectrosc.* 11
[14] Qasim M, Anwar-Ul-Haq M, Sher Afgan M, Haq S U and Baig M A 2020 *Plasma Sci. Technol.***22** 074007
[15] Aguilera J A, Aragón C, Cristoforetti G and Tognoni E 2009 *Acta Part B At. Spectrosc.***64** 685–9

The effect of co-dopants (Cu^{3+}, Sm^{3+}-ions) on the optical properties of Sodium-Zinc-Borate glasses

S. Hemalatha[1], M. Nagaraja[1, $], Madhu.A[2,*], Vinayak Pattar K[3,4], K. Suresh [5], Srinatha N[6]

[1]*Department of Physics, SSIT, Sri Siddhartha Academy of Higher Education, Tumakuru-572 105, India*
[2]*Department of Physics, Bangalore university, Bengaluru-560 056, India.*
[3]*Jawaharlal Nehru Centre for Advanced Scientific Research, Bangalore – 560064, India*
[4]*Department of Physics, B.G.S R&D Centre, SJC Institute of Technology, Chikkaballapur, India*
[5]*Department of Physics,Sri Venkateswara Engineering College, Tirupati – 517 502, India.*
[6]*Department of Physics, RV Institute of Technology and Management, Bengaluru, 560 076, India*
*Corresponding author: *mmathi.33@gmail.com; [$]madihally.nagaraja@gmail.com*

Abstract

In this article, we describe the synthesis and optical characteristics of (Cu^{2+}, Sm^{3+}-ions) doped sodium zinc borate glasses made using the conventional melt quenching technique. The produced glasses were investigated using XRD, UV-Visible-NIR, and Photoluminescence spectroscopy. XRD analysis reveals the amorphous nature of produced glass samples. The presence of both Cu^{2+} and Sm^{3+} ions may be seen in UV-Visible-NIR absorption spectra in the 200-2000 nm range. Refractive index, density, and energy band gaps were all examined as physical and optical variables.In addition, photoluminescence emission spectra for two excitation wavelengths, 290 nm and 400 nm, were obtained to confirm Cu^{2+} and Sm^{3+}-ions-induced photoluminescence in sodium zinc borate glasses, respectively. When using a 290 nm excitation, the spectra exhibited one wide emission peak about 460 nm, however when using a 400 nm excitation, the spectra revealed three separate peaks in the visible region.The spectral properties of all of the samples were identical. The drop in PL intensity with Sm3+ concentration might be attributed to energy transfer between copper and samarium ions with 400 nm excitation. Surprisingly, the glass sample with 1 mol% Sm^{3+} exhibits the highest visible PL intensity / emission. The investigation's findings show that the produced samples are viable candidates for photonic and LED applications.

1. Introduction

In recent decades, the various applications of high-performance photoluminescence (PL) materials have generated considerable research interest. Inorganic crystals with unique coordination have long been believed to be a potential host for tunable emissions because they provide good transfer efficiency for luminescence centres. Complex manufacturing processes and high fabrication costs are disadvantages that restrict their use [1-3]. PL glasses, on the other hand, are a low-cost option for ensuring high transfer efficiency and a simple preparation process. Glasses are also suitable hosts for PL materials because of their high Ultra-Violet and visible transparency, chemical stability, and relative ease of doping with different activators [2, 3]. Glass is a good candidate for transition metal or rare earth ions to penetrate because it has a lot of interconnected routes and a lot of specific surface area. In recent years, copper has received a lot of attention as one of the most important TMs. The copper cluster, Cu^{+} ions, and Cu^{2+} ions in particular have exceptional optical properties [4,5].Because of its energy levels, which govern fluorescence characteristics, Sm^{3+} is a major RE ion utilised as dopants for glasses with optical applications. Glasses doped with Sm^{3+} ions may benefit fibre lasers and planner waveguides. Due to its 4f electronic band structure, Sm^{3+} shows distinct and strong absorption bands in the UV to blue region [6]. The presence of Cu^{2+} ions in octahedral locations in the samples suggested that dipole–dipole interactions between the Cu^{2+} ions were the most prevalent, according to a previous research. In the UV-Visible region, spectra of Sm^{3+} ions revealed numerous f-f electronic transitions.Cu^{2+} ions with an octahedral-system surrounded by six oxygen's caused a part of them to overlap the peak, implying the presence of Cu^{2+} ions at octahedral-system surrounded by six oxygen's sites [7, 8]. Based on the data and current study, an attempt is being made to optimise the Cu^{2+} and Sm^{3+} ion co-dopants concentration in the sodium-zinc-borate glass. In earlier work [9,10], we discussed the effect of individual dopants on the same host.

2. Composition and method of preparation

The glass composition (70-x) B_2O_3 - $20Na_2O$ -10ZnO – 0.5 (CuO)-xSm_2O_3 was made using the conventional melt quenching method, with x = 0.1, 0.5, 1.0, and 2.0 mol percent marked as NZBCu05Sm01, NZBCu05Sm05, NZBCu05Sm1, and NZBCu05Sm2. As described in earlier work [9,10], the comprehensive synthesis and experimental approaches were used. The amorphous character of the co-doped NZB series is confirmed by XRD analysis of the samples. The sample thickness ranged from 1mm to 1.5mm, and the density went from 2.963 to 3.183 g/cm^3.

3. Results and discussion

3.1 UV-Vis-NIR absorption

Fig. 1 shows the absorption spectra of NZB co-doped with Cu^{2+} and Sm^{3+}-ions, which depicts distinct peaks in the range 300-1800 nm wavelength range. The absorption peaks centred at both lower wavelength and higher wavelength region are 360, 373, 388, 402, 417, 456, 475, 770, 1080, 1226, 1372,1474,1533,1492,1645 nm have been assigned to $^4D_{3/2}$,$^6P_{7/2}$,$^6P_{3/2}$,$^6P_{3/2}$,$^4G_{9/2}$,$^4I_{13/2}$, $^4M_{15/2}$+$^4I_{11/2}$,$^2B_{1g}$-$^2B_{2g}$,$^6F_{11/2}$,$^6F_{9/2}$,$^6F_{7/2}$,$^6F_{5/2}$,$^6F_{3/2}$,$^6H_{15/2}$,$^6F_{1/2}$Fig.1.(b) respectively [6,9].In the near infrared range, the interaction of Cu^{2+} and Sm^{3+} ions resulted in high-intensity peaks. Urbach energy and indirect band gap energy are computed when the absorption spectra are further expressed in terms of Tauc's plot. Our prior work [9,10] may be used to calculate and estimate band gaps and Urbach energies. The band gap reduced from 3.24 to 2.90 eV as seen in Fig. 2(a), resulting in an increase in Urbach energy values from 0.14 to 0.38 eV. This shows that the glass' structural properties are altered, resulting in the induction of prohibited gaps and a reduction in the optical band distance [5-10]. This function may be observed in all samples and is accompanied by a red shift (323 to 338 nm) in the cut-off wavelength [6-10].

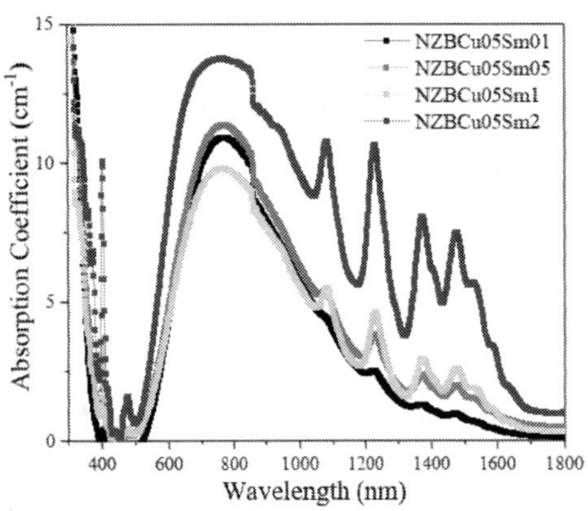

Fig.1(a)UV-Visible-NIR spectra of NZB Samples.

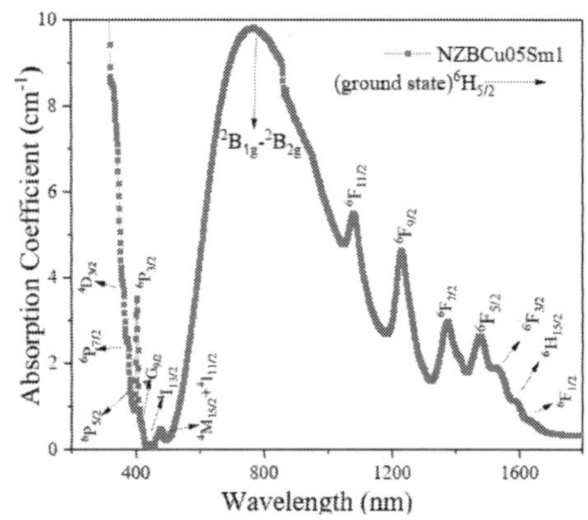

Fig.1(b) Assignments of peak in absorption Spectra.

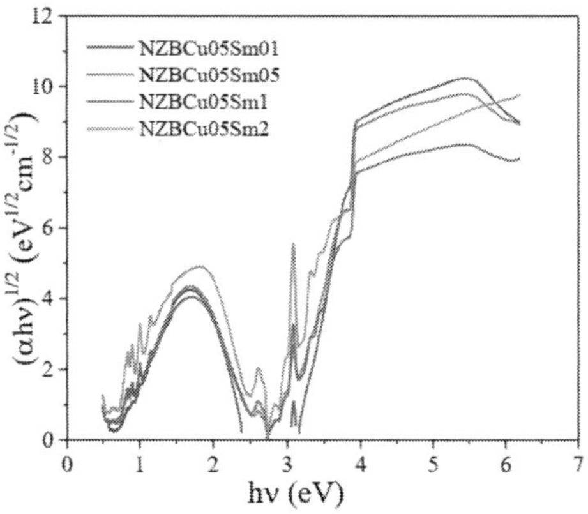

Fig.2(a) Indirect band gap from Tauc's plot.

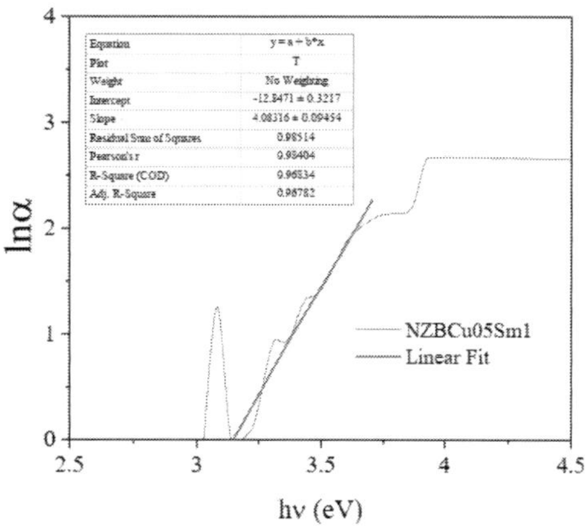

Fig.2(b) Typical Urbach energy plot

3.2 Photoluminescence

Fig. 3(a) and (b) show the emission spectra for 290 and 400 nm excitation. The key criterion used to determine energy transfer is the spectrum overlap between the luminescence spectra of Sm^{3+} donor ions and the absorption spectra of acceptor Cu^{2+}. In earlier publications [6,11], this was likewise found and adequately explained. The highest intensity for Sm^{3+} ions 0.5mol percent and 0.5mol percent Cu^{2+}-ions was obtained in the 290 nm excitation for the NZB series, whereas the maximum intensity for 400 nm excitation was observed in Sm^{3+} ions 1mol percent and 0.5mol percent Cu^{2+}ions.We can clearly see that energy transfer occurred in the Co-doped NZB series, resulting in an increase in intensity in the 1mol percent NZBCu05Sm1 sample, based on the aforementioned observation.

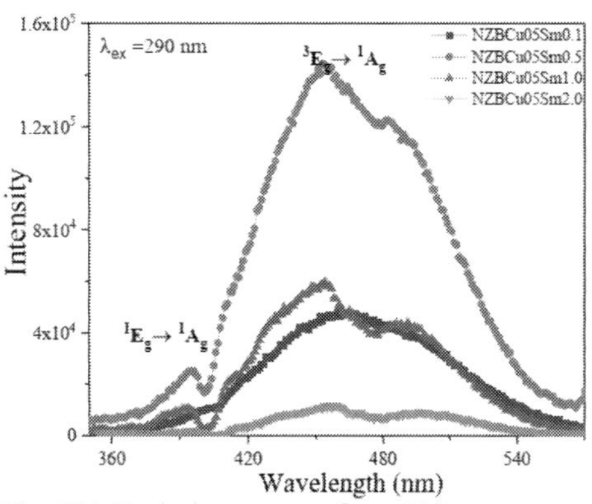

Fig. 3(a) Emission spectra from 290 nm Excitation.

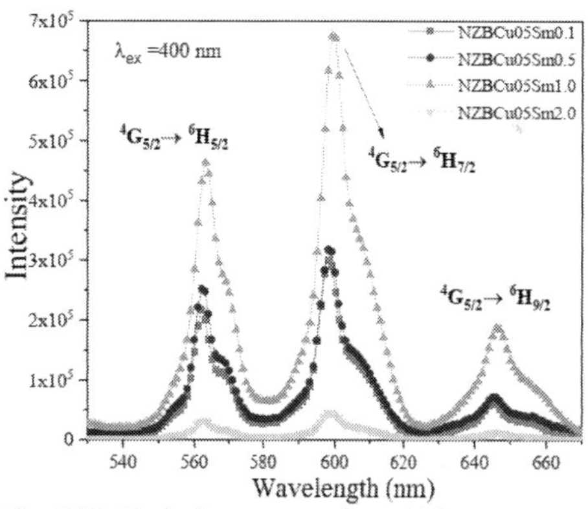

Fig. 3(b) Emission spectra from 400 nm Excitation.

3.3 CIE co-ordinates and CCT values

The CIE diagram of the NZB series is produced using the OSRAM software for both 290 and 400 nm excitation. The purity of NZBCu05Sm1 is higher, as shown in Fig. 4 (a) and (b), and it emits blue and orange-red emission, respectively. Table 1 lists the remaining relevant parameters. The importance of luminous effectiveness of radiation (LER), which is denoted by the letter K by the International Commission on Illumination (CIE), Color Rendering Index (CRI), and x,y co-ordinates is detailed in earlier reports[12]. The reported LER and CRI for 400 nm excitation is within the range of commercially available devices, according to the studies.

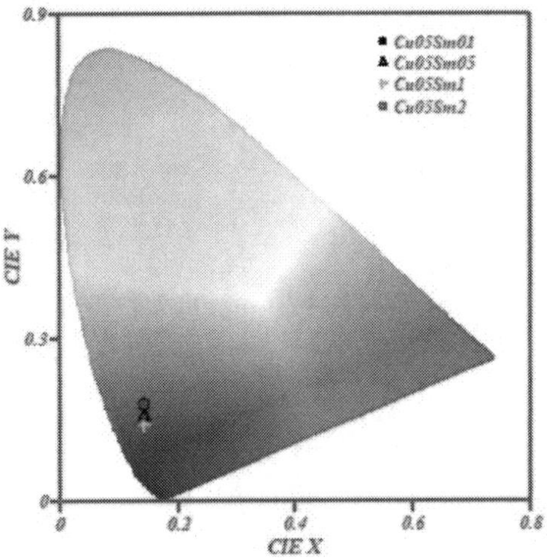

Fig.4(a) CIE diagram from 290 nm Excitation.

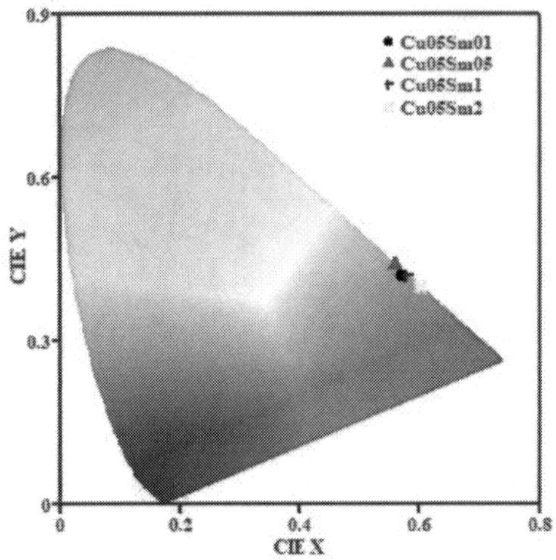

Fig.4(b)CIE diagram from 400 nm Excitation.

Table1. Colour coordinates(x,y),CRI,CCT,LER and other parameters of CIE diagram.

	Samples	x	y	u'	v'	CCT	CRI	LER	Duv	Du'v'	Ldom(nm)	Purity(%)
400 nm Excitation	NZBCu05Sm01	0.595	0.403	0.358	0.545	1499	44	339	0.003	0.005	593.7	99.7
	NZBCu05Sm05	0.579	0.417	0.338	0.548	1657	49	360	0.005	0.007	591.1	99.4
	NZBCu05Sm1	0.556	0.439	0.310	0.552	1927	39	405	0.008	0.013	587.2	99.1
	NZBCu05Sm2	0.575	0.422	0.332	0.549	1709	39	350	0.006	0.008	590.2	99.5
290 nm Excitation	NZBCu05Sm01	0.137	0.180	0.112	0.331	N/A	N/A	160	0.080	0.093	480.8	80.0
	NZBCu05Sm05	0.138	0.142	0.125	0.288	N/A	N/A	126	0.090	0.120	477.3	84.1
	NZBCu05Sm1	0.139	0.121	0.133	0.261	N/A	N/A	137	0.100	0.141	475.1	86.4
	NZBCu05Sm2	0.139	0.175	0.115	0.326	N/A	N/A	151	0.079	0.094	480.2	79.8

4. Conclusions

Using the traditional melt quenching technique, we have effectively prepared (Cu^{2+}, Sm^{3+}-ions) doped sodium zinc borate glasses. XRD, UV-Visible-NIR, and Photoluminescence spectroscopy were used to characterise the produced glasses. The amorphous properties of the produced glasses are shown by the XRD findings. Optical absorption spectra obtained using UV-Visible-NIR in the 200-2000 nm range show the presence of both Cu^{2+}- and Sm^{3+} ions. Several physical and optical factors, including as refractive index, density, and energy band gaps, were also investigated. Photoluminescence emission spectra for two excitation wavelengths, 290 nm and 400 nm, were also obtained to confirm Cu^{2+} and Sm^{3+}-ions generated photoluminescence in sodium zinc borate glasses, respectively. When 290 nm excitation was used, the spectra exhibited one wide emission peak about 460 nm, but when 400 nm excitation was used, the spectra revealed three separate peaks in the visible region.The decrease in PL intensity with Sm^{3+} concentration might be due to energy transfer between copper and samarium ions while using 400 nm excitation. Surprisingly, the glass sample with 1 mol% Sm^{3+} had the highest visible PL intensity/emission. According to the findings, the produced samples are suitable for photonic and LED applications.

REFERENCES

1. Yang W, Ilaria D Sa, Alessia C, Stefania B andGuorong C, 2019 *Opt. Mater.* **87** 80.
2. Wang Y, Liu S, Mao J, Li X, Li L, Zeng H and Chen G, 2017 *J. Lumin.* **185** 241.
3. Wang Y, Zhang Z, Lu J, Lu J, Zeng H and Chen G, 2017 *J. Lumin.* **192** 105.
4. Richter M, Fait M J G, Eckelt R,Schneider M, Radnik J, Heidemann Dand Fricke R, 2007 *J. Catal.* **245** 11.
5. Zhou S, Zhu B, Lin G, Yang H, Xu S, Wang Zand Qiu J,2007 *J. Phys. D: Appl. Phys.* **40** 7339.
6. Herrera A, Fernandes RG, de Camargo ASS,Hernandes AC, Buchner S, Jacinto C and Balzaretti NM., 2016 *J. Lumin.* **171** 106.
7. Petru P, Razvan S, Loredana E O, Liviu Ca B,and Eugen C, 2020 *Mater.* **13**5040.
8. Jose AJ,2016*J.Non-Cryst. Solid.* **432**227.
9. Hemalatha S, Nagaraja M, Madhu A, Suresh K and Srinatha N, 2021 *J. Non-Cryst. Solid.* **554** 120602.
10. Hemalatha S, Nagaraja M, Madhu A, Suresh K, Srinatha N, 2021 *Mater. Today: Proc.* (In press); DOI: 10.1016/j.matpr.2021.08.075.
11. Rudramamba K S, Taherunnisa S K and Rami Reddy M, 2020 *Opt. Mater.* **109** 110340.
12. MICHAEL ROYER,2019*OSA Continuum.* **2(8)** 2308.

IOP Conf. Series: Materials Science and Engineering 1221 (2022) 012022 doi:10.1088/1757-899X/1221/1/012022

Role of Scanning Speed in Morphological and Optical Properties of Nanosecond Laser Textured Silicon Surfaces

Rajeev Rajendran[1], Lekha Mary John[1] and K. K. Anoop[1,*]
[1]Department of Physics, Cochin University of Science and Technology, Kochi-22
*Email: anoopkk@cusat.ac.in

Abstract. Laser surface texturing (LST) is a rapid single-step method for surface functionalization. Different micro and nanoscale structures can be fabricated by controlling the laser parameters, ambient conditions, and material properties. In this work, we investigated the role of laser scanning speed in the morphological and optical properties of nanosecond laser textured silicon surfaces. Keeping the laser flounce just above the threshold, we controlled the laser scanning speed in the range 50–100 μm/s to monitor the morphological and optical changes. Morphological analysis shows various randomly arranged microstructures in the processed area, and the induced structures strongly depend on the scanning speed. The total reflection spectrum (specular+diffuse) from laser textured silicon shows that a significant reduction (≈60%) in reflectivity is possible with the decrease in scanning speed from 100 μm/s to 50 μm/s. These highly absorptive and large-area microstructures have numerous possibilities in photovoltaics, optoelectronic devices, and other material science applications.

1. Introduction

Laser surface texturing (LST) is a novel and versatile technology for surface functionalization. LST excels over other conventional techniques, like mechanical, chemical, and electric discharge texturing. It is due to its capacity to process a wide range of materials, like metals, polymers, semiconductors and glasses[1]. LST is a single-step, maskless method that can process non-planar surfaces with a high degree of control over the shape and size of the structures. It can be carried out in the open air without the use of a clean roomenvironment with a minimum material processing time[1]. Past few decades, it has shown numerous applications in the field of industry, material science and biomedical applications due to its enhancement in surface properties such as optical, wetting, mechanical, chemical, biological and other properties[2].For solar cells, sensors, and optoelectronic devices, silicon is the most commonly utilised material. The capacity to transform incident light into an electrical response is critical in the majority of these applications. Due to the high refractive index of silicon, the reflection loss of silicon at the air-material interface is very large[3]. There are many techniques developed for reducing the reflection of silicon surface like wet chemical etching[4,5], electrochemical etching[6], and reactive ion etching[7,8] etc. The recent developments in laser material processing led to the creation of highly absorptive silicon by direct laser surface texturing and laser-assisted chemical etching in the presence of various background gases[9–13].

In this work, we performed surface texturing on the silicon (100) surface at various scanning speeds using a Q-switched nanosecond laser.The effect of scanning speed on the morphological and optical features of the treated surface was investigated.Our study shows a strong a dependency of induced micro and nanoscale structures on the laser scanning speed. A fine-tuning of the reflectivity of processed surfaces as a function of scanning speed was also observed in our investigations

2. Experimental Details

Surface texturing of silicon (100) wafer is done by irradiating the surface with Q-switched Nd:YAG nanosecond laser pulses ($\lambda = 532$ nm, $\tau = 10$ ns, repetition rate = 10 Hz) at normal incidence in the air as shown in Figure 1. The laser source's spot profile is elliptical, and the laser pulse energy was kept constant at 12 mJ throughout the experiment. The corresponding laser intensity on the silicon target is 22×10^{10} Wcm^{-2}. An electromechanical shutter (Thorlabs SH05) is used to control laser irradiation. The beam is focused on the target using a lens of focal length-25 mm, and the target is placed on a motorised XY translation stage (Holmarc MTS-90115). The texturing is done by line-by-line scanning over an area of 25 mm^2 on the target surface, the scan line separation is set to 30μm, and the scanning speed is varied for creating different types of surface structures. Three samples are created by setting the scanning speed at 50 μm/s, 75 μm/s and 100 μm/s.

Images taken with a Field Emission Scanning Electron Microscope(Carl ZEISS) are used to study the surface morphology of the structured area. A stylus profilometer (Dektak 6M) is used to acquire the groove depth profile. Using an integrating sphere connected spectrophotometer(JASCO V-570), the reflectivity of complex surfaces is investigated by gathering specular and diffusive reflections.

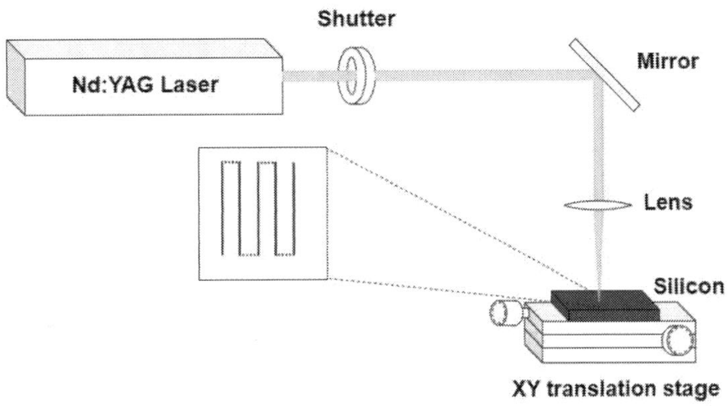

Figure 1.Experimental setup for laser surface structuring of silicon

3. Result and Discussion

The structured area in all three samples is found to be blackened in their appearance. The surface morphology of the structured area was studied using a scanning electron microscope. The SEM images in Figure 2 (a), (e) and (i) clearly shows the multiple parallel grooves engraved on the silicon surface with a periodicity of 30 μm that corresponds to the scan line separation, and this period of grating is larger than the wavelength range (0.2 μm-0.8 μm) chosen for reflectance measurement. Also, the higher magnification SEM images in Figure 2 ((c), (d),(g),(h),(k) and (l)) shows that the valleys and the ridges of the microgrooves are covered by a large number of nano and micro-structures. The concentration of nano and micro-structures over the structured area increases as the scan speed decreases. The decrease in scan speed causes to increases in the number of laser pulses per unit length along the scanning line. This results in more overlapping of the laser pulse as shown in Figure 3(a). This increase in the overlapping area causes more deepening of the crater formed by the last pulse by the upcoming next pulse. Also, the decrease in scan speed results in the increase of the width of the groove, which can be seen in the SEM images.

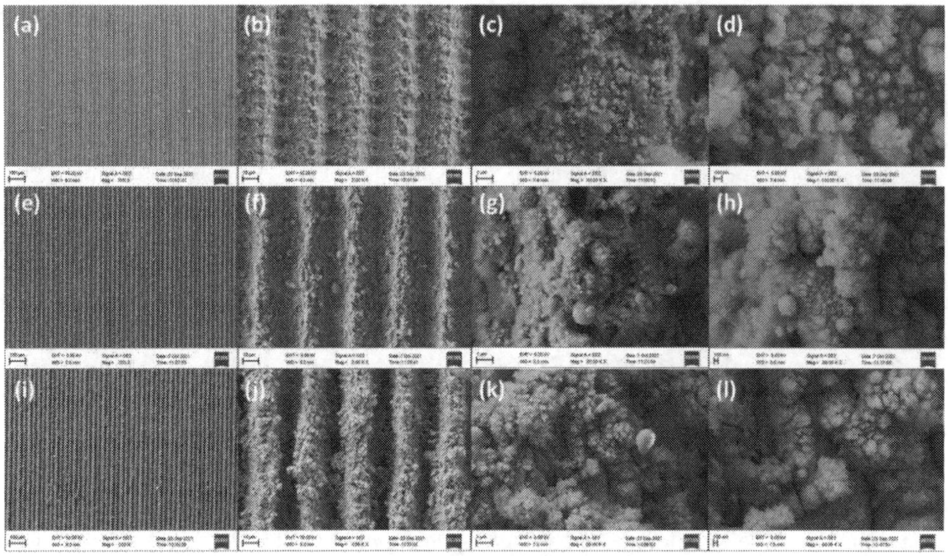

Figure 2.FESEM images of structured silicon surface with various laser scanning speed: (a) 100 μm/s, (e) 75 μm/s and (i) 50 μm/s (magnification 100 μm)and their higher magnification (10 μm, 1 μm, 100 nm) images to the right side.

The depth profile of the grooves measured using the stylus profilometer is shown in Figure 3(b). The average depth of the groove in each structured sample decreases as on increasing the scanning speed. The average depth of grooves are 8.28±2.70 μm, 7.57±1.67 μm and 6.13±0.90 μm for scan speed 50 μm/s, 75 μm/s and 100 μm/s respectively. This result validates the observations from the SEM images that the scan speed increase causes the number of laser pulses per unit length to decrease, leading to less deepening of the grooves.

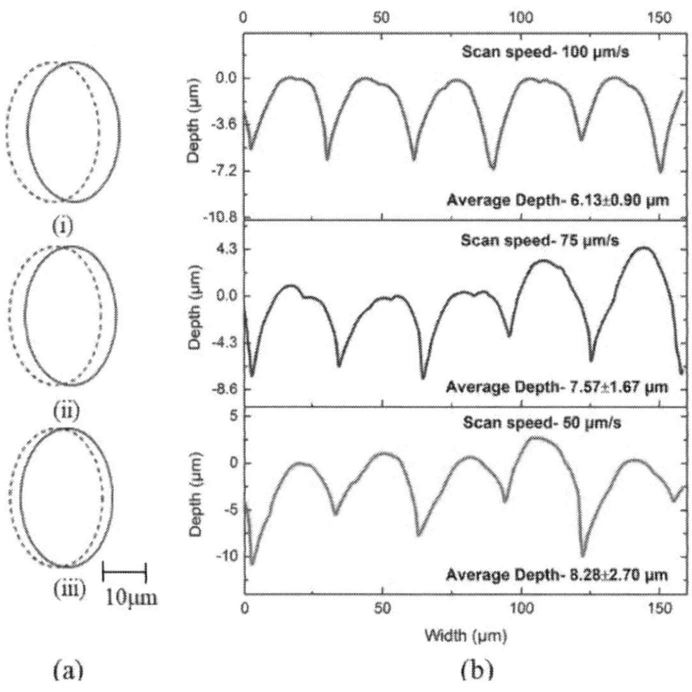

Figure 3. (a) Schematic representation of overlaps between successive laser pulses (b) depth profile of the surface structures created.

The total reflectance (diffuse + specular) spectra of the structured area for the wavelength shorter than the bandgap wavelength of silicon are shown in Figure 4. The results show that all three structured silicon surfaces significantly reduce reflectivity (≈60 % to 75 %) over the unstructured silicon wafer surface. Also, it shows a reduction in reflectivity (≈3 %) when the scan speed is reduced from 100 μm/s to 75 μm/s and the same is observed when the scan speed is reduced from 75 μm/s to 50μm/s.

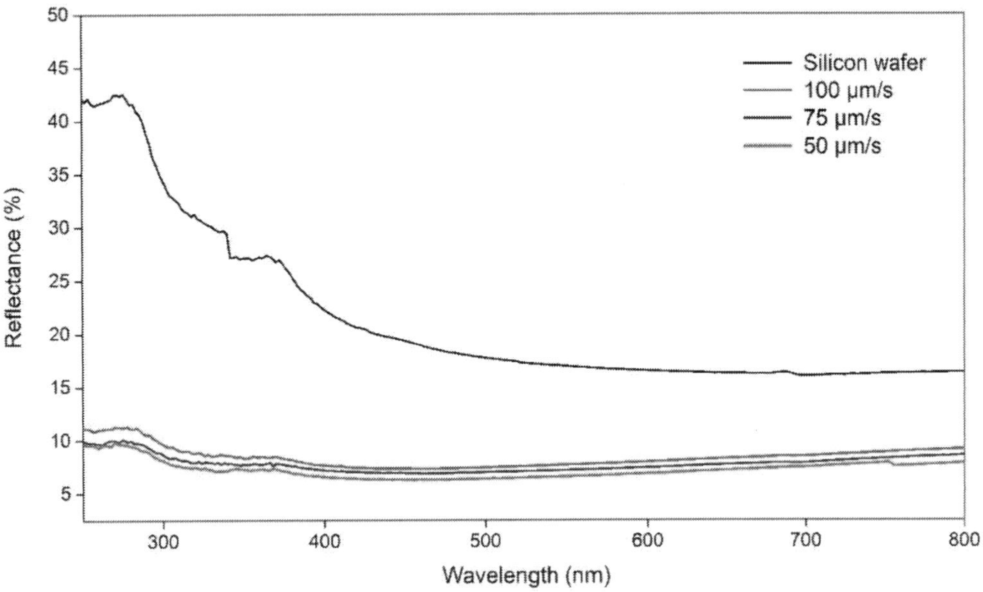

Figure 4. Reflectance spectra of the unstructured silicon surface and structured silicon surfaces at various scan speeds.

The absorption of light by silicon surface involves four different mechanisms: intrinsic absorption due to band transitions, absorption due to free carriers, absorption due to impurities or lattice defects, and intrinsic absorption due to lattice vibration[13]. When the silicon surface is irradiated with laser pulse,structures are created on the surface and these surface structures reduces its reflectance.The reduction in reflection of the laser surface textured silicon surface can be due to multiple reflection, scattering and moth-eye effect of structures having of variable sizes. The protruding structures of dimensions greater than several wavelength enhances the coupling to the material by light trapping due to multiple reflections. Also the absorption is enhanced by the increase in the effective optical path length due to the refraction of light at oblique angles. The structures with dimension of the order of the incident wavelength successively scatters the light results in increase of optical path length and enhances the absorption.Moth-eye effect is caused due to the presence of periodic array of structures with dimensions smaller than the incident wavelength[3]. The FESEM images in Figure 2 clearly reveal that large number of sub-micro and nanoparticles are deposited back when we reduce the scanning speed. Also the microgrooves are more deepened when the scan speed is reduced. This variation in the concentration of nano and sub-micro structures and the microgroove depth causes the reduction in the reflectivity of the textured surface when the scan speed is varied.

4. Conclusion

In this paper, we perform a detailed study of surface structures produced by laser surface texturing on silicon surfaces. The effects of scan speed on morphological and optical (reflectivity) properties of the silicon surface are studied. We found that the structured surfaces are covered with large number of micro and nano structures. Also the surfaces shows a significant reduction in the total reflectance over the unstructured silicon surface. The reduction in reflectivity of the structured surface is found to be due to the presence of nano and sub-micro structure covered grooves and subwavelength structures. The concentration of these structures and the depth of the micro grooves are found to be strongly depended on the scanning speed of laser surface texturing. Our results suggests that highly absorptive silicon surfaces can be created by optimising the scanning speed in laser surface texturing.

References

[1] Vorobyev A Y and Guo C 2012 Direct femtosecond laser surface nano/microstructuring and its applications *Laser Photonics Rev.* **7** 385–407
[2] Ion J 2005 Laser Processing of Engineering Materials *Elsevier Butterworth-Heinemann, Oxford*
[3] Brown M.S. A C B 2010 *Fundamentals of Laser-Material Interaction and Application to Multiscale Surface Modification. In: Sugioka K., Meunier M., Piqué A. (eds) Laser Precision Microfabrication.* (Springer Series in Materials Science, vol 135. Sp)
[4] Koynov S, Brandt M S, Stutzmann M, Koynov S, Brandt M S and Stutzmann M 2006 Black nonreflecting silicon surfaces for solar cells Black nonreflecting silicon surfaces for solar cells **203107** 88–91
[5] Branz H M, Yost V E, Ward S, Jones K M, To B, Branz H M, Yost V E, Ward S, Jones K M and To B 2009 Nanostructured black silicon and the optical reflectance of graded-density surfaces Nanostructured black silicon and the optical reflectance of graded-density surfaces **231121** 98–101
[6] Ma L L, Zhou Y C, Jiang N, Lu X, Shao J, Ma L L, Zhou Y C and Jiang N 2006 Wide-band " black silicon " based on porous silicon **171907**
[7] Huang Y, Chattopadhyay S, Jen Y, Peng C, Liu T, Hsu Y, Pan C, Lo H, Hsu C, Chang Y, Lee C, Chen K and Chen L 2007 Improved broadband and quasi- omnidirectional anti-reflection properties with biomimetic silicon nanostructures **2** 1–5
[8] Zaidi S H, Ruby D S and Gee J M 2001 Characterization of Random Reactive Ion Etched-Textured Silicon Solar Cells **48** 1200–6
[9] Wu C, Crouch C H, Zhao L, Carey J E, Younkin R, Levinson J A, Mazur E, Farrell R M, Gothoskar P and Karger A 2001 Near-unity below-band-gap absorption by microstructured silicon *Appl. Phys. Lett.* **78** 1850–2
[10] Carey J E, Crouch C H, Shen M and Mazur E 2005 Visible and near-infrared responsivity of femtosecond-laser microstructured silicon photodiodes *Opt. Lett.* **30** 1773
[11] Sheehy M A, Winston L, Carey J E, Friend C M and Mazur E 2005 Role of the Background Gas in the Morphology and Optical Properties of Laser-Microstructured Silicon **95** 3582–6
[12] Zorba V, Boukos N, Zergioti I and Fotakis C 2008 Ultraviolet femtosecond , picosecond and nanosecond laser microstructuring of silicon : structural and optical properties
[13] Vorobyev A Y and Guo C 2011 Direct creation of black silicon using femtosecond laser pulses *Appl. Surf. Sci.* **257** 7291–4

Influence of substrate temperature and laser wavelength on the structural, optical and electrical properties of laser ablated tin oxide thin films

Harisha Kumar K[1,2,*], Prasad Kumar[2] and S M Dharmaprakash[2]

[1]Department of Physics, University College of Science, Tumkur University, Tumkur, Karnataka 572 103, India
[2]Department of Studies in Physics, Mangalore University, Mangalagangothri, Mangalore, Karnataka 574 199, India
[*]harishakumark@gmail.com

Abstract: The pulsed laser deposition (PLD) of tin oxide thin films on glass substrates has been achieved by ablating tin oxide (SnO_2) pellet with Nd-YAG laser. The effects of laser wavelength (532nm, 1064nm) on the deposition as well as substrate temperature (room temperature, 250°C, 450°C) were investigated. The XRD studies were carried out and observed the growth of tetragonal rutile structure of SnO_2 films. The FESEM micrographs show changes in surface morphology as a function of laser wavelength as well as the deposition temperature. The optical property investigations using UV-Vis spectra reveal an increase in the band gap with increasing temperature. The investigation of electrical properties using four-point probe method revealed the lower electrical resistivity of the SnO_2 films. Thus, the as grown optimized SnO_2 conducting thin film electrodes find its potential application in supercapacitor devices.

Keywords: Tin oxide; pulsed laser deposition; optical properties; four-point probe

1. Introduction

Tin oxide (SnO_2) is an important transparent conducting oxide (TCO)material.It is a large bandgap (3.6eV) semiconductor with plenty of technological applications. They are used in optoelectronic devices,solar cells, gas sensors, electrochromic devices, supercapacitor/lithium battery electrodes and flat panel displays[1-4]. SnO_2 films have been prepared by variety of deposition techniques[5-7]. Among various techniques the pulsed laser deposition (PLD) technique has been employed successfully in the growth of desired SnO_2 surfaces[8-9]. PLD has several advantages when compared to other deposition techniques. PLD is very flexible deposition method and multicomponent epitaxial films with smooth surface [10-11] can be grown. In PLD technique SnO_2 films are deposited either by ablating a SnO_2 target or Sn target under oxygen atmosphere. Thus, the characteristics of deposited films depends upon deposition parameters.Supercapacitors and lithium ion batteries are widely used as energy storage devices. Supercapacitors provide high energy density and high-power density than conventional capacitors and batteries respectively. The electrodes are the most important part of electrochemical capacitors. Among various electrode materials (MnO_2, NiO, SnO_2 etc.) SnO_2 attracted a considerable attention due to its low resistivity, low costand high-power density [12-14].To study the effective use of SnO_2 layers finds their immediate requirements in the application of supercapacitors.

The structural, electrical and optical properties vary with deposition conditions like substrate temperature, oxygen pressure and laser fluence. A good crystalline SnO_2 film can be grown above 200^0C substrate temperature and at 20 Pa oxygen pressure [15].In this paper, we report the influence of laser wavelength (532nm, 1064nm) and substrate temperature (room temperature, 250°C, 450°C) on the structural, optical and electrical properties of pulsedlaserablatedSnO_2 thin films.

2. Experimental

The tin oxide films were produced in a standard PLDsetup. The laser (Q1 DNA, Quanta Systems) with wavelengths 532nm and 1064nm were used to ablate the SnO_2 pelletized targets. Pelletized targets were prepared from finely ground tin(IV) oxide powder (99.9% purity). The powder was pressed at 1000kgm^2 to form pellet and sintered at 1000°C for 4hours. Thus, the target prepared have high density and porosity. The laser beam was focused on the 1-inch diameter SnO_2 target at an incident angle of 45°. The target was continuously rotated to ensure uniform ablation.Ultrasonically cleaned glass plates were used as substrate and kept at a distance of 5cm from the target.All depositionswere carried out at 20 Pa of oxygen atmosphere[15]. The SnO_2 films were deposited at three different substrate temperatures (room temperature, 250°C, 450°C).The structure of the SnO_2 films was investigated by Rigaku miniflex-600 X-ray diffractometer with CuK_α as an X-ray source. Surface morphology of SnO_2 films were examined using Ultra 55 Carl-Zeiss field emission scanning electron microscope (FESEM). The optical transmittance was measured over 200-950nm wavelength range using Fluoro-Max-Plus-C-P spectrophotometer. Current-voltage (I-V) characteristics were studied using X-plore I-V source and measurement unit ofQuazar Technologies.

3. Results and discussion

Fig. 1 shows the XRD pattern of PLD SnO_2 films deposited using Nd:YAG laser wavelengths 532nm and Q-switched second harmonic laser wavelength 1064nm. Fig. 2 shows the XRD pattern of SnO_2thin films deposited at three different deposition temperatures. The diffraction peaks detected are matching with the tetragonal rutile structure of SnO_2 (JCPDS #88-0287). We observed a well-defined peak at 26.6°correspondingto(110) plane in all samples. Also, there are weak peaks at 33.9° (101) and 51.8° (211) for all samples. The XRD pattern of the film also contains peaks corresponding to SnO (JCPDS #78-1913) for the films deposited using laser wavelength 1064nm as well as at 450°C substrate temperature. In this case, SnO and SnO_2 both phases are present.The oxygen deficiency within the grain, in grain boundary or at both could have assisted the growth of SnO phase[16].

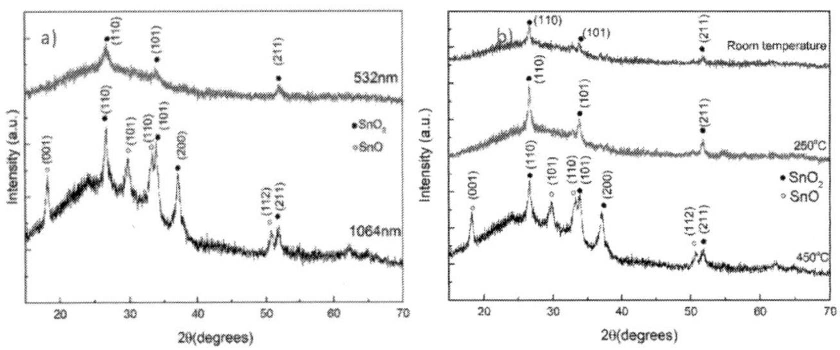

Fig. 1XRD patterns of SnO_2 films deposited at a) different wavelengths and b) different substrate temperatures

Fig. 2 shows SEM micrograph images of SnO_2 films. The microstructure of PLD SnO_2 films change significantly with laser wavelength and substrate temperatures. Grain size is found to be increasing with increase in deposition substrate temperature. This is due to the agglomeration of smaller grains [18]. Grain sizes are uniform and have a definite pattern when deposited at 450°C and when using 532nm wavelength.

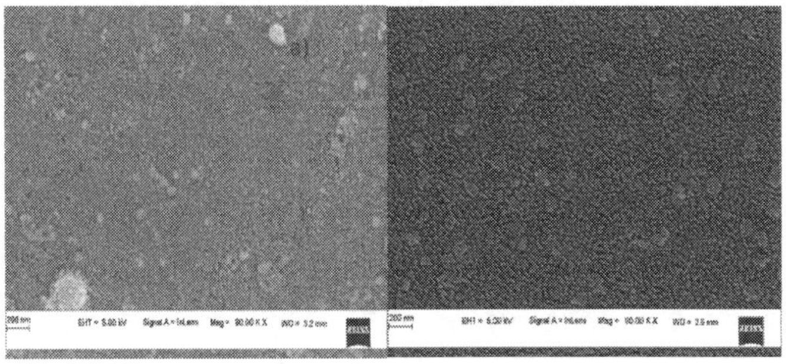

Fig. 2SEM micrographs of PLD SnO_2 films deposited using laser wavelengthsa) 1064nm and b) 532nm

Fig. 3SEM micrographs of PLD SnO_2 films deposited at different substrate temperatures a) room temperature b) 250° C and c) 450°C

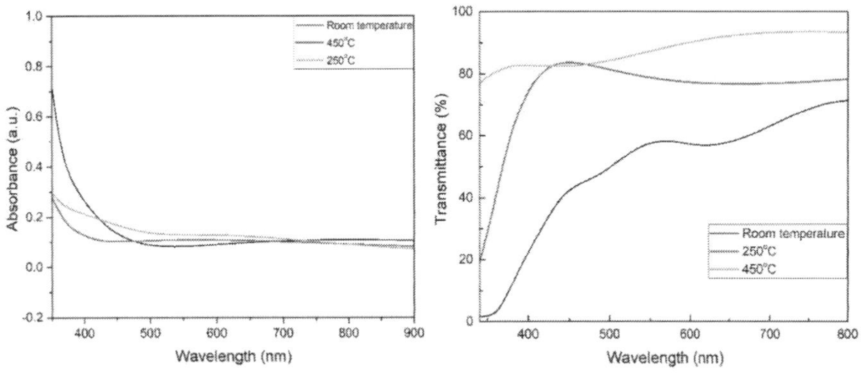

Fig. 4 The optical absorbance and transmittance spectra at different substrate temperatures.

The optical band-gap of the films calculated from the transmittance spectra. The optical band-gap (E_g) is calculated using the Tauc relation,

$$\alpha h\nu = C\left(h\nu - E_g\right)^n$$

Where C is constant, ν is the transition frequency and n is the nature of band transition. Since SnO_2 is direct band-gap material, n is taken as 0.5. Fig. 3 shows the Tauc plot and the optical band-gap was determined by extrapolating the straight regions of the plot. The optical band-gap is found to be 3.54eVand which agrees with the reported value (3.6eV)[13].The observed band-gap valueremains almost same in all samples and a similar trend wasobserved by Stanimirova et. al.[12]. Optical transmittance of the films is below 80% when deposited at room temperature and at 250^0C.

IOP Conf. Series: Materials Science and Engineering 1221 (2022) 012023 doi:10.1088/1757-899X/1221/1/012023

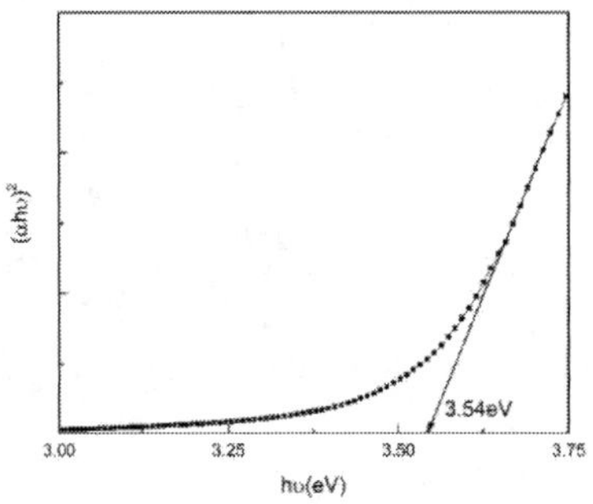

Fig.5 Typical Tauc plot of 400nm thick SnO_2 film deposited at 450°C substrate temperature.

The room temperature electrical resistivity of the films was determined and the values are presented in table 1 and table 2. The resistivity of the film afound to be sensitive to deposition temperature and drastically decreases from 53.74 to 0.432 mΩm. This suggests acorrelation between the resistivity and grain size. The conductivity in prepared SnO_2 films most likely influenced by carrier scattering at the grain boundaries and oxygen vacancies[17-18]. The resistivitymeasuredis minimum for the film deposited at 450°C substrate temperature. This could be due to densely packed microstructure and oxygen vacancies [18].

Table 1. Variation of SnO_2 film resistivity with substrate temperature.

Substrate Temperature	Resistivity ($10^{-3}\Omega m$)
Room temperature	53.74
250°C	4.115
450°C	0.432

Table 2. Variation of SnO_2 film resistivity with deposition laser wavelength.

Laser wavelength	Resistivity ($10^{-3}\Omega m$)
532nm	2.388
1064nm	0.432

4. Conclusion

Tin oxide thin films were deposited using Nd-YAG laser wavelengths 532nm and 1064nm and at different substrate temperatures employing pulsed laser deposition technique. The effect of substrate temperature on the surface morphology has been investigated and we observed uniform film deposition when the deposition wavelength is 532nm. The optical band-gap we obtained is 3.5eV and the lowest electrical resistivity of 0.432 mΩm is achieved for the film deposited with laser wavelength 1064nm.ThephysicalparametersobtainedinthepresentstudyaresuitabletoutilizeSnO_2thinfilmsforelectrodeapplicationsinsupercapacitors.

References

[1] B C Lewis and B D Paine, *MRS Bull.* 25, 22 (2000).

[2] M Chen, Z L Pei, C Sun, L S Wen and X Wang, *J. Crystal Growth,* 220. 254 (2000).

[3] A Kar, M A Stroscio, M Datta, J Kumar and M Meyyappan, *App. Phys. Letters* 94, 101905 (2009).

[4] I Aruna, F E Kruis, S Kundu, M Muhler, R Theissmann and M Spasova, *J. Appl. Phys.,* 105, 064312 (2009).

[5] L Bruno, C Pijolat and Lalauze, *Sens. Actuators B* 18, 195 (1994)

[6] A Dieguez, A Romano Rodriguez, J R Morante, J Kappler, N Barsan and W Gopel, *Sens. Actuators B* 60, 125 (1999).

[7] A. Cricenti, R. Generosi, M. A. Scarselli, P. Perfetti, P. Siciliano, A. Serra,A. Tepore, C. Coluzza, J. Almeida, and G. Margaritondo, J. Phys. D: *Appl.Phys.* **29**, 2235 (1996).

[8] M. A. El Khakani, R. Dolbec, A. M. Serventi, M. C. Horrillo, M. Trudeau,R. G. Saint-Jacques, D. G. Rickerby, and I. Sayago, *Sens. Actuators B* **77**,383 (2001).

[9] W. S. Hu, Z. G. Liu, J. G. Zheng, X. B. Hu, X. L. Guo, and W. Göpel, *J.Mater. Sci.: Mater. Electron.* **8**, 155 (1997).

[10] H. Kim, A. Pique ', J. S. Horwitz, H. Mattoussi, H. Murata, Z. H. Kafafi,and D. B. Chrisey, *Appl. Phys. Lett.* **74**, 3444 (1999).

[11] D. B. Chrisey and G. K. Hubler, *Pulsed Laser Deposition of Thin Films,* Wiley, New York (1994).

[12] Seng, N. K., Ahmad, M. K., Jasmin, N. S., Zainal, N. B., Nayan, N., Fhong, C., … Mohamed, A. (2016), *ARPN Journal of Engineering and Applied Sciences*, *11*(14), 8915–8920.

[13] Nam, S. C., Yoon, Y. S., Cho, W. I., Cho, B. W., Chun, H. S., & Yun, K. S. (2001), *Electrochemistry Communications*, *3*(1), 6–10.

[14] Velmurugan, V., Srinivasarao, U., Ramachandran, R., Saranya, M., & Grace, A. N. (2016). *Materials Research Bulletin*, *84*, 145–151. https://doi.org/10.1016/j.materresbull.2016.07.015.

[15] Stanimirova, T. J., Atanasov, P. A., Dimitrov, I. G., & Dikovska, A. O. (2005). *Journal of Optoelectronics and Advanced Materials*, *7*(3), 1335–1340.

[16] Vispute R D, Godbole V P, Chaudhari S M, Ogale, S. B., S M Kanetkar(1988). *J. Matter. Res.*, *3*(6), 1180–1186.

[17] Nomura, K. (2015). *Croatica Chemica Acta*, *88*(4), 579–590.

[18] Dolbec, R., El Khakani, M. A., Serventi, A. M., Trudeau, M., & Saint-Jacques, R. G. (2002). *Thin Solid Films*, *419*(1–2), 230–236. https://doi.org/10.1016/S0040-6090(02)00769-1.

IOP Conf. Series: Materials Science and Engineering 1221 (2022) 012024 doi:10.1088/1757-899X/1221/1/012024

Physico-Chemical Key Parameters, Freundlich and Langmuir Isotherm, Rate Constant Studies on the Removal of Divalent Nickel using *Delonix regia* Fruit Pod as Low Cost Bioadsorbent

M M Mendiretta[1], S L Mudliar[2] and A V Bharati[3]
[1] Assistant Professor, Yeshwantrao Chavan College of Engineering, Nagpur, India
[2,3] Assosocate Professor, ShriRamdeobaba College of Engineering and Management Nagpur, India
E-mail: khurmamycc@gmail.com

Abstract. The exploitation of *Delonix regia* fruit pod (DRFP) as the low-cost biodsorbents is verified as an alternative for current pricey techniques for elimination of nickel ions from aqueous solutions. Batch adsorption studies is conducted to examine the effects of physico-chemical key parameters such as the initial metal ion concentration , pH, agitation time, stirring rate and adsorbent dosage on the adsorption from *Delonix regia* fruit pod (DRFP). The % adsorption, Langmuir constants, Freundlich constant for [Ni(II)] were determined for the adsorption system as a function of sorbate concentration. Equilibrium data were analyzed using the Langmuir and Freundlich isotherms whereas the adsorption kinetics data is also evaluated.

1. Introduction

Biosorption as a novel treatment tool is acquiring its standing in remediation of wastewaters from heavy metals [1]. Biosorption being a capable, clean and viable technique is evolving as an potent alternative for wastewater treatment specially in developing countries, as the cost involved in different treatment methods many time surpasses the requirement to treat water [2]. Heavy metal pollutions get out as effluents from different industries, like mining, electroplating, batteries, metal plating, stabilizers, pigments, sewage sludge and alloy industries where the chief constituent includes lead, nickel, copper, chromium, and other heavy metals. Heavy metals being non-biodegradable have a tendency to get mount up in living organisms through food chains. Few metals like Ag, Hg, Pb and Cd could be enormously toxic for to living beings. Other metals such as Zn, Cu, Fe, Mn, Co and Ni, when exist in excess concentrations above definite limits, could be very damaging to living organisms but are important for animals and plants in minute quantity as well.

Nickel is a silvery metal and Ni(0) and Ni(II) are the two oxidation states which occur in nature. Divalent nickel, Ni(II) has been identified as a common toxic pollutant. Nickel (II) and finds its way to the waterways from industries such as nickel plating, coins and jewellery making, electroplating, porcelain enamelling etc. At elevated concentrations nickel may cause cancer of nose, lungs and bones. Acute poisoning of Ni(II) results Allergic skin reactions, asthma, conjunctivitis, lung fibrosis, carcinogenic.

2. Materials and Methods

2.1 Selection of Adsorbent

Before selection of a biomaterial as an adsorbent material it should be made sure that selected sample is locally and conveniently available, is non-toxic, cost wise reasonable, and should have surface characteristics appropriate for proficient and swift adsorption. Based on the above criteria profusely available biomaterial with proven medicinal value was tapped for its potential as an alternative to the available techniques for treating waste waters.

2.2 Preparation of Bioadsorbent

The starting material has been collected from Nagpur, Maharashtra; it takes a position between 21.1458° N latitude and 79.0882° E longitude (elevation 311 m), washed with double distilled water for elimination of dirt and other impurities. The material was then boiled in double distilled water through varying the water repetitively till water turn into colorless, which is the indication that the water-soluble color compounds are removed off. The boiled and thoroughly washed material was dried in oven for 24 hours at 80°C and stocked in desiccator to avoid adsorption of moisture. The material first got crushed into small pieces before pulverization then sieved to a particle size (0.20-0.30 mm).

2.3 Preparation of Metal Ion Solutions

AR grade chemicals are employed in the study from Merck, India. The calculated quantity of salts, $NiCl_2.6H_2O$ respectively were added in double distilled water for preparing the metal ion stock solution of Ni(II).NaOH, HCl and buffer solutions ACS reagent grade (E. Merck) were used to adjust the solution pH. For measurement of pH, an Elico (LI-129) pH meter was utilized. FTIR spectrum of adsorbents and metal loaded adsorbents in KBr pellet were recorded in 450-4000 cm^{-1} range on computer-controlled Fourier transform infrared spectrophotometer. SHIMADZU UV-VIS Spectrophotometer Model UV1201 is used to determine the concentration of metal ions.

2.4 Experiments on Batch Adsorption

Several factors affecting extent of adsorption have been calculated in order to attain the best conditions for the whole process. For determining the ratio essential for optimum biosorption, the concentration of DRFP was varied between 0.5 to 5g. For the adsorption isotherm studies, at pH 5.The influence of pH on biosorption was measured ranging from pH 2.0 to 7.0. At five different temperatures, i.e., 25, 30, 40, 50 and 60°C, the percentage removal of metal ions was measured. The outcome of agitation speed on the adsorbent material was studied from 50 to 300 rpm.

3. Results and Discussion

3.1 Characterization of Bioadsorbents: The Physicochemical Characters are shown in Table 1.

Table 1. *Physicochemical Characters of DRFP.*

Physicochemical Characters	Apparent density (g ml^{-1})	Moisture content (%)	Ash content (%)	Zero point charge (pH_{zpc})	Fixed Carbon (%)
(DRFP)	0.44	7.52	7.93	3.7	54.96

3.2 Scanning Electron Microscopic (SEM) Studies

Evidence regarding the textural and morphological characteristics of the surface of the adsorbent can be obtained from Scanning Electron Microscopic studies. Scanning which is shown in Figure 1, it is evident from the analysis that the surface area is uneven, heterogeneous and pores are present on the surface of the powdered biomass.

Figure 1. SEM image of DRFP

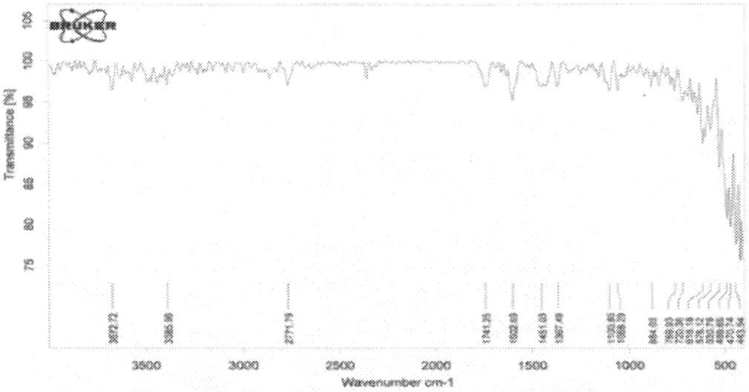

Figure 2. FTIR spectrum of DRFP

3.3 Infrared Spectroscopy

FTIR spectroscopy is a vital means to identify the different functional groups existing in an adsorbent. FTIR analysis of prepared biomaterial samples DRFP is presented in Figure 2. The peak assignments of various vibration bands of the spectra are given in Table 2. It is inferable from the spectra and peak assignment that the adsorbents contain aromatic, phenolic, alcoholic, aldehydes, alkanes, carboxyl and amine (primary/secondary) groups.

Table 2. Functional group assignment for FTIR absorption bands of DRFP.

Wavelength (cm^{-1})	Possible Assignment	Functional Group
3672	free hydroxyl,O-H stretching	Alcohols, phenols
3385	H-Bonded, O-H stretching	Alcohols, phenols
2771	H-C=O; C=H stretching	Aldehydes
1741	C=O stretching	Esters, saturated aliphatic

3.4 Effect of the Solution pH on Adsorption

A test was performed at increasing initial pH from 2 to 7. From figure 3, in the pH range i.e. 2 to 7, it is seen that the adsorption percentage of nickel rise with augment in initial pH of the medium. This behavior could be illustrated using pH_{zpc} of the adsorbent. The pH_{zpc} of DRFP is found to be 3.7 in this research module. According to Shukla and Pai when the pHof the solution exceeds, the pH_{zpc} extent of adsorption rises [3] [4]. The utmost value of adsorption of metal ions on bioadsorbents was established at pH 5 for Nickel ions. Above this pH value biosorption significantly decreased. The reduction in adsorption of metal ions at small and high pH is credited to the interaction between OH^- and H^+ with nickel ions. From figure 3, it is clear that at higher pH, the extent of adsorption is little. Percentage removal of nickel by adsorption onto 0.5 to 5 g activated carbon sample of DRFP with 100 mg/l of nickel with agitation time of 60 min and at pH 6 was carried out. The outcomes percentage removal of copper versus adsorbent dose is presented in figure 4.

3.5 Effect of Adsorbent Dosage

Observation made show that the percent removal of metal increases with increase in the adsorbent dose from 0.5 to 5g as the surface area of the adsorbent increases. Thus, better metal recovery is synonymous to a high biomass dose for availability of more active sites. This existence could be illustrated because of variety of reasons, like electrostatic interactions, availability of solute, reduced integration at elevated biomass densities and interference between binding sites.

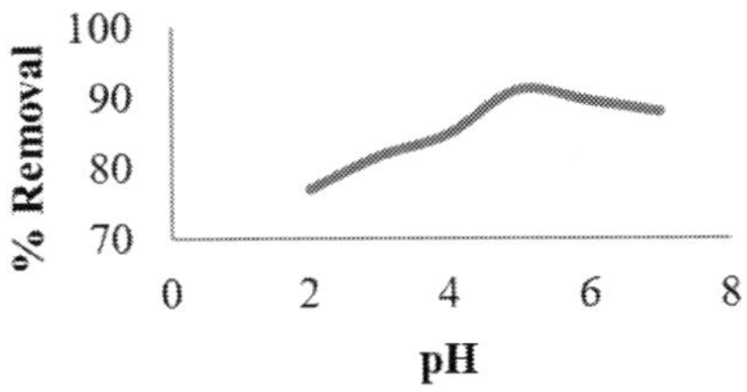

Figure 3. Effect on nickel removal by with pH(Initial metal nickel ion concentration = 100 mg/l, adsorbent dose 0.5g, and time = 40 min, at 200 rpm).

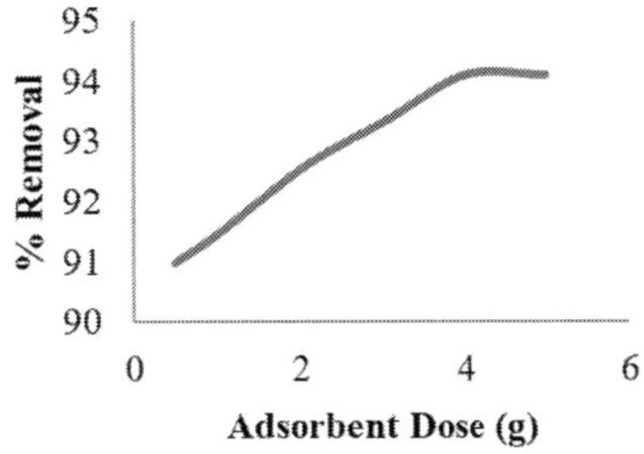

Figure 4. Effect of adsorbent dose on nickel removal(Initial metal ion concentration = 100 mg/l, pH = 5, time = 40 min, at 200 rpm).

3.6 Effect of the Initial Metal ion Concentration

The adsorption data analysis is important for designing purposes and to develop an equation that precisely signifies the results. The theoretical equilibrium isotherm: the Langmuir, Freundlich, Dubinin-Radushkevich and Temkin are the frequently used isotherms [5]. Correlation coefficient (R^2) and the isotherm constants of the models for sorption of metal ion onto DRFP are presented in Table 3.

Table 3. Isotherm parameters of Ni adsorption of DRFP.

Model	Parameter	DRFP Ni(II)	Model	Parameter	DRFP Ni(II)
Langmuir Isotherm	q_{max}(mg/g)	48.17	*Tempkin Isotherm*	A_T(L/mg)	0.58
	K_L(L/mg)	0.07		b_T	292.6
	R^2	0.98		R^2	0.97
	χ^2	2.7128		χ^2	3.6154
Freundlich Isotherm	N	1.26	*Dubinin Radushkevich Isotherm (DRK)*	q_m(mol/g)	17.06
	K_F(mg/g)	2.11		K (mol^2/kJ^2)	0.04
	R^2	0.99		R^2	0.84
	χ^2	0.9854		χ^2	13.257

3.7 Effect of Biosorption on thermodynamics &Temperature

The percentage removal of Ni(II) ions was studied at five different temperatures, i.e., 25, 30, 40, 50 and 60°C. Figure 5depicts the temperature dependence of adsorption. The adsorption of metal ions by the adsorbent is observed to be high at 60°C. The metal ions removal increased with a rise in temperature. Such may be due to either reduced thickening of boundary layer immediate to adsorbent. Also, there is a scope of solute diffusion within the pores of adsorbent at higher temperatures. At higher temperatures, greater adsorption will be observed as diffusion is an endothermic process [6].The thermodynamic parameters were computed from the following equations [7]:

$$\Delta G^0 = -RT \ln K_c \qquad (1)$$

$$\log K_c = \frac{\Delta S^0}{2.303\,R} - \frac{\Delta H^0}{2.303\,RT} \qquad (2)$$

Where,K_c is the equilibrium constant[8]. The intercepts and slope provides values of ΔH^0 and ΔS^0 from the plot of lnK_cversus *1/T* (7). Finally, the ΔG^0 value is computed using the equation:

$$\Delta G^0 = \Delta H^0 - T\Delta S^0 \quad (3)$$

Values of ΔH^0, ΔS^0, and ΔG^0 are listed in Table 4. The spontaneous nature of adsorption is predicted for the negative ΔG^0 values obtained.

Table 4. Thermodynamic Parameters of Ni(II) Adsorption on DRFP.

Temp (K)	*ΔH* (kJ/mol)	*ΔS* (J/mol K)	*ΔG* (kJ/mol)
298			-33.02
303			-33.59
313			-34.70
323	0.0030	110.90	-35.81
303			-10.39
313			-10.74
323			-11.08

Elovich plots of *qt* versus *ln(t)* are shown in Figures 8, for the prepared variants of adsorbent materials. The corresponding results are compiled in Table 5 showing good *R* square values, hinting toward the possibility that the degree-determining stage being diffusion in nature [9].
The fitting of the kinetic data w.r.t correlation coefficient and SSE illustrate that the dynamics of sorption can be best demonstrated by the pseudo 2[nd]order model which indicates that a chemisorptive rate restricting for all the selected adsorbent. Intra-particle diffusion model plots indicates that the rate of sorption is limited at more than one mechanism.

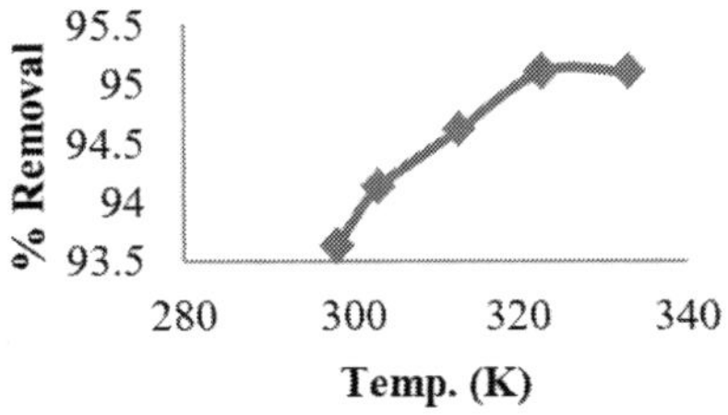

Figure 5. Effect on removal by bioadsorbents w.r.t temperature (Initial metal copper ion concentration = 40 mg/l, nickel ion concentration = 100 mg/l, adsorbent dose 0.5g, time = 40 min, pH = 5, at 200 rpm).

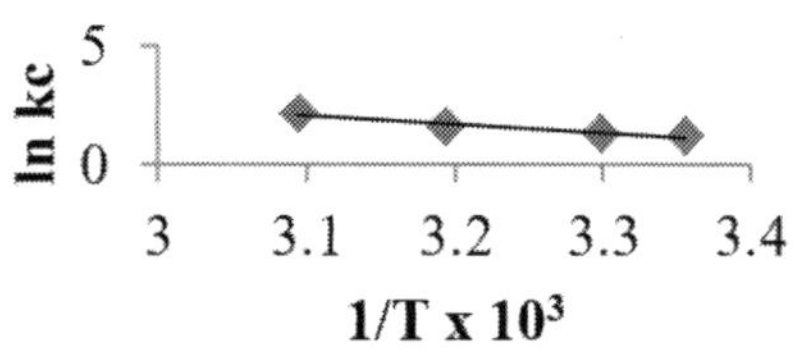

Figure 6. Van't Hoff plots for Biosorption of nickel onto DRFP.

Table 5. Kinetic parameters of Ni(II) Adsorption on DRFP.

Model	*II order kinetics*				*Intraparticle Diffusion Model Weber Moris*			*Elovich equation*		
Parameter	K_2(g mg^{-1} min^{-1})	*qe*(mg g^{-1})	R^2	*SSE*	K_i(mg g^{-1} min$^{-0.5}$)	R^2	*SSE*	*α* (mg g^{-1} min^{-1})	R^2	*SSE*
DRFP/ Ni(II)	0.001	23.81	1	0.34	0.43	0.8	47	2.24	0.9	12.8

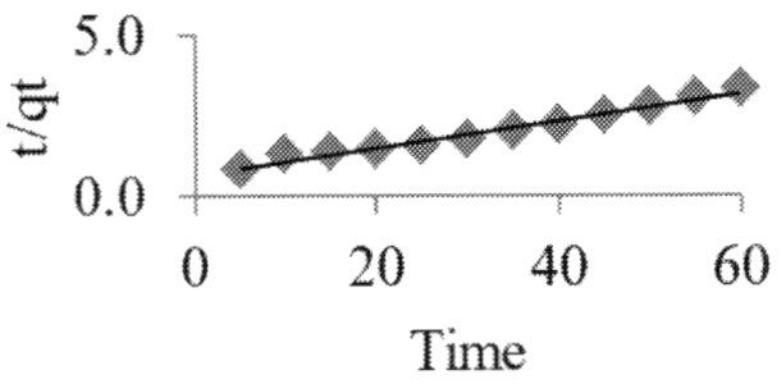

Figure 7. Second order kinetic modeling of nickel onto DRFPF.

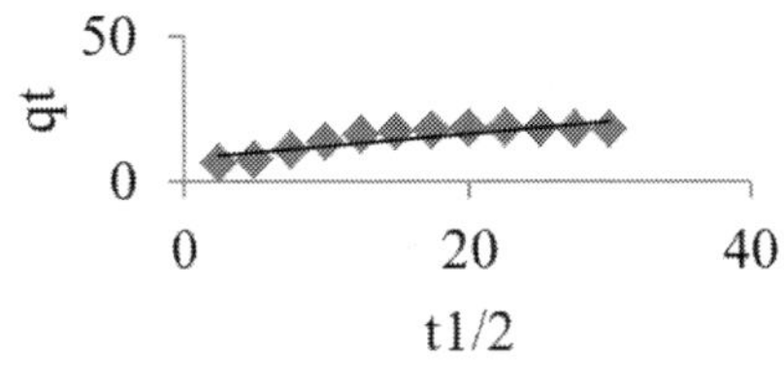

Figure 8. Intraparticle Diffusion Model kinetic modeling nickel onto DRFP.

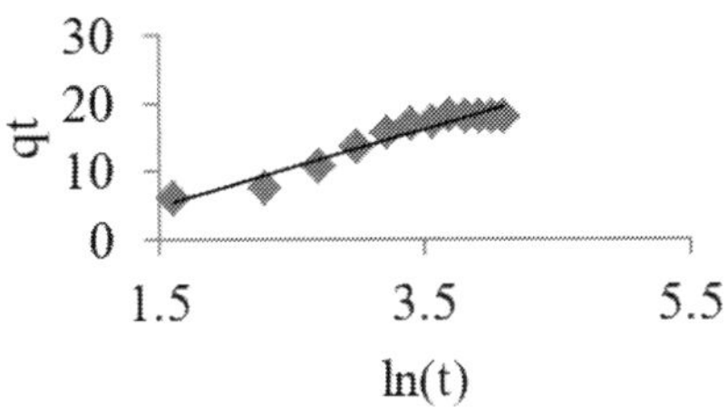

Figure 9. Elovich Model kinetic modelling of nickel onto DRFP.

4. Conclusion

The experimentalstudy concluded that for the elimination of Ni(II) from aqueous solution,DRFP could be utilized as a potential sorbent. The pH of the solution,batch study parameters, temperature,contact time andbiomass concentration were found to be valuable on the biosorption techniques. The kinematic studies shows that the pseudo-second-order kinetic modelis followedby the biosorption technique. The measured thermodynamic parameters showed the practicability, spontaneousand endothermicnature of the biosorption of Ni(II) onto DRFP biomass. Thus, we can say thatfor the treatment of wastewaters containing Ni(II) ions, the DRFP biomasscould be used as analternative biosorbent.

References

[1] Gurel L., 2017, Applications of the Biosorption Process for Nickel Removal from Aqueous Solutions – A Review. Chem. Engg. com. p. 711-722.

[2] Musayev S. MAa. 2017, Multicomponent Biosorption of Heavy Metals from Aqueous Solutions: A Review. J. of Envir. Studies. 26: p. 1433-41.

[3] Das N JRK. 2006, Adsorption of some bivalent heavy metal ions from aqueous solutions by Mn nodule leached residues. Journal of Colloid and Interface Science. 293: p. 253-262.

[4] Sharma A BKG. 2004, Adsorption of Pb (II) from aqueous solution by azadirachta indica(Neam) leaf powder. J. Hazard. Mater. 113: p. 97-109.

[5] Nisha Gaur AKMYAT, 2018, Adsorptive removal of lead and arsenic from aqueous solution using soya bean as a novel biosorbent: equilibrium isotherm and thermal stability studies. 98(3).

[6] Dundar M NCNY 2008, Biosorption of Cu(II) ions onto the litter of natural trembling poplar forest. J. Hazard. Mater. 151: p. 86–95.

[7] Kumar P S RSARV, 2011, Adsorption of Metal Ions onto the Chemically Modified Agricultural Waste. Clean Soil Air Water. 40: p. 188–197.

[8] Reddy D H K RDKVSKRAVR, 2011, Biosorption of Ni(II) from aqueous phase by Moringa oleifera bark, a low cost biosorbent. Desalination. 268: p. 150–157.

[9] Babu Ch S CEKAKSR, 2014, Adsorption Kinetics of A Cationic Dye onto Indigenously Prepared Activated Kaza's Carbon. J. Appl. Chem. 3: p. 2462-69.

IOP Conf. Series: Materials Science and Engineering 1221 (2022) 012025 doi:10.1088/1757-899X/1221/1/012025

Structural, capacitive and impedance properties of graphene oxide-PVDF composites for flexible electronics

Richa Pandey[1,2], Naveen Singhal[1], Parveen Kumar[2] and Sajjan Dahiya[3]

[1]Department of Chemistry, School of Physical Sciences, DIT University, Dehradun-248009, India
[2]Materials and Nano Engineering Research Laboratory, Department of Physics, School of Physical Sciences, DIT University, Dehradun-248009, India
[3]Department of Physics, MaharshiDayanand University, Rohtak-124001, India
parveenpaliwal@gmail.com, parveen.kumar@dituniversity.edu.in

Abstract. Flexible composite thick films of polyvinylidene fluoride 'PVDF' loaded with graphene oxide 'GO' as reinforcement were prepared using solution casting method. The PVDF:GO ratio was selected as 100:0 (R0), 50:50 (R5) and 40:60 (R6). The structural properties were analysed using X-ray diffraction investigations. Dielectric constant 'ε' and tangent loss 'tanδ' was measured as a function of frequency (1kHz to 1MHz) at different temperatures from 30 to 150°C for all the samples. Impedance 'Z' and capacitive modulus 'M' was calculated. The compositional effect on the electrical properties was investigated and a correlation was developed between the composition and property. The evaluated results suggest that the prepared composites for flexible electronic applications.

1. Introduction

Over past few decades, polymer based nanocomposites 'PNC' due to their high dielectric permittivity and low tangent loss have shown potential as an advance material for energy storage devices, generators, lithium ion batteries etc [1-3]. These are the lightweight materials with good chemical stability and binding capability. Due to its excellent piezoelectric and pyroelectric properties, polyvinylidene difluoride (PVDF) have been used as potential candidate for electronics applications [4]. PVDF is a fluoropolymer which is chemically non-reactive produced by polymerization of vinylidene difluoride. It found in five different polymorphic crystal structures (α, β, γ, δ, ε) which have different electrical and mechanical property. Out of these α and β are the most common phases, where α being a primary phase is thermally stable at room temperature and pressure in comparison to all other phases. But it is a non-polar phase. On the other hand, β phase is having a polar structure and is kinetically more stable under ambient temperature. It exhibits extraordinary pyro and piezoelectric properties [5]. However, this phase is the most commonly used phase, but its electronic and mechanical properties can be further modified using different fillers in optimized quantity such as metal nano particles, carbon nano tubes 'CNT', graphene etc. The carbon based fillers such as CNTs, graphene oxide 'GO' and reduced graphene oxide 'rGO' helps the investigators to overcome the low performance of polymers such as PVDF for energy and other electronic applications. The properties of PVDF loaded with C-based material like GO, rGO and MWNTs are strongly dependent on various factors like particle size, homogeneity, dispersion level and synthesis techniques [6-8]. Furthermore, GO is has also shown its potential for various electronic applications [9,10]. Many reports have been made on the use of GO's due to its very high electrical conductivity and its potential use as in fabrication of nano-sensors, transparent electrodes and many other applications. In the present study we synthesized nanocomposites of PVDF as matrix and GO as fillers. The structural, dielectric and impedance properties were investigated and discussed.

2. Experimental

PVDF (Sigma Aldrich), graphite powder (Molychem), N,N,-dimethyl formamide (S.D. Fine), sodium nitrate ($NaNO_3$), potassium permanganate ($KMnO_4$), sulphuric acid (H_2SO_4), hydrogen peroxide (H_2O_2), hydrochloric acid (HCl), dimethyl formamide (DMF) were used as raw materials. Graphene oxide was prepared via the modified Hummer's method, as described in literature [9,10]. 10wt% of PVDF dissolved in N, N-dimethyl formamide (DMF) and stirred at 60°C for 2 hours. A mixture of GO and DMF with concentration of 1mg\ml is prepared and sonicated for 1 hour to get homogenous solution. Nanocomposite films of PVDF and GO mixed in ratio 100:0, 50:50 and 60:40 (named as R0, R5 and R6) were fabricated by solution casting process. For this, the prepared GO-DMF solution was added drop-wise in PVDF-DMF solution in different ratios and continuously stirred for 2 hours. Prepared mixture was then poured on a petri dish and placed in a hot air oven for solvent evaporation at 40°C. The nanocomposite films were then cut into pieces for XRD measurements from $10° \leq 2\theta \leq 30°$. Silver paste was used to make the ohmic contacts to measure the capacitance and tangent loss at different temperatures (30 to 150°c) at different frequencies.

3. Results and discussion

3.1. Structural analysis

Figure 1 shows the XRD patterns of R0, R5 and R6. Two common phases α and β of polyvinylidene fluoride (PVDF) can be observed. Peaks at 2θ angles of 18.4 and 26.6 with planes (100) and (021) confirms the presence of α phase [21]. While the peak with 2θ angle 20.4 confirms the presence of β phase. It indicates that α-phase contents in initial crystallization process are dominant in the PVDF. By increasing the ratio of GO, the peak intensity of α- phase of PVDF decreases, whereas the intensity of β phase get increased which may be due to the compatibility between =CO group in GO and fluorine in PVDF. As a result of which no individual characteristic peak of GO is observed in prepared composites [10].

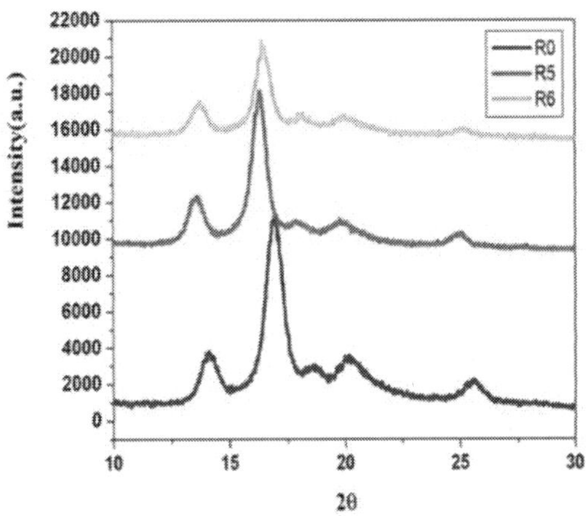

Figure 1. X-ray diffraction pattern of R0, R5 and R6 samples.

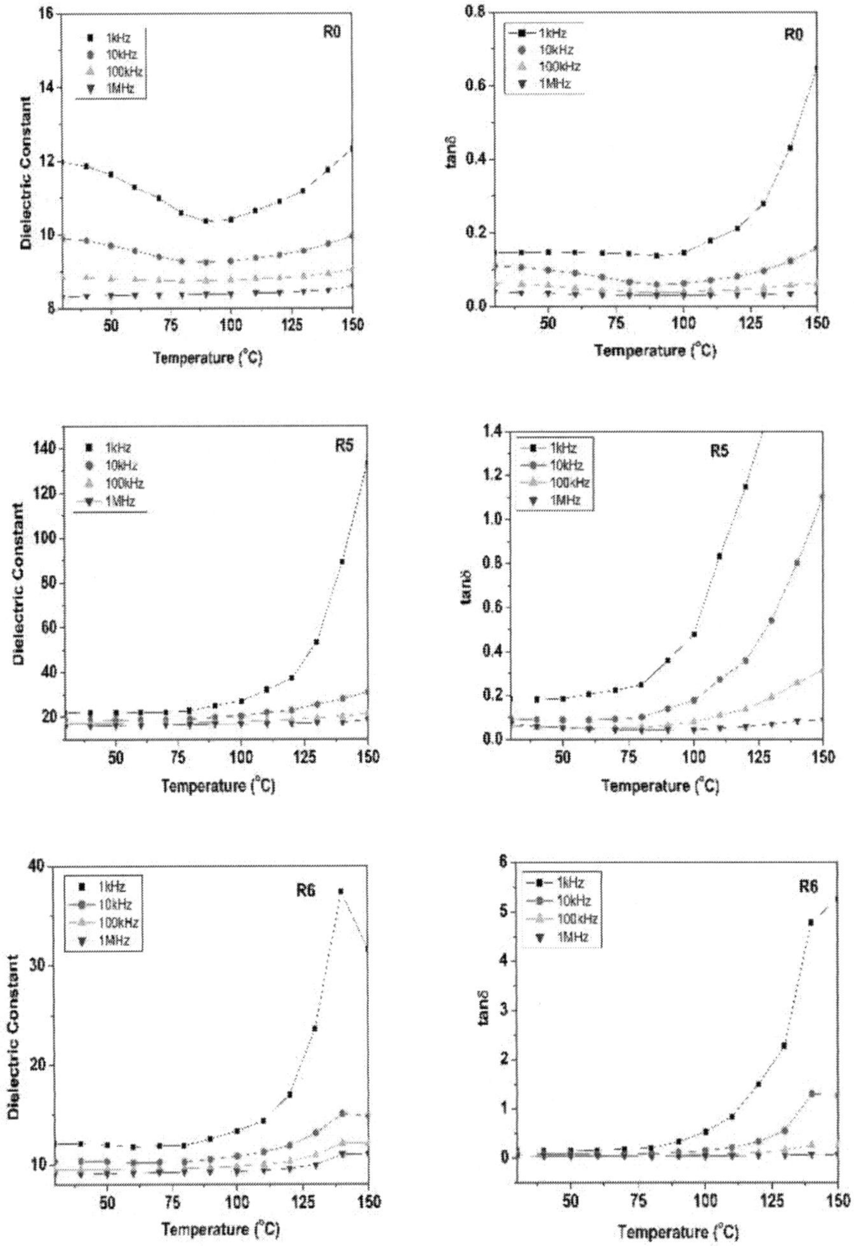

Figure 2. Variation of dielectric constant and tanδ as a function of temperature

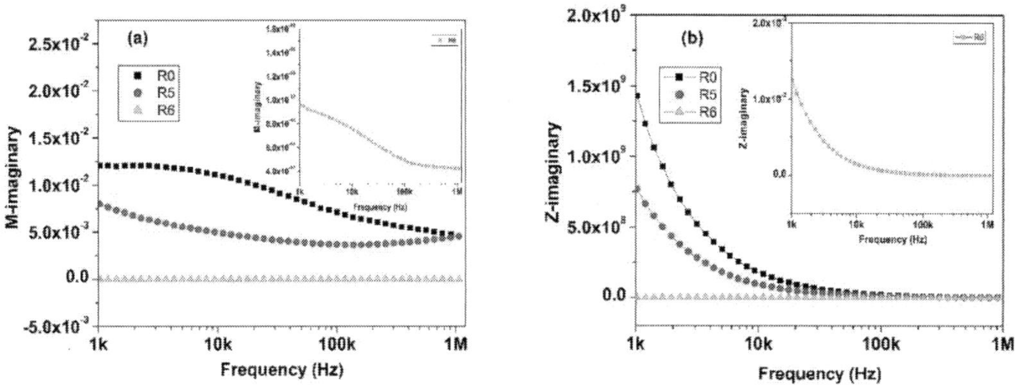

Figure 3. Variation of imaginary M and Z for R0, R5 and R6 samples

3.2. Electrical properties

Dielectric constant and tangent loss 'tanδ' was measured over a range of temperature from 30°C to 150°C in steps of 10°C. Figure 2 shows the variation of dielectric constant and tanδ as a function of temperature at 1kHz, 10kHz, 100kHz and 1MHz. Dielectric dispersion can be observed in all the samples which prominent at higher temperatures. It can be attributed to the presence of the small percolated structure of particles. In these types of composites, the dielectric constant is mainly due to the orientation and interfacial polarization. At low frequencies these polarizations result in the increase of charge on the surface of dielectric material which get restricted to contribute at relatively higher frequencies. As a result of which the dielectric constant decreases. A high value of dielectric constant can be observed in R5 sample as compared to R0 and R6 samples. This increase in dielectric constant can be mainly attributed to formation of micro/nano-capacitor structures in the PVDF due to increase in the mass fraction of conductive phase. Furthermore, the dielectric dispersion is also small in it in lower temperature range as compared to other samples.

Tangent loss also shows the similar variation with temperature for all samples. It can be observed in R5 sample that initially the dielectric constant increases slowly with temperature for all the frequencies. However, it increases with higher rate at elevated temperatures especially at lower frequency. This behavior shows the thermal stability of the dielectric constant which makes the material more suitable for thermally stable capacitive applications. Therefore, it can be concluded that dielectric properties improved with GO filling up to a PVDF:GO ratio equal to 50:50. The value of tanδ is also quite small for this sample [11].

From the measured dielectric constant, complex impedance 'Z' and capacitive modulus 'M' were calculated. Being a more informative component, the imaginary part of M and Z are very important to study. Figure 3 shows the variation of imaginary part of M and Z as a function of frequency at room temperature (30°C). Like dielectric parameters, the value of these parameters are out of trend for sample R5 as compared to R0 and R6. The variation of imaginary M with frequency is quite similar for R0 and R6 samples. It is nearly constant over a small frequency range and then decreases, while a continuous decrease in its value can be noticed in sample R5. This also represents the formation of micro capacitors due to filling of conductive material GO in PVDF. The variation of imaginary Z is almost similar for all the three samples, which shows a regular decrease with frequency. The improved dielectric properties, impedance and capacitive modulus makes the material suitable for capacitive applications.

IOP Conf. Series: Materials Science and Engineering 1221 (2022) 012025 doi:10.1088/1757-899X/1221/1/012025

4. Conclusion

Flexible composite of polyvinylidene fluoride 'PVDF' loaded with graphene oxide 'GO' as reinforcement were prepared using solution casting method. The structural properties were analysed using X-ray diffraction investigations. Dielectric constant 'ε' and tangent loss 'tanδ' was found to be improved for sample R5. The evaluated results suggest that the prepared composites for flexible electronic applications.

References

[1] Pazhamalai P, Mariappan V K, Sahoo S, Kim W Y, Mok Y S and Kim S Je 2020 Micromachines 11 198.
[2] Patodia T, Sharma K B, Dixit S, Srivastava N, Katyayan S, Jain S K, Tripathi B 2020 AIP Conference Proceedings 2265 030630.
[3] Abdelhamid E H, Jayakumar O D, Kotari V, Mandal Balaji P, Rao R, Naik V M, Naik R and Tyagi A K 2016 RSC Adv. 6 20089.
[4] Moradi R, Karimi-Sabet J, Shariaty-Niassar M and Koochaki Mohammad A 2015 Polymers 7 1444.
[5] Hasegawa R, Takahashi Y, Chatani Y and Tadokoro H 1972 J. Polym. 3 600.
[6] Levi N, Czerw R, Xing S, Iyer P and Caroll D L 2004 Nano Letters 4 1267.
[7] Takashima K, Horie S, Mukai T, Ishida K and Matsushige K 2008 Sensors and Act. A 144 90.
[8] Islam A, Khan A N, Shakir M F and Islam K 2020 Mat. Res. Exp. 7 015017.
[9] He Z Z, Yu X, Yang Z-hui, Zhang N, Huang T, Wang Y and Zhou Z-wan 2018 Composites Part A: App. Sci. Man. 104 89.
[10] Wu W, Tu J, Li H, Zhan Z, Huang L, Cai Z, Li Q, Jiang M and Huang J 2020 J. of Mat. Sci: Mat. Elect. 31 807.
[11] Kumar KS, Pittala S, Sanyadanam S and Paik P 2015 RSC Adv. 5 14768.

IOP Conf. Series: Materials Science and Engineering 1221 (2022) 012026 doi:10.1088/1757-899X/1221/1/012026

Fabrication of Janus Liquid marbles With Tunable Particle Coverage by Controlling the Surface Tension of the Core Liquid

Bindhu Sunil Kumar Lekshmi[1] and Subramanyan Namboodiri Varanakkottu*[1]

[1]Department of Physics, National Institute of Technology Calicut, 673601, INDIA
* varanakkottu@nitc.ac.in

Abstract. Janus Liquid marbles (JLMs) are next-generation liquid marbles enwrapped with multi-particles of different sizes, hydrophobicity, and/or chemical properties. The functionally distinct surface of Janus Liquid marble enables the selective and precise manipulation of liquidmarble in a controllable manner. Here, we present an efficient method to tune the particle coverage of Janus Liquid marble by controlling the surface tension of the core liquid. The impact-based fabrication strategy is used for the continuous productionof JLMs of different volumes. The tunability in particle coverage is achieved for JLMs of volumein the range 5-10 μL. A simple energy balance analysis is employed for the theoretical estimation of the maximum spread parameter (β_m) and found to be in reasonableagreement with the experimentally measured value.

1. Introduction

Liquid marbles (LMs) are millimeter-sized liquid drops encapsulated by hydrophobic particles of submicron to micron dimensions.[1,2] The particles envelope entraps air among particle aggregates which completely isolates inner liquid from the underlying substrate, which facilitates the manipulation of LMs like a soft solid.[3] The air-cushioning enables LMs to be completely non-wetting (including liquid substrates) soft solids but also facilitates reduced resistance to the motion,[4,5] endowing LMs the functionalities like for cargo transport and delivery.[6] LMs are emerging as a potential candidate in various applications such as gas sensors,[7] micro-reactors,[8] micro pumps,[9]and micro-bioreactors.[10] Control over actuation and manipulation of LMs can further be improved by multi-particle wrappings, which are generally referred to as Janus Liquid Marbles (JLMs).[11]

JLMs are capable of providing precise and improved manipulability than pristine LMs. The control over the particle coverage on JLM will enhance performance in selective manipulability. Several methods have been developed to fabricate JLM, including coalescence,[11] impact-based,[12], and magnetic-induced methods.[13] In the coalescence method, the tunability in particle coverage is realized by changing the volume of coalescing LMs, whereas, in the impact-based fabrication method, the impacting droplet volume and Weber number dictate the particle coverage tunability. Though the impact-based method and coalescence method facilitate the tuning of particle coverage of JLM by varying the droplet volume, the fabrication of JLM having different particle coverage for a fixed volume of the LMis not explored yet. Herein, we propose a simple, impact-based fabrication of Janus LMs of the same volume with tunable particle coverage by controlling the surface tension of the inner liquid. It is a single-step process yielding tunable particle coverage by varying the surface tension of inner liquid drop.

IOP Conf. Series: Materials Science and Engineering 1221 (2022) 012026 doi:10.1088/1757-899X/1221/1/012026

2. Working Principle

The working principle of the impact-based Janus fabrication technique relies on the particle redistribution on a pre-coated LM due to the internal and surface flows created upon its impact on a solid surface at a We ≥30.[14] The impact dynamics of particle-laden droplets on a solid surface could be explained based on forces acting on it, the capillary, inertial and viscous forces.[15] Before the impact, the droplet possesses inertial energy and surface energy. When the particle-laden liquid drop hits the solid surface at a moderate Weber number (≥30), the droplet spreads in the radial direction, which modifies the distribution of the adsorbed particles. Under this condition, the interplay of capillary forces set off a vortical motion inside the droplet, and the surface flow distribution directs towards the droplet-squeezing zone, as shown in figure 1. This result in the assembly of particles adsorbed at the liquid-air interface near the drop-squeezing zone at the maximum spreading stage. Beyond this point, the capillarity dominates and enables the droplet to retract and regains its shape leading to the redistribution and assembly of particle ring towards the center of the drop surface.

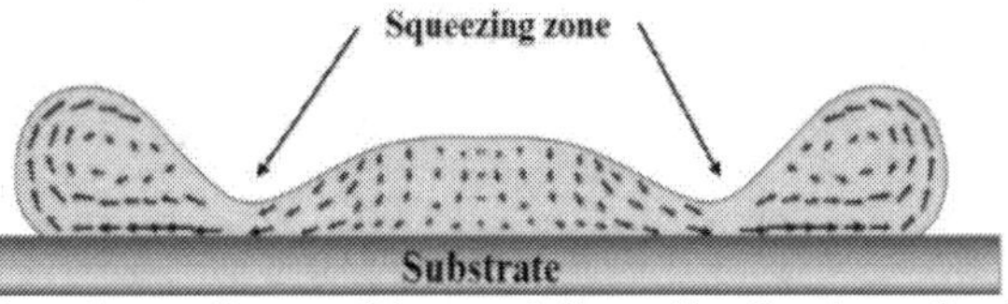

Figure 1: Schematic showing the surface and internal flow distribution of a droplet upon impact on a hydrophobic substrate at a Weber number We = 30 and Reynolds number Re = 500.[15]

3. Experimental method

In this work, the impact-based fabrication method was employed for the fabrication of stable JLMs(See SI for more details). The liquid droplet of volume 5-10 μLis dispensed using a micropipette on the PVDF particle bed. The droplets collect particles while rolling and form a sparsely covered LM. The subsequent impact of sparsely coated LM on the second particle bed results in the formation of JLM. To fabricate JLM of inner liquid with low surface tension, a surfactant solution (0.4 mM) was prepared with anionic surfactant Sodium dodecylbenzenesulfonate (SDBS) (Sigma-Aldrich). A high-speed camera (Promon 501, AOS) is positioned vertically above the BTP bed to capture the impact dynamics at 1374 and 1453 fps. A USB camera (Celestron44308) was used to capture the formed JLMs. The surface tension of the droplet was measured using the pendant drop plugin in ImageJ software.

4. Results and Discussion

First, we investigated the impact dynamics of a particle-laden droplet with deionized water and surfactant solution as inner liquid on the BTP pellet (contact angle 105^0). A 10 μL deionized liquid droplet is dispensed on the PVDF particles bed. Owing to their hydrophobicity, the particles adhere to the droplet while rolling and form a sparsely covered liquid droplet. This sparsely covered droplet is allowed to impact on the BTP pellet at a Weber number of 35±2.

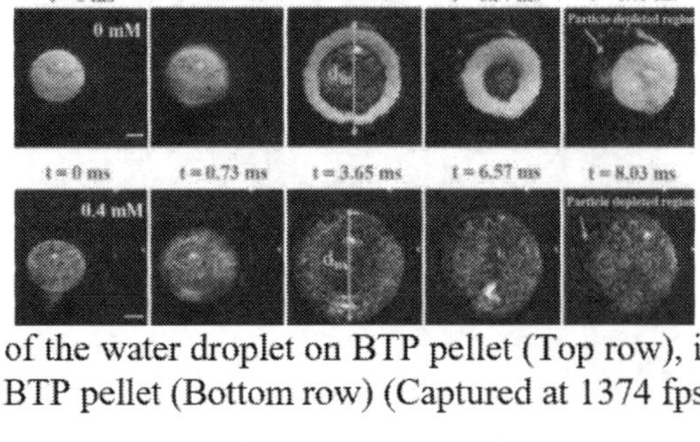

Figure 2: Impact dynamics of the water droplet on BTP pellet (Top row), impact dynamics of 0.4 mM SDBS surfactant droplet on BTP pellet (Bottom row) (Captured at 1374 fps)(Scale bar-1 mm).

After the impact, the liquid starts spreading in the radial direction and reaches maximum spread at t = 3.65 ms, where momentarily inertial energy vanishes, as shown in figure 2. The particle adsorbed on the liquid-air interface follows the surface and internal flows and forms a ring-like assembly. Beyond this point, the capillary force dominates, and with the onset of surface energy pull, the liquid starts retracting, and the particles ring collapses at t = 8.03 ms. Then the experiment is carried out using 0.4 mM of SDBS surfactant solution as inner liquid. The surface tension corresponds to 0.4 mM SDBS solution was measured to be 57.8±1 mN/m. Similar to the water droplets, the impact dynamics of surfactant solution also showed the assembly of the particle at the droplet crust. Figure 2shows the impact dynamics of both particle-laden water droplets and surfactant droplets.

Further, for the fabrication of JLM, we performed the experiment with BTP bed. A 10 µL deionized liquid droplet sparsely covered with PVDF particles is allowed to impact on the BTP bed at Weber number 35±2, as shown in figure 3(top row). Upon impact, the droplet deforms, and the liquid starts spreading radially outwards along with particles and reaches maximum spread. At the maximum spreading stage, at t = 2.72 ms, a raised rim is formed at the periphery with particles assembled on the rim. The liquid in contact with black toner particles collects particles from the bed. On retraction and rebound, the droplet regains its spherical shape and forms a Janus liquid marble.

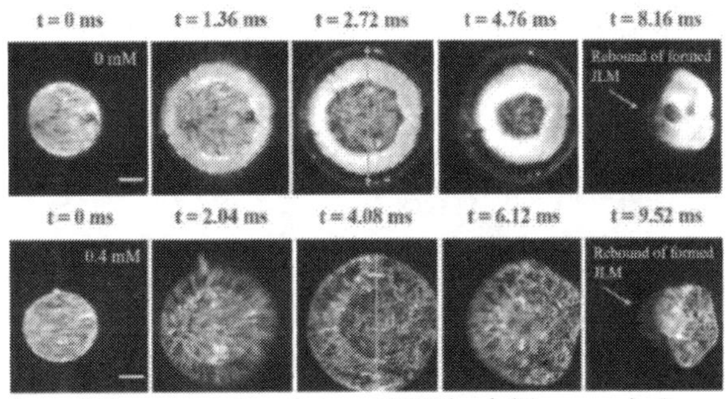

Figure 3: Impact dynamics of the water droplet on BTP bed (Top row), Impact dynamics of 0.4 mM SDBS surfactant droplet on BTP bed (Bottom row) (Captured at 1453) (Scale bar-1 mm)

To investigate the effect of surface tension on the impact dynamics of the particle-laden droplet, 0.4 mM of SDBS aqueous solution of surface tension 57.8±1 mN/m is prepared. A 10 µl of 0.4 mM aqueous solution is dispensed on the PVDF particles bed. The addition of surfactant to the liquid droplet reduces the droplet's surface tension, thereby minimizing the total energy. Besides, the reduction in the surface tension of the liquid droplet increases the wettability of the particles, thereby allowing the particles to submerge into the bulk liquid. Thus the surfactant droplets collect fewer particles from the particles bed compared to the water droplet, and this attachment further minimizes the total free energy of the system. This facilitates controlling the particle coverage of the Janus liquid marble by adding a surfactant to the inner liquid. Upon impact, the aqueous droplet starts spreading. At t = 4.08 ms, the submerged particles form a ring like assembly of particles. On retraction, the submerged PVDF particles are assembled to the droplet crest. Compared to the pristine particle-laden droplet, the surfactant droplet collects fewer particles from the first particle bed since the droplet is already in its minimum energy state.

Further, we characterized the PVDF particle ring formation after the droplet impactby plotting the normalized radial intensity profile of the droplets at their maximum spreading stage, as shown in figure 4. We have chosen the image that corresponds to t=2.72 ms for the water droplet and t=4.08ms for the surfactant added droplet. The contrast of images was enhanced by 10%, and the profile is plotted using the radial profile plugin in ImageJ software. The radial profile of both the droplets showed an amplified intensity at the periphery, confirming the assembly of particles at the rim.

However, the peak intensity is diminished in the case of surfactant solution, indicating low particle coverage compared to pure water.

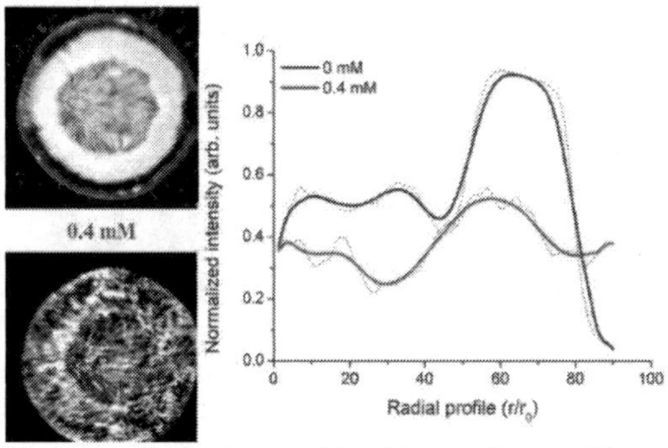

Figure 4: Contrast enhanced images (left). Normalized intensity profile at maximum spreading stage. The blue and red line corresponds to the 10 point averaging of the original data points represented in dotted lines (right).

To characterize the impact dynamics, the maximum spread parameter $\beta_m$is estimated experimentally as well as theoretically. The maximum spread parameter is the ratio of maximum spreading diameter to the diameter of the spherical droplet prior to the impact. The dimensionless maximum spread parameter can be estimated theoretically using different approaches. The commonly employed methods are energy balance analysis and momentum balance analysis. Here, we use the model formulated by Mozhi et al. on the basis of energy balance analysis for the calculation of the maximum spread parameter.[16]
The model balances the inertial and surface energy prior to the impact to the total energy at the maximum spreading stage, assuming that the drop forms a cylindrical shape during the maximum spreading stage.

$$\mathrm{Energy}_{\text{Prior to impact}} = \mathrm{Energy}_{\text{At the maximum spreading stage}}$$

$$E_{ki} + E_{si} = E_{Smax} + E_d^V + E_d^C + E_d^H \quad (1)$$

Where E_{ki} = mgZ represents kinetic energy of the droplet (m- mass of the droplet, Z- impact height), $E_{si} = \pi D_0^2 \gamma_{LV}$, the surface energy of the droplet prior to impact (D_0 – diameter of the spherical droplet,γ_{LV} – liquid- vapor interfacial energy), $E_{Smax} = \frac{\pi D_0^2}{4}(1 - cos\theta)\gamma_{LV}$he surface energy of liquid at maximum spreading stage, $E_d^V = \frac{6\beta_m^4}{Re} E_{ki}$, $E_d^C = k\left(\frac{B_0}{We}\right)^{\frac{1}{3}} E_{ki}$, energy loss due to viscous dissipation and $E_d^H = \frac{1}{2} E_{ki}$, energy loss due to internal flows.
Simplifying equation (1) gives,

$$\frac{6.36}{Re}\beta_m^4 + \left[\frac{6}{We}\right]\beta_m^2 - \left[\frac{12}{We} + \frac{1}{2} - k\left(\frac{Bo}{We}\right)^{\frac{1}{3}}\right] = 0 \quad (2)$$

Solving equation (2) for β_m^2 gives the value forβ_m. Here $Re = \frac{\rho v d}{\mu}$ represents Reynolds number (ρ- density of the liquid, v- impact velocity, d- diameter of the droplet), $We = \frac{\rho v^2 d}{\gamma}$ represents Weber number(γ- surface tension the liquid), $k = 0.39\left(\frac{\varphi \rho_p}{\rho_w}\right)^{\frac{1}{3}}$ represents the crater dissipation constant (φ- packing fraction, ρ_p- density of the particles, ρ_w- density of liquid), $B_0 = \frac{\Delta\rho d^2 g}{\gamma}$ represents the bond number. The values of density, dynamic viscosity, and surface tension of water and 0.4 mM surfactant solution used are 997 Kg/m^3, 10^{-4} Pa s, 0.072 N/m and 0.057 N/m. Solving equation (2) gives the

IOP Conf. Series: Materials Science and Engineering 1221 (2022) 012026 doi:10.1088/1757-899X/1221/1/012026

theoretical value of maximum spread for 5µL, 7 µL, and 10 µL liquid droplets with surface tension values 0.072 N/m and 0.057 N/m for each volume. The theoretical and experimental values of $\beta_m$are listed in table 1.

Table 1: The theoretical and experimental value of maximum spread **(β_m)**on particle bed for volume 5-10 µL. Each measurement represents the mean of five individual measurements with the standard deviation as the error bar.

Volume (µL)	**Spreading diameter on BTP bed (mm)**		**Maximum spread on BTP bed(β_m)**			
			0 mM		**0.4 mM**	
	0 mM	**0.4 mM**	**Theoretical**	**Experimental**	**Theoretical**	**Experimental**
5	3.52±0.17	3.34±0.04	2.00	1.66±0.1	1.99	1.57±0.02
7	4.04±0.07	3.84±0.13	2.01	1.70±0.03	1.99	1.62±0.05
10	4.75±0.22	4.54±0.12	2.02	1.77±0.1	1.98	1.7±0.05

The discrepancy between the theoretical and experimentally estimated values of $\beta_m$can be attributed to the cylindrical shape assumption employed in the model. In the real scenario, a thick rim is formed near the edge with the accumulation of particles, which is not considered in the model.
Since the particle coverage of JLM depends on the weight percentage of particles collected from both beds, we were able to tune the particle coverage by lowering the surface tension of the inner liquid. As the surface tension of water was reduced from 72 mN/m to 57.8 mN/m, the weight percentage of particles collected from the PVDF bed decreased significantly. To demonstrate the tunability of particle coverage using the presented method, we fabricated JLMs of different volumes, and the results are presented in figure 5.

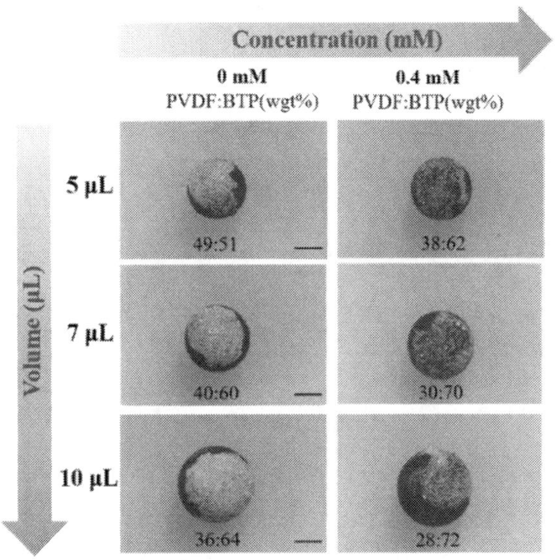

Figure 5: Janus liquid marbles (5-10µL)showing particle tunability.(Scale bar-1 mm)

Table 2: The weight percentage represents the mean of five individual measurements with the standard deviation as the error bar.

Volume (µL)	**Weight percentage PVDF : BTP**	
	0 mM	**0.4 mM**
5	49±4:51±4	38±2:62±2
7	40±4:60±4	30±3:70±3
10	36±2:64±2	28±3:72±3

5. Conclusion

In conclusion, we developed an efficient strategy for tuning the particle coverage of JLMs by controlling the surface tension of the inner liquid. We used the impact-based fabrication method for the production of JLMs with water and 0.4 mM surfactant as the inner liquid. The addition of surfactant led to the minimization of total energy, thus collecting lesser particles from the first PVDF particles bed compared to the water droplet. This allowed us to tune the particle coverage. To show the capabilities of the developed method and realize the tunability in particle coverage, the JLMs of different volumes are fabricated. Energy balance analysis was employed to estimate the value of maximum spread (β_m) theoretically, and it was found to be in a reasonable agreement with the experimentally measured value.The study of LMs is now at the forefront of soft matter research, and we believe these functionally distinct JLMs can significantly contribute to the progressing fields of microfluidics.

6. Acknowledgment

SNV sincerely acknowledges the funding from the Department of Science and Technology, India, through an INSPIRE Faculty award (Faculty Award/2016/DST/INSPIRE/04/2015/000544). The authors acknowledge Yarra Siddhartha and Vankudoth Naveen Kumarfor their preliminary experiments.

References

[1] Aussillous P and Quéré D 2001 *Nature***411** 924
[2] Aussillous P and Quéré D 2006 *Proc. R. Soc. A***462** 973–99
[3] Bormashenko E 2011 *Current Opinion in Colloid & Interface Science***16** 266–71
[4] McHale G and Newton M I 2011 *Soft Matter***7** 5473–81
[5] Fullarton C, Draper T C, Phillips N, Mayne R, de Lacy Costello B P J and Adamatzky A 2018 *Langmuir***34** 2573–80
[6] Singha P, Swaminathan S, Yadav A S and Varanakkottu S N 2019 *Langmuir***35** 4566–76
[7] Tian J, Arbatan T, Li X and Shen W 2010 *Chem. Commun.***46** 4734–6
[8] Arbatan T, Li L, Tian J and Shen W 2012 *Advanced Healthcare Materials***1** 80–3
[9] Bormashenko E, Balter R and Aurbach D 2010 *Appl. Phys. Lett.***97** 091908
[10] Arbatan T, Al-Abboodi A, Sarvi F, Chan P P Y and Shen W 2012 *Advanced healthcare materials***1** 467–9
[11] Bormashenko E, Bormashenko Y, Pogreb R and Gendelman O 2011 *Langmuir***27** 7–10
[12] Lekshmi B S, Yadav A S, Ranganathan P and Varanakkottu S N 2020 *Langmuir***36** 15396–402
[13] Xu Z, Zhao Y, Dai L and Lin T 2014 *Particle & Particle Systems Characterization***31** 839–42
[14] Wildeman S, Visser C W, Sun C and Lohse D 2016 *J. Fluid Mech.***805** 636–55
[15] Mao T, Kuhn D C S and Tran H 1997 *AIChE* Journal**43** 2169–79
[16] Mozhi Devan Padmanathan A, Sneha Ravi A, Choudhary H, Varanakkottu S N and Dalvi S V 2019 *Langmuir***35** 6657–68

IOP Conf. Series: Materials Science and Engineering 1221 (2022) 012027 doi:10.1088/1757-899X/1221/1/012027

Simulation of Optical Emission Spectra of Cu-Sn-Zn Alloy Plasmas for Laser-Induced Breakdown Spectroscopy Applications

M S Suryan Sivadas1, Lekha Mary John1 and K K Anoop1, *
[1]Department of Physics, Cochin University of Science and Technology, Kochi-682 022
E-mail: *anoopkk@cusat.ac.in

Abstract. Atomic emission spectroscopy (AES) is a potential tool for qualitative and quantitative elemental analysis of multi-element materials. The emission spectrum is composed of the characteristic lines of the constituent species, and their intensities are proportional to the population of their energy states and transition probability. We simulated the emission spectrum of copper-tin-zinc alloys under typical laser-induced plasma conditions (e.g.T_e = 1 eVand $N_e = 10^{17}\,\mathrm{cm}^{-3}$). Saha LTE equation determines the population of neutral and ionised species for each of these constituent elements. The Boltzmann distribution is used to calculate the strength of their probable transitions, and each line is given a broadened Gaussian profile. Their sum will approximate an ideal plasma spectrum of the alloy. Spectra are generated for varying temperatures, electron density, and compositions and show variations based on these parameters. This simulation provides a simple and convenient way to perform laser-induced breakdown spectroscopy (LIBS) applications on any alloy sample.

1. Introduction

Atomic emission spectroscopy (AES), is a widely used tool for qualitative and quantitative analysis of multi-element samples like alloys. In AES, samples are excited to a high-temperature plasma state, through electrical discharge or laser ablation. This plasma contains its constituent species' in their excited states and they will emit radiation upon returning to their ground state. This emitted light corresponds to the characteristic emission of lines of these species' and constitute the plasma emission spectrum. Intensity of these lines are proportional to the population of the excited energy state and transition probability. This makes the emission spectra useful for the qualitative and quantitative analysis of these samples.Laser-Induced Plasma (LIP) is a transient micro plasma that shows spatial and temporal inhomogeneity. Typical laser produce plasma contains 10^{17} to 10^{19} electrons per unit volume (cm^{-3}), and plasma temperature ranges from 6000 K to 20000 K[1]. And they mainly depend on laser pulse characteristics (pulse width, pulse energy, and wavelength), ambient conditions (pressure), and sample (physical and chemical properties). Electron number density and plasma temperature vary in the different temporal and spatial windows; due to this nature, a complete thermodynamic equilibrium (CTE) state cannot be explained for laser-produced plasma; therefore, we use the Local Thermodynamic Equilibrium (LTE) assumption[2]. To give a correct interpretation of plasma spectra and for the meaningful application of theoretical Boltzmann–Maxwell and Saha–Eggert expressions, the existence of LTE is essential[3].

In this work, we simulated emission spectrum of some copper-tin-zinc alloys under typical laser-induced plasma conditions. We have used Saha's LTE equation and Boltzmann statistics to estimate the intensity of possible transitions and gave a broadened Gaussian profile to each line.

2. Simulation Details

In this section, we describe the steps in the plasma spectrum simulation in detail. There are three input variables, which specifies the condition of the plasma. These are the plasma temperature (T_e) number density of electrons in the plasma (n_e) and the fractional composition of each constituent element in the plasma. This composition is expressed as *weight fractions*, (denoted as w_X for an element $X$and$\sum_X w_X = 1$). For calculations, we need the number *fraction*, which can be computed from the

IOP Conf. Series: Materials Science and Engineering 1221 (2022) 012027 doi:10.1088/1757-899X/1221/1/012027

weight fraction, knowing the atomic mass of the element as and the proportionality is set by the normalization. Besides these inputs, we need a database of all possible transitions of the constituent elements and related information like transition probability, level energies and corresponding degeneracies. These data were taken from NIST atomic spectroscopy database[4] and Kurucz database[5].

2.1. Step 1: Estimating Contributions from Different Species

In the plasma, an element X can exist in its neutral state or any of its ionized states. All these states will have their contribution to the plasma spectrum, based on their abundance. So, the first step is to quantify the abundance of each ionized species. For this, we assume *the plasma to be in LTE*. Then, we can use *Saha's ionization equation*, which relate the abundance of a certain ionization state to the next higher state of the same element in a plasma at LTE, at some temperature[6,7]. It is given by

$$\frac{n_{X,i+1}}{n_{X,i}} = 6.009 \times 10^{21} \frac{T_e^{3/2}}{n_e} \cdot \frac{U_{X,i+1}(T_e)}{U_{X,i}(T_e)} \cdot \exp\left(-\frac{I_{X,i}}{T_e}\right) \quad (1)$$

Where $n_{X,i+1}$ is the number fraction, $U_{X,i}$ is the partition function and $I_{X,i}$ is the ionization energy of the i-th ionization state of element X. (In this section, we adopt the notation X^{i+} to denote the *i*-th ionization state of an element X. The 0-th ionization state is the neutral species.) With these ratios and the relation $\sum_i n_{X,i} = n_X$,we can estimate the fractional contribution of each state of a particular element. In our simulation, we only considered the first two ionization states of each element and their abundances can be found to be.

$$n_{X,0} = \frac{n_X}{1+y} \text{ and } n_{X,1} = y n_{X,0}, \quad \text{where } y = \frac{n_{X,1}}{n_{X,0}} \quad (2)$$

2.2. Step 2: Estimation of Peak Line Intensities

As mentioned earlier, each constituent species in the plasma has their own contribution to the spectrum. Each of their characteristic transitions appears in the spectrum, with intensities depending on their population and the probability of that transition. Let us consider the speciesX^{i+}. If it has a transition$u \to l$, with a probabilityA_{ul}, resulting in a line at wavelengthλ_{ul}, then its intensity can be estimated as[6].

$$I_{ul} \coloneqq I(\lambda_{ul}) = \frac{1}{4\pi} n_{ul} A_{ul} \epsilon_{ul} \quad (3)$$

Here, $\epsilon_{ul} = E_u - E_l$ is the energy of the emitted photon, which is the difference between the upper and lower-level energies and n_u is the fractional population of the upper level u of that species. The latter depends on the specific statistics the particles (here, atoms and ions) follows. For a fairly high temperature plasma, one could use the *Boltzmann statistics*, which gives the population to be

$$n_u(T) = \frac{n}{U(T)} g_u \exp\left(\frac{-E_u}{T}\right), \quad \text{where } U(T) = \sum_j g_j \exp\left(\frac{-E_j}{T}\right) \quad (4)$$

Here n is the total population of the species and $g_u$is the degeneracy for energy level E_u. Required spectroscopic data are collected from NIST Atomic spectra database[4]and Kurucz atomic line list[5]. With this, intensities of all the available lines at a specific set of temperature and electron density are pre-computed so that spectra can be generated at various wavelength ranges as desired.

2.3. Step 3: Generating the Final Spectrum

Last step in the simulation is to generate the final spectrum in the desired wavelength range. For this, we give each line a broadened shape and sum over all available lines to produce the output spectrum.

Known mechanisms of spectral broadening are the *Doppler* and *Stark* mechanisms. But, for simplicity, we only consider a broadening that is attributed to the *resolving power* of the spectrometer, given by $R = \lambda_0/w$. Here, w is *the full width at half maximum*(FWHM) of the broadened line and λ_0 is its central wavelength[8]. We have chosen a Gaussian line shape function for each line, given by (see, for example,[9].

$$I(\lambda) = I_0 \exp\left[-\frac{1}{2}\left(\frac{\lambda-\lambda_0}{s}\right)^2\right] \tag{5}$$

Where $w = 2s\sqrt{2\ln(2)}$ and I_0 is the peak intensity of the line, as given by Eqn. (3). We have also assumed the plasma to be optically thin, so that effects such as self-absorption can be avoided. Final spectrum will be the sum of all the available line profiles. The result will resemble the plasma spectrum in ideal conditions.

3. Results and Discussion

Using the method described in the former section, we have simulated spectra for three different alloys, namely *brass* (65% copper and 35% zinc by weight), *bronze* (88% copper and 12% tin) and *admiralty brass* (69% copper, 30% zinc and 1% tin). Simulations are done at temperatures of 0.5 and 1 eV, with an electron density of 10^{17}cm^{-3}. Simulated spectra are presented in Fig. (1a), (1b) and (1c).

We got more intense emission in the shorter wavelength (UV) region at higher temperatures. This is because the higher energy states becomes more populated at higher temperatures, and this results in more intense UV emissions. Also, as a result of more ionizations, transitions of higher ionized species becomes intense at these temperatures.

The reported emission lines of LIP spectra of copper are 324.754 nm, 327.396 nm, 465.112 nm, 510.554 nm, 515.324 nm, 521.820 nm, and 578.213 nm; of zinc are 330.293 nm, 334.501 nm, and 472.215 nm, and 481.05 nm; and of tin are 270.65 nm, 283.997 nm, 286.331 nm, 300.913, 303.411nm and 326.233 nm[10,11]; and they are present in the simulated spectrum as well. Spectroscopic information on these lines are given in Table (1). There are also some lines which are closely packedand combined into a single line and are practically indistinguishable. Therefore, we can infer that simulations in AES give prior knowledge of intense characteristics emission of any samples.

These simulated plasma spectra are useful for identifying possible lines we could observe in a real life spectrum. Then we could choose which lines are useful for the qualitative and quantitative analysis of multi-element samples in Laser-Induced Breakdown Spectroscopy (LIBS). In LIBS, observed intensities of known lines of the component elements are used to estimate plasma conditions and composition of the sample.

IOP Conf. Series: Materials Science and Engineering 1221 (2022) 012027 doi:10.1088/1757-899X/1221/1/012027

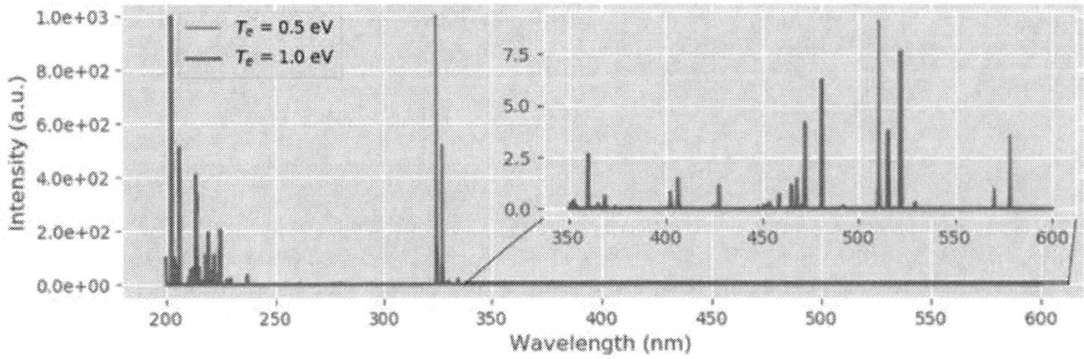

(a) Brass

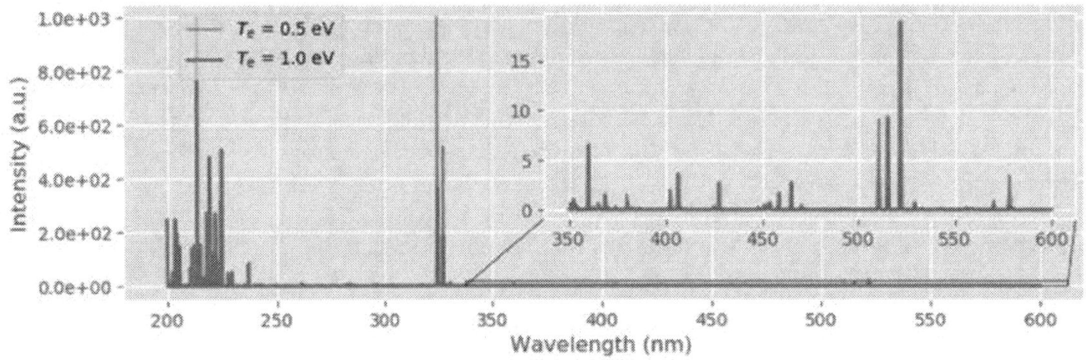

(b) Bronze

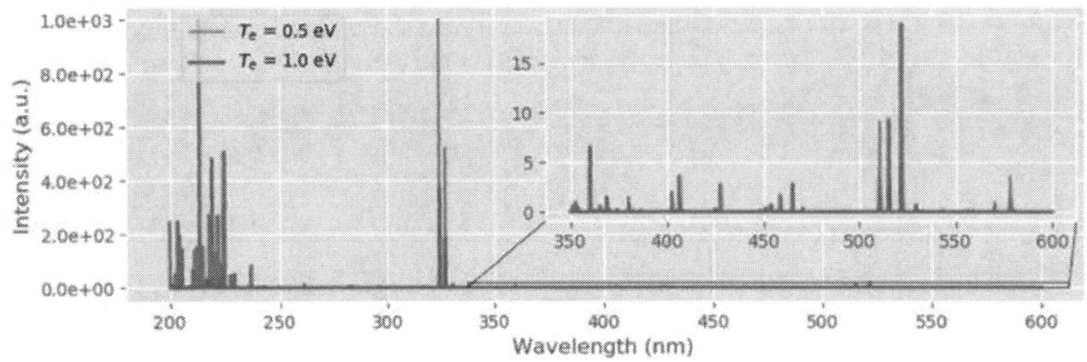

(c) Admiralty Brass

Figure 1: Simulated spectrum of the three alloys at temperatures of 0.5 and 1 eV ($n_e = 10^{17} cm^{-3}$ and R = 1000). An enlarged view of 350-600 nm region in inset. All spectra are normalised so that the strongest line have intensity 1000.

Table 1:Persistent lines that are observed in the spectra.Lines of a specific species will bepresentonlyifthatisacomponentofthealloyconsidered.

Species	λ(nm)	$A_{ul}(s^{-1})$	E_l(eV)	E_u(eV)	g_l	g_u
CuI	324.754	1.395e+08	0.00000	3.81669	2	4
CuI	327.396	1.376e+08	0.00000	3.78590	2	2
CuI	465.112	3.800e+07	5.07209	7.73703	10	8
CuI	510.554	2.000e+06	1.38895	3.81669	6	4
CuI	515.323	6.000e+07	3.78590	6.19118	2	4
CuI	521.820	7.500e+07	3.81669	6.19203	4	6
CuI	578.213	1.650e+06	1.64223	3.78590	4	2
ZnI	330.294	6.700e+07	4.02966	7.78233	3	3
ZnI	334.501	1.700e+08	4.07788	7.78335	5	7
ZnI	472.216	4.576e+07	4.02966	6.65451	3	3
ZnI	481.053	7.004e+07	4.07788	6.65451	5	3
SnI	270.650	6.600e+07	0.20976	4.78937	3	5
SnI	283.998	1.700e+08	0.42498	4.78937	5	5
SnI	286.331	5.400e+07	0.00000	4.32882	1	3
SnI	300.913	3.800e+07	0.20976	4.32882	3	3
SnI	303.411	2.000e+08	0.20976	4.29491	3	1
SnI	326.233	2.700e+08	1.06787	4.86725	5	3

The spectrum generated here is an ideal one, assuming ideal conditions. Simulating the exact spectrum requires more theoretical considerations and additional parameters like absorptions, broadening mechanisms etc. has to be considered. But this spectrum could be a useful tool to LIBS applications. It can be used to identify strong lines that could be present in the spectrum. It can also be used to choose which spectral window is to be sampled for specific conditions of the plasma, to get enough emission lines that are useful for analysis.

4. Conclusion

We have generated the plasma spectrum for three alloys of copper, tin and zinc under different plasma conditions. These spectra has many of the characteristic lines of the constituent species' and has shown a strong dependence on the plasma temperature. More intense lines were observed in the UV region as temperature increased. The spectrum simulated here is not a replica of an exact one. But it is a useful approximation under ideal conditions. Spectral lines and their intensities encodes the qualitative and quantitative information about the plasma. In LIBS, such a simulated spectrum can be used for identifying lines that are useful for the analysis. It can also be used to tell which spectral region has to be chosen for LIBS study, for a specific plasma condition.

Acknowledgement
K. K. Anoop acknowledges the Science and Engineering Research Board (SERB) for the financial assistance through a TARE Grant (No. TAR/2018/000340).

References

[1] Capitelli M, Casavola A, Colonna G and De Giacomo A 2004 *Acta Part B: Atomic Spectroscopy***59** 271–89
[2] Cristoforetti G, Tognoni E and Gizzi L A 2013 *Spectrochimica Acta Part B: Atomic Spectroscopy***90** 1–22

[3] Cristoforetti G, De Giacomo A, Dell'Aglio M, Legnaioli S, Tognoni E, Palleschi V and Omenetto N 2010 *Spectrochimica Acta Part B: Atomic Spectroscopy* **65** 86–95

[4] Kramida A and Ralchenko Y 1999 NIST Atomic Spectra Database, NIST Standard Reference Database 78

[5] Kurucz R and Bell B 1995 Atomic Line Data *Atomic Line Data (R.L. Kurucz and B. Bell) Kurucz CD-ROM No. 23. Cambridge* **23**

[6] Griem H R 1997 Principles of Plasma Spectroscopy, Cambridge Uni *Press, Cambridge*

[7] Noll R 2012 Laser-Induced Breakdown Spectroscopy *Laser-Induced Breakdown Spectroscopy: Fundamentals and Applications* ed R Noll (Berlin, Heidelberg: Springer) pp 7–15

[8] International Union of Pure and Applied Chemistry 1996 Compendium of Chemical Terminology, 2nd ed. (the "Gold Book") (Blackwell Scientific Publications, Oxford)

[9] Weisstein E W Gaussian Function from MathWorld-A Wolfram Web Resource, [Online]. Available: https://mathworld.wolfram.com/GaussianFunction.html [2021, October 2].

[10] Achouri M, Baba-Hamed T, Beldjilali S A and Belasri A 2015 *Plasma Phys. Rep.* **41** 758–68

[11] Elhassan A, Giakoumaki A, Anglos D, Ingo G M, Robbiola L and Harith M A 2008 *Spectrochimica Acta Part B: Atomic Spectroscopy* **63** 504–11

IOP Conf. Series: Materials Science and Engineering 1221 (2022) 012028 doi:10.1088/1757-899X/1221/1/012028

Investigation on Electrical properties of PEO/CMC solid polymer blend electrolyte

Samartharama B N[1], Nagaiah N[2], Demappa T[3], Ambika M R[4] and Devendrappa H[5]

[1]Department of Physics, S.J.V.P College, Harihara, Karnataka, India
[2]Department of Physics, Jnanabharati Campus, Bangalore University, Bengaluru, Karnataka, India
[3]Department of post-graduate studies and research in Polymer science, Sir. M Vishweswaraiah PG centre, Mysore University, Tubinakere, Mandya, Karnataka, India
[4]Department of Physics, PES University, Bengaluru, Karnataka, India
[5]Department of Physics, Mangalagangotri, Mangalore University, Konaje, Karnataka, India
Email: ramaforever2007@gmail.com

Abstract: The solid polymer blend sample films of Poly (ethylene Oxide) (PEO)/Sodium Carboxy methyl cellulose (CMC) with 10/90, 30/70, 50/50, 70/30 and 90/10 compositions were prepared by solution cast technique and electrical properties were investigated by Dielectric studiesassociated with dielectric constant (ε') and dielectric loss (ε'')including AC conductivity (σ_{ac}) measurements. Dielectric constant and dielectric loss are found to decrease with the increase of frequency, and laterremains constant for all blend compositions in higher frequency range. Conductivity studyreveals that,the blend of 10/90 composition exhibits higher electrical conductivity of 2.707X10^{-5} Sm^{-1} and is found to be decreased with increase inwt.% of CMC in the blend matrix. The deviation in the trend of conductivity has been observed at 70/30 and 90/10 composition due to the miscibility.

Key words: Polymer blends, PEO, CMC, Solid polymer electrolytes, and electrical conductivity.

1. Introduction:

Polymer blending is the process by which it is possible to obtain the tailored properties of desired interest,which are not possible to achieve from individualpolymers. Industrial applications considerably increase the interest in the study ofthe polymer blends [1]. Among enormous industrial applications, production of energy storage device is afascinating and highly demanded technology utilized in batteries, capacitors, fuel cells and solar cells applications [2]. Now days, among various invented energy storage devices, the usage of solid polymer electrolytes is a promising method employed in large-scale electricity storage batteries due to its flexibility, good mechanical property, adequate conductivity, electrochemical stability, high temperature resistance, less difficulties in transportation, ease of fabrication &processing and simple maintenance features[3]. Because of these practical applications, usage of solid polymer electrolytes has gained attention of many researcherssince more than three decades [4]. Polymer blending technology offers an efficient way to fulfil such novel requirements in designing solid polymer electrolyte materialsdue potential properties and applications.

It has been found that, many researchershave worked on several host polymers like poly (ethylene oxide) (PEO), poly (aniline) (PANI), poly (vinyl chloride) (PVC), poly (vinyl acetate) (PVAc) etc. [5-8]. These investigations reveal that, among various constituent polymers used in polymer blend electrolyte preparation, poly (ethylene oxide) (PEO) exhibits excellent ion transport properties because of its special feature of high concentration of electron pairs and rapid segmental dynamics [9-11]. CMC being another constituent polymer which is a water soluble polysaccharide and is a potential candidate in view of its low cost, good abundance in nature, ability to provide good electrode-electrolyte contact and the most significant characteristic of biodegradable property [12-15].In the proposed research work, PEO/CMC blend has been considered as an interesting candidate for the preparation of solid polymer electrolyte which plays an important role in preparation of various electronic storage devices by increasing the mobility of ions which in turn enhances the conductivity.

2. Materials and Experimental methods

2.1. Materials:

In the proposed research work, we used Poly (Ethylene Oxide) with average molecular weight 200,000 procured from Sigma Aldrich and Carboxymethyl cellulose (Na-CMC) with average molecular weight of 295,188 g/mol (calculated by Mark-Houwink Sakudara equation $[\eta] = kM_w^a$, where [η] is intrinsic viscosity, k is 5.37×10^{-4} and α=0.73 are Mark-Houwink constants for CMC in water solvent) procured from SD Fine-Chem Ltd.,

2.2. Preparation:

The polymer blend films of PEO/CMC with varied compositions were prepared by solution cast method. Here, specified amount of PEO and CMC in terms of weight percentage were dissolved in double distilled water and stirred about 24 hours continuously using magnetic stirrer. The obtained solution was poured intoclean petri dish kept on well levelled table top. The same was dried at room temperature about 48 hours. During the whole process, care was taken to avoid the entry of dust or other foreign particles. Dried, free-standing films were peeled off carefully from the petri dish and stored in desiccators to avoid moisture absorption [16].

2.3. Experimental methods:

Dielectric studies: Dielectric constant, Dielectric loss and AC conductivity of PEO/CMC blend samples with varied compositions were investigated using Electrical Impedance Spectroscopy (EIS) The dielectric properties of PEO/CMC blend with varied compositions were determined by complex dielectric constant equation

$$\varepsilon_r(\omega) = \varepsilon' + i\varepsilon''$$

Where $\varepsilon_r(\omega)$ is complex dielectric permittivity,ε' and ε'' are real and imaginary components called dielectric constant thatmeasures the ability of the sample to store electric charges and dielectric lossmeasures the energy dissipated during migration or orientation of dipole respectively. Dielectric constant (ε') and dielectric loss (ε'') have been calculated using the relations

$$\varepsilon' = \frac{tC_p}{A\varepsilon_0}$$

$$\varepsilon'' = \varepsilon' tan\delta$$

Where, t is the thickness of the sample film, A is the area of cross section of the electrode used, C_p is parallel capacitance of electrolyte sample, ε_0 is dielectric constant of free space and tan δ is dielectric loss tangent or energy dissipation factor. AC conductivity has beencalculatedusing dielectric constant and dielectric loss by the relation

$$\sigma_{ac} = \varepsilon' \varepsilon_0 \omega \tan\delta$$

3. Result and Discussion:

Dielectric property: Dielectric property of materials is associated with two quantities dielectric constant and dielectric loss.As PEO and CMC were regarded as dielectric materials, PEO/CMC blend is an interesting candidate and essential to study the dielectric constant and dielectric loss of these materials. Generally, polymer exhibits higher value of dielectric constant and dielectric loss in lower frequency range and follows exponential decay upon increasing frequency of applied electric field and finally reaches saturation. The accumulation of space charges at the interface of electrode-electrolyte in low frequency range in the direction of electric field decreases the transfer of chargeswhich contributeshigh value of dielectric constant and dielectric loss. Further by increasing the frequency, the dipoles have less time to orient themselves in the direction of the field due to rapid reversal of applied electric field [17]. The same behaviour can be observed for dielectric constant and loss in studied polymer samples.

Fig (1.a & 1.b) revealthat, the dielectric constant and loss in low frequency region are high for 10/90 composition and gradually decreases upto 50/50 composition and further increases for 70/30 and decreases for 90/10 composition. The decreasing nature of dielectric constant from 10/90 to 50/50 composition is due to decreasing the wt.% of CMC in the blend matrix as it possesses high dielectric constant compared to PEO [18]. The increase of dielectric constant for 70/30 composition is due to the blend is miscible when wt% of PEO is greater than 50 wt. % as reported in previous work [19]. Further it decreases for 90/10 composition even though miscible, because of the blend consisting higher wt% of PEO which is less dielectric in nature compared with CMC. This higher value of dielectric constant signifies the presence of a greater number of charges and confirms the higher conducting nature of the specific sample [20]. Further by increasing frequency, dielectric constant and lossesexponentially decayand reaches saturation for all the blend samples.

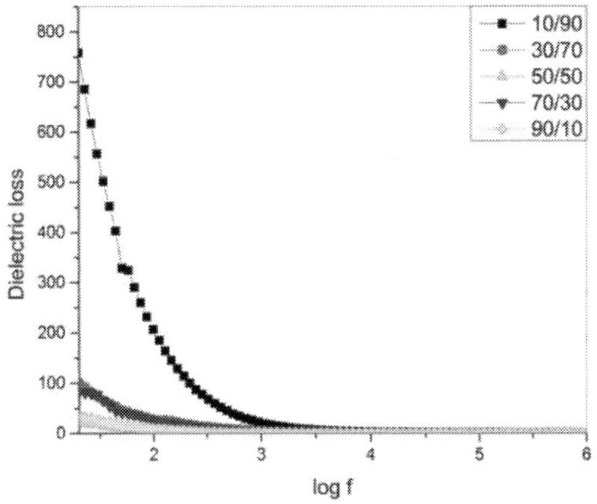

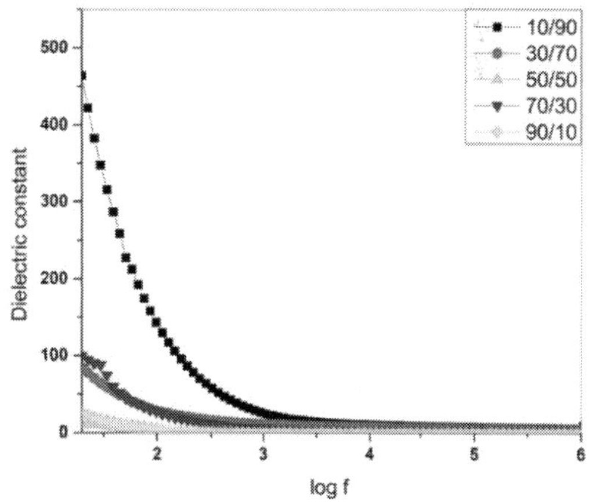

Fig (1): Variation of a) dielectric constant and b) dielectric loss with logarithmic frequency of PEO/CMC polymer blends with varied compositions.

AC Conductivity studies: Fig (2.a) reveals that, conductivity increases with increase in frequency of applied electric field. According to the literature, the electrode polarization plays a dominant role in increasing the conductivity due to acquiring sufficient time to orient the dipoles in low frequency region. The plateau region associated with intermediate frequency range contributes saturation region and signifies long range ion migration. The dispersion region contributes due to ion migration by hopping mechanism in higher frequency region, signifies the short-range ion migration.[21].

Variation of AC conductivity with increasing weight percentage of CMC in PEO matrix can be observed in fig (2.b). It is o (a) d that, for prepared blend samples, 10/90 c (b) tion shows higher value of conductivity 2.707 ... Sm^{-1} and it gradually decreases upto 50/5... position. For 70/30 composition it increases and decreases for 90/10 composition. The increase of conductivity at 70/30 composition may be attributed to the miscibility nature when it contains greater than 50 wt.% of

IOP Conf. Series: Materials Science and Engineering 1221 (2022) 012028 doi:10.1088/1757-899X/1221/1/012028

PEO[22]. Though 90/10 composition is miscible, conductivity decreases due to the presence of lower wt% of CMC.

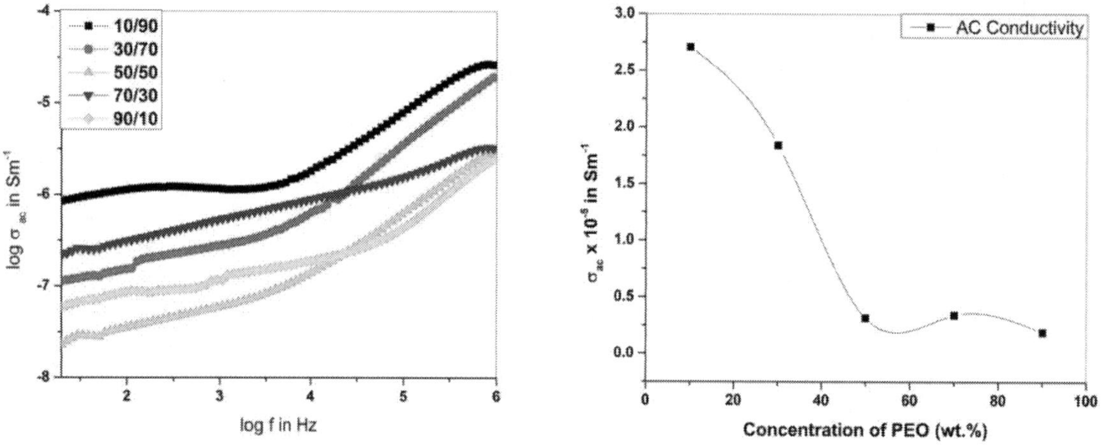

Fig (2): Variation of a) logarithmic AC conductivity with logarithmic frequency for PEO/CMC polymer blend with varied compositions b) AC conductivity with increasing concentration of CMC in PEO polymer matrix

4. Conclusion:

PEO/CMC polymer blend films with varied compositions were suc (a) ly prepared and electrical properties were investigated. Dielectric studiesshow usual behaviour ymers and in present case, dielectric constant and loss are decreasedbecause of decreasing wt.% of CMC in the blend. In addition, upon addition of 70 weight percentage of PEO in the blend, the conductivity raises because of miscibility. Further for 90 wt.% of PEO, the conductivity again decreases though its miscible composition as it consistsof minimum wt.% of CMC. This proves that, even miscibility plays an important role in enhancing conductivity, minimum wt.% of CMC causes to decrease the conductivity. Therefore 10/90 composition of PEO/CMC blend exhibit higher ionic conductivity and can be used in the preparation of solid polymer electrolyte in battery applications.

References:

[1] Elashmawi I S, Abdelrazek E M and Yassin A Y2014*Br J Appl Sci Technol.***4** 4263-4279.
[2] Rahul Singh, Anji Reddy Polu, Bhattacharya B, Hee-Woo Rhee, Canan Varlikli and Pramod K Singh2016*Renew.Sustain. Energy Rev.***65** 1098–1117
[3] Hashmi S A, Latham R J, Linford R G and Schlindwein W S1997*J. Chem. Soc. Faraday Trans***93** 4177–4182
[4] Block H and North A M1970*Advan. Mol. Relaxation Processes***1** 309-374.
[5] Kulasekarapandian K, Jayanthi S, Muthukumari A, Arulsankar A and Sundaresan B 2013*Int. j. eng. sci. invention res. dev.***5** 30-39
[6] Prajakta Joge, Dinesh Kumar Kanchan, Poonam Sharma and Nirali Gondaliya 2013*Indian J. Pure Appl. Phy.***51** 350-353
[7] Safenaz Reda M and Sheikha Al-Ghannam M2012*Adv. Materials Phy. Che.***2** 75-81
[8] Anil Arya and Sharma A L*Ionics* DOI 10.1007/s11581-016-1908-6.
[9] Manoratne C H, Rajapakse R M G and Dissanayake M2006*Int. J. Electrochem. Sci.***1** 32-46
[10] Young W S and Epps T H2009 *Macromolecules***42** 2672-2678.
[11] Chiu C Y, Chen H W, Kuo S W, Huang C F and Chang F C2004 *Macromolecules***37** 8424-8430.
[12] Ahmad, N H and Isa M I N2016 *Carbohydr. Polym.* **137** 426–432.
[13] Rani M S A, Mohamed N S and Isa M I N2015 *Int. J. Polym. Anal. Charact.***20** 491–503.

[14] Samsudin A S and Isa M I N 2014*J. Korean Phys. Soc.*, **65**, 1441–1447.
[15] Isa M I N and Samsudin A S 2016*Int. J. Polym. Mater. Polym. Biomater.* **65** 561–567.
[16] Poonam Sharma, Kanchan D K, Nirali Gondaliya, Meenakshi Pant and Manish Jayswal S 2013*Ionics***19** 301–307
[17] Kenneth S. Cole,1941*J. Chem. Phys.***9** 341-351
[18] Rania Badry, Sherif El-Khodary, Hanan Elhaes, Nadra Nada and Medhat Ibrahim 2021*Opt. Quantum Electron***3**1-15
[19] Chun-Yi Chiu, Hsien-Wei Chen, Shiao-Wei Kuo, Chih-Feng Huang, and Feng-Chin Chang 2004*Macromolecules.*, **37**, 8424-8430
[20] Azli A A, Manan N S A and Kadir M F Z 2015*Adv. Mater. Sci. Eng.*1-10
[21] Anil Arya and SharmaA L, 2019*Ionics.* 1-16
[22] Samartha Rama Banavathi Nagaraj, Nagaiah Ningaiah, Ambika Madalakote Rajanna, Hadadi Yallappa and Demappa Thippaiah2021*Int. J. Polym. Anal. Charact.***26** 561-572

One-Pot Synthesis of Biocompatible Glycine Protected Chromium Doped ZnS Nanoparticles and their Optical Properties

Vighneshwar S Bhat[1], T S Tilakraj[1], Mallikarjun K Patil[1], Vikram Pujari[1] and Sanjeev R Inamdar[1]*

[1] Laser spectroscopy programme, Dept. of Physics, Karnatak University, Dharwad 580003, India

*Corresponding author: him_lax3@yahoo.com

Abstract. Here in, we report the synthesis and characterization of Chromium doped Zinc Sulfide nanoparticles (ZnS NPs). Initially, ZnS NPs are synthesized by bio-compatible glycine cap using simple one-pot co-precipitate method, and further it is doped by Chromium. The structure and morphology of these ZnS NPs was confirmed by X-Ray Diffractometry (XRD) and Scanning Electron Microscope with Elementary Dispersive Spectrum (SEM with EDS) techniques. The optical characterization techniques reveal that the Chromium doping affected the absorption and photoluminescence properties of the NPs. Photoluminescence of these NPs shifts from 384 nm to 428 nm upon Chromium doping. By using Tauc plot we obtained the energy band gap of 4.7 eV and it reduces to 3.9 eV for Chromium dope. The resultant ZnS NPs have size of 2.17nm and 1.86nm (with Chromium doping), also it gives Cubic Zinc blend phase as proved by XRD. The instrumentation techniques SEM with EDS, XRD, FTIR confirms that high purity Chromium doped ZnS NPs can be obtained by the proposed simple, low cost and highly effective method.

Keywords: Nanoparticles (NPs); Synthesis; Biomolecule; Photoluminescence; X-Ray Diffractometry (XRD)

1.Introduction

Since the last couple of decades, nanomaterials-related research has been the most fascinating and most-pursued field among the research community. Each and every research on it exposes to work in a futuristic way. There is tremendous scope for research on the nanomaterials and their applications in various fields including biomedical, biophysics, and so on [1-2]. Nanomaterials exhibit extraordinary properties compared to their bulk counterparts and differ in many aspects; with the size, shape, and crystal structure deciding their properties and consequent stability for different applications. The nanomaterials show a high optical property, catalytic activity, electrical and magnetic properties due to their increased surface to volume ratio, charge carrier spatial confinement effect, and quantization of electronic band structure. So large number of its application-related work is going on medical, energy harvesting, solar energy, food, and communications fields [3-11]. The group II-IV semiconductor nanomaterials having the ability to shows size-dependent electrical and optical properties due to the quantum confinement effect of the carrier. In this group among all semiconducting materials like CdS, CdSe and ZnO, the Zinc Sulfide (ZnS) is the novel and most significant compound because of its direct band gap with the widest energy band gap nature. The ZnS has bandgaps 3.7eV and 3.4eV for viz. cubic structure and wurtzite crystal structure respectively. ZnS nanomaterials shows good fluorescent nature. Based on its optical properties many papers were published. Mn-doped ZnS NPs are used as effective nanoscale sensors [12]. Histidine in biological fluid can be detected using phosphorescence property of doped ZnS and so on [13-14].

In the present work we tried to synthesize the ZnS NPs in such a way that we can utilize this work in a biological approach in the future. So we have to take bio capping agent Glycine and Chromium doping during the synthesis. The Glycine is more stable, nontoxic, biofriendly and has the simplest molecular structure compare to the conventional amino acids. It can minimize the complications in the surface characterization works on the surface of the nanomaterials. The amino acids are employed for biological applications like Drug delivery, Biosensors, Imaging, and so on. Thus the building blocks of protein called amino acids may be deliberated as ideal biocompatible stabilizing agents. The physical property of the side chain may influence the interaction properties of the amino acid residues with the other compound. Bio-capped NPs synthesis work was done by some research community. Tilakraj et. al. reported Lactic Acid capped ZnS QDs shows good fluorescent behavior than normal ZnS QDs [15]. Jinwoo Park et.al. worked on photosensor and photocatalytic activities of L-Glycine capped ZnS:Mn NCs [16]. Cysteine capped ZnS NPs is an excellent fluorescent probe in the detection of nucleic acids was reported by Youngxi Li et.al. [17]. Dhrubjyoti S et.al. reported Glycine capping gives an good photo catalyst property to the host NPs [18]. We used Chromium doping to improve the optical, electronic and textural properties of the ZnS. The doping effects the grain size, structural defect and contact surface area. There are few research articles on Cr doped ZnS were reported. Palvindar kaur et.al reported as ZnS is doped by Cr^{2+} ions the ferromagnetism increases [19]. Reddy et.al. reported [20] Cr doped ZnS NPs with 2-mercaptoethanol shows band gap increases with increasing Cr concentration. Then we adopted the colloidal co-precipitation synthesis method which is the simplest and gives good results [21-24]. From literature, the obtained sample may be utilizing in biosensors, FRET and biomedical applications in future [25-26].

In this paper, we reported the synthesis protocol of the biocompatible Glycine capped undoped ZnS and Cr doped ZnS NPs. The impact of Cr ions as a dopant on structural and optical properties was investigated. The synthesized materials characterization was followed using different experimental techniques. The used techniques are Fourier Transform Infrared Spectroscopy (FTIR), X-Ray Diffraction (XRD), Scanning Electron Microscope with Elementary Dispersive Spectrum (SEM-EDS). The optical properties were studied using UV-Visible absorption and Photoluminescence measurements, which allows us to study how the Glycine cape and Cr ions doping influences the optical behavior of ZnS NPs.

2. Experimental procedure

2.1. Materials

We procured all chemicals which are used in the synthesis procedure from HiMedia©. In the experiment we used salts that are of analytical grade and without further purification. We took Zinc sulfate [$Zn(SO)_4$] as a source of Zinc ions and Sodium sulfide [NaS_2] as a source of Sulfur ions. In this reaction biocompatible organic molecule Glycine [$C_2H_5NO_2$] is used as capping agent. Chromium nitrate [$Cr(NO_3)_3$] is used as a dopant in the experiment. Pure Double distilled De-ionized water was used in all synthesis steps.

2.2. Synthesis of NPs

Glycine capped ZnS and ZnS:Cr NPs were synthesized by the chemical co-precipitation method. All synthesis steps were carried out at room temperature and throughout the experiment, we used double distilled water as a solvent. First 39.6ml of Zinc Sulfate solution in water with 0.2M was prepared in a conical flask. Taken this solution in a beaker with magnetic stirring and heated up to 70℃. The freshly prepared 0.4ml (0.2M) Chromium nitrate solution and 20ml (0.05M) Glycine solution were added to the beaker with continuous stirring and maintain temperature at 70℃. Later add 40ml (0.2M) Sodium sulfide solution added dropwise to the beaker to retain temperature and to consume all the reactants. Keep the reaction on a hot plate for 1 hour with continuous stirring. Using a centrifuge machine the formed precipitate is separated from the solution. The precipitate is washed repeatedly by using deionized water which helps to remove the unwanted water soluble byproducts. Finally using the oven we slowly dried the obtained white colored gel and make it a fine powder. The preparation of undoped Glycine capped ZnS NPs were follows the same synthesis procedure but only doping agent Chromium nitrate was not used. Finally, the powder sample of Glycine capped ZnS and ZnS:Cr NPs were obtained and preserved at room temperature for further characterization.The corresponding chemical reactions are shown in equation (1) and (2)

$$ZnSO_4 + Na_2S \xrightarrow{Glycine(C2H5NO2)\ @\ 70^{\circ}C} ZnS + Na_2SO_4 \longrightarrow \quad (1)$$

$$ZnSO_4 + Na_2S + [Cr(NO_3)_3] \xrightarrow{Glycine(C2H5NO2)\ @\ 70^{\circ}C} ZnS{:}Cr + Na_2SO_4 + NaNO_3 \longrightarrow \quad (2)$$

3. Instrumentation and methods of Characterization

The present work involves various characterization techniques to confirm the optical and physical properties of the synthesized materials. UV-Visible absorption photometry was carried out using a JASCO spectrophotometer (model V-670) which is a double beam instrument. We have used Horiba Fluoromax-4 a spectrofluorometer that is present in our lab for the study of fluorescence measurement. P L emission

IOP Conf. Series: Materials Science and Engineering 1221 (2022) 012029 doi:10.1088/1757-899X/1221/1/012029

(330nm-550nm) spectra were recorded for synthesized Glycine capped ZnS and ZnS:Cr NPs in deionized water as solvent. Fourier Transform Infrared (FTIR) spectra of synthesized materials were performed on a Nicolet FTIR Spectrometer with KBr pellet technique ranging from 500cm^{-1} -5000cm^{-1}. X-ray diffraction (XRD) pattern were obtained on a Rigaku Smart Lab SE powder XRD measurement unit. In XRD the used source having wavelength 1.54Å and scan speed was 5°/min. The microscopic image of synthesized NPs is obtained by JEOL Scanning Electron Microscope and EDS spectrum study confirms the presence of the target sample.

4. Results and discussions

4.1. Material characterization: X-Ray Diffraction studies

XRD is a powerful tool to confirm the structure and composition of any materials. The XRD study gives information on d-spacing, phase purity, crystal structure, crystal size and even material composition. Figure 1 shows the XRD patterns of Glycine capped ZnS and ZnS:Cr NPs along with the miller indices. After analyzing data from software we found that the peaks matches with the ICDD file numbers 01-080-4443 of Zinc Sulfide and 01-080-4444 of Zinc Chromium Sulfide both shows Cubic Zinc Blende structure. There are three major peaks with lattice planes (111), (220) and (311) are observed. The necessary XRD data are display in Table 1. Broadening of the diffraction peak indicates the very small sized NPs are formed. There were no additional peaks observed which means the absence of impurity phases. In the present work we observed that the diffraction peaks position shifted towards higher 2Θ values and the FWHM of ZnS:Cr NPs are large then pure ZnS, it reflecting a decrease in lattice parameters with adding Cr dopant. This happens due to the smaller radius of the Cr ions (0.63Å) than the Zn ions (0.74Å). This indicates the Cr ions has entered the ZnS lattice substitutially. The grain size of the NPs can be calculated using the Scherrer equation [i.e. equation (3)] . Due to uniform broadening the mean crystal size of undoped ZnS NPs is a (2.07nm) and that of Cr doped ZnS NPs is (1.86nm). Thus Cr doping can appreciably influence the ZnS crystallinity.

$$D_{hkl} = \frac{k_{hkl}\,\lambda}{\beta cos\theta} \longrightarrow \quad (3)$$

Where D_{hkl} is the mean crystallite size, k_{hkl} is a geometric factor and value is consider to be 0.9, λ is the wavelength of the X-ray , β be the FWHM of diffraction peak and Θ is the diffraction angle.

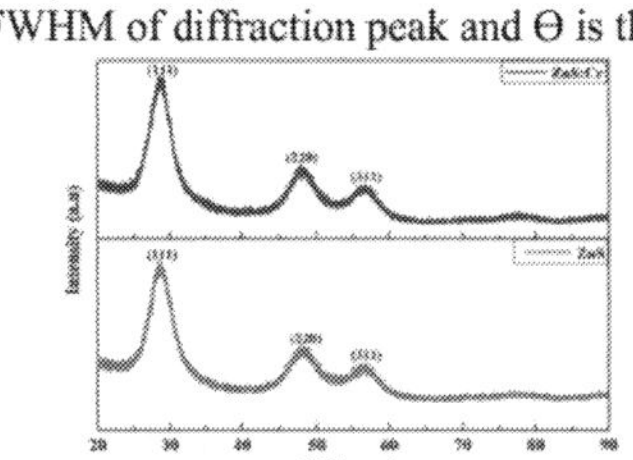

Figure 1 A powder XRD pattern of Glycine capped ZnS and ZnS:Cr NPs

Table 1 Results and calculations originated from powder XRD of Glycine capped ZnS and ZnS:Cr NP

Sample	2Θ (degree)	d (Å)	FWHM(Å)	Particle size(nm)	(h k l)	ICDD file
ZnS	28.70	3.107	3.529	2.427	(1 1 1)	01-080-4443
	47.93	1.896	4.421	2.054	(2 2 0)	01-080-4443
	56.62	1.624	4.670	2.005	(3 1 1)	01-080-4443
ZnS:Cr	28.72	3.105	3.831	2.236	(1 1 1)	01-080-4444
	48.04	1.892	4.791	1.896	(2 2 0)	01-080-4444
	56.65	1.623	5.527	1.705	(3 1 1)	01-080-4444

4.2. SEM-EDS

The morphology and chemical compositions of synthesized NPs were studied by using a fully equipped Scanning Electron Microscope with Elementary Dispersive Spectrum [SEM-EDS] analyzer. The corresponding SEM images of Glycine capped ZnS:Cr NPs were shown in Figure 2(a). As we observed that the undoped ZnS NPs show agglomeration with random size, after Cr ion doping ZnS NPs morphology changes and shows spherical shapes. The SEM image when we magnified at 15,000 it reveals the formation of nano-sized structure and morphology. The image shows that there was an agglomerated crystal whose diameter approximately less than 80nm.

To get elemental compositions in a synthesized sample, EDS analysis is one of the best tools. The amount of Cr ions substituted into ZnS crystal lattice is inspected by EDS. The EDS study of Figure 2(b) gives the presence of Zn, S and Cr elements with mass% of 41.97, 15.82 and 0.28 respectively. Here trace of the Al atom due to aluminium foil used in the experiment and also presence of O and H due to the considerably existence in the capping agent Glycine.

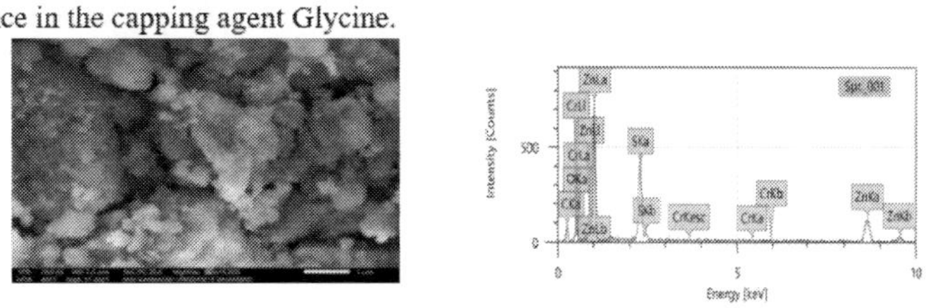

2(a) **2(b)**

Figure 2(a), (b) Scanning Electron Microscope (SEM) image and X-Ray Energy Dispersive Spectrum (EDS) of Biocompatible Glycine capped ZnS:Cr NPs

4.3. FTIR studies

FTIR is a powerful tool in understanding the structural formation of materials and gives an idea about the involvement of functional groups in it. FTIR spectra of glycine capped ZnS and ZnS:Cr NPs are as shown in Figure 3. There is a broad intense absorption in higher energy region at 3405.65 cm^{-1} and 3409.06 cm^{-1} appeared due to the O-H vibration of glycine and water molecules [27]. The peaks at 2075.93 cm^{-1} and 2075.95 cm^{-1} can be found, it should be assigned to the existence of CO_2 molecules [29]. The 1615.88 cm^{-1} and 1616.15 cm^{-1} peaks attributed to C–O stretching of carboxylic acid and the transmission bands 1118.12 cm^{-1} and 1117.45 cm^{-1} is characteristic of NH^{+3} group of glycine molecule [28]. The peak appearing at 618cm^{-1} and 618.90cm^{-1} are assigned to the Zn-S stretching vibration [29] which confirms the formation of ZnS NPs. Other peaks observed are due to the vibrations due to hydrocarbons and O-H entities present in glycine.

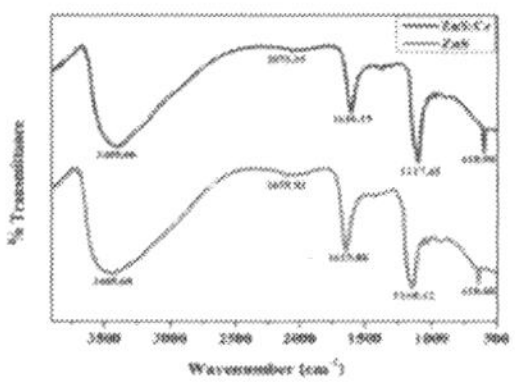

Figure 3 FTIR spectrum of Glycine capped ZnS and ZnS:Cr NPs

4.4. UV-Visible absorption study

The basic optical properties of nanomaterials can be studied using UV-Visible absorption spectroscopy technique which is a simple and well-known spectroscopy technique in the research area. The data obtained from this technique helps to study the absorption spectra and also calculate the energy bandgap of the nanomaterials. The optical absorbance spectra from 200 nm to 450 nm of Glycine capped ZnS and ZnS:Cr NPs are shown in Figure 4. The absorption peak for pure ZnS NPs is at 265nm and ZnS:Cr NPs is at 262nm. The Cr doped ZnS shows sharp peak with absorbance maxima shift towards a shorter wavelength than the pure ZnS. This is due to the quantum confinement effect of small particles and narrow shape is evidence of the formation of the very small size of the dispersed particles [30]. The energy band gap of the undoped ZnS NPs and ZnS:Cr NPs can be calculated using the Tauc plot. Figure 5 shows the Tauc plots of the NPs and its corresponding energy band gap is found to be 4.7eV (ZnS) and 3.9eV (ZnS:Cr). The obtained energy band gap is higher than the bulk ZnS crystals band gap, which indicates the formation of small crystals. The dimensions of the ZnS and their corresponding optical properties are controlled by glycine capping and the Cr dopant will affect its confinement properties.

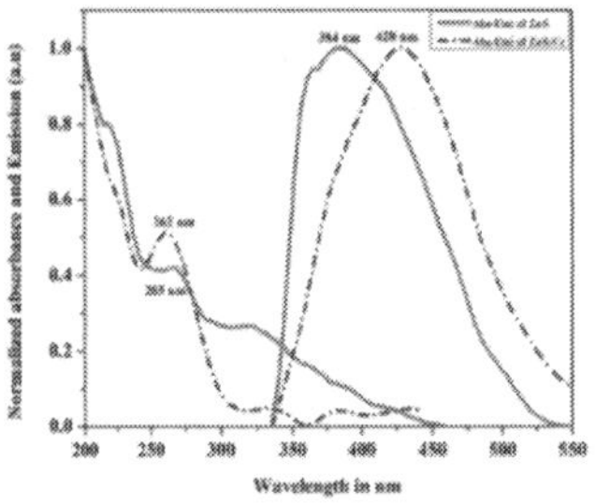

Figure 4 UV-Visible absorption and PL spectrum of ZnS and ZnS:Cr NPs

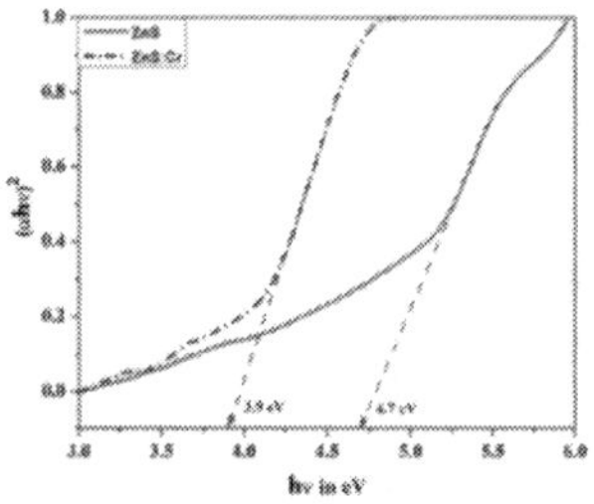

Figure 5 Tauc plots of ZnS and ZnS:Cr NPs

4.5. Photoluminescence study

The room temperature PL spectra of Glycine capped ZnS and ZnS:Cr NPs with excitation at 315nm are shown in Figure 4. It can be seen that intense PL emission peak for Glycine capped ZnS NPs is at 384nm and that for ZnS:Cr NPs is 428nm, which is shifted towards blue emission. The peak shift is related to native defects (e.g., Sulfur vacancy). In general UV emission at room temperature is associated with the modulation of the electron concentration, where shifting towards blue emission (i.e., towards visible emission) connect directly with the materials intrinsic defect states. When Cr ions are doped into ZnS NPs more defect states will be introduced. Therefore it is reasonable with peaks appearing on the longer wavelength side. When Cr ions are incorporated in the ZnS host lattice the luminescence centers of Cr ions are formed. Hence the PL emission spectra confirm that the Cr ions are incorporated successfully into the ZnS host lattice. Xu [31] studied the PL properties of Cu^{2+} doped ZnS nanopaticles and observed blue (460nm) and green (570nm) emission. Reddy et al [32] reported at 438nm PL emission peak obtained due to intrinsic defect states of the materials. Here another factor the biocompatible capping agent glycine also affects the luminescence property of the ZnS NPs. In our previous reported work Lactic acid capped ZnS NPs shows PL spectrum at 377nm [15]. Compared to Lactic acid capped ZnS nanoparticle work here the spectrum is enhanced because of agglomeration which enhances the PL property of the ZnS NPs. Thus capping agent Glycine and Cr doping will affects the ZnS NPs morphology and optical properties during the synthesis.

5. Conclusion

In the present work, ZnS and ZnS:Cr NPs have been synthesized successfully through a simple one-pot co-precipitation method. The biomolecule Glycine effectively acts as a capping agent around NPs. The optical and physico-chemical properties were investigated using UV-Visible absorption photometry, Photoluminescence study, FTIR, XRD and SEM-EDS. The optical study reveals that the synthesized undoped ZnS NPs shows high fluorescence in the violet region and after doping the spectra shifted towards the blue region. The Cr doping highly affected the optical properties of the Glycine capped ZnS. FTIR analysis confirms that there was an O-S bonding occurring between the capping molecule and ZnS nanoparticle. X-Ray Diffraction study reveals that the synthesized doped and undoped NPs having a Zinc Blende structure. Using the Scherer equation we calculated the particle size of undoped and doped ZnS NPs. The sizes of undoped ZnS and Cr doped ZnS NPs are 2.07nm and 1.86nm respectively, which confirms that Cr ions are incorporated into ZnS ions. Using SEM-EDS techniques sample composition and images were obtained. The present investigation demonstrates that biocompatible Glycine capped Chromium Doped ZnS NPs give rise to appreciable high fluorescence properties near the blue region in water solvent, which may lead to work on future biosensors and drug delivery systems in biomedical fields.

Acknowledgements

The authors would like to acknowledge University Scientific Instrumentation Centre (USIC), Karnatak University, Dharwad for providing instrumentation facility. TST would like to thank UGC for NET-JRF fellowship (Sr. no. 2061651259 dt: 26/10/2017).

References

[1] Wen Long Liu, Mei Zhen Zou, Si Yong Qin, Yin Jia Cheng, Yi Han Ma, Yun Xia Sun, and Xian-

Zheng Zhang 2020, *Adv. Funct. Mater* 2003559

[2] Simon D Anderson, Vanessa V Gwenin and Christopher D Gwenin, 2019, *Nanoscale Research Letters*,14:188.

[3] Kulkarni S K 2015, *Nanotechnology: Principles and Practices,* Springer International Publishing

[4] Xia Y 2008, Nanomaterials at work in biomedical research. *Nat. Mater.* 7 (10), 758–760.

[5] Lee J Mahendra S and Alvarez P J J 2010 *ACS Nano,* 4 (7), 3580–3590.

[6] Rao C N R and Cheetham A K 2001, *J. Mater. Chem.* 11 , 2887–2894.

[7] Conte M Prosini P P Passerini S 2004, *Mater. Sci. Eng. B,* 108, 1–2, 25, 2-8.

[8] Gogos A Knauer K and Bucheli T D 2012, *J. Agric. Food Chem.* 60 (39), 9781– 9792.

[9] Thibeault S A Kang J H Sauti G Park C, C C Fay and G C King, 2015, *MRS Bulletin* 40 (10) , 836–841.

[10] Kumar A Sharma G M Naushad, A H Al Muhtaseb, A García Peñas, G T Mola, C Si, and F J Stadler, 2019, *Chemical Engineering Journal.*

[11] Fatima Mustafa and Silvana Andreescu, 2020, *RSC Adv.*, 10, 19309,

[12] Jyoti Patela, Bhawana Jaina, Ajaya K. Singha,, Md Abu Bin Hasan Susanb, and Lellouche Jean Paulc, 2020, *Microchemical Journal* 155, 104755

[13] Wei Bian, Fang Wang , Yanli Wei , Li Wang , Qiaoling Liu , Wenjuan Dong , Shaomin Shuang and Martin M F Choi , 2015, *Analytica Chimica Acta* 856 , 82–89.

[14] Nirmal M Brus L 1999 Luminescence Photophysics in Semiconductor Nanocrystals, *Acc.Chem. Res.* 32 (5), 407–414.

[15] Tilakraj T S Mallikarjun K Patil, Vighneshwar S Bhat, Vikram Pujari and Sanjeev R Inamdar, 2021, *Materials Today: Proceedings,* Volume 47, Part 14 , Pages 4189-4193.

[16] Jinwoo Park, Mi Choi and Cheong Soo Hwang, 2018, *Journal of Nanoscience and Nanotechnology* Vol. 18, 6137–6141.

[17] Yongxin Li Jinlong Chen Changqin Zhu, Lun Wang, Danhua Zhao, Shujuan Zhuo, Yuqin Wu, 2004, *Spectrochimica Acta Part A* 6 , 1719–1724.

[18] Parita Basnet , Dhrubajyoti Samanta, T Inakhunbi Chanu, Satadru Jha and Somenath Chatterjee, 2019 *Journal of Materials Science: Materials in Electronics.*

[19] Kaur P Kumar S Singh A Chen C L Dong C L Chan T S Lee K P Srivastava C Rao S M and Wu M K, 2015, *Superlattices and Microstructures.*

[20] Li Y, Ding Y I Zhang Y Qian Y 1999 *J. Phys.Chem. Solids* 60 (1) , 13–15.

[21] Mahamuni S Khosravi A A Kundu M Kshirsagar A Bedekar A Avasare D B Singh P Kulkarni S K Ali Azam Khosravi Manisha Kundu, and Anjali Kshirsagar, 1993 *J. Appl. Phys.* 73 (10) , 5237–5240.

[22] Li H Shih W Y Shih W H 2007 *Nanotechnology* 18 (20) , 205604.

[23] H Labiadh, K Lahbib, S Hidouri, S Touil, T BEN Chaabane and Tahar Ben Chaabane, 2016, *Asian Pac. J. Trop.Med.* 9 (8) , 757–762.

[24] Z Zheng et al, 2021, *J. Phys.: Conf. Ser.* 1775 , 012008.

[25] I L Medintz, A R Clapp, H Mattoussi, E R Goldman, B Fisher, J M Mauro, 2003, *Nature Mater* 2 (9) , 630–638.

[26] Jianwei Bai, Longqi Chen, Yawei Zhu, Xiqing Wang, Xiaodan Wu and Yujie Fu, 2021, *Coll. Surf. A: Physicochem. Eng. Aspec.,* 610, 125696, ISSN 0927-7757.

[27] M Aqeel, M Ikram, A Asghar, A Haider, A Ul Hamid, M Naz, M Imran and S Ali, 2020, *Applied Nanoscience.*

[28] S Suresh, A Ramanand, D Jayaraman, S M Navis Priya and K. Anand, 2011, *International Journal of the Physical Sciences* Vol. 6(16), pp. 3875-3878.

[29] M Jothibasa, C Manoharanb, S Johnson Jeyakumara, P Praveenc, I Kartharinal Punithavathya and J Prince Richarda, 2017, *Solar Energy* 159 , 434–443.

[30] M Kuppayeea, G K Vanathi Nachiyara and V Ramasamyb, 2011, *Applied Surface Science* 257 , 6779–6786.

[31] X Xu, 1996, *Solid Luminescence,* Chinese Academy of Science and University of Science and Tech. China.

[32] D Amaranatha Reddy, G Murali, R P Vijayalakshmi, B K Reddy, and B Sreedhar,2011, *Cryst.Res.Technol.* 46, No. 7,731 – 736.

IOP Conf. Series: Materials Science and Engineering 1221 (2022) 012030 doi:10.1088/1757-899X/1221/1/012030

Optically Controlled Liquid Marble Transport Platform

Chetteente Meethal Ragisha[1] and Subramanyan Namboodiri Varanakkottu*[1]

[1]Department of Physics, National Institute of Technology Calicut, 673601, INDIA
* varanakkottu@nitc.ac.in

Abstract. Transport of liquid marbles is gaining significant attention owing to its application in various fields, including digital micro fluidics and drug delivery. Though various strategies exist for the transport of liquid marbles over the liquid substrates, on-demand transport over solid substrate remains an unexplored challenge. Most of the existing transport strategies rely on the properties of the droplet, limiting their applicability for specific liquids. Herein, we demonstrate an on-demand liquid marble transport platform based on local surface deformation of a polymer layer controlled by the thermal expansion of the confined air cavity. The platform is made of acrylic with a 5 mm hole, and a thin super hydrophobic polymer layer over it. To obtain maximum surface deformation, we have optimized the parameters such as hole diameter and thickness of the polymer layer. The hydrophobicity of the polymer layer is enhanced by coating the surface with Polyvinylidene difluoride (PVDF) nanoparticles for efficient liquid marble transport. We demonstrate the transport of liquid marble of volume 3 μL with a laser source (450 nm). A localized temperature rise of about 0.6 °C at the surface is sufficient to realize the transport of liquid marbles. The demonstrated strategy is simple and could be used for the on-demand transport of liquid marble, irrespective of the properties of the liquid.

1. Introduction

On-demand transport of liquid marbles (LMs) over solid/liquid substrate is of particular interest in view of their potential application in digital microfluidics.[1] The conventional transport methods over liquid substrates are based on magnetic,[2–4] light,[5,6] chemical actuation,[7] capillary interactions,[8] and Marangoni flow.[9–12] Though the transport of liquid marble over the liquid substrate is well explored, the on-demand transport over solid substrate remains a concern. For instance, the transport of ferrofluidic liquid marbles over polymer substrate was achieved by magnetic actuation.[13] Further, multi-particle liquid marbles have been developed to achieve better control over the transport of liquid marbles.[14–16] However, the applicability of such methods is limited to specific liquid marbles only. Methods based on optical actuation are advantageous as it ensures additive-free and contactless manipulation. Moreover, it does not rely on the properties of the core liquids or substrate. In this work, we demonstrate a simple and cost-effective optofluidic platform for liquid marble transport based on local surface deformation generated by the thermal expansion of air.

The method relies on the optothermal expansion of the air cavity confined in a Polymethyl methacrylate (PMMA) substrate. Optothermal expansion of the air cavity (at constant pressure) results in the deformation of the attached flexible polymer layer, as shown in Figure 1. The experimental setup consists of Polymethyl methacrylate (PMMA) (dimension-25 mm×25 mm×4 mm) with a hole (diameter = 5 mm) at the center as the substrate and a Polydimethylsiloxane (PDMS) layer (thickness=0.2 mm) on top of it. The bottom of the PMMA substrate was sealed using a copper foil to trap the air inside the cavity (see materials and method section for more details on the chip fabrication). The PDMS layer is undeformed at ambient temperature (Laser OFF), as shown in Figure 1(left). A laser beam (wavelength 450 nm, power 850 mW) was focussed into a spot size of 1 mm using a planoconvex lens to induce localized heating of the air cavity. The copper foil attached to the bottom of the PMMA substrate absorbs the incident beam and generates heat inside the cavity. Thus, the air expands, resulting in an upward deformation (convex deformation) at the PDMS layer (Figure 1(right)). We exploit this laser-induced deformation to realize the transport of liquid marble. Switching OFF the laser decreases the temperature inside the cavity, and thereby the system regains its initial state (undeformed state). Further reduction in the temperature of the air cavity results in thecontraction of air which in turn creates a downward deformation (concave deformation) in the PDMS layer. Such a system can find application in the trapping of liquid marble/droplets (reservoir).

2. Materials and Methods

Preparation of PDMS layer

PDMS layer of desired thickness were made by mixing PDMS precursor (Sylgard 184 silicon elastomer) with a curing agent in a 10:1 ratio by weight in a petri dish. After that, it is cured by placing in a vacuum oven at 70 ^{0}C and 0.06 mPa for one hour. PDMS layer is carefully peeled off and placed on the PMMA substrate.

Preparation of PVDF coated PDMS layer

PDMS precursor is mixed with curing agent in the ratio 10:1 by weight. The mixture is partially cured (80-90%) by placing it in a vacuum oven at 70 ^{0}C and 0.06 mPa for 45 minutes. The partially cured PDMS is dip-coated in the PVDF-ethanol solutiontogetauniformcoatingalloverthesurface. The surface is dried and then rinsed with water to remove excess PVDF particles.

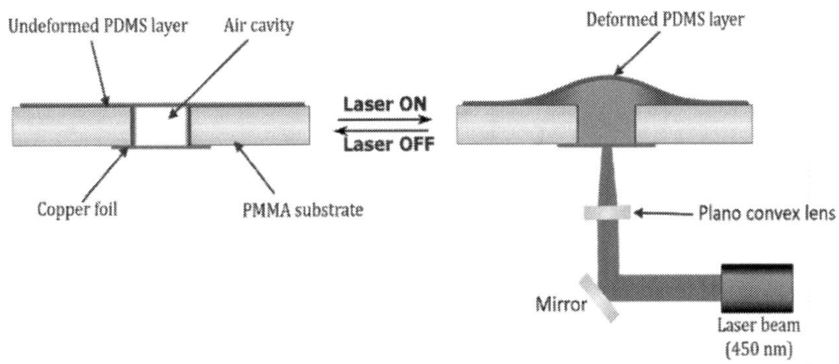

Figure 1. Schematic showing the principle of laser-induced deformation of a flexible PDMS layer.

3. Results and discussion

First, we performed the experiments using a Peltier thermoelectric cooler attached to the bottom of the PMMA substrate to control the temperature. Temperature variations of the substrate were monitored using a thermocouple connected to a temperature data logger (DIGI SENSE 20250-44). The side view images of the deformation were captured using a USB camera (USB microscope 2.0). We optimized the hole diameter (5 mm) and PDMS thickness (0.2 mm) to obtain maximum deformation at the PDMS layer. Initially, we kept the temperature at 25 °C. Then the temperature was increased to 45 °C in steps of 2°C. Figure 2a represents the change in the height of deformation (Δh) as a function of temperature rise (ΔT). Each data point corresponds to an average of 3 individual measurements with standard deviation as the error bar. Initially, the PDMS layer was undeformed. The air inside the cavity expands upon increasing the temperature, and the PDMS layer gets deformed, as shown in Figure 2b. With a temperature variation of 20 °C, the height of deformation reaches 0.12±0.02 mm (Figure 2a).

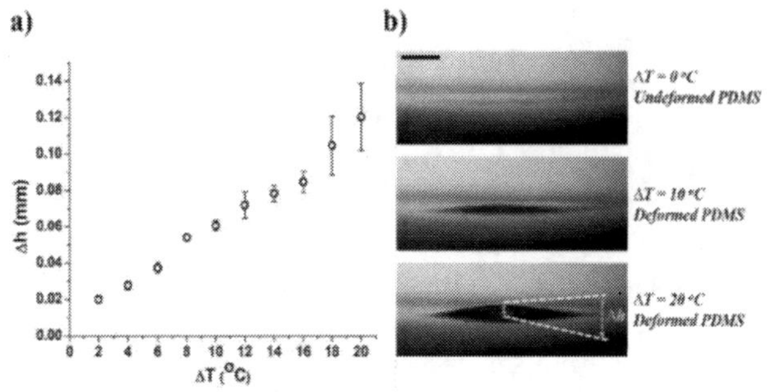

Figure 2. (a) Temperature-dependent variation of the height of the deformation of the polymer film. (b) Side view images of PDMS layer at different temperatures upon heating (Scale bar corresponds to 0.5 mm).

Cooling the substrate from 25 °C to 15 °C resulted in the downward deformation of the PDMS layer due to the contraction of air. Figure 3a shows the variation of depth of deformation (Δd) upon reducing the temperature. The side view images of the PDMS layer at different temperature intervals are shown in Figure 3b.

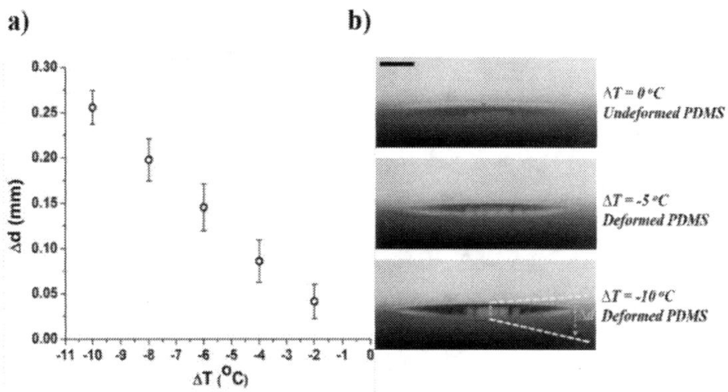

Figure 3. (a) Temperature-dependent variation of depth of deformation of the polymer film. (b) Side view images of PDMS layer at different temperatures upon cooling (Scale bar corresponds to 1 mm).

To demonstrate the liquid marble transport over the PDMS layer, we enhanced the hydrophobicity of the PDMS surface by incorporating Polyvinylidene Fluoride (PVDF) particles (see materials and method section). The top view images correspond to normal PDMS, and PVDF-coated PDMS are shown in Figure 4a (left) and Figure 4b (left), respectively.

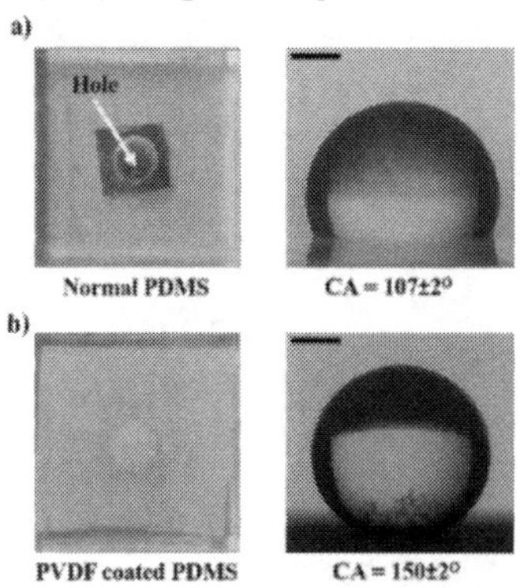

Figure 4. (a) Top view image of normal PDMS layer (left) and the side view image of a water droplet on PDMS layer (right). b) Top view image of PVDF coated PDMS layer (left) and the side view image of a water droplet on PDMS layer. Scale bars correspond to 0.5 mm.

The contact angle (CA) of the PDMS surface was measured using a contact angle meter(HO-IAD-CAM-01A). The PVDF-coated surface shows a significant enhancement in the water contact angle compared to that of a normal PDMS surface. Since there is a trapped air layer within the PVDF-coated surface, the Cassie-Baxter regime comes into effect, which causes a substantial increase in the contact angle.[17]The contact angle changes from 107±2° to 150±2° by the incorporation of the PVDF particles. Then we performed the experiment by replacing the Peltier thermoelectric cooler with a laser (450 nm) source, and realized light-controlled upward deformation on the PDMS layer upon switch ON the laser. To monitor the surface temperature distribution due to laser irradiation, we performed IR imaging. The temperature profile of the PDMS surface was recorded using an IR camera (FLIR E75),setting the emissivity of the PDMS layer as 0.86.[18] At t = 0 s, the surface shows a nearly uniform temperature distribution (T_{max}= 24.3 °C), as shown in Figure 5a. T_{max} represents the maximum temperature in a frame. As the irradiation proceeds, the temperature rises to 25 °C (Figure 5b) at an irradiation time of 20 s. A temperature rise of 0.6 °C (ΔT_{max}) was realized in the hole region after 20 s laser irradiation (Figure 5a).

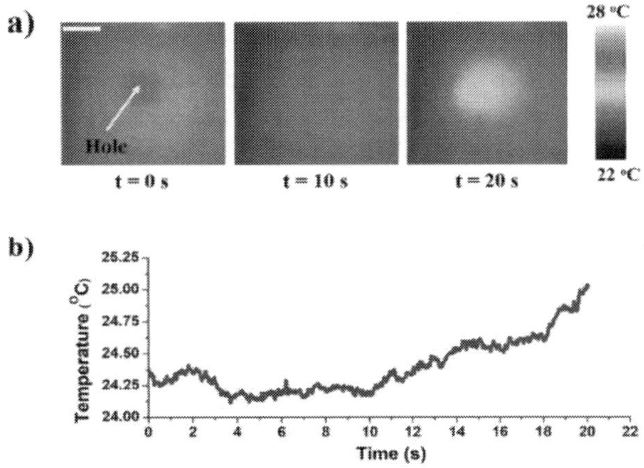

Figure 5. Thermal characterization of PDMS layer (a) Thermal images (top view mode) of the PDMS layer upon laser illumination. (b) Time-dependent temperature rise at the PDMS layer. Scale bar corresponds to 5 mm.

Figure 6 depicts the transport of liquid marble over the PVDF-coated PDMS substrate. For the fabrication of liquid marble, we prepared a lycopodium particle bed in a petri-dish, and a deionized water drop of volume 3 μL was rolled over the particle bed until the liquid drops were fully encapsulated by lycopodium particle. The fabricated liquid marble was carefully placed on top of the PDMS layer, slightly off-center with respect to the center of the hole.

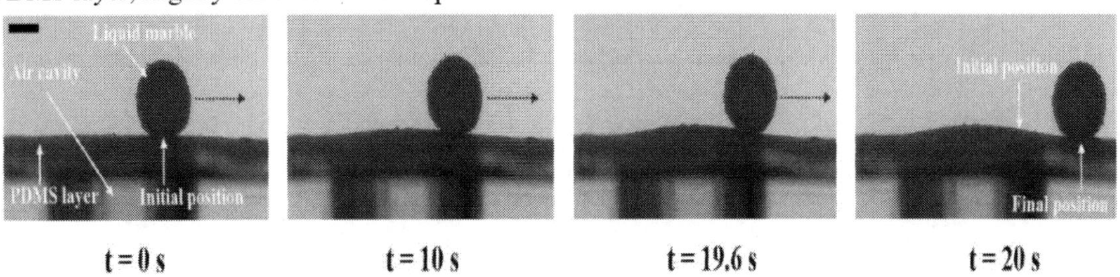

Figure 6. Optical images (side view) of the transport of liquid marble over the PDMS layer upon laser irradiation (Scale bar corresponds to 1 mm).

Upon switching ON the laser, the liquid marble moves in response to the upward deformation generated at the PDMS surface. A snapshot of the optical images corresponds to different laser illumination time is shown in Figure 6. At t=1.8 s, the initial movement of liquid marble was observed. The surface deformation increases over time, which pushes the liquid marble away from the initial position. Within 20 s of laser irradiation, the liquid marble moves out of the deformation regime, with no further movement. The directionality of the transport could be further improved by exploiting the downward deformation (concave deformation) created on the PDMS layer upon cooling the substrate. Further, the single-hole optofluidic platform can be extended to a multi-hole chip array by creating an array of holes on the substrate to realize the long-range transport of the liquid marbles.

4. Conclusion

In conclusion, we have developed an optofluidic platform for the transport of liquid marbles. The strategy is based on the optothermal expansion of the air cavity confined in a PMMA substrate and the corresponding deformation of a flexible polymer layer (PDMS layer). We have optimized the hole diameter (5 mm) and the thickness of the PDMS layer (0.2 mm) to obtain maximum deformation at the PDMS layer. The localized heating of the air cavity resulted in the upward deformation (convex deformation) of the PDMS layer, and the localized cooling resulted in the downward deformation (concave deformation). Further, we exploit the upward deformation to realize liquid marble transport over the PDMS layer. The hydrophobicity of the surface was enhanced by coating it with Polyvinylidene Fluoride (PVDF) particles to demonstrate liquid marble transport. The transport of liquid marble (3μL) was achieved by the localized heating of the air cavity upon laser irradiation (450 nm). This work is expected to enhance the capabilities of light-actuated microfluidic systems for the manipulation of liquid marbles.

Funding Sources

Funding agency: SNV sincerely acknowledges the funding from the Department of Science and Technology, India, through an INSPIRE Faculty award (Faculty Award/2016/DST/INSPIRE/04/2015/000544) and the National Institute of Technology Calicut (FRG/2019/PHY/52).

Acknowledgment

SNV sincerely acknowledges the funding from the National Institute of Technology Calicut (FRG/2019/PHY/52) and the Department of Science and Technology, India, through an INSPIRE Faculty award (Faculty Award/2016/DST/INSPIRE/04/2015/000544). The authors acknowledge Aarood U R and P Aravind for their preliminary experiments on the fabrication part.

References

[1] Bormashenko E 2017 Liquid Marbles, Elastic Nonstick Droplets: From Minireactors to Self-Propulsion *Langmuir* **33** 663–9

[2] Vialetto J, Hayakawa M, Kavokine N, Takinoue M, Varanakkottu S N, Rudiuk S, Anyfantakis M, Morel M and Baigl D 2017 Magnetic Actuation of Drops and Liquid Marbles Using a Deformable Paramagnetic Liquid Substrate *Angew. Chem. Int. Ed.* **56** 16565–70

[3] Khaw M K, Ooi C H, Mohd-Yasin F, Nguyen A V, Evans G M and Nguyen N-T 2017 Dynamic behaviour of a magnetically actuated floating liquid marble *Microfluid. Nanofluidics* **21** 110

[4] Zhao Y, Xu Z, Parhizkar M, Fang J, Wang X and Lin T 2012 Magnetic liquid marbles, their manipulation and application in optical probing *Microfluid. Nanofluidics* **13** 555–64

[5] Kavokine N, Anyfantakis M, Morel M, Rudiuk S, Bickel T and Baigl D 2016 Light-Driven Transport of a Liquid Marble with and against Surface Flows *Angew. Chem. Int. Ed.* **55** 11183–7

[6] Paven M, Mayama H, Sekido T, Butt H-J, Nakamura Y and Fujii S 2016 Light-Driven Delivery and Release of Materials Using Liquid Marbles *Adv. Funct. Mater.* **26** 3199–206

[7] Frenkel M, Dombrovsky L, Multanen V, Danchuk V, Legchenkova I, Shoval S, Bormashenko Y, Binks B P and Bormashenko E 2018 Self-Propulsion of Water-Supported Liquid Marbles Filled with Sulfuric Acid *J. Phys. Chem. B* **122** 7936–42

[8] Singha P, Swaminathan S, Yadav A S and Varanakkottu S N 2019 Surfactant-Mediated Collapse of Liquid Marbles and Directed Assembly of Particles at the Liquid Surface *Langmuir*

[9] Bormashenko E, Bormashenko Y, Grynyov R, Aharoni H, Whyman G and Binks B P 2015 Self-Propulsion of Liquid Marbles: Leidenfrost-like Levitation Driven by Marangoni Flow *J. Phys. Chem. C* **119** 9910–5

[10] Kawashima H, Mayama H, Nakamura Y and Fujii S 2017 Hydrophobic polypyrroles synthesized by aqueous chemical oxidative polymerization and their use as light-responsive liquid marble stabilizers *Polym. Chem.* **8** 2609–18

[11] Kawashima H, Paven M, Mayama H, Butt H-J, Nakamura Y and Fujii S 2017 Transfer of Materials from Water to Solid Surfaces Using Liquid Marbles *ACS Appl. Mater. Interfaces* **9** 33351–9

[12] Ooi C H, van Nguyen A, Evans G M, Gendelman O, Bormashenko E and Nguyen N-T 2015 A floating self-propelling liquid marble containing aqueous ethanol solutions *RSC Adv.* **5** 101006–12

[13] Bormashenko E, Pogreb R, Bormashenko Y, Musin A and Stein T 2008 New Investigations on Ferrofluidics: Ferrofluidic Marbles and Magnetic-Field-Driven Drops on Superhydrophobic Surfaces *Langmuir* **24** 12119–22

[14] Bormashenko E, Bormashenko Y, Pogreb R and Gendelman O 2011 Janus Droplets: Liquid Marbles Coated with Dielectric/Semiconductor Particles *Langmuir* **27** 7–10

[15] Lekshmi B S, Yadav A S, Ranganathan P and Varanakkottu S N 2020 Simple and Continuous Fabrication of Janus Liquid Marbles with Tunable Particle Coverage Based on Controlled Droplet Impact *Langmuir* **36** 15396–402

[16] Xu Z, Zhao Y, Dai L and Lin T 2014 Multi-Responsive Janus Liquid Marbles: The Effect of Temperature and Acidic/Basic Vapors *Part. Part. Syst. Charact.* **31** 839–42

[17] Simpson J T, Hunter S R and Aytug T 2015 Superhydrophobic materials and coatings: a review *Rep. Prog. Phys.* **78** 086501

[18] Byun I, Ueno R and Kim B 2014 Micro-heaters embedded in PDMS fabricated using dry peel-off process *Microelectron. Eng.* **121** 1–4

IOP Conf. Series: Materials Science and Engineering 1221 (2022) 012031 doi:10.1088/1757-899X/1221/1/012031

Coulomb drag rate in a coupled electron-electron quantum wire system

Vishal Verma[1], Devi Puttar[1], Vinayak Garg[1,*] and R. K. Moudgil[2]

[1]Department of Physics, Punjabi University, Patiala-147 002, India
[2]Department of Physics, Kurukshetra University, Kurukshetra-136 119, India

[*]E-mail:vinayak2000@gmail.com

Abstract. In this paper, we calculate the Coulomb drag rate in a coupled electron-electron (*e-e*) quantum wire system over a wide range of temperature *T*, and some selected values of electron number densityandinter-wire spacing. The intra- and inter-wire exchange-correlations are treated within the self-consistent mean-field approximation of Singwi et al (the so called STLS theory) suitably generalized to the coupled system at finite-*T*. At a fixed *T*, the drag rate is found to increase with decrease in inter-wire separation and/or electron number density due to increased intra- and inter-wire coupling among electrons. Further, we note that the magnitude of drag rate is more in the STLS theory as compared to the random-phase approximation (RPA), and the difference consistently increases with decrease in inter-wire spacing.

1. Introduction

There has been extensive experimental [1-5] and theoretical [6-8] activity on the Coulomb drag in coupled quantum wire systems in recent years. When a current is allowed to pass in one of the wires, the charge carriers in the other get dragged due to inter-wire momentum and energy transfer of the carriers *via* inter-wire scattering process. In such coupled systems, the inter-wire separation *d* is large enough so that tunnelling effects are not significant at all.

Debray et al [1] have experimentally studied the Coulomb drag in coupled electron-electron (*e-e*) quantum wire system and found that the drag resistance $R_D$decreases rapidly with increase in inter-wire separation *d*. Later, Yamamoto et al [2] reported negative Coulomb drag between parallel quantum wires for both the zero and non-zero magnetic field cases. They noted that under the condition of strong particle correlations *i.e.*, low carrier density and *T*, and high magnetic field the electrons flow in opposite directions in the coupled wires. Positive and negative Coulomb drag was further experimentally observed by Laroche et al [3] in vertically integrated quantum wires at different values of electron number density. Recently, Zhou and Guo [4] studied the Coulomb drag effect in coupled wires using the non-equilibrium Green's function formalism. Very recently Mitra et al [5] have investigated the Coulomb drag in a novel drag system consisting of graphene and *InAs* nanowire (NW) heterostructure.

Theoretically, Tanatar [6-8] have studied the Coulomb drag in parallel cylindrical electron quantum wires in the random-phase approximation (RPA) and Hubbard approximation (HA). The momentum transfer rate (Coulomb drag rate) was calculated as a function of *T* for different values of *d*. It was found that the drag rate increases with decrease in *d* and it peaks around $T \sim 0.2 - 0.3T_F$ (T_F being the

Fermi temperature) for all the selected d-values [6]. Moreover, the drag rate was found to exhibit a non-monotonous behaviour*w.r.t.* the strength of disorder present in the coupled system [7,8].
Notably, the Coulomb drag has mainly been studied in the RPA which completely ignores the intra- and inter-wire exchange-correlations among electrons and is valid only in the high electron number density regime. Motivated by this, we here compute the Coulomb drag rate in a coupled e-e quantum wire system using the self-consistent mean-filed approximation of Singwi Tosi Land and Sjölander (the STLS theory) [9] for a wide range of T, and some selected values of d and electron number density.

2. Theoretical Formalism

We consider a system of two parallel and coupled quantum wires having b as the transverse width of each wire and separated by a centre-to-centre distance d. In this coupled quantum wire system, electrons are assumed to be the charge carriers in both the wires and their confinement is modelled in such a way that they can move only in one spatial direction. The Fourier transform of intra- and inter-wire Coulomb interaction potential is taken as

$$V_{ll'}(q) = \frac{2e^2}{\varepsilon_0} K_0\left(q\sqrt{b^2 + (1-\delta_{ll'})d^2}\right), \tag{1}$$

with l and l' as the wire indices (l =1, 2for first and second wire, respectively). In Eq. (1), $d > b$ to avoid the tunnelling effects and $K_0(x)$the zeroth-order modified Bessel's function of the first kind. The background dielectric constant ε_0 and the electron number density of both the wires is assumed to be same in the system.
To study Coulomb drag in the coupled e-e quantum wire system, we use the expression for the drag rate derived through the Boltzmann equation as [6-8]

$$\tau_D^{-1} = \frac{\hbar^2}{4\pi m^* n k_B T} \int_0^\infty dq\, q^2 \int_0^\infty d\omega\ |W_{ll'}(q;T)|^2 \times \frac{Im\chi_l^0(q,\omega;\mu_l,T) Im\chi_{l'}^0(q,\omega;\mu_{l'},T)}{\sinh^2[\hbar\omega/(2k_BT)]}. \tag{2}$$

Here, m^* is the electron effective mass, T the temperature, n the linear electron number density, μ_l the chemical potential of electrons in wire l and k_B the Boltzmann constant. In Eq. (2), $W_{ll'}(q;T)$ is the statically screened inter-wire Coulomb interaction potential defined as

$$W_{ll'}(q;T) = \frac{V_{ll'}(q)[1 - G_{ll'}(q;T)]}{\epsilon(q;T)}. \tag{3}$$

$\epsilon(q;T)$is the static dielectric function and can be obtained by substituting $\omega = 0$ in

$$\varepsilon(q,\omega;T) = \left[1 - V_{ll}(q)\left(1 - G_{ll}(q;T)\right)\chi_l^0(q,\omega;\mu_l,T)\right]\left[1 - V_{l'l'}(q)\left(1 - G_{l'l'}(q;T)\right)\chi_{l'}^0(q,\omega;\mu_{l'},T)\right] \\ -\left[V_{ll'}(q)\left(1 - G_{ll'}(q;T)\right)\right]^2 \chi_l^0(q,\omega;\mu_l,T)\chi_{l'}^0(q,\omega;\mu_{l'},T). \tag{4}$$

$\chi_l^0(q,\omega;\mu_l,T)$ is the T-dependent non-interacting density response function of electrons in wire l and expressed in terms of its zero-T counterpart as [10]

$$\chi_l^0(q,\omega;\mu_l,T) = \begin{cases} \int_{e^{(-\mu_l/k_BT)}}^1 \chi_l^0(q,\omega;\mu_l + k_BT\ln|x|, T=0)\frac{dx}{(1+x)^2} \\ \quad + \int_0^1 \chi_l^0(q,\omega;\mu_l - k_BT\ln|x|, T=0)\frac{dx}{(1+x)^2};\ \mu_l > 0 \\ \int_0^{e^{(\mu_l/k_BT)}} \chi_l^0(q,\omega;\mu_l - k_BT\ln|x|, T=0)\frac{dx}{(1+x)^2};\ \mu_l < 0. \end{cases} \tag{5}$$

In Eq. (5), $\chi_l^0\ (q,\omega;\mu_l,T=0)$is the Lindhard function obtained as

$$\chi_l^0\ (q,\omega;\mu_l,T=0) = \frac{m^*}{\pi q} ln \left| \frac{\omega^2 - \left(\frac{qk_F}{m^*} - \frac{q^2}{2m^*}\right)^2}{\omega^2 - \left(\frac{qk_F}{m^*} + \frac{q^2}{2m^*}\right)^2} \right|. \tag{6}$$

$k_F(=\pi n/2)$ is the one-dimensional (1D) Fermi wave vector.

In our work, we employ the dielectric formulation to calculate the components of density response function $\chi_{ll'}(q,\omega;T)$of coupled *e-e* quantum wire system. In the STLS approach [9], $\chi_{ll'}(q,\omega;T)$takes the form as

$$\chi_{ll'}(q,\omega;T)=\frac{\chi_l^0(q,\omega;\mu_l,T)\left[\delta_{ll'}+(-1)^{\delta_{ll'}}\Psi_{\bar{l}\bar{l}'}(q;T)\,\chi_{\bar{l}}^0(q,\omega;\mu_{\bar{l}},T)\right]}{\epsilon(q,\omega;T)}, \tag{7}$$

where $\bar{l}$ denotes the wire index opposite to l and $\Psi_{ll'}(q;T)=V_{ll'}(q)[1-G_{ll'}(q;T)]$ is the effective Coulomb interaction potential accounting for the exchange-correlations among electrons. $G_{ll'}(q;T)$ appearing in Eq. (3), (4) and $\Psi_{ll'}(q;T)$ is commonly referred to as the local-field correction (LFC) factor and in the STLS approach is given as

$$G_{ll'}(q;T)=-\frac{1}{n}\int_{-\infty}^{\infty}\frac{dq'}{2\pi}\frac{q'}{q}\frac{V_{ll'}(q')}{V_{ll'}(q)}[S_{ll'}(|q-q'|;T)-\delta_{ll'}]. \tag{8}$$

Notably, $G_{ll'}(q;T)=0$ in the random-phase approximation (RPA) employing the absence of exchange-correlation effects in the system. In Eq. (8), $S_{ll'}(q;T)$is the static structure factor related to $\chi_{ll'}(q,\omega;T)$through the finite-T fluctuation dissipation theorem [11] as

$$S_{ll'}(q;T)=-\frac{k_BT}{n}\sum_{m=-\infty}^{m=+\infty}\chi_{ll'}(q,z_m;T), \tag{9}$$

where $z_m=i\omega_m$ and $\omega_m=2\pi mk_BT/\hbar$ with $m=0,\pm1,\pm2,\pm3,\ldots$.

It can be seen that the $S_{ll'}(q;T)$, $\chi_{ll'}(q,\omega;T)$ and $G_{ll'}(q;T)$can only be obtained numerically by solving Eqs. (7)-(9) in an iterative manner. The self-consistently obtained $G_{ll'}(q;T)$is put in $W_{ll'}(q;T)$ [Eq. (3)] which in turn is used to calculate the Coulomb drag rate τ_D^{-1} as expressed by Eq. (2).

3. Results and discussion

For computations, it is convenient to work in dimensionless system of units. We take wave vector q in the units of Fermi wave vector k_F, inter-wire spacing d and wire width b in the effective Bohr atomic radius $a_0^*[=\varepsilon_0\hbar^2/(m^*e^2)]$. ω and μ_l (l being the wire index) are expressed in Fermi energy E_F. The linear electron number density $n$in both the wires is taken to be equal and described by a dimensionless parameter r_s measured in the units of a_0^*; $r_s=1/(2na_0^*)$. Throughout our calculations, we take ϵ_0, k_B and $\hbar$equal to unity.

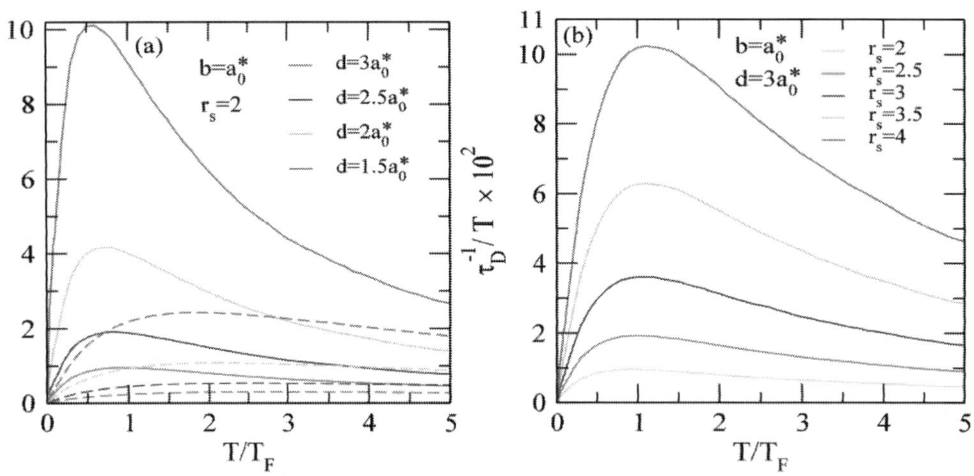

Figure 1. The scaled drag rate τ_D^{-1}/T plotted as a function of T/T_F at $b=a_0^*$ for r_s=2 and indicated d-values in panel (a), and $d=3a_0^*$and indicated r_s-values in panel (b) for the coupled *e-e* quantum wire system in the STLS theory. The corresponding RPA results are shown by dashed lines in panel (a).

The set of Eqs. (7), (8) and (9) is solved in an iterative manner to obtain static structure factor $S_{ll'}(q;T)$and local-field correction (LFC) factor $G_{ll'}(q;T)$within a predefined tolerance of 10^{-7} at the chosen q-points. $G_{ll'}(q;T)$thusobtained is put in Eq. (3) and (4) to calculate the statically screened inter-wire interaction potential $W_{ll'}(q;T)$ which in turn is used in Eq. (2) to compute the Coulomb drag rate τ_D^{-1}. In Fig. 1(a), we plot the scaled drag rate τ_D^{-1}/T as a function of T at $b = a_0^*$, r_s=2 and some selected values of d. We find that for $d = 3a_0^*$, there starts to appear a peak in τ_D^{-1}/T which grows in magnitude with further decrease in d and becomes quite prominent at $d = 1.5a_0^*$. The peak in τ_D^{-1}/T attains maximum height at T~$0.5T_F$ for all selected d-values. For comparison, the random-phase approximation(RPA) results are also shown. It can be clearly seen that τ_D^{-1}/T is more in the STLS theory as compared to theRPA and RPA curves of τ_D^{-1}/T are almost flat with no peaked structure. Thisis due to the fact that the RPA completely neglects the exchange-correlation effects and the STLS theory captures them through $G_{ll'}(q;T)$. Qualitatively, a similar behaviour of τ_D^{-1}/T in the RPA has already been reported by Tanatar [7,8] for a coupled *e-e* quantum wire system containing disorder.
In Fig. 1(b), we show τ_D^{-1}/T as a function of T at $d = 3a_0^*$andsome selected values of r_s. We observe that there starts to develop a peak in τ_D^{-1}/T at T~T_F for r_s=2.5. The peak height increases with further increase in r_s and attains maximum height at $r_s = 4$. This implies that the coupling among intra- and inter-wire electrons significantly increases with increase in r_s which appreciably enhance τ_D^{-1}/T. However, we could not obtain self-consistent $G_{ll'}(q;T)$and hence τ_D^{-1}/T beyond $r_s = 4$.

4. Summary and conclusions
In this paper, we have presented a theoretical calculation of Coulomb drag rate in a coupled electron-electron (*e-e*) quantum wire system using the self-consistent mean-field approximation of Singwi et al (the STLS theory). The inter-wire interaction potential is assumed to be statically screened, and the temperature, electron number density and inter-wire separation dependence of the drag rate has been computed. The drag rate is found to increase with decrease (increase) in d (r_s) at a fixed T. The STLS results of drag rate have been compared with the RPA and it is noted that its magnitude is more in the STLS theory. This is due to the fact that STLS incorporates the exchange-correlation effects among electrons which are completely ignored by the RPA.

5. Acknowledgments
Vishal Verma and Devi Puttar are grateful to DST and CSIR (New Delhi) for providing financial assistance.

References
[1] Debray P,Vasilopoulos P,Raichev O,Perrin R,Rahman M andMitchel W C 2000 *Physica* E 6 694;Debray P,Zverev V,Raichev O,Klesse R,Vasilopoulos P andNewrock R S 2001 *J. Phys.: Condens. Matter* 13 3389
[2] Yamamoto M,Stopa M,Tokura Y,Hirayama Y andTarucha S 2002 *Physica* E 127 26
[3] Laroche D, Gervais G,Lilly M P andReno J L 2011*Nat. Nanotechnol.*6 793; Laroche D,Gervais G,Lilly M P andReno J L 2014*Science* 343 631
[4] Zhou C and Guo H 2019*Phys. Rev.* B 99 035423
[5] Mitra R,Sahu M R,Watanabe K,Taniguchi T,Shtrikman H, Sood A K and Das A 2020*Phys. Rev. Lett.*124 116803
[6] Tanatar B 1996*Solid State Commun.*991
[7] Tanatar B 1998*Phys. Rev.* B 58 1154
[8] Tanatar B 2000*Turk. J. Phys.*24143
[9] Singwi K S,Tosi M P,Land R H andSjölander A 1968*Phys. Rev.*176589
[10] Hu B Y-K and Das Sarma S 1993*Phys. Rev.* B 485469
[11] Pines D and Nozières P 1966 *The Theory of Quantum Liquids* vol 1 (New York: Benjamin)

IOP Conf. Series: Materials Science and Engineering 1221 (2022) 012032 doi:10.1088/1757-899X/1221/1/012032

Study of optical properties of photo annealed solution based As_2S_3 chalcogenide glass thin films

Prince[*1] **and M Zulfequar**[2]

[1] Dept. of Physics, Dayanand Subhash National P.G. College, Unnao, U.P., India- 209801
[2]Dept. of Physics, Jamia Millia Islamia, New Delhi, India-110025
*Email: princevish2@gmail.com

Abstract: Chalcogenide glasses have wide range of applications in Mid- IR region due to its attractive optical properties. In this paper, we study the influence of UV light on the optical properties of solution based A_2S_3 chalcogenide thin films by examining transmission spectraand further we compare optical properties of photo annealed solution based thin films with the optical properties of thermally deposited As_2S_3 thin film annealed at 150°C for one hour. Illumination removes the excess amount of solvent which results in high refractive index, lower absorption coefficient and optical band gap ranging from 2.32 eV to 2.27eV.

1. Introduction

Chalcogenide glasses exhibit many attractive optical properties, including a high refractive index, nonlinear optical behavior, wide transmission window and photosensitive[1,2]in nature. These glasses and its thin films are used to fabricate optical components for IR applications[3,4]. However, traditional method of depositing chalcogenide thin films such as thermal vapor deposition and sputtering often suffer stoichiometry deviation, complex patteringand essentialhigh vacuum system requirements which result the change in optical properties of deposited thin films and cost escalation of IR optical components[5,6].In order to maintainnatural optical propertiesof chalcogenide glass thin films and to produce lowcostIR optical components, we explore easy and simple technique of solution based depositionwhichinvolves dissolution of bulk chalcogenide glasses in appropriate solvent. The solution based deposition technique also provides a new approach towards easy patterning in chalcogenide glasses which make this technique very suitable in terms of stability and cost effectiveness of optical components for IR applications[7,8]. Moreover, solution based patterning methods could be a low temperature, additive in nature and etch free process [9,10].

In order to use solution based chalcogenide glass thin films in IR optical components, the associated optical properties should be known and how these properties change with the illumination of light. Therefore,in this paper, we analyzethe influence of ultra violet light to the optical properties of solution based As_2S_3 films and comparethese properties with the thermally deposited As_2S_3 thin film thermally annealed for 1 hour. UV-visible transmission spectra are measured to characterize the optical properties of films exposed to UV light of duration from 1 hour to 3 hours. Finally, we compare all calculated optical constant i.e. optical band gap, refractive index, absorption coefficient and extrinsic coefficient for different IR applications.

2. Experimental Details

2.1. Bulk Preparation

Bulk As_2S_3 glass, used as reference material is prepared by Melt Quenching method. Constituent of high purity elements(As: Alfa-Aesar 99.999%, S: Cerac 99.999%) are weighted according to their

IOP Conf. Series: Materials Science and Engineering 1221 (2022) 012032 doi:10.1088/1757-899X/1221/1/012032

atomic weight percentage. Pre-cleaned quartz ampoules are used to prepare batches which are evacuated under a vacuum of 10^{-5} Torr for half an hour and sealed using gas-oxygen flame torch. These batches of sealed quartz ampoules are then placed in to muffle furnace in which temperature is increased up to the melting point of bulk A_2S_3 glass sample accordingly at the rate of 3-4°C/minute heating cycle. To get homogenous melt, ceramic rod is rotated in which ampoules are fixed and placed into muffle furnace for 12 hours.Quartz ampoules containing homogenous melt are cooled air quenched. Further, these quartz ampoules are broken to get bulk sample of As_2S_3 glass. Further prepared samples are annealed at 40°C well below the glass transition temperature (T_g) for 15 hours to aviodoxides or oxygen that might come during the breaking of ampoules.The prepared bulks are characterized to check the stoichiometry and nature of the samples and it is found the amorphous nature of the samples with the same stoichiometry.

2.2. Thin film Preparation

As_2S_3 thin films have been prepared by using twodifferent ways of deposition i.e. Thermal deposition method and Spin coating method.

2.2.1. Film Deposition by Thermal evaporation method. In this method, Chalcogenide thin films are deposited on glass substrates. To avoid any external element on the surface of substrate, these substrates are cleaned with deionized water properly and placed inside the chamber of coating unit(HHV, India, model 12A4D-T).Evaporation process is carried out with a pressure of $\sim 2\times10^{-6}$ Torr at room temperature through molybdenum boat. The film deposition rate is kept at~10-15 nm/second.Uniform films are produced by rotating substrate holder at the frequency of 3 Hertz during the process of deposition. Thicknesses of prepared thin films are found to be in range of 1-2 μm by using quartz crystal monitor. Nature of prepared thin films are checked by using X-ray diffraction measurements (PAN alytical, The Netherlands) with source wavelength λ = 1.540 Å (Cu $K_{\alpha 1}$). Chalcogenide glass thin films are found amorphous in nature. To use these thin films for further optical measurements, we anneal (thermal annealing) these films at 150°C for 1 hour.

2.2.2. Film Deposition by Spin Coating Unit. To deposit solution based thin film on the pre-cleaned glass substrates (2.54cm ×2.0cm), solution of bulk As_2S_3 is prepared using ethanolamine as a solvent at concentration of 0.5gm/ml. To achieve a true transparent solution, solution is continuously stirred for 24 hours using a magnetic stirrer in appropriate atmosphere inside the wooden box. Color of the solution changes with time and finally prepared solution gives orange color. Further, this prepared solution is accelerated on rotating substrate at 3000rpm for 5minutes using a spin coater unit (Model: SPIN 150-V3). The thickness of these solution based thin films are measured to be ~1-2 μm using optical profilometer.

For optical measurements, these thin films are annealed by exposing UV light (λ=375nm) using excimer UV (ultra-violet)lamp (Model: Quark QEF100)with illumination power 10mW/cm^{2}for 1 hour, 2 hours and 3 hours respectively.Transmission spectra are measured for all thermally and photo (UV light) annealed thin films to calculate optical constants using double beam UV/Vis spectrophotometer (Model: Camspec M550) in wavelength range of 190-1100nm.

3. Result and Discussion

3.1. Structural and compositional Study

X-Ray diffractometer is employed for studying the nature of bulk / thin film samples of As_2S_3chalcogenide glass. A copper target is used as an X-ray source with Cu$K_{\alpha 1}$ radiation (λ=1.540Å). Scanning angle is recorded in the range of 10° to 90°. A scan speed of 2°/minute and chart speed of 1 cm/minute is maintained. X-ray diffraction patterns, are taken for bulk As_2S_3, thermally depositedand spin coated As_2S_3 films, are shown in figure 1. The absence of sharp peaks in X-ray diffraction confirms the amorphous nature of bulk and thin films.

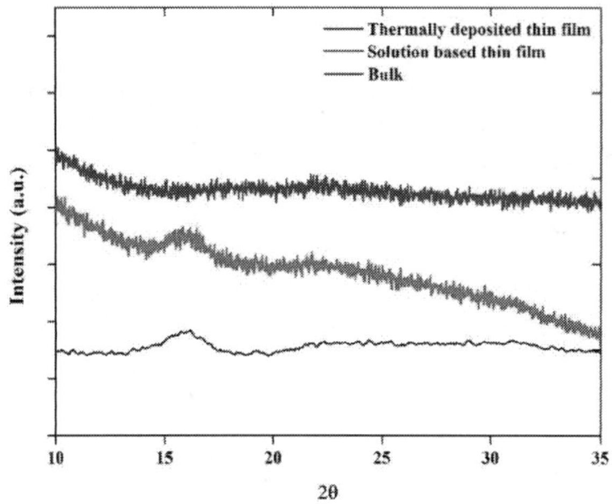

Figure 1. XRD patterns for bulk, thermally deposited and solution derived thin films of As_2S_3

The elemental compositions of 3 hours photo annealedspin coated and 1 hour thermally annealed thin films of As_2S_3 have been confirmed by Energy-Dispersive X-ray spectroscopy (EDX). Table 1 shows the presence of different atomic compositions of As (72.87%), S (27.13%) for thermally deposited thin films and As (70.17%), S (29.83%) for solution derived thin films which are foundapproximately similar with the actual composition of bulk As_2S_3 respectively.

Table 1. Atomic Composition of As_2S_3 films prepared by different techniques

Type of Films	As(weight %)	S (weight %)
Thermally Deposited thin film(Thermal annealing at 150°C for 1 hour)	72.87	27.13
Solution Derived thin films (Photo annealed for 3 Hours)	70.17	29.83

3.2. Optical measuremnts

Optical parameters are the significant feature for the fabrication of IR optical components. In order to understand the photo (UV)-annealing behavior of solution based thin films of As_2S_3, we compare its optical parameters withthe thermally deposited thin films annealed at 150°C for 1hour.To calculate different optical parameters, transmission spectra of solution based thin films (photo annealed for 1 hour, 2 hours and 3 hours) and thermally deposited thin films (thermally annealed at 150°C for 1 hour) are recorded in wavelength region of 190-1100nm. The variation of optical band gap are measured,as a function of photo annealing duration for solution derived thin films and,thermal annealing at 150°C for thermally deposited thin films using Tauc's extrapolation [11,12]. Using Tauc's equation i.e equation 1.

$$\alpha h\nu = B(h\nu - E_g)^{1/2} \quad (1)$$

Where, B is the slope of the Tauc edge called the band tail parameter.In order to determine the optical band gap, intercepts of the plot of $(\alpha h\nu)^{1/2}$ against photon energy are measured as shown in figure 2.

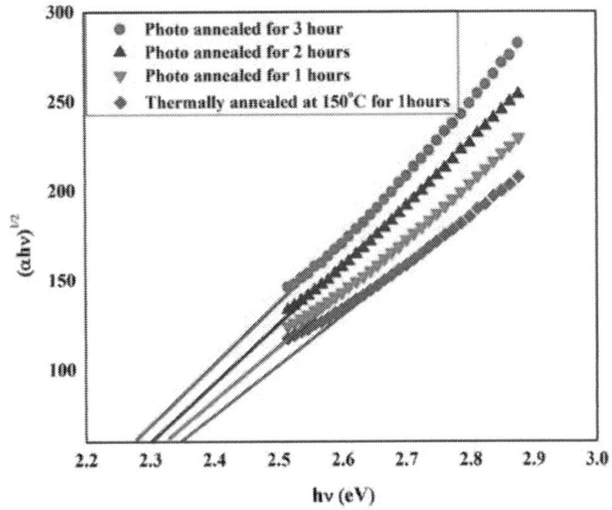

Figure 2. Variations of $(\alpha hv)^{1/2}$ vs photon Energy (hv) with different annealing methods for As_2S_3 thin films

It is observed from the figure 2 that optical bandgap for thermally deposited thin film (annealed at 150°C) is 2.35eV and for solution based thin films (photo annealed for 1hour, 2hours and 3hours) it varies from 2.32eV to 2.27eV which indicates the redshift of optical band gap with increase of photo annealing time. The corresponding shift of optical bandgap is may be due to reduction of bond energy of As_2S_3. Thus, it is observed that increase of photo annealing time of solution based thin films reduces the optical band gap for As_2S_3.

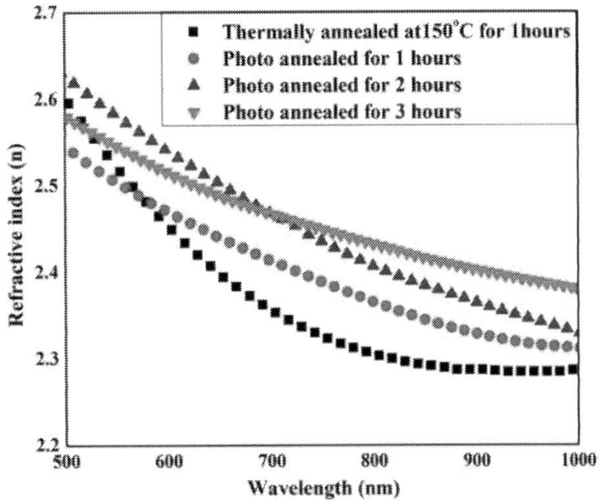

Figure 3. Variation of refractive index with wavelength for solution based thin films and thermally deposited thin films

Swanepoel's method [13,14]has been used to calculate the refractive index of prepared thin films.The calculated value of refractive index for thermally deposited thin films is found 2.30 and for solution based thin films it varies from 2.32 to 2.40. The increase in refractive indexwith photo annealing duration is may be due to the densification of solution based thin films. The Cauchy's dispersion relationship is also used to for extrapolation of all wavelengths [15]. The variation of refractive index with the wavelength is shown in figure 3 which verifies the normal dispersion law. The bang gap and refractive index value obtained after annealing at 150 °C for 1 hour is quite similar to the values reported by Shanshan song et.al. [16] for As_2S_3 films.

Table 2. The observed and calculated value of all optical constants for solution derived and thermally deposited thin films

As_2S_3 Chalcogenide glass	Thermally Annealed at 150°C	Photo Annealed for		
		1 Hour	2 Hours	3 Hours
Refractive Index **(n)**	2.30	2.32	2.36	2.40
Optical Band gap**(E_g)**	2.35	2.32	2.30	2.27
Extrinsic Coefficient **(K)**	0.238	0.224	0.217	0.214
Absorption Coefficient **(α)**	3.33	3.13	3.03	2.99

The value of absorption coefficient (α) and extrinsic coefficient (K) has also been calculated which found to be in decreasing order with increasing time of exposing of UV light as tabulated in table 2. Results clearly indicates that the prepared thin films deposited by spin coating can be used for the fabrication of any optical components as an alternative to the thermally deposited films.

4. Conclusion

The effect of photo annealing in optical properties has been studied for the thin films of As_2S_3 glassy alloys prepared by spin coating method and further all optical properties have been compared with thermally deposited thin film of As_2S_3. Optical band gap is found decrease with increase in photo annealing time to the solution derived thin films. The value of optical band gap for 3 hours photo annealing of solution derived thin films is found to be 2.27eV whereas for 1 hour thermally annealed thin film prepared by thermal deposition technique, its value is found to be 2.35eV. Solution based thin films photo annealed for different time results to increase in the value of refractive index (from 2.32 to 2.40) as compared to thermally deposited thin films, its value found to be 2.30. Hence, we can say that the solution driven films can be used for optical device fabrication, solar cells, wave guide, electrical switching and other opto-electronic devices.

References

[1] Kanamori T, Terunuma Y, Takahashi S, and Miyashita T 1984 *J. Light wave Tech.* **2** 607–13
[2] Owen A E, Firth A P, and Ewen P J S 1985 *Philos. Mag. B***52** 347–362
[3] Song S, Howard S S, Liu Z, Dirisu A O, Gmachl C F, and Arnold C B 2006 *Appl. Phys. Lett.* **89** 041115
[4] Lee M W, *et.al* 2007 *Opt. Express***15** 1277–85
[5] Balan V, Vigreux C, and Pradel A 2004 *J. Optoelectron. Adv. Mater.* **6**, 875
[6] Song S, Carlie N, Boudies J, Petit L, Richardson K, and Arnold C B 2009 *J. Non-Cryst. Solids***355** 2272-78.
[7] Mitzi D B, Copel M and Chey S J 2005 *Adv. Mater.***17** 1285–89
[8] Zha Y, Waldmann M and Arnold C B 2013 *Opt. Mat. Exp.***3**1259-72
[9] Gonzá-Leal J M, Stuchlik M, Vlcek M, Jiménez-Garay R, and Márquez E 2005 *Appl. Surf. Sci.* **246** 348–355
[10] Santiago J J, Sano M, Hamman M, and Chen N 1987 *Thin Solid Films***147** 275–284
[11] Wood D L, Tauc J, 1972 *Phys. Rev. B.* **5** 3144–51
[12] Hafiz M M 2001 *J.of Mat. Sci.: Materials in Electronics***12** 395–401

IOP Conf. Series: Materials Science and Engineering 1221 (2022) 012032 doi:10.1088/1757-899X/1221/1/012032

[13] Swanepoel R2000 *J. Phys. E.* **16** 1214–1222
[14] Marquez E, Ramirez-Malo J, Villares P, Jimenez-Garay R, Ewen P J S and Owen E 2000 *J. Phys. D. Appl. Phys.* **25** 535–541
[15] Moss T S 1959 *Optical Properties of Semiconductors* (London: Buttenworths)
[16] Song S, Dua J and Arnold C B 2010 *Opt. Express* **18** 5472–80

IOP Conf. Series: Materials Science and Engineering 1221 (2022) 012033 doi:10.1088/1757-899X/1221/1/012033

Impact of electron-phonon interactions on Wigner crystallization in an electron quantum wire

Devi Puttar[1], Vishal Verma[1], Vinayak Garg[1, a)]and R. K. Moudgil[2]

[1]Department of Physics, Punjabi University, Patiala-147 002, India
[2]Department of Physics, Kurukshetra University, Kurukshetra-136 119, India

[a)] Email: vinayak2000@gmail.com

Abstract. In this paper, we explore the impact of electron-phonon (*e-ph*) coupling on the Wigner crystallization occurring due to strong electron-electron (*e-e*) interactions in a semiconductor-based electron quantum wire. The *e-ph* coupling is represented by the Fröhlich interaction potential and electron correlations are included within the dynamic version of self-consistent mean-field approximation of Hasegawa and Shimizu. Numerical results are reported for the static structure factor and pair-correlation function for fixed wire width and some selected values of electron number density parameterr_s. A strong peak in the static structure factor and pronounced periodic oscillations in pair-correlation function indicate the formation of Wigner crystal in the quantum wire at a critical electron number density r_s^c.We find that the inclusion of*e-ph* interactions in addition to the usual *e-e* interactions significantly enhance r_s^c, at a fixed wire width.

1. Introduction

In semiconductor-based quantum wire structures, electrons move freely along the length of the wire while their motion is restricted along the remaining two transverse directions. The strong confinement of electrons and dynamics of electron correlations lead to a variety of interesting phenomena in quantum wires. The Wigner crystallization [1] is among one of fascinating effect observed in quantum wires wherein the electrons undergo the formation of a periodic crystalline array and get localized at sufficiently low electron number density. This situation crop up as the electrons energetically favour an ordered state due to the dominance of Coulomb potential over the kinetic effects. Many researchers have already studied the Wigner crystallization theoretically [2-4] and in the simulation study [5] with consideration of only the usual electron-electron (*e-e*) interactions in quantum wires. Experimental evidence also supports the idea of formation of one-dimensional (1D) Wigner crystal (WC) in a *GaAs* based quantum wire structure [6]. However, in doped polar semiconductor-based quantum wire structures the electron-phonon (*e-ph*) coupling is also important and cannot be ignored. The *e-ph* coupling is in addition to the usual electron-electron (*e-e*) interactions present in the system.
Motivated by this, we here investigate the impact of *e-ph* coupling on the Wigner crystallization at absolute zero occurring due to the dynamics of strong *e-e* interactions in an electron quantum wire. Our focus will be to explore as how the Wigner crystallization in the wire get affected by including the *e-ph* interactions. The *e-ph* interactions will be considered *via* the Fröhlich coupling and the dynamics of electron correlations will be included through the approach of Hasegawa and Shimizu commonly known as the quantum STLS (qSTLS) theory[7]. It is important to mention here that the qSTLS approximation had been quite successful in predicting the Wigner crystallization in a quasi-one-dimensional electron gas (electron quantum wire) at absolute zero [2,3] and finite-temperature [4] for

the situation in which only the *e-e* interactions are considered. In the next section, we discuss briefly the theoretical formalism related to our work.

2. Theoretical Formalism

To explore the Wigner crystallization in an electron-phonon *(e-ph)* coupled quantum wire, we calculate the static structure factor and pair-correlation function using the dielectric formulation within the linear response theory. Here, the dynamic density response function $\boldsymbol{\chi(q,\omega)}$ is the main physical quantity as the correlation functions of the quantum wire can easily be obtained from it.In theqSTLS approximation, the density response function $\boldsymbol{\chi(q,\omega)}$is defined as

$$\chi(q,\omega) = \frac{\chi_0(q,\omega)}{1-V_T(q,\omega)[1-G(q,\omega)]\chi_0(q,\omega)}. \quad (1)$$

Here, $\chi_0(q,\omega)$ is the non-interacting density response function of one-dimensional (1D) electrons at absolute zero and $G(q,\omega)$ is the dynamic local-field correction (LFC) factor accounting for the exchange-correlations among the wire electrons given as

$$G(q,\omega) = -\frac{1}{n}\int_{-\infty}^{\infty}\frac{dq'}{2\pi}\frac{\chi_0(q,q',\omega)}{\chi_0(q,\omega)}\frac{V_{e-e}(q')}{V_{e-e}(q)}[S(q-q')-1]. \quad (2)$$

$\chi_0(q,q',\omega)$ in Eq. (2) is the inhomogeneous non-interacting density response function of 1D electrons and at absolute zero obtained as

$$\chi_0(q,q',\omega) = -\frac{2}{\hbar}\int_{-\infty}^{+\infty}\frac{dp}{h}\frac{f^0\left(p+\frac{\hbar q'}{2}\right)-f^0\left(p-\frac{\hbar q'}{2}\right)}{\omega-\frac{pq}{m^*}+i\eta}. \quad (3)$$

Here, $f^0(p)$ is the Fermi-Dirac distribution function and η is the small positive infinitesimal. For$q'=q$, $\chi_0(q,q',\omega)$ reduces to $\chi_0(q,\omega)$ which is used in Eq. (1).

In Eq. (1), $V_T(q,\omega)$ is the total interaction potentialwhich includes the usual *e-e* and *e-ph* interactions and given as

$$V_T(q,\omega) = V_{e-e}(q) + V_{e-ph}(q,\omega). \quad (4)$$

Here, $V_{e-e}(q)$ is the Fourier transform of bare *e-e* interaction potential and for the harmonic confinement model of electrons in a cylindrical quantum wire derived as [8]

$$V_{e-e}(q) = \frac{e^2}{2\varepsilon_\infty}f(qb);f(x) = 2E_1(x^2)\exp(x^2), \quad (5)$$

with b the transverse wire width and $E_1(x)$ as the exponential-integral function. $V_{e-ph}(q,\omega)$ is the Fröhlich interaction potential which describes the longitudinal optical (LO)phonon-mediated*e-e* interactions and is defined as [9]

$$V_{e-ph}(q,\omega) = M_q^2 D_0(\omega). \quad (6)$$

Here, M_q is the effective 1D Fröhlich *e-ph* coupling matrix element and$D_0(\omega)$ is the unperturbed (free) bare LO-phonon propagator given as

$$M_q^2 = V_{e-e}(q)\frac{\omega_{LO}}{2}\left(1-\frac{\varepsilon_\infty}{\varepsilon_0}\right); D_0(\omega) = \frac{2\omega_{LO}}{\omega^2-\omega_{LO}^2}. \quad (7)$$

ω_{LO} is the LO-phonon frequency and $\varepsilon_0(\varepsilon_\infty)$ is the static (high frequency) background lattice dielectric constant.

Using Eqs. (5) and (6) in Eq. (4), the final form of the total effective interaction potential *i.e.*Eq. (4) yields as

$$V_T(q,\omega) = \frac{e^2}{2\varepsilon_\infty} f(qb)\left[1 + \frac{\omega_{LO}^2}{\omega^2-\omega_{LO}^2}\left(1 - \frac{\varepsilon_\infty}{\varepsilon_0}\right)\right]. \quad (8)$$

In Eq. (2), $S(q)$ is the static structure factor related to $\chi(q,\omega)$ through the fluctuation-dissipation theorem [10] as

$$S(q) = -\frac{\hbar}{\pi n}\int_0^\infty d\omega\, \chi(q, i\omega). \quad (9)$$

Here, the ω-integration has been transformed to theimaginaryω-axis *i.e.*$\omega \to i\omega$in $\chi(q,\omega)$through the Wick's rotation [11] to avoid the plasmon poles.

Further, $S(q)$ can be used to calculate the pair-correlation function $g(r)$ which gives the probability of finding an electron at a distance *r* away from the other at the origin as

$$g(r) = 1 + \frac{1}{n}\int_{-\infty}^{+\infty} \frac{dq}{2\pi} e^{-iqr}\,[S(q) - 1]. \quad (10)$$

Notably in the limit $\hbar \to 0$, the qSTLS dynamic LFC factor $G(q,\omega)$ reduces to

$$G(q) = -\frac{1}{n}\int_{-\infty}^{\infty} \frac{dq'}{2\pi}\frac{q'}{q}\frac{V_{e-e}(q')}{V_{e-e}(q)}[S(q-q') - 1], \quad (11)$$

which is simply the static LFC factor of the STLS approximation [12] at absolute zero.

In the next section, we present the numerical results of static structure factor $S(q)$and pair-correlation function $g(r)$ for the *e-ph* coupled quantum wire at some selected values of electron number density parameter r_s and fixed wire width *b*.

3. Results and discussion

For numerical calculations, we take q in the units of Fermi wave vector k_F, r in k_F^{-1} and wire width *b* in the effective Bohr atomic radius $a_0^*[= \epsilon_\infty \hbar^2/(m^* e^2)]$.$\omega$ and $\omega_{LO}$are expressed in Fermi energy E_F, and linear electron number density *n* as$r_s = 1/(2na_0^*)$. Throughout our calculations, we take$b = 2a_0^*$ and$\hbar = 1$. We use the physical parameters corresponding to a *GaAs* based electron quantum wire system *i.e.*$\omega_{LO} = 36.8\, meV$, $1\, Ryd^* = 7.67 meV$,$m^* = 0.067 m_e$ and$\varepsilon_0\left(\varepsilon_\infty\right)$ =12.9 (10.9).

The numerical results of $S(q)$ are plotted in Fig. 1(a) at wire width $b = 2a_0^*$ and r_s = 2, 4, 6, 7 and 8.7 by including the *e-ph* interaction effects in the electron quantum wire. We find that a peak start evolving in $S(q)$ for $r_s \geq 6$ whose magnitude grows with further increase in r_s. Eventually, the peak height becomes maximum for $r_s = r_s^c \approx 8.7$ at$q \approx 3.1k_F$ and beyond r_s^c the convergence in $S(q)$ could not be achieved. The strong peak in $S(q)$ at r_s^c signals the onset of an ordered state *i.e.*Wigner crystal in the electron quantum wire. To make a comparison, we also plot the results of $S(q)$ without considering the *e-ph* interactions for r_s=2, 4, 6 and 7 in the system. It is observed that there is a considerable difference in the results for both the situations *i.e.* with and without inclusion of the *e-ph* interactions which becomes more visible with the increase in r_s.
It is to be noted that Garg *et at*[3] have earlier predicted the Wigner crystallization in quasi-one-dimensional electron gas (Q1DEG) at $b = 2a_0^*$ for $r_s = r_s^c \approx 7.5$considering only the usual*e-e*

interactions and completely ignoring the*e-ph* coupling. Our present study demonstrates that if the *e-ph*interactions are taken into account (along with *e-e* interactions), then r_s^c increases by a factor of ~16% at $b = 2a_0^*$ as r_s^c changes now to 8.7 from 7.5. This is due to the fact that *e-ph* interactions tend to weak the exchange-correlations. To highlight the importance of the dynamics of correlations, we compare the qSTLS $S(q)$ with the corresponding STLS result at $r_s = 8.7$ and note that it remains a smooth function of q having no peak in it. This is due to the reason that the STLS theory deals electron correlations in a static manner and consequently do not detect the Wigner crystallization in the system.

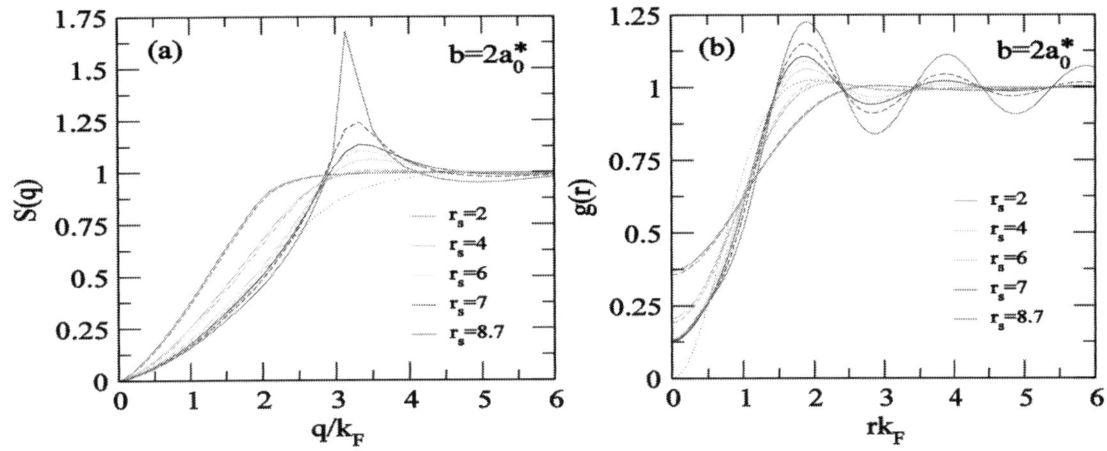

Figure 1.$S(q)$as a function of q/k_F in panel (a) and $g(r)$as a function of r/k_F^{-1} in panel (b) at $b = 2a_0^*$ in the qSTLS approximation with *e-ph* interactions (solid lines) for r_s = 2, 4, 6, 7 and 8.7, and without *e-ph* interactions (dashed lines) for r_s = 2, 4, 6 and 7. The dotted line is the STLS curve at r_s = 8.7 with the *e-ph* interactions.

Further, the numerically computed $S(q)$ is used in Eq.(10) to obtain the corresponding pair-correlation function $g(r)$ as presented in Fig. 1(b) and is shown at the same wire parameters for which we depicted $S(q)$ in Fig. 1(a). We find that $g(r)$exhibits oscillatory behaviour as one moves towards the higher r_s and strong periodic oscillations are observed at $r_s^c \approx 8.7$ which again reaffirm the formation of a Wigner crystal. The STLS result of *g(r)* is also shown at $r_s = 8.7$ and it is noted that it remains a smooth function of *r* and does not contain any sort of periodic oscillations.

4. Conclusions

In conclusion, we have studied the role of *e-ph* coupling along with the usual *e-e* interactions on the Wigner crystallization in an electron quantum wire at absolute zero. The *e-ph* coupling is included by the Fröhlich interaction potential and electron correlations have been treated within quantum (dynamic) version of the mean-field theory of Hasegawa and Shimizu (the qSTLS approximation) for different values of electron number density parameter $\boldsymbol{r_s}$. Our study reveals that the wire electrons may undergo Wigner crystallization at a critical electron number density parameter $\boldsymbol{r_s^c}$ whose value get enhanced by a factor of 16% at $\boldsymbol{b = 2a_0^*}$, if we include the *e-ph* coupling in addition to the usual *e-e* interactions. This implies that the inclusion of *e-ph* coupling significantly affects the Wigner crystallization by weakening the electron correlations in the electron quantum wire. On the other hand, the STLS theory does not predict formation of the Wigner crystal in electron quantum wire due to neglect of the dynamics of electron correlations.

Acknowledgments

DP and VV are grateful to CSIR (New Delhi) and DST for providing financial assistance.

References

[1] Wigner E 1934 *Phys. Rev.*461002
[2] Tanatar B andBulutay C 1999 *Phys. Rev. B*5915019
[3] Garg V *et al* 2008 *Phys. Rev. B*78045406
[4] Kaur K*et al*2020 *J. Phys.:Condens. Matter*32335403
[5] Casula M *et al* 2006 *Phys. Rev. B*74245427
[6] Steinberg H *et al* 2006 *Phys. Rev. B*73113307
[7] Hasegawa T andShimizu M 1975 *J. Phys. Soc. Japan*38965
[8] Friesen W I andBergersen B 1980 *J. Phys. C: Solid State Phys.*136627
[9] Mahan G D 1990 *Many Particle Physics 3rd Edition* (New York: Plenum)
[10] Pines D andNoziers P 1966 *The Theory of Quantum Liquids Vol 1* (New York: Benjamin)
[11] Pines D 1964 *Elementry Excitations in Solids*(New York: Benjamin)
[12] Singwi K S *et al* 1968 *Phys. Rev.*176589

Numerical Simulation of Silicon Crystal Growth: A 2-DOpen FOAM implementation

S G Nagarajan[1], Balram Panjwani[2], Eivind Johannes Øvrelid[2], M P Gururajan[1], and N N Viswanathan[1]

[1]Department of Metallurgical Engineering and Materials Science, Indian Institute of Technology Bombay, Mumbai-400076 India.
[2]SINTEF Industry, Trondheim-7465, Norway.
Email ID: sgnagarajan@ymail.com, gururajan.mp@gmail.com

Abstract. The main objective of this work is to develop a numerical solver to simulate the silicon solidification process using an open source software named open FOAM.We describe ways in which the existing solvers in open FOAM such as buoyant Boussineq PimpleFoam and buoyant PimpleFoam can be extended /modified to study solidification problems. Specifically, in this paper, we describe the development of the solidification solvers based on (a) the buoyant Boussineq PimpleFoam along with solidification Melting Source and, (b) bouyantPimpleFoam along with temperature dependent thermo-physical properties using the latest openFOAM2012 version. It is known that the solidification solver using bouyantBoussineqPimpleFoamalong with solidification Melting Source works well in cases where the solidification process can be modelled using a source term; we are also able to reproduce this textbook case satisfactorily. On the other hand, when we use the bouyantPimple Foam solver with temperature dependent properties for silicon solidification process, we are not able to satisfactorily match our results with those in the literature for the flow patterns. We present these preliminary results and indicate the possible reasons for the mismatch between our results from open FOAM and the results from a global model. The work on the benchmarking of the flow patterns and the incorporation of one more scalar equation to obtain the oxygen distribution in the solidification domain are in progress.

1. Introduction

The performance of Photovoltaic (PV) panels is significantly affected by the quality of the silicon crystals from which the wafers are prepared for the solar cell fabrication. Directional solidification (DS) process is one of the main methods to produce thesilicon ingots. The poly-silicon feedstock is melted in a quartz crucible and solidified from the bottom to top direction in the same crucible in a controlled heat transfer atmosphere. The quality of the mc-Si ingots is greatly affected by the heat transfer properties of the DS furnace. Nowadaysthe numerical simulations have become an important tool to analyse and optimize the process parameters of the solidification process which involves conduction and convection heat transfer. Computation fluid dynamics (CFD)is a powerful technique that is being used to solve a wide range of fluid flow and heat transfer problems using the basic principles of the conservation of mass, momentum and energy. The first step in the CFD simulation is the definition of the geometry and the grid generation, which sub-divides the domain into a number of smaller and non-overlapping control volumes.The conservation of relevant properties is established through the finite volume method (FVM) that involves the integration of relevant governing partial differential equations over all the control volumes of the simulation domain. The discretization of the integral equations leads to a system of algebraic equations, which are than solved by proper iterative methods. Due to the buoyancy force, the convection current is developed in the melt domain because of density difference that actually alters the shape of the solid-melt interface and species distribution during the solidification process. This convection in the fluid domain and the time dependent moving interface are the main features of the phase change problems that require the development of suitable governing equations and their numerical implementations. The governing equations with classical Stefan formulation have difficulties due to the involvement of transforming grids that need careful development of tailor-made codes. On the other hand, the enthalpy formulations lead to the fixed grid approach, which is more appropriate to be used when the phase change system has only one component. The zero velocity condition when the liquid turns into solid is established through number of alternative ways such as switch-off technique, variable viscosity technique, and Darcy source technique. Among these, the Darcy source technique has more physical significance for modelling the flow in the mushy zone; and,it only requires the addition of source terms in the momentum and energy equations [1].

OpenFOAM (Open Source Field Operation and Manipulation) is a free open source software developed using C++ to solve complex fluid flow problems.In open FOAM, there is a precompiled utility called solidificationMeltingSource [2] in the fvOption directory through which the additional source terms can be added to implement the Darcy source technique for phase change simulation. Besides, the pre-existing solvers can becustomized by directly adding the additional source terms in the source codes. In this work, we useboth the methods to develop the solver for phase change problemsby adding additional source terms. For developing phase change solver, wetake two existing solvers from openFOAM such as buoyantBoussinesqPimpleFoam and bouyantPimpleFoam among which we use bouyantPimpleFoam to develop the solver for silicon solidification process.The basic formulation, implementation and development of phase change solver is described in the nextsection. This is followed by some preliminary results obtained using the solvers. We conclude the paper with a brief summary of the status of our current numerical simulation efforts.

2. Formulation

The phase change problems are modelled based on three conservation principles, namely, mass, momentum and energy. The mass and momentum conservation equations are solved for a given geometry and boundary conditions in order to obtain the velocity field. Using the velocity (also called flow field), the convection and diffusion of the scalar quantities in the domain is calculated.

In the absence of volumetric dilatation, for an incompressible fluid, the mass conservation equation takes the following simple form:

$$div\,\vec{u} = 0 \tag{1}$$

where $\vec{u}$ is the velocity field of the fluid.

Equation 2 is the momentum conservation equation for y direction, which is nothing but the Navier-Stokes equation with two additional source terms. According to the fixed gird approach suggested by Voller and Prakash [3,4], there is no need to represent separate momentum and energy equations for the two phases. The equation 2 is valid for both solid and melt phases.

$$\frac{\partial}{\partial t}(\rho v) + div(\rho\vec{u}v) = \frac{-\partial P}{\partial y} + div(\mu grad v) + S_b + S_d \tag{2}$$

where ρ is the density, $\mu$is the dynamic viscosity. S_b is the buoyancy source term and $S_d$is an additional source term which is the Darcy source term. Note that in this formulation due to the presence of Darcy source term which dominates, the dynamic viscosity can be taken to have the same numerical value in both the solid and the melt even though the solid is infinitely viscous.

Equation 4 represents the temperature equation which is derived from the conservation of internal energy (equation 3). When substituting the expression for internal energy, the additional terms that appear (that is, terms except the temporal, convection and diffusion) are collected and expressed as an enthalpy source term (S_h).

$$\frac{\partial}{\partial t}(\rho h_{tot}) + div(\rho\vec{u}h_{tot}) = div(kgradT) \tag{3}$$

$$h_{tot} = c_P T + f_l L_f \tag{3a}$$

$$\frac{\partial}{\partial t}(\rho C_P T) + div(\rho C_P \vec{u} T) = div(kgradT) - S_h \tag{4}$$

$$S_h = \rho L_f\left[\frac{\partial f_l}{\partial t} + div(\vec{u} f_l)\right] \tag{4a}$$

where C_P is the specific heat capacity, L_f is the Latent Heat of Fusion, k is the thermal conductivity. The one more scalar equation can be added for analysing the species distribution in the solidification domain. The convection and diffusion of the species can be analysed by using the equation 3 by replacing the enthalpy by a scalar variable which will give the species (oxygen) concentration. In the results and discussion session, we describe the boundary conditions for the different simulations that are to be used along with the above partial differential equations.

3. Implementation

The blockMesh and snappyHexMesh utilities in the openFOAM are the available grid generation utilities for creating simple and complex geometries. In this work, the blockMesh utility is used to generate square and rectangle geometries with hexahedron control volumes where the front and back faces are assigned as empty patch type in order to implement 2D calculations. Two adjacent control volumes share each faces of the interior control volumes, whereas the face of control volume that is on the boundary of the domain is shared by only one control volume where the appropriate boundary conditions for relevant variables are assigned. The gradient and divergence terms in the governing equations are discretized based on the Gauss linear scheme. The temporal and Laplacian terms are discretized based on Euler(implicit) and Gauss linear orthogonal schemes.

As indicated in the previous section, we use two different approaches. In the first solver, the buoyancy driven flow is included using the Boussinesq approximation, which ignores the density variation, expect in the source term S_b where the gravity (g) term appears.

$$S_b = -\rho_o \vec{g}\beta(T - T_m) \tag{5}$$

where $\vec{g} = -9.8\hat{j}$, β is the thermal expansion coefficient, T_m is the melting temperature, ρ_o is the reference density.

In the second solver, the temperature dependent density variation has been used where the buoyancy force is calculated directly from the density variations.

$$S_b = \vec{g}(\rho - \rho_{ref}) \tag{6}$$

The additional source term S_d is added to ensure that the velocity field in the solid region is zero. In the isothermal phase change process, the mushy region is not present and the phase front is a finite line.

$$S_d = -C\frac{(1-f_l)^2}{f_l^3+q}v \qquad f_l(T) = \begin{cases}0, if\ T > T_m\\ 1, if\ T < T_m\end{cases} \tag{7}$$

where f_l is the liquid fraction, C and q are computational constants. If the temperature is higher than the melting temperature (T_m), the S_d will become zero and the equation 2 reduces to the general Navier-Stokes equation. If the temperature is lower than the melting temperature, the term S_d will dominate all other term in the momentum equation which will lead to $\vec{u} = 0$ in the solid region. So that constant C in the S_d term should be sufficiently large in order to extinguish the velocity in the solid region. The constant q is avery small number and is used to avoid the zero denominator when the liquid fraction is zero.

Equations 8(a),(b) are used to transform liquid fraction f_l from 1 to 0 when the temperatures become lower than the melting temperature (T_m). The superscripts n+1 and n refer to the present and last iterations. γ is the relaxation factor which is 0.9. Initially the temperature of the entire domain is higher than the melting temperature ($T > T_m$) and f_l^n=1. So according to 8(a) f_l^{n+1}>1 and equation 8(b) will reset the fraction back to f_l^{n+1}=1. When $T < T_m$, according to 8(a) f_l^{n+1}<1 and the liquid fraction will decrease until it reduces to zero [5].

$$f_l^{n+1} = f_l^n + \frac{\gamma C_p}{L_f}(T - T_m) \quad 8(a)$$

$$f_l^{n+1} = \max(0, \min(1, f_l^{n+1})) \quad 8(b)$$

The solidificationMeltingSource has links for all the source terms and calculations for the liquid fraction that can be used along with the buoyantBoussinesqPimpleFoam solver to deal with the phase change problems. Through the fvOptions the necessary entries required for the simulation can be given. In the other method, we edit the source files to include the additional source terms and to include the formulations for the liquid fraction calculation. The bouyantPimpleFoam source files (UEqn.H, EEqn.H) were edited to include the source terms. The further details about the implementation and the properties are given in the results and discussion. The input requirements, the suitable schemes and solutions for the corresponding solvers are available at the openFOAM site [5].

4. Results and Discussions

We have taken two different test problems for the implementation of the developed phase change solvers. The geometry, boundary conditions and simulation parameters for the first problem were taken from the paper by Voller et al.[1]. This problem is also implemented and solved using open FOAMby Torabi [6]; our implementation of this problem is a benchmark exercise and we are able to reproduce the flow patterns reported by Torabi by following the steps given in [6].In the second test problem, we take the geometry and boundary conditions from a global model for silicon solidification. For solver,we edit buoyant PimpleFoam source code to include the terms shown in the equation 6, 7 and 8. We implement the temperature dependent properties and make the necessary modifications by following the steps given by Hummel [7](for the n-octadecane melting process). Thus, in our implementation, the only modification we make is to use the parameters relevant for silicon solidification and start with a liquid fraction (f_l) initialized to unity on the overall domain. For the boundary conditions prescribed by us, the solidification starts from the bottom of the domain. In the following sections, we describe these simulations and the results from them in greater detail.

4.1. Phase Change Solver -1

In this approach, the existing buoyant Boussinesq Pimple Foam solver is used along with fvOption utility to simulate the phase change problem. We assume that the melt is an incompressible fluid and that the thermo-physical properties are constant; the fluid flow is considered to be laminar; the density variation due to temperature difference is incorporated through the Boussinesq approximation.This solver is generally used for solving the natural convection problems due to buoyancy driven flow. As we discussed in the previous section, the addition of Darcy and enthalpy source terms in the momentum and energy equations leads to the implementation of phase change solver which is carried out by using the derived solidificationMeltingSource utility from the fvOption directory. The fvOption file is added in the system case directory. The geometry, boundary conditions and simulation parametersaretaken from the paper by Voller et al [1]. The rectangle geometry of 1m X 1m with 40X40 grids is generated using blockMesh utility and the time step is 1. According to equation 5 the density variations are taken into account in terms of temperature difference that sets up convection current in the fluid domain. According to the boundary conditions, the left side wall is the colder side and the right side wall is the hotter side. The fluid near the hot wall moves upward and the fluid near the interface comes down, meanwhile the fluid from the hot wall carries heat and melts the top side of the solid-melt interface which is shown in the figures where the figure 1(a) shows the flow pattern reported by Voller et al and the figure 1(b) shows the flow pattern generated by the developed model in the openFOAM using solidificationMeltingSource utility. The addition of Darcy source term leads to zero velocity in the solid domain. The figure 1(c) shows the simulated temperature profile in the solidification domain.The solid-melt interface is also as expected.The limitations of using this solver to model the silicon solidification process are the following: (i) In this solver the thermo-physical properties of solid and liquid are same whereas they are different for Si solid and Si melt, (ii) Constant thermo-physical properties were considered here, whereas most of the thermo-physical properties of silicon are dependent on temperature (iii) The Boussinesq approximation used in this simulation is only valid when the temperature difference is small.

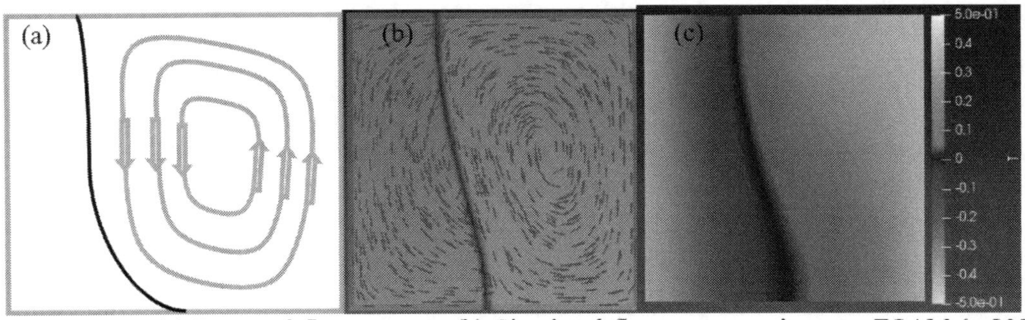

Figure 1. (a) Reported flow pattern, (b) Simulated flow pattern using openFOAM (t=2000s), (c) temperature distribution.

In the next part we take the bouyantPimpleFoam solver which is a solver for compressible flow and customize the solver to include all the above mentioned points to develop the solver for the silicon solidification process. Having said that, the implementation of this text book problem is a good starting point for openFOAMers to gain experience with openFOAM solvers for phase change problems [6]; and, our purpose in implementing the solver is also the same, namely to gain experience and comfort level in using openFOAM for solidification problems.

4.2. Phase change solver II – phaseChangeCavityFoam

In this approach, we consider the different thermo-physical properties for Si solid and Si melt.The density of the silicon solid is lesser than the silicon melt and the density of the silicon melt decreases as the temperature is increased which is shown in table 1. So it is necessary to take temperature dependent thermo-physical properties to carry out the numerical simulation. The equation 6 is used as the buoyancy source term instead of using the Boussinesq approximation. Instead of adding the derived utility, here the additional source terms in the momentum (UEqn.H), energy equations (EEqn.H), the phase fraction variable, and its implementation are directly added in the source file of the preexisting solver bouyantPimpleFoam. The detailed implementation steps are found in Ref [7]. OpenFOAM has a utility using which the temperature dependent properties can be defined in the form of polynomial expressions, however in the case of solidification, due to the presence of discontinuity at the melting temperature, the temperature dependent properties are given in the tabulated format. In order to access the intermediate values given in the table, we use linear interpolation. Table 1. Simulation parameters of silicon solidification [8]

Tmelt (K)	1685	Fluid Si: Dynamic viscosity (Pa s)	0.0008
L (Latent heat) (J/Kg)	1800000	q	0.02
Solid Si: Density (Kg/m^3)	2530	ρ_{ref} (Kg/m^3)	2551
Fluid Si: : Heat conductivity (W/mK)	66.5	Pr number	0.011
Solid Si: C_p (Specific heat)(J/Kg.K)	915	g (gravity) (m/s^2)	9.81
Fluid Si: C_p (Specific heat)(J/Kg.K)	1000	Cu	1X10^5
Fluid Si: Density (Kg/m^3)	P(T)=3193-0.3701T		
Solid Si: Heat conductivity (W/mK)	P(T)= 110.6122402-0.1507227384 T + 0.0001093579825 T^2 – 4.009416795 X 10^{-8} T^3 + 5.66839358 X 10^{-12} T^4		

The geometry of the computational domain, which consists crystal and melt region has been described using the default *blockMesh* utility with the dimensions of 876 mm x 245 mm, which is the size of G5 multi-crystalline silicon ingot (approx. 450 kg). The hexagonal mesh option has been used to define the 2D mesh (130 X 120) and the third direction has been set as empty. The geometry, grid size, and boundary conditions were taken from a global model. The position dependent temperatureon the boundary has been established using *codedFixedValue* option. Due to the addition of the Darcy source terms in the momentum equation and enthalpy source term in the energy equation, the zero velocity in the solid region is established and the solid and liquid fraction is identified using equations

8. The figure 2(a) shows the solid-liquid fraction from a global model developed using CGSim[8,9]. The tare 2(b) shows the solid liquid fractiondeveloped using customized openFOAM phaseChangeCavityFoam solver. The preliminary results generated using the openFOAM model fairly agree with the global model except near the edges where the openFOAM solution has a bump. However, as can be seen from the flow patterns shown in Fig 2(c) and 2(d), the openFOAM solver does not agree well with global model solutions. We believe this is because of the differences in the velocity boundary conditions on the surface of the melt; the boundary condition implemented by the global model is not known; on the other hand,we have implemented no-slip or zero velocity on the melt surface. We are exploring the other available boundary conditions in openFOAM to identify the appropriate boundary condition, which can give better match with the global mode. We are also in the process ofadding one scalar variable to the model and obtaining the parameters associated with such a variable from literature in order to track the distribution of oxygen in the phase change domain.

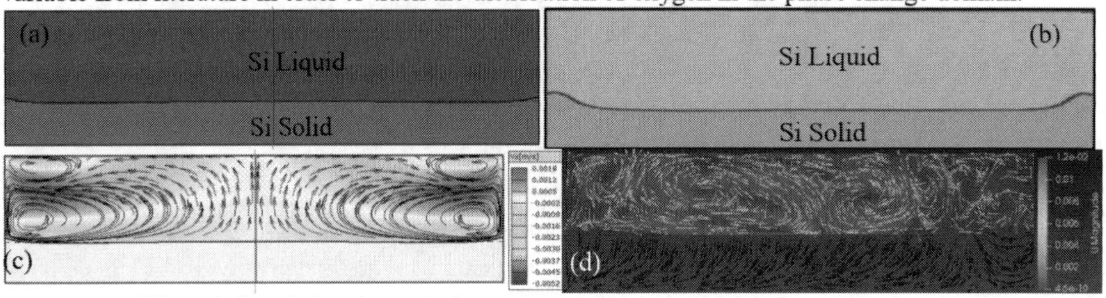

Figure 2. Solid and liquid fraction (a) from the global model, (b) from the openFOAM model; Velocity flow pattern (c) from the global model (d) from the openFOAM model.

5. Conclusions

We have described the development of the solidification solvers using the existing openFOAM solvers such as bouyantBoussineqPimpleFoam, and bouyantPimpleFoam.We are able to reproduce the results in the literature using our implementation of the solidification solver using bouyanBoussineqPimple Foam with fv Options utility, which uses Boussinesq approximation and considers constant thermo-physical properties for the phase change simulation. This exercise has given us confidence in using openFOAM for solidification problems. However, such a simple solver is inadequate for modelling the silicon directional solidification. Hence, we have used the bouyantPimpleFoam solver with temperature dependent thermo-physical properties. We have shown the preliminary result such as the solid-liquid fraction obtained from the developed openFOAM model agrees fairly well with the global model. Our preliminary results also show that the identification of appropriate boundary conditions is essential to obtain meaningful flow patterns and such an exercise is in progress.

Acknowledgements: One of us (SGN) thank IIT Bombay for the Institute Post-Doctoral Fellowship; we thank DST for funding this project work and C-DAC, Pune, and IIT Bombay for the computational facilities

References

[1] Voller VR, Cross M, Markatos NC, 1987, *Int. J. Num. Met. Eng.,* vol 24, pp 271-284.
[2] solidificationMeltingSource.html Accessed on 2021.
[3] Voller VR, Prakash C,*Int. J. Heat Mass Transf.,* 1987, vol 30, pp 1709-1719
[4] Voller VR, Swaminathan CR,*Numer. Heat Tr. B-Fund.,* 1991 vol 19, pp 175-189
[5] Torabi Rad M. In: Nobrega J., Jasak H. (eds) OpenFOAM, *Springer, Cham.*
[6] https://www.openfoam.com/documentation/guides/latest/doc/guide-applications-solvers-heat-transfer-buoyantPimpleFoam.html Accessed on 2021.
[7] DanielHummel/hummelReport.pdf Accessed on 2021.
[8] Nagarajan SG, Srinivasan M, Aravinth K, Ramasamy P, 2016. *Int. J. Mater. Res.,* vol 107, pp 525-533
[9] Smirnova OV, Vasif M, Mamedov, Kalaev VV, 2014, *Cryst. Growth Des.* vol 14, pp 5532–5536

Shape Control of Au Nanoparticles Using *Flacourtia Inermis* Extract

Rani M Pattabi[1, a)], Krishnaprabha M[2] and Manjunatha Pattabi[3]
[1,3] Department of Materials Science, Mangalore University, Mangalagangothri – 574199
[2]Department of Physics, Dr. Dayananda Pai- P. Satish Pai Government First Grade College Carstreet, Mangaluru - 575001, India and St. Aloysius College (Autonomous), Mangalore- 575003, Indi
[a)]Corresponding author: ranimpattabi@rediffmail.com

Abstract. Spherical Au nanoparticles and triangular nanostructures were synthesized with ease using the extract of the fruit of Flacourtia inermis. This sour fruit, colloquially called "lololikka" is found extensively in some regions of South India and it is reported to possess good antimicrobial activity. We have succeeded in controlling both the size as well as the shape of the particles by merely changing the reaction temperature or the extract concentration thereby rendering the synthesis protocol extremely useful for designing them for medical or catalytic applications. The effect of the nanoparticle shape in its catalytic activity has been studied by following the reduction reaction of 4- nitrophenol.

1. Introduction

Nanoparticles of both gold and silver find enormous applications in the field of catalysis, sensors, therapeutic and biomedical applications. Noble metal nanoparticles possess special optical and electronic properties which make them ideally suited for various applications such as anti-cancer agents, biocatalysts etc [1]. Although there are a large number of physical and chemical methods reported for the shape-controlled synthesis of metal nanostructures [2,3], the synthesis of anisotropic metallic nanoparticles and their shape control through green routes is still a challenge. Developing protocols for the synthesis of such particles using non-toxic, biological agents is desirable not only for their eco-friendliness but also in view of their applications in the bio-medical field. Thus, the use of biological agents such as fungi[4], bacteria [5]and plant extracts are being widely accepted as an alternate method, with plant extracts being one of the largest studied methods [6,7]. Extracts prepared from aloe vera, *Cinnamomum camphora*, *Carica papaya* [1] etc. have been reported where they act as reducing agents for the rapid synthesis of gold nanoparticles.

The present paper reports the synthesis and shape control of gold nanoparticles using the fruit extract of *Flacourtiainermis* belonging to *Flacourtiaceae* family. The fruits of this plant are edible and used in preparation of jams and pickles. The fruit isreported to possess good antifungal and antioxidant activities [8].

2. Experimental Details

A weighed amount of the fresh fruit is thoroughly washed, chopped into small pieces, boiled with known volume of deionized water and then filtered to get the extract. Gold nanoparticles were synthesized by mixing an aqueous solution of chloroauric acid ($HAuCl_4$) of the desired molarity with different amounts of the fruit extract, under continuous stirring. The synthesis is carried out at room temperature, elevated temperature and ice-cold condition.

The absorbance spectra of the Au nanoparticle colloids (AuNP) so obtained were recorded using UV–1800 Shimadzu UV spectrophotometer in the wavelength range of 200- 1000 nm. Morphological

studies were carried out, in a Carl Zeiss Field Emission Scanning Electron Microscope (FESEM) model Σigma operated at an accelerating voltage of 3 kV, from AuNP colloids dried on carbon tape.
The catalytic activity of the synthesized nanoparticles was assessed by studying the reduction of 4-nitrophenol to 4- aminophenol. For this, 1 ml of a 5 mM solution of 4-nitrophenol and 2 ml of a 0.25 mM $NaBH_4$ solution is added to 25 ml de-ionized water. To this, 2 ml of the nanoparticles are added, and the absorption spectra recorded as a function of time.

3. Results and Discussions

Localized surface plasmon resonance (LSPR) endows AuNPs with the ability to absorb light in the visible region which makes them useful in medical applications such as hyperthermic cancer treatment and imaging [6]. Changing the LSPR peak position can thus make these particles capable of absorbing light of different wavelengths, resulting in modifying their photothermal properties. Anisotropic metal nanoparticles exhibit LSPR behaviour different from spherical particles and hence in order tomake use of thisproperty, it is essential to find means to control both the size and shape of the particles. For medical/biological applications, it is especially desirable that the methods adopted for their synthesis and the reagents used are biocompatible [9]. In our studies, we demonstrate that control over particle size could be achieved by controlling the *concentration* of the fruit extract of *Flacourtiainermis*and control over the shape could be achieved by changing the reaction *temperature*.
Due to the size and shape dependent surface plasmon resonance (SPR), both silver and gold nanostructures exhibit attractive colours. On mixing different amounts of the fruit extract with $HAuCl_4$, a colour change was observed in the resulting solution. The colour of the solution turned from light pinkish to a turbid pinkish brown at low and room temperatures, whereas at 100 °C a dark violet coloured solution was obtained within 15 minutes after the addition of fruit extract.
Figure 1a shows the UV-Vis spectra of the particles synthesized by using the fruit extract prepared from 10 g of the fruit in 30 ml of de-ionized (DI) water. FE is the absorption spectra of the fruit extract which showed a peak at ~ 512 nm, corresponding to the absorption by the light pinkish coloured extract.

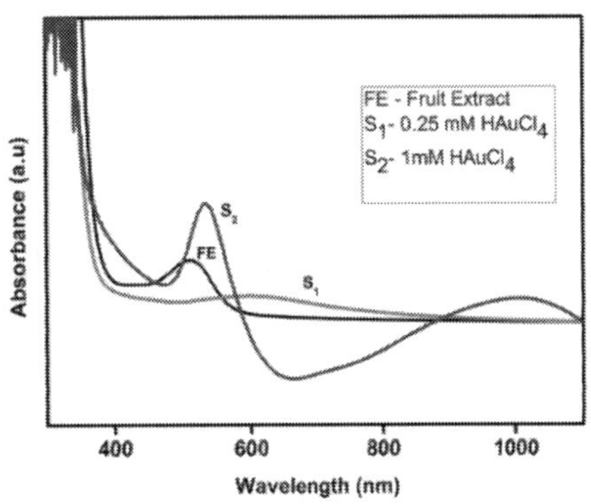

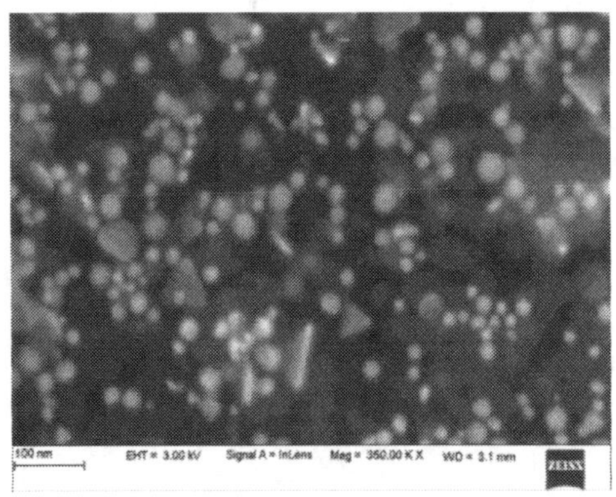

Figure 1. (a) UV-Vis spectra of Fruit extract (FE), 0.25 mM $HAuCl_4$ (S_1) and 1mM $HAuCl_4$ (S_2) reduced with4 ml of fruit extract and (b) FESEM micrograph of sample S_2.

S_1 and S_2 are the absorption spectra of Au nanoparticles obtained by adding 4 ml of this extract to 16 ml of 0.25 mM and 1 mM $HAuCl_4$ respectively. The spectra of S_1 exhibit a weak and broad SPR peak at 597 nm. Surface plasmon resonance peak from spherical AuNPs in aqueous media occurs at ~520 nm for particles of size approx. 10nm[7]. The growth of colloidal nanoparticles is by a nucleation growth mechanism and the peak of very low intensity may be because of insufficient nucleation [10]. The concentration of $HAuCl_4$ was increased and S_2 is the spectra obtained from a 16 ml, 1 mM

$HAuCl_4$ solution reduced with the same amount of fruit extract. The spectra exhibit two peaks, one at 533 nm and a second one at 1005 nm. The appearance of two peaks, one in the visible and another in the near infra-red region of the electromagnetic spectrum is known to be due to the formation of anisotropic particles [11,12,13].

Figure 1(b) shows the FESEM picture of this sample in which nanotriangles having edge length varying from ~ 25 nm to 75 nm, along with spherical particles are visible. This confirms that the appearance of two peaks in the absorption spectra of S_2 is indeed due to the formation of nanotriangles.

It has been reported that the position of the long wavelength peak in the absorption spectra of nanotriangles depends on their size - triangles with smaller edge lengths produce a blue shifted in-plane SPR peak [11]. In Figure 2 we reproduce the UV-Vis spectra of particles synthesized with two different concentrations of the extract. S_2 is the spectra of the sample given in figure 1 and S_3 is the spectra from a sample prepared using a higher concentration of extract i.e, 15 gm of fruit in 30 ml DI water. It can be observed that as the Au^{3+}: fruit extract concentration decreases, the out-of-plane SPR band is slightly red shifted whereas there is a large blue shift of the in-plane SPR band from 1005 nm to 865 nm. This indicates that control over the size of the triangles formed could be achieved by changing the concentration of the extract.

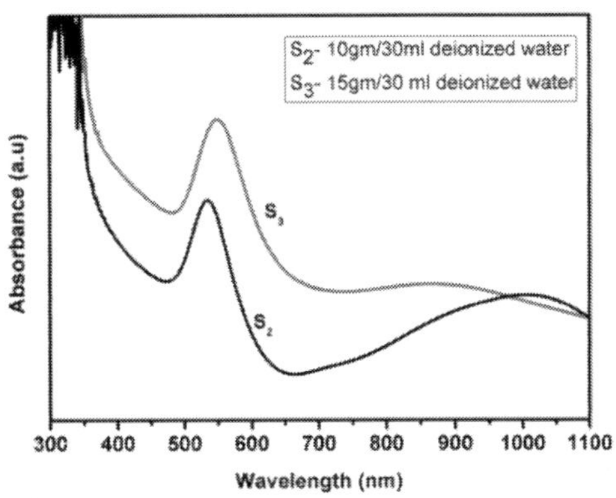

Figure2 UV-Vis spectra of 1 mM $HAuCl_4$ reduced with 4 ml extract of two different concentrations.

Finally we also demonstrate that a mere change in temperature of the reaction helps in producing particles of different shapes which is effectively revealed in the absorption spectra, figure 3, of the specimen synthesized at room temperature (S_3), 0°C (S_4) and at 100°C (S_5).

IOP Conf. Series: Materials Science and Engineering 1221 (2022) 012035 doi:10.1088/1757-899X/1221/1/012035

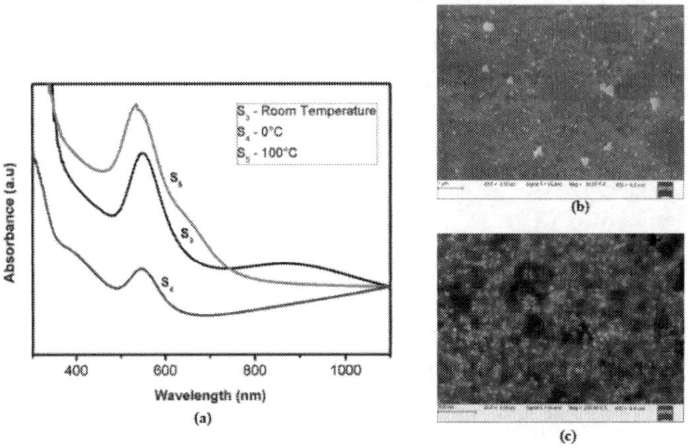

Figure 3 (a) UV-Vis spectra of nanoparticles synthesized at room temperature (S_3), 0°C (S_4) and 100°C(S_5), (b) FESEM micrograph of S_4 (scale bar 1μm) and (c) FESEM micrograph of S_5 (scale bar 200 nm).

The room temperature spectrum is of the same sample S_3 shown in figure 3. When synthesized at 0°C (S_4) the in-plane SPR peak is beyond the measuring range of the instrument. However, the spectra shows an increasing trend at higher wavelengths which indicates that the in-plane SPR peak does exist beyond 1200 nm whereas at room temperature this peak is blue shifted to 865 nm. This is also accompanied by an increase in the intensity of the out-of-plane SPR peak and a decrease in the in-plane SPR peak. At 100°C, the in-plane peak is only faintly distinguishable at 651 nm whereas the out-of plane SPR peak is blue shifted and is of larger intensity. Taken together, it can be understood that the nanostructures synthesized at lower temperatures using the extract of*Flacourtiainermis* produces larger anisotropic Au nanoparticles.
The FESEM pictures of S_4 and S_5, displayed in figure 3 (b) and (c) respectively, corroborates the finding of the UV-Vis spectra. S_4 consists of large triangular particles having an edge length ranging from ~125 to 500 nm consistent with the observation that the in-plane SPR peak of S_4 lies beyond 1200 nm. Comparing the size of the triangular particles in samples S_2 (shown in figure 1b) and S_4, it is evident that smaller triangles exhibit a blue shift in the position of the in-plane SPR peak. Hence, S_3 exhibiting a blue shifted peak at 865 nm suggests the formation of triangles of edge length smaller than that of S_4. Also, we observe that for the sample S_3 synthesized at room temperature, the intensity of the low wavelength peak is enhanced while that of the high wavelength peak decreases compared to that of S_4 synthesized at 0°C. It can be concluded that with increase in temperature, the number of sphĕrical particles increases with a concomitant decrease in the number of triangular particles. This aspect is further strengthened from the UV-Vis and FESEM of sample S_5 synthesized at 100°C, both reaffirming the presence of only spherical particles.
Shankar et al. [14] and Rai et al..[15] have bio-synthesized anisotropic Au nanostructures using lemon grass extract. In their studies, Au nanotriangles were obtained by varying the extract concentration [14] or temperature [15]. They conclude that a higher reaction temperature increases the reduction and consequently the nucleation rate of Au ions, leading to the formation of spherical particles. At lower temperatures, the reduced reduction rate allows the aggregation and rearrangement of smaller particles which act as nucleation centers for the anisotropic triangular structures.

The catalytic activity of sample S_2 was studied for its capacity in the reduction of 4-nitrophenol to 4-aminophenol. 4-nitrophenol is an organic pollutant present in industrial wastewater. They can cause adverse effect to the blood, liver and central nervous system. However, the reduced product, 4-aminophenol, is employed in the production of analgesics, corrosion inhibitors etc. [16]. It has been well reported that an aqueous solution of 4- nitrophenol exhibits two peaks in its absorption spectra,one at 226 nm corresponding to *π-π transitions from the phenolic ring and another at 317 nm corresponding to *π-n transitions from the lone pair of electrons in the oxygen and nitrogen atoms [17,18]. On the addition of $NaBH_4$, the 317 nm peak shifts to 400 nm due to the formation of 4-nitrophenolate ions. The intensity of this peak remains unaltered for as long as a few days. However, on addition of suitable catalysts the intensity of the peak at 400 nm reduces and a new peak at 300 nm emerges due to the formation of 4- aminophenol, which is an indication of the conversion of 4-nitrophenol. We have monitored the intensity of the peak at 400 nm, to assess the conversion of 4-nitrophenol as a function of time, for the sample S_2 consisting of nanotriangles and the result is presented in figure 4.

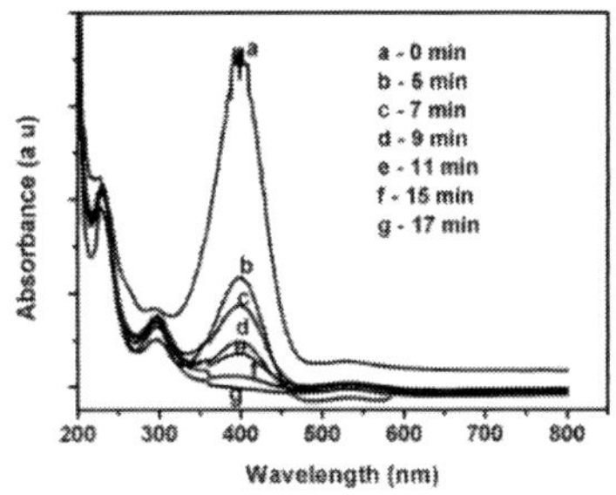

Figure 4. Reduction of 4- nitrophenol by sample S_2 as a function of time.

It can be seen that the intensity of the 400 nm peak decreases with a concomitant increase in the peak at 300 nm indicative of gradual transformation of 4- nitrophenol to 4 – aminophenol. Complete reduction of 4- nitrophenol is accomplished at 17 minutes, proving the efficient catalytic activity of these particles.

In conclusion, we have demonstrated a simple method for the size and shape control of gold nanoparticles by a bio-friendly method using the fruit extract of *Flacourtiainermis*. The synthesized nanotriangles exhibit very good catalytic activity towards the reduction of 4- nitrophenol. The reported simple protocol for the size-controlled synthesis of anisotropic gold nanotriangles may be promising since such nanostructures could find applications especially in the field of bio-medicine.

References

[1] Lee S Y Krishnamurthy S Cho C-W and Yun Y-S 2016 *ACS Sustainable Chem. Eng.*4 2651
[2] Abedini A Bakar A AA LarkiF Menon P S Islam M S and Shaari S *Nanoscale Research Letters* 11 287
[3] Kawamura G Nogami M and Matsuda A 2013 *Journal of Nanomaterials* 2013 1
[4] Kitching M, Ramani M and Marsili E 2015 *Microbial Biotechnology* 8 904
[5] Pourali P, Badiee S H, Manafi S, Noorani T, Rezaei A, Yahyaei B 2017 *Electronic Journal of Biotechnology*
29 86
[6] Vines J B, Yoon J-H, Ryu N-E, Lim D-J and Park H 2019 Front. Chem. **7** 167
[7] Kimling J, Maier M, Okenve B, Kotaidis V, Ballot H and Plech A 2006 *J. Phys. Chem. B* 110 15700
[8] Alakolanga A G A W, Kumar N S, Jayasinghe L and Fujimoto Y 2015 *J Food Sci Technol.*52 8383
[9] Huang X, Jain P K, El-Sayed I H and El-Sayed M A 2007 *Nanomedicine***2** 681
[10] Thanh N T K, Maclean N and Mahiddine S 2014 Chem. Rev. 114 7610
[11] Chandran S P, Chaudhary M, Pasricha R, Ahmad A and Sastry M 2006, *Biotechnol. Prog.*577
[12] Huang X and El-Sayed M A 2010 *Journal of Advanced Research* 1 13

[13] Altaf M, Jaganyi D 2016 *Synthesis and Reactivity in Inorganic, Metal-Organic and Nano-Metal Chemistry*
46 1332
[14] Shankar S S, Rai A, Ahmad A and Sastry M 2005 *Chem. Mater.* 17 566
[15] Rai A, Singh A, Ahmad A and Sastry M 2006 *Langmuir* 22, 736
[16] Krishna R, M. Fernandes D M, Domingos V F, Ribeiro E S, Gild J C, Diase C and Ventura J Freire C,Titus E 2015 *RSC Adv.*5 60658
[17] Seoudi R, Said D A 2011 *World J. Nano Sci. Eng.*1 51
[18] Krishnaprabha M and Pattabi M 2016 *Int J Nanosci.* 15 1660015

IOP Conf. Series: Materials Science and Engineering 1221 (2022) 012036 doi:10.1088/1757-899X/1221/1/012036

An Investigative Study on the effect of Dopants on Manganese Based Mixed Nano Ferrites Synthesized through Sol-Gel Method

Dhanya Jose[1], Sruthy Mildred[2], Rintu Mary Sebastian[3] and Sheena Xavier[4]

[1,2,4] St. Xavier's College for Women, Aluva, Kerala, India, 683101,[3]Morning Star Home Science College, Angamaly, Kerala, India, 683573
Email: sheena.xavier1@gmail.com

Abstract. The present article investigates the effect of different dopants on manganese based mixed nano ferrites synthesized through sol-gel method.MnF, MnCoF and MnCoMgF ($MnFe_2O_4$, $Mn_{0.5}Co_{0.5}Fe_2O_4$and$Mn_{0.5}Co_{0.25}Mg_{0.25}Fe_2O_4$) were synthesized by Sol-gel method. The structural, magnetic and optical properties of the samples were analysed through XRD(X-RayDiffractometer), VSM (Vibrating Sample Magnetometer,) and UV-Visible Spectroscopy. The phase identification and lattice parameter determination were carried out by using X-RayDiffraction and hence the crystallite size and X-Ray density were calculated. Peaks (220), (311), (400),(511), and (440) were obtained from the XRDpattern of the samples, which confirmed their single-phase cubic spinel structure. It was observed that because of the large ionic radius of Co^{2+} (0.745 Å) ions as compared to that of Fe^{3+} (0.645 Å) ions, lattice parameter of MnCoF increased in comparison with that of MnF but, the substitution of Mg^{2+} to MnCoF sample decreased the lattice constant. By investigating the hysteresis loops produced by the samples in the VSM analyses, it was found that the MnF have low coercivity and moderate saturation magnetization amongst the three. The substitution of Mg^{2+} in MnCoF sample decreased the value of saturation magnetization. UV-Visible analysis was used to determine bandgap energy of three samples.From the Tauc plot, we can observe bandgap energies of 3.202 eV, 3.180 eV and 3.239 eV, for MnF, MnCoF and MnCoMgF respectively.On substituting with the Co^{2+} in MnF the band gap of sample decreases indicating quantum confinement phenomenon.When substitution was done both Co^{2+} and Mg^{2+}, the band gap energy value increases.

1. Introduction

Nano ferrites have been of huge interest due to their wide spread applications in biomedical and technological applications. Their tunable electromagnetic properties make it a favorite for scientists all around the world. Manganese doped ferrites are one of the most explored amongst the ferrites for their high permeability, high magnetization, high resistivity and low core losses [1-4]. Manganese ferrite nanoparticles are used as bio-analytical tools for immunoassay and protocol for protein separation [5]. To ensure the purity and to control the size of the particles, novel preparation techniques and proper characterization techniques are prerequisites. The proper analyzation techniques help us to develop samples with suitable properties that we are looking for.In this study the sol-gel auto-combustion process is preferred in this study for the synthesis of the ferrite nanoparticles with the highest purity [6]. To maintain a good stoichiometric control, Sol-gel method is best, also we can produce ultrafine particles with narrow size distribution with very low expenses [7].
The objective of our study is to develop Manganese based mixed nano ferrites substituted with Co and Mg, synthesized through Sol-Gel method. The substitution plays part in tuning the properties of the magnetic nano ferrite under study. XRD, VSM and UV-Visible spectroscopy were used to conduct a proper characterization of the samples prepared [6].

2. Experimental

2.1. Synthesis of nanoparticles

In order to prepare manganese ferrite nanoparticles, analytical reagent (AR) grade manganese nitrate, ferric nitrate, magnesium nitrate and cobalt nitrate were used as chemical precursors.Firstly, $MnFe_2O_4$ (MnF) sample was prepared in which the ratio of Ferric nitrate to Manganese nitrate was 2:1. Then, ions ofCo^{2+} and Mg^{2+} were substituted into MnF to obtain the mixed ferrite nanomaterials.
The nitrates were dissolved in minimum amount of ethylene glycol at room temperature as per the required stoichiometric ratio. With the help of the magnetic stirrer the mixture is stirred till all the particles dissolve and the sol was heated at 60°C to obtain a wet gel. Further heating of the gel at higher temperatures led to self-combustion. The obtained dry powder is ground well using agate mortar to form ultrafine particles of manganese ferrite.Thus,$MnFe_2O_4$, $Mn_{0.5}Co_{0.5}Fe_2O_4$and$Mn_{0.5}Co_{0.25}Mg_{0.25}Fe_2O_4$ were synthesized [3].

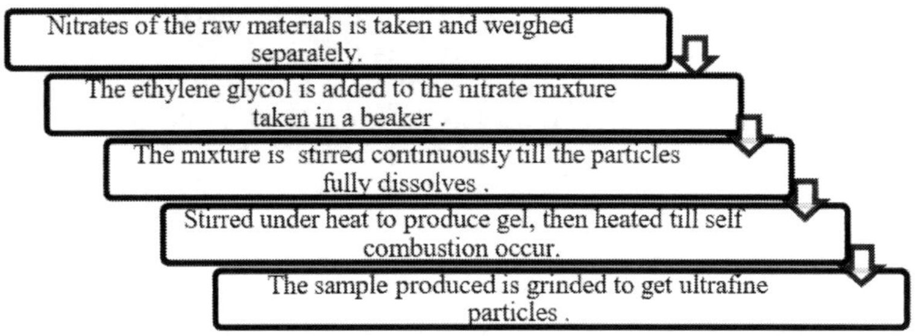

Figure 1: Schematic representation of the synthesis procedure

2.2. Characterization

To get the structural information of a material we have utilised the X-ray diffraction analysis, using the Bruker AXS D8 Advance Diffractometer. The phase identification and lattice parameter determination were carried out by using X-Ray Diffraction and hence the crystallite size and X-ray density were calculated.
We have the Bragg's condition,

$$2d\sin\theta = n\lambda$$

where λ is the wavelength of X-Ray radiation, n is an integer, d is the spacing between atomic planes in the crystalline planes and θ is known as diffraction angle. The intensity of the diffracted X-rays is measured and plotted as a function of diffraction angle 2θ. From the 2θ values of the peaks, with the assistance of the above equation, the lattice spacing (d) is calculated. From the d values we can calculate the lattice parameter a, hence the crystallite size D and X-ray densityρ.

3. Results and discussion

3.1. XRD analysis

In order to understand the crystalline structure and their behaviour we opt for X-Ray Diffraction analysis. For XRD analysis, we have used powder samples of MnF, MnCoF and MnCoMgF calcinated at 400°C for two hours. In this case, X-ray Diffractometer is equipped with Cu-Kα radiation (λ=1.5406 Å) in the range of 2θ=20° to 80°. The figure 2 show the X-ray Diffraction patterns of manganese ferrite and its metal substituted manganese ferrites. The XRD patterns from Figure 2 show peaks indexed as (220), (311), (400), (511) and (440), which were in agreement with the standard peaks of spinel ferrites as per the data provided by the ICDD (International Centre for Diffraction Data) [20].
From the XRD data, the lattice constant (a) for each peak of each sample is calculated from the knowledge of the interplanar spacing (d), and the miller indices (hkl). The crystalline size (D) is

calculated using the Scherer's formula. The X-ray density (ρ_x) is determined from the calculated values of the lattice constant.

Table 1: Position of the X-Ray peaks and corresponding miller indices

Sample	Miller indices (*hkl*) of the X-ray peaks and their position 2θ in degree				
	(220)	**(311)**	**(400)**	**(511)**	**(440)**
MnF	30.099	35.523	43.279	57.107	62.772
MnCoF	30.099	35.420	43.100	56.977	62.595
MnCoMgF	30.135	35.420	43.150	57.061	62.681

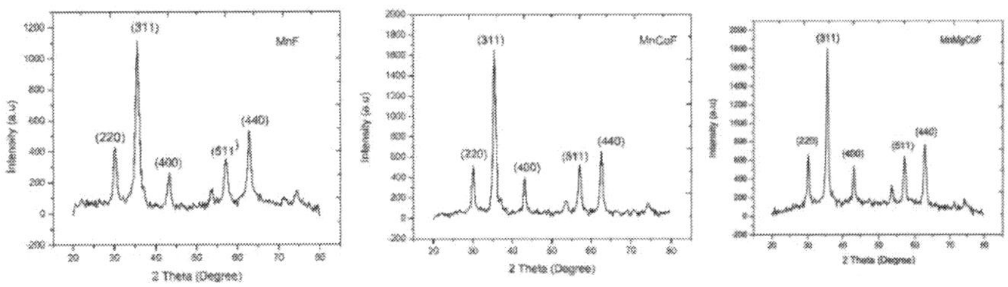

***Figure 2**:X-Ray Diffraction pattern for MnF, MnCoF and MnCoMgF*

From the calculated values of the crystalline size of the nanocrystalline ferrites under study it can be seen that the size ranges between 12 -17 nm. The comparison of the structural parameters of the sample givenis in Table 2.

It can be observed that the nanocrystalline manganese ferrite as well as its substituent nanoparticles showed a single-phase cubic spinel structure due to the presence of peaks at the hkl indices of (220), (311), (400), (511) and (440), which agrees with the standard reported values.Also, it is evident that there are no impurity crystal phase present in the sample. Thus, the formation of magnetic manganese ferrites is confirmed. The XRD lines are broad and the breadth decreases with the substitution of transition metals, which indicates gradual increase in the crystalline size as a function of the concentration of the Co^{2+} and Mg^{2+}.

Table 2: Comparison of structural parameters

Sample	$D(nm)$	$a(Å)$	$\rho(\frac{gm}{cm^3})$
MnF	12.442	8.376	5.302
MnCoF	15.957	8.396	5.265
MnCoMgF	16.092	8.378	5.300

The increase in the crystalline size can be considered as a good indication of improved crystallinity and great chemical homogeneity [9-10].The lattice constant of MnCoF is observed to be larger than that of MnF. The substitution of Mg^{2+} to MnCoF sample further decreases the lattice constant.

An increase in the lattice parameter with increase in Co^{2+} ion content is expected because of the large ionic radius of Co^{2+} (0.745 Å) ions when compared to that of Fe^{3+} (0.645 Å) ions. Thus, it is seen that the lattice constant is affected by the cationic stoichiometry[1]. It can be inferred that the values of the lattice parameter exhibit an almost linear dependence, which indicates that they are obeying Vegard's law [11-12].

Co^{2+} ions are expected to enter into the octahedral sites in place of Fe^{3+} ions which could result in an internal stress to make the lattice distorted and an expansion of unit cell [13-14]. From the table 2 we can see that there is a decrease in X-ray density when compared to manganese ferrite nanoparticle

VSM Analysis

The magnetic property is investigated with a vibrating sample magnetometer at room temperature. The observed hysteresis loops from vibrating sample magnetometer of the three samples at room temperature are shown in Figure 3and the saturation magnetization (M_s), coercivity (H_c), retentivity (M_r) and remnant ratio of the three samples are given in the table 3.

Table 3: Magnetization, Coercivity, Retentivity & Remnant ratio of the samples.

Sample	Saturation Magnetization (M_s)emu/g)	Coercivity (H_c) (Oe)	Retentivity (M_r) (emu/g)	Remnant Ratio(M_r/M_s)
MnF	64.038	6.37	0.420	0.0065
MnCoF	73.228	1028.90	23.837	0.3255
MnCoMgF	54.094	446.16	13.497	0.2495

The MnF sample have low coercivity and moderate saturation magnetization. The retentivity and remnant ratio of this sample is low when compared with other two samples. It is found that the coercive force of MnF is low compared to other samples. The magnitude of coercive force increases with the substitution of cobalt ions in MnF nanoparticles. The substitution of Mg ions in MnCoF sample further decreases the magnitude of coercive force. The low coercive force confirm that the MnF is a soft ferrite[15].

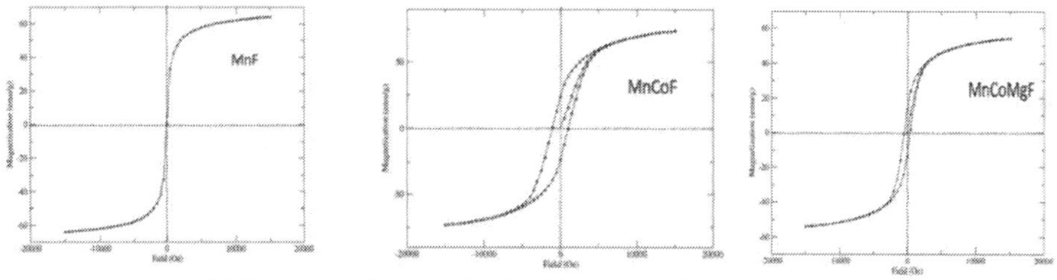

Figure 3: Hysteresis loop of MnF, MnCoF and MnCoMgF

The substitution of cobalt ions in manganese ferrite nanoparticles increased the value of saturation magnetization (M_s) and coercivity (H_c). This may be caused by the substitution of Co^{2+} ions by Mn^{2+} ions in the octahedral sites. The MnCoF sample have high retentivity and high remnant ratio compared to other samples. Therefore,the presence of Co^{2+} ions on octahedral sites may result in an increase in magnetic moment and equivalently an increase in magnetization.It is also observed that as the particle size increases the coercivity also increases[16-17]. The substitution of magnesium ions in MnCoF sample decreased the value of saturation magnetization. The decreasing concentration of Co^{2+} ions on octahedral sites may result in a decrease in magnetic moment and hence leads to a decrease in magnetization. The MnCoMgF has moderate coercivity compared to other samples. The MnCoF sample exhibit hysteresis loops of typical magnetic behavior, indicating the presence of ordered magnetic structure in the spinel system. The various hysteresis parameters are dependent on grain size, domain state, stresses and temperature [18].

Thus, from the investigation of the hysteresis loops produced by the samples in the VSM analysis, it was found that the MnF have low coercivity and moderate saturation magnetization amongst the three. The magnitude of coercive force increases with the particle size of the nanoparticles. The substitution of magnesium ions in MnCoF sample decreased the value of saturation magnetization. The decreasing concentration of cobalt ions on octahedral sites of MnCoMgF, may result in a decrease in magnetic moment and hence leads to a decrease in magnetization [10] [13].

3.2. *UV-Visible Analysis*

The samples are scanned within the frequency range 200-800 nm and graphs are plotted by taking hv along the x-axis and $(\alpha hv)^2$ along the y-axis. The obtained graph for various samples is given in the figure 4.1, 4.2 & 4.3. Tauc plot method is used to determine bandgap energy from absorption data. Extrapolation of linear regions of these plots of hv versus $(\alpha hv)^2$ gives the direct band gap values. The band gap energy values are tabulated in the table.

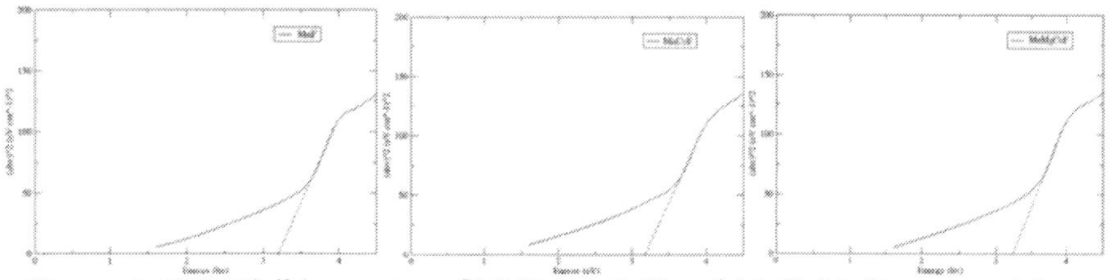

Figure 4: UV-Visible spectra of MnF, MnCoF and MnCoMgF nanoparticles.

Table 4: Comparison of Band Gap Energies

Sample	Band gap energy, E_g (eV)
MnF	3.202
MnCoF	3.180
MncoMgF	3.239

The estimated band gap values of MnF, MnCoF and MnCoMgF are tabulated in the above table. By analyzing the table, it is clear that on doping with the Co^{2+} ions the band gap decreases with increase in Co^{2+} ions concentration in manganese ferrite. This decrease in the bandgap energy from MnF to MnCoF can be attributed to quantum confinement phenomenon [19]. But when we make doping with both Cobalt and Magnesium, the bandgap energy value increases. From the Tauc plot, we can observe bandgap energies of 3.202 eV, 3.180 eV and 3.239 eV, for MnF, MnCoF and MnCoMgF respectively.

4. Conclusion

Nanocrystalline $MnFe_2O_4$,$Mn_{0.5}Co_{0.5}Fe_2O_4$and$Mn_{0.5}Co_{0.25}Mg_{0.25}Fe_2O_4$samples were synthesized by Sol-Gel auto combustion method and their structural, magnetic and optical properties were studied. Peaks (220), (311), (400), (511) and (440) was obtained from the XRD pattern of the samples, which confirmed their single-phase cubic spinel structure. The substitution of magnesium to MnCoF sample decreased the lattice constant. By investigating the hysteresis loops produced by the samples in the VSM analyses, it was found that the MnF have low coercivity and moderate saturation magnetization amongst the three. The substitution of magnesium ions in MnCoF sample decreased the value of saturation magnetization. From UV-Visible analysis we were able to determine bandgap energies of the samples. It was found to be 3.202 eV, 3.180 eV and 3.239 eV, for MnF, MnCoF and MnCoMgF respectively. On substituting with the cobalt ions in MnF the band gap of sample decreases indicating quantum confinement phenomenon. But when we substitute with both Co and Mg, the band gap energy value showed as increase.

Acknowledgments

Authors wish to acknowledge the help provided by STIC, CUSAT and SAIF, IIT Chennai for providing facilities for measurements and DST-FIST for the financial support to St. Xavier's College for Women, Aluva.

References

[1] Bhandare S V, Kumar R, Anupama A V, Mishra M, Kumar R V, Jali V M and Sahoo B 2020

*Mat. Chem.And. Phys.*vol **251**123081
[2] Aravind G, Raghasudha M, Ravinder D, Raja M M, Meena S S, Bhatt P and HashimM 2015 *Cera. Int.* vol **42(2)B** p 2941-50
[3] Sekhar B C, Rao G S N, Caltun O F, Lakshmi B D, Rao B P and Rao P S V S 2016 *J. of. Mag.And. Mag. Mat.* vol **398**p 59-63
[4] Dippong T, LeveiEA and CadarO 2021, Recent Advances in Synthesis and Applications of MFe2O4 (M = Co, Cu, Mn, Ni, Zn) Nanoparticles*Nano.* vol **11(6)** 1560
[5] Hung LY, Chang JC, Tsai YC, et al 2014, *Nanomed.* vol **10(4)** p 819-29
[6] Rahim S, Iftikhar F J and Malik M I 2019, Biomedical applications of magnetic *Nanoparticles for Drug Delivery and Diagnostic App. (Elsevier)* chapter 6 p 301-28
[7] Brinker C J and Scherer G W 1990, *Sol-Gel Science* (Boston: Academic Press)
[8] B. D. Cullity 1959, *The Elements of X-ray Diffraction*(London: Addison-Wesley)
[9] Desai I. Nadagouda M N, Elovitz M, Marc M and BoulangerB 2019 *Mat. Science for Energy Tech.*vol **2(2)** p150-60
[10] Rao B P, Caltun O, Cho W S, Kim C O and Kim C G 2007 *J. of. Mag. And. Mag. Mat.* vol **310 (2-part3)** pe812-14
[11] Köseoğlu Y, Alan F, Tan M, Yilgin R and Öztürk M 2012, *Cera. Int.* vol **38(5)** p3625–34
[12] Gabal M and Ata-Allah SS 2004, *Mat. Chem. and Phys.* Vol **85(1)** p 104-12
[13] Ati A A, Abdalsalam A H andHasan A S 2021, *J. of Materials Science:* **32(3)**p 3019–37
[14] Kumar R and Kar M 2016, *J. of. Mag. And. Mag. Mat.* Vol **416** p 335-41
[15] Kotnala R K and Shah J 2015, Ferrite Materials: Nano to Spintronics Regime ed KHJ Buschow *Handbook of Magnetic Materials (Elsevier)* chapter 4 vol **23**p 291-379
[16] Ishizaki T, Yatsugi K and Akedo K 2016, *Nanomaterials (Basel)* vol **6(9)** 172
[17] KnellerE F and Luborsky F E 1963, *J. Appl. Phys.*vol **34(3)** 656
[18] Sobhani A and Salavati-Niasari M 2015, *J. of Alloys and Comp.*vol 625 pg 26-33
[19] Race T D 2014, Nanodevices: fabrication, prospects for low dimensional devices and applications ed Martin Feldman*Nanolithography(Woodhead Publishing)*p 399-423.
[20] Xavier S, Jose D , George S and Alekha K V 2020, *AIP Conf. Proc.***2263** 060007

IOP Conf. Series: Materials Science and Engineering 1221 (2022) 012037 doi:10.1088/1757-899X/1221/1/012037

Diamond-Like Carbon Thin Films from Low-Pressure and High-Density CH_4 Plasma

Sucharita Saha and Debajyoti Das*
Energy Research Unit, School of Materials Sciences
Indian Association for the Cultivation of Science,
Jadavpur, Kolkata – 700 032, INDIA

*E-mail: erdd@iacs.res.in

Abstract. DLC thin films were grown using a low-pressure high-density CH_4/Ar plasma on glass substrates in a planar RF (13.56 MHz) inductively coupled plasma (ICP) CVD system. Using a plasma triggered by RF power to a mixture of 20 sccm CH_4 and 50 sccm Ar gas maintained at a low pressure of ~30 mTorr, a series of samples were prepared by varying the substrate temperature and then changing the RF power. The optimised DLC films, obtained at 500 °C and 900 W demonstrated a high optical transmission of ~97 % corresponding to awideoptical band gap of ~3.58 eV; and a minimum of I_D/I_G ~0.61 along witha maximum ofI_{Dia}/I_D ~1.32 in Raman analysis established the maximum sp^3/sp^2 content in the carbon network.

1. Introduction

Diamond-like carbon (DLC) films are comprisedof C atoms arranged in different co-ordinations e.g., sp^3, sp^2 and sp[1,2]. DLC films possess high hardness, electrical insulation, low friction coefficient and are highly transparent, which make them useful in opto-electronic devices. DLC films are prepared by magnetron sputtering, capacitively coupled RF plasma CVD and mostly by high-pressure microwave plasma CVD [3–6].In thepresent work we intend to prepare DLC films on inexpensive glass substrates from alow pressure (~mTorr) and high-density (CH_4 + Ar) plasma of planar inductively coupled plasma (ICP) CVD.

2. Experimental

DLC thin films were prepared in an ICP CVD system with an RF (13.56 MHz) power source. A series of films were deposited on Corning® Eagle™ glass substrates at 30 mTorr deposition pressure, using 20 sccm CH_4 gas as the main precursor and 50 sccm Ar as the diluent gas. The substrate temperature (T_S) was varied from 200 °C to 500 °C at RF power of 500 W. At an optimised substrate temperature, the effect of changing the RF power (from 500 W to 900 W) on the sp^3/sp^2 content of the films was studied. The samples were characterized by Raman data obtained from a J-Y Horiba Raman spectrophotometer equipped with 532 Ar^+ laser excitation at ~2 $mWcm^{-2}$ power density. Optical transmission of the films was measured by Hitachi 330 double-beam spectrophotometer. Surface morphology of the films was studied using JEOL JSM-6700F field effect scanning electron microscope.

3. Results and discussion

Figure 1(a) shows the deconvoluted Raman spectrum of the film prepared at 500 °C and 500 W RF power. The deconvoluted Raman spectrum of each of the films showssixsatellite peaks. 'Diamond-peak' at 1332 cm^{-1} arises from the crystalline orientation of sp^3 hybridized C-C bonds of diamond phase. The disorder present in sp^2 bonded carbon atoms originates the D-peak at ~1350 cm^{-1}. The G-peak at ~1600 cm^{-1} evolves from the stretching vibrations of sp^2 (graphitic) C=C bonds.

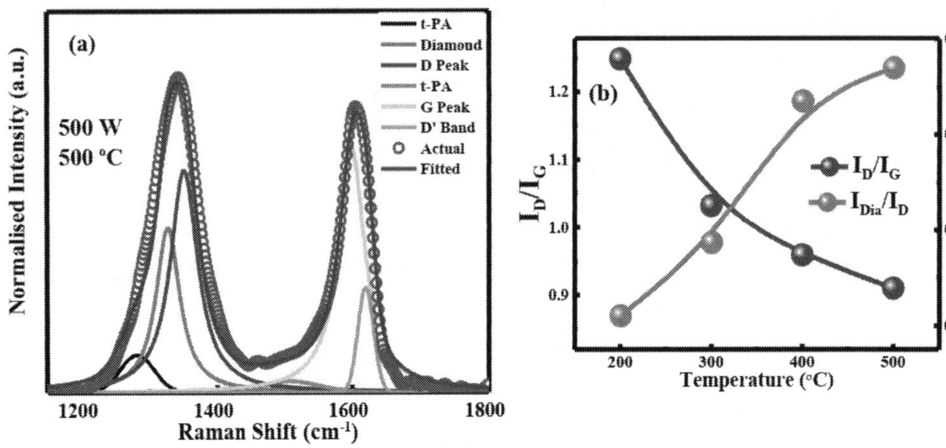

Figure1.(a)The deconvoluted Raman spectra of the film grown at 500 Wand 500 °C, (b) the changes in I_D/I_G and I_{Dia}/I_D with increasing substrate temperatures.

The D′-band at 1620 cm^{-1} arises from disorders in the orientation of graphitic lattice. Two companion peaks at 1150 cm^{-1} and 1480 cm^{-1} arise due to the presence of trans-polyacetylene (t-PA). A small shift of the t-PA peaks from the standard positions may occurdue to the contributions from the grain boundaries [7].

The decrease in the I_D/I_G (intensity of the D peak with respect to the G peak) ratio with increasing temperature [Figure 1(b)] indicates that there is a decrease in disorder within the network. The simultaneously increased intensity of the diamond peak with respect to the D peak (I_{Dia}/I_D) signifies that the diamond content in the film increases at high temperatures. Accordingly, it is inferred that the film prepared with 20 sccm CH_4 at 500 °C and 500 W RF power is the optimum DLC film with characteristic minimum I_D/I_G~0.91 and maximum I_{Dia}/I_D ~0.74 which are related to the increased sp^3/sp^2 content in the carbon network [8].

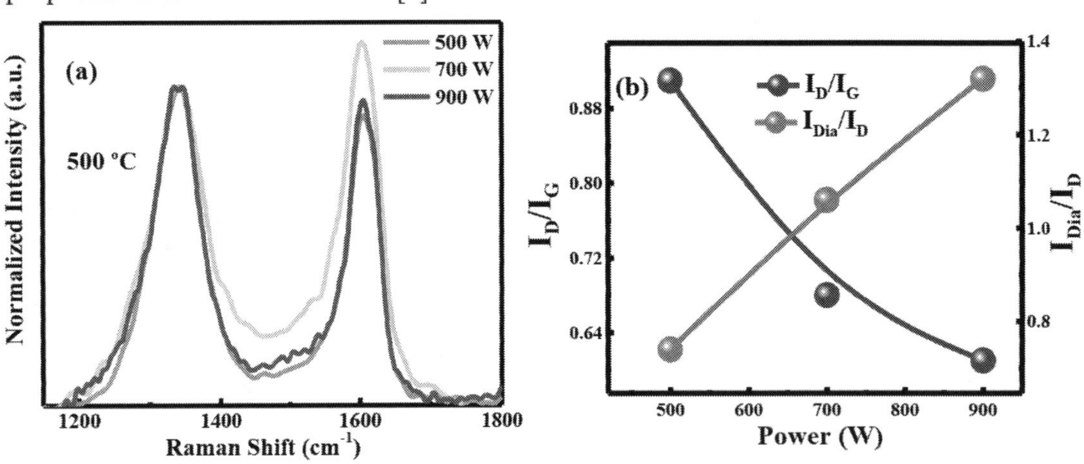

Figure 2.(a) Raman spectra of the films grown at varying RF powers, (b) variation of I_D/I_G and I_{Dia}/I_D with RF power.

In order to further increase the sp^3/sp^2 content of the DLC films, the effect of varying RF power from 500 W to 900 W was studied at the previously optimised condition. Figure 2(a)represents the Raman spectra of the films deposited with 500, 700 and 900 W. From Figure 2(b), the increasing I_{Dia}/I_D along with decreasing I_D/I_G ratios at elevated RF powers signifies that the films with enhanced diamond content and lower defects are formed[9,10]. Thus, at an optimum RF power of 900 W and 500 °C, the DLC film possesses the least I_D/I_G value ~0.61 and a maximum I_{Dia}/I_D value of 1.32.

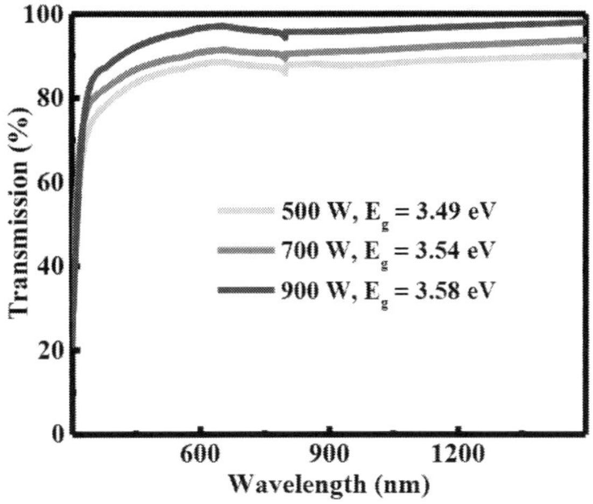

Figure 3. Transmission spectra of films prepared at 500–900W.

The high transparency of the DLC films makes them useful as optical coatings for optoelectronic devices. The transmission spectra of the films prepared at 500, 700 and 900 W, obtained from UV-Vis spectroscopy are shown in Figure 3. All the films in this series have high transmission above ~80 % in the wavelength range 450-1500 nm. The optical band gaps are calculated from the Tauc's plots, obtained from the above transmission spectra. These are tabulated at the inset of Figure 3. The DLC films, prepared at 900 W, shows a maximum transmission of ~97 %and an optical band gap of ~3.58 eV.

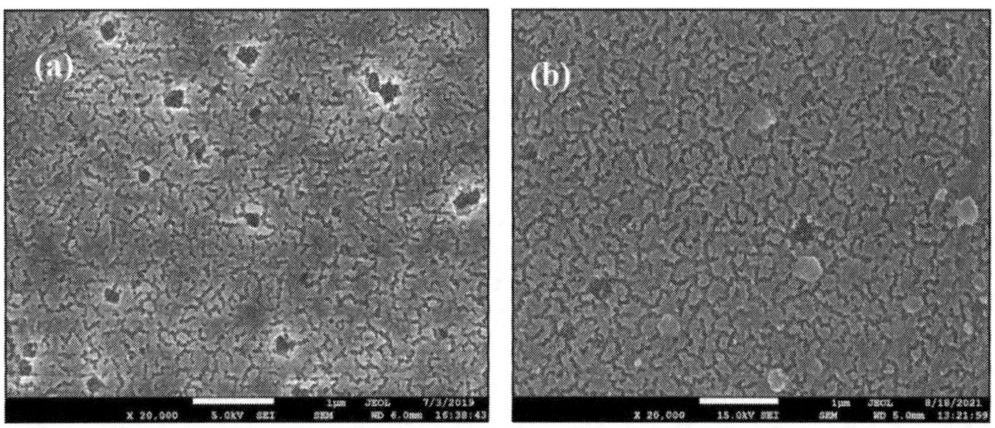

Figure 4. SEM micrograph of the film grown at (a) 700 W and 500 °C, (b)900W and 500 °C.

The SEM images of the DLC filmsprepared at 700 W and 900 Wat substrate temperature of 500 °Care shown in Figure 4(a)and4(b). Both images demonstratedense and uniform matrix with patches and pores. The holes present in the 700 W film are more abundant as compared to that in film grown at 900 W RF power.

In ICP CVD, a high ion-density in the plasma (~10^{11}–10^{12}) is obtained via inductive discharges, enabling low pressure depositions [11].The formation of DLC films is achieved by the dissociation of the CH_4 molecules into lower hydrides CH_x, which act as the precursor ions for film formation [12]. Here, DLC films with increased sp^3/sp^2 content are formed at higher substrate temperatures and RF powers. At elevated substrate temperatures, the mobility and hence reactivity of the precursor ions increases that facilitates in reducing the defect densities in the network. Increased RF power helps in effective dissociation of CH_4 into suitable precursors for increased growth rate of the DLC films.

4. Conclusion

DLC films were grown by a low pressure and high-density plasma in a planar ICP CVD system with 20 sccm CH_4 and 50 sccm Ar at 30 mTorr and 500 W RF power, by varying the substrate temperature (200–500 °C) and further by increasing the RF power (500–900 W) at 500 °C. Optimum DLC films obtained at 500°Cand 900 W RF power demonstrated an optical band gap of ~3.58 eV and a related high optical transmission of 97 % in the vis-NIR region. A minimum of I_D/I_G~0.61 and a maximum of I_{Dia}/I_D~1.32 in the Raman spectrum demonstrated the superior sp^3/sp^2 content in the carbon network.

Acknowledgements
The work has been done under projects funded bythe Department of Science and Technology(DST) and Council of Scientific and Industrial Research (CSIR), Government. of India. SS acknowledges DST, for providing Senior Research Fellowship (SRF).

References

[1] Das D 2019, *Carbon-Based Nanofillers and their Rubber Nanocomposites*, Editors: Yaragalla, Mishra, Thomas, Kalarikkal and Maria (Amsterdam, Netherlands: Elsevier), 123–181
[2] Reisel G,Wielage B,Steinhauser S and Hartwig H 2003, DLC for tools protection in warm massive forming, *Diam. Relat. Mater***12**1024–1029
[3] Das D and Banerjee A 2015, Anti-reflection coatings for silicon solar cells from hydrogenated diamond like carbon, *Appl. Surf. Sci* **345** 204–215
[4] Caschera D, Federici F, Kaciulis S, Pandolfi L, Cusmà A and Padeletti G 2007, Deposition of Ti-containing diamond-like carbon (DLC) films by PECVD technique, *Mater. Sci. Eng.* C **27** 1328–1330
[5] Xu J, Fan H, Kousaka H, Umehara N, Diao D and Liu W 2007 *Diam. Relat. Mater***16** 161–166
[6] May P W 2000, Diamond thin films: a 21st-century material,*Phil. Trans. R. Soc. London***A358** 473–495
[7] Klauser F, Steinmu¨ller-Nethl D, Kaindl R, Bertel E and Memmel N 2010, Raman studies of nano- and ultrananocrystalline diamond films grown by hot-filament CVD, *Chem. Vap. Deposition* **16** 127–135
[8] Tai F C, Lee S C, Wei C H and Tyan S L 2006, Correlation between I_D/I_G ratio from visible Raman spectra and sp^2/sp^3 ratio from XPS spectra of annealed hydrogenated DLC film,*Mater. Trans* **47** 1847–1852
[9] Das D and Banerjee A 2015, Further improvements of nano-diamond structures on unheated substrates by optimization of parameters with secondary plasma in MW-PECVD, *Surf. Coat. Technol* **272** 357–365
[10] R. Paul, Das S N, Dalui S, Gayen R N, Roy R K, Bhar R and Pal A K 2008, Synthesis of DLC films with different sp^2/sp^3 ratios and their hydrophobic behaviour, *J. Phys. D: Appl. Phys* **41**055309
[11] Raha R and Das D 2013, Nanocrystalline silicon thin films prepared by low pressure planar inductively coupled plasma, *Appl. Surf. Sci***276** 249–257

[12] Riccardi C, Barni R, Fontanesi M and Tosi P 2000 Gaseous precursors of diamond like carbon films in Ar/CH_4 plasmas,*Chem. Phys. Lett***329** 66–70.

IOP Conf. Series: Materials Science and Engineering 1221 (2022) 012038 doi:10.1088/1757-899X/1221/1/012038

Influence of lanthanum doping on the structural and optical properties of cerium tungstate nanocrystals

Anjaly Jose[1,2], Hitha H[1,2], Soumya Kuriakose[2,3], Mathew John[1,2] and Thomas varghese[1,2*]

[1]Nanoscience Research Centre (NSRC), Dept. of Physics, Nirmala College, Muvattupuzha, 686 661, Kerala, India

[2]Dept. of Physics, Newman College, Thodupuzha, 685584, Kerala, India

[3]Dept. of Physics, St. Peters College, Kolencherry, 682311, Kerala, India

*Email: gigithomas08@gmail.com

Abstract

In this study, the effect of lanthanum (La) doping on the structural and optical properties of cerium tungstate nanocrystals synthesized by the chemical precipitation method is investigated. The X-ray diffraction studies confirmed the crystalline phase formation and observed no phase change with doping. It is interesting to note that the crystallite size is decreased with La doping. The characteristic vibrational peaks of W-O bonds are confirmed from the FTIR and Raman spectroscopic studies and found no appreciable changes in the spectral positions with doping. The optical properties are studied using UV-Vis spectroscopy and photoluminescence spectroscopy. The characteristic broad blue PL emission from the tetrahedral tungstate group is observed for the pure and doped samples, and the appreciable variation in PL intensity can be associated with the formation of defects that occurred while doping.

1. Introduction

Metal tungstates have been effectively used for scintillation detectors. The defects in metal tungstates that arose/developed during the synthesis can exert a significant influence on the scintillation performance of the material [1]. The radiation properties can be enhanced by doping with suitable ions [2, 3]. There have been many studies to explain the dependence of photoluminescence on metal vacancies, oxygen vacancies and oxygen interstitials [4].

The $Ce_2(WO_4)_3$ possess distorted Scheelite structure crystallized in a monoclinic crystal system with a distorted tetrahedral WO_4 [5]. The $Ce_2(WO_4)_3$ has many defects due to the easy oxidation of Ce^{3+} into Ce^{4+} in $Ce_2(WO_4)_3$ causing many oxygen vacancies in the structure. There are many reports on the increase in photoluminescence with lanthanum doping in metal oxides and tungstates [4, 6]. Many authors have reported that with lanthanum doping in metal oxides, the enhancement in PL intensity is due to the formation of more oxygen vacancies and surface defects [6]. In metal tungstates, the doping mechanism of La^{3+} ions especially at high concentration on the effect of optical properties is a matter of debate [7-10].

In this paper, we have synthesized lanthanum doped cerium tungstate nanoparticles by the chemical precipitation method and studied their structural and optical properties in detail.

2. Experimental details

2.1. Materials
The chemicals used for the synthesis are cerium (III) nitrate hexahydrate, $Ce(NO_3)_3.6H_2O$ (Sigma Aldrich, 99.9%) and lanthanum (III) nitrate, $La(NO_3)_3$. $6H_2O$ hexahydrate (Merck, 98%) as cation sources and sodium tungstate, $Na_2WO_4.2H_2O$ (Merck, 99.9%,) as anion source.

2.2. Synthesis Route
The undoped cerium tungstate and a series of lanthanum doped cerium tungstates were synthesized using the chemical precipitation method. The concentrations of La^{3+} taken for doping at 2, 4 and 6 mol% of cerium. The solutions of required molar concentrations of cerium nitrate and lanthanum nitrate mixed well in a magnetic stirrer. After the thorough mixing, sodium tungstate solution poured into this mixture solution, stirred for half an hour and precipitate was obtained. The precipitate was washed several times with distilled water and then with ethanol to remove any unreacted products and impurities. The product was then dried at 70°C for 20 h in a hot air oven. The dried powder crushed well using a mortar and pestle and calcined at 600 °C in a muffle furnace.

2.3. Characterization
The X-ray diffraction (XRD) patterns of pure and lanthanum doped samples were recorded by a Bruker D8 advance X-ray diffractometer with CuKα radiation (λ=1.5406 Å, X–ray tube voltage = 35 kV and current = 35 mA). Diffraction data was taken in the 2θ range from 10 to 60° at increments of 0.02° with a step time of 29.1s. Fourier transform infrared (FTIR) spectra were recorded by Thermonicolet Avatar 370 FTIR spectrophotometer in the mid-IR range 4000 to 400 cm^{-1}. FT-Raman spectra were recorded using Bruker RFS-27 Raman spectrometer using Nd-YAG Laser of wavelength λ = 1064 nm. UV-Vis absorption spectra were recorded using Shimadzu 2600 UV–Visible spectrophotometer. The photoluminescence spectra were recorded using Horiba Flourolog 3spectrometer keeping the entrance and exit slit width 4 nm.

3. Results and Discussions

3.1. XRD analysis
The crystal structure, phase purity and average crystallite size of lanthanum doped cerium tungstate nanoparticles are determined from the X-ray diffraction patterns. 'Figure 1' shows the X-ray diffraction patterns of pure and doped samples. All diffraction peaks can be indexed with the monoclinic type structure of $Ce_2(WO_4)_3$ with space group C2/c, confirmed from ICDD file No. 98-040-1920 [5]. Interestingly, no additional phases are observed with lanthanum doping. As $Ce_2(WO_4)_3$ possess both Ce^{3+} (ionic radius, r=1.143 Å) and Ce^{4+} (ionic radius, r=0.97 Å) oxidation states, the possibility of substitution of both sites with La^{3+} (ionic radius, r =1.16 Å) exists.

The crystallite sizes of the nanoparticles are calculated from the line broadening of XRD patterns using the Scherrer equation [11],

$$D=k\lambda/\beta \cos\theta, \quad (1)$$

where D is the crystallite size, k=0.9, λ= 0.15406 nm for Cu-Kα radiation, β the full width at half maximum (FWHM) of the peak and θ the Bragg's angle. The average crystallite sizes obtained are tabulated in Table 1. The crystallite size is found to be decreased with lanthanum doping ,that is due to the distortion in the host lattice, which decreases nucleation and growth of $Ce_2(WO_4)_3$ [12].Williamson

IOP Conf. Series: Materials Science and Engineering 1221 (2022) 012038 doi:10.1088/1757-899X/1221/1/012038

Hall analysis is carried out to identify the line broadening contributions from crystallite size and microstrain values [13]. The W-H equation is

$$\beta \cos\theta = \frac{k\lambda}{D} + 4\varepsilon \sin\theta, \quad (2)$$

where β, θ, λ and D represent the same parameters as in Scherrer equation and ε is the micro-strain value. The crystallite sizes obtained from W-H plots are bigger than that obtained from Scherrer equation. The micro-strain values are also increased with increase in lanthanum doping concentrations which indicates more defects are created due to lanthanum incorporation in the lattice.

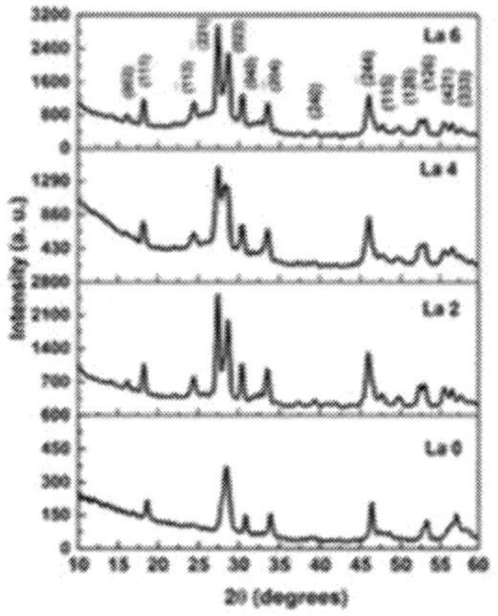

Figure 1. XRD patterns of pure and lanthanum doped cerium tungstate nanoparticles

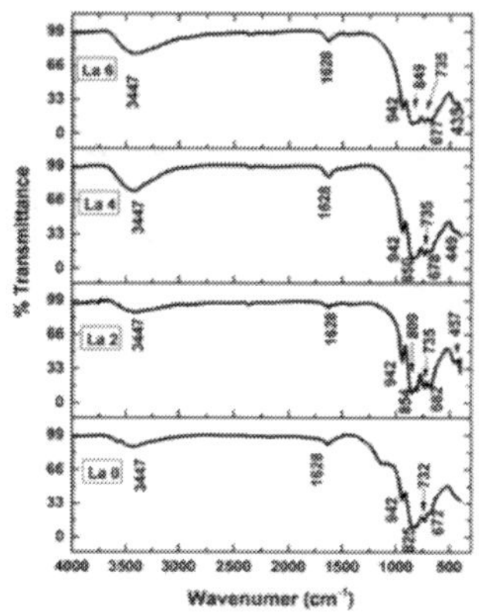

Figure 2. FTIR spectra of pure and lanthanum doped $Ce_2(WO_4)_3$ samples

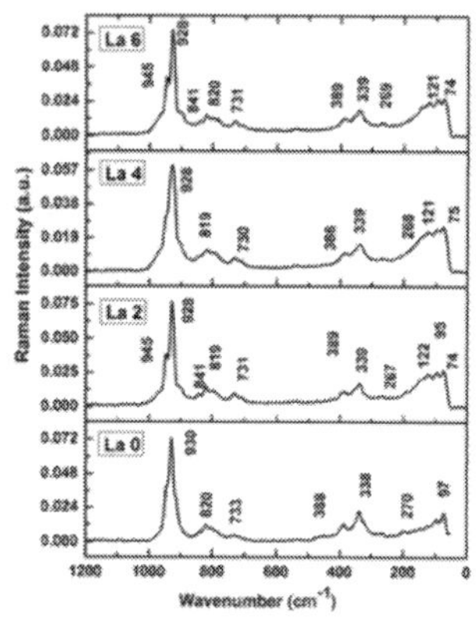

Figure 3. Raman bands of pure and lanthanum doped $Ce_2(WO_4)_3$ samples

Table 1. Comparison of crystallite sizes from Scherrer equation and W-H plots

Sample Name	Intercept	Slope (microsrain)*10^{-3}	D (nm) Scherrer equation	D (nm) from W-H plot
La 0	0.00621	2.1	20.52	22.3
La 2	0.00808	3.1	15.94	17.3
La 4	0.00812	3.27	15.30	17.07
La 6	0.00686	6.86	15.88	20.2

3.2. FTIR spectroscopy

'Figure 2' shows the FTIR spectra of pure and lanthanum doped cerium tungstate nanoparticles. The weak absorption bands observed at 3447 and 1628 cm^{-1} for all the samples are due to the presence of adsorbed water molecules and are attributed to the symmetric stretching and symmetric bending modes of vibration of hydroxyl groups respectively [14]. The characteristic vibrational bands of tetrahedral WO_4^{2-} groups observed at 942 cm^{-1} corresponds to $A_1(v_1)$ vibrations, the bands at 850, 825, 809, 735 and 677 cm^{-1} are of $F_2(v_3)$ and $E(v_2)$ vibration band is observed at around 450 cm^{-1} [15]. There is no visible change in the absorption band positions of tungstate groups due to doping.

3.3. Raman Spectroscopy

The low frequency stokes lines obtained in the Raman spectra of the cerium tungstate samples are presented in 'figure 3'. The Raman bands occurred at 945 and 928 cm^{-1} corresponds to $A_1(\nu_1)$ vibrations, the bands at 841, 820, 731 cm^{-1} are of $F_2(\nu_3)$ vibrations and those at 389,339 cm^{-1} belongs to $F_2(\nu_4)$ vibrations [15]. The low frequency bands present in the spectra below 300 cm^{-1} are due to the external vibrational modes.

3.4. UV-Visible Spectroscopy

The recorded diffused reflectance spectra of La doped samples and the corresponding optical absorption spectra is shown in 'figure 4'.

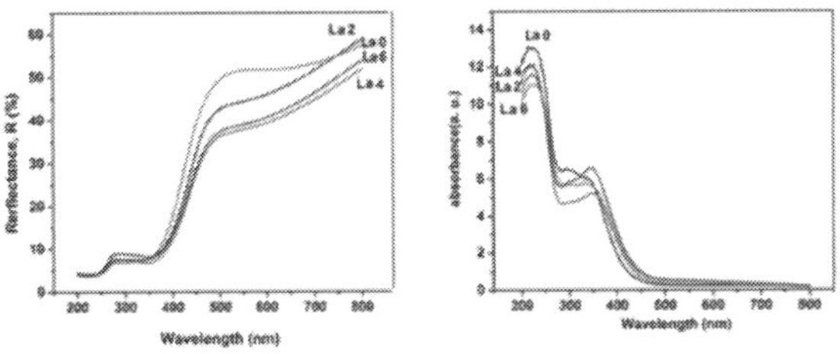

Figure 4. Reflectance and absorbance spectra of pure and La doped $Ce_2(WO_4)_3$ samples

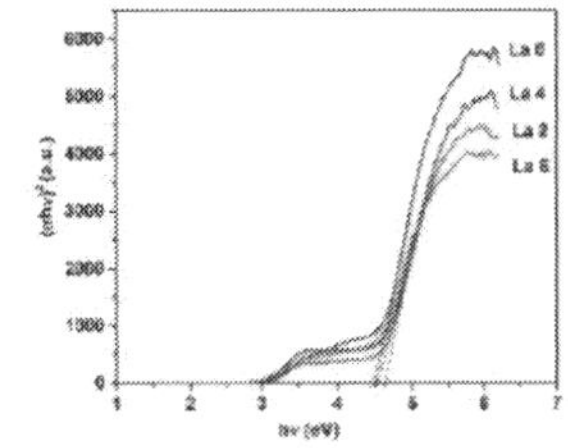

Figure 5. Tauc plots of pure and La doped $Ce_2(WO_4)_3$ samples

There is a clear strong and broad absorption peak below 400 nm. The peaks corresponding to 4f-5d transition in cerium is occurred at 300 nm and 340 nm [16]. The strong absorption peak below 240 nm is due to the absorption peak of WO_4^{2-} group [17].

The absorption coefficient, α due to inter-band transition near the band gap is well described by Tauc relation [18, 19] for direct transition,

$$\alpha h\upsilon = C_1(h\upsilon - E_g)^{1/2} \quad (3)$$

where α is the absorption coefficient, hν is the photon energy, Eg is the optical band gap and C_1 is the proportionality constant. From the Tauc plots ('figure 5'), the optical band gaps are 4.5, 4.58, 4.54 and 4.47 eV for pure and La doped samples La 2, La 4 and La 6 respectively.

The shift of the absorption edge to higher energy is observed in the case of doping at low concentrations (2 and 4%). But, the red-shift of the cut-off edge is found for doping at higher concentration (6%). The similar trend has been observed in lanthanum doped $PbWO_4$ crystals [1]. At low concentrations, the possibility of the replacement of Ce^{3+} mostly occurs, since the ionic radius of Ce^{3+} ions is matched with La^{3+} ions. The cerium vacancies (V_{Ce}) and lanthanum in the cerium sites (La_{Ce}) at low concentrations do not introduce new states deep in the band gap. At high La doping, the possibility of the substitution of Ce^{3+} and Ce^{4+} ions might be occurred. As with increase in La concentrations, more defects will be created [1]. These defects might be caused by oxygen vacancies created by La^{3+} substitution in place of Ce^{3+} and Ce^{4+} [4]. The defect energy bands interact with the conduction band and/or valence band of the crystal and hence the reduction in optical band gap.

3.5. Photoluminescence Spectroscopy

The photoluminescence emission spectra of pure and lanthanum doped samples are shown in 'figure 6' at an excitation wavelength of 380 nm.

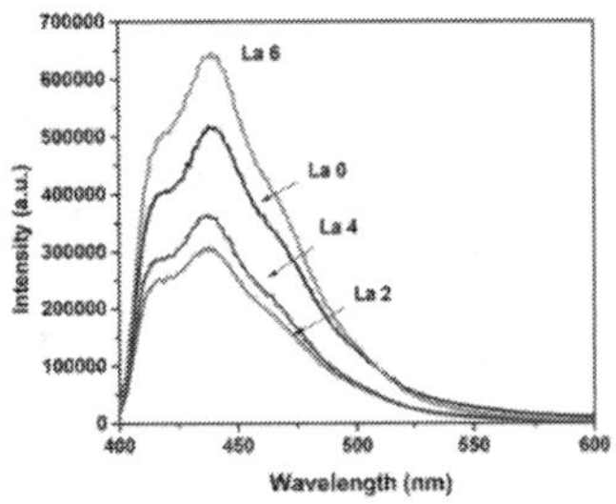

Figure 6. PL emission spectra of pure and La doped $Ce_2(WO_4)_3$ samples

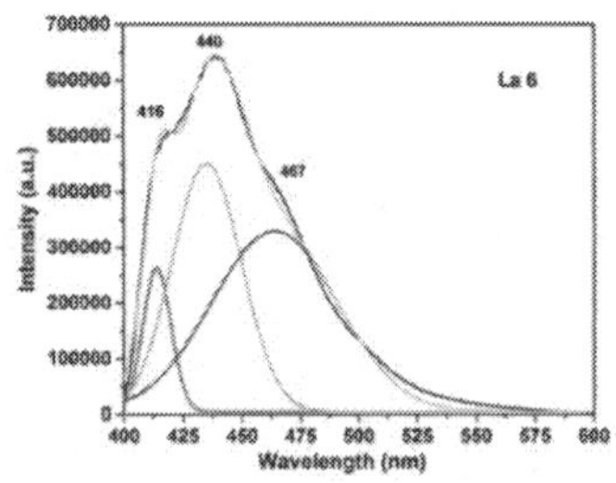

Figure 7. Deconvoluted PL emission spectrum of sample La 6

Spread-eagle shape of PL emission is observed for the pure and doped $Ce_2(WO_4)_3$ samples ('figure 6') [20]. The main broad emission peak is a recombination of three different Gaussian peaks. It is due to the influence of the Jahn-Teller effect on the degenerated excited state of WO_4 tetrahedra [21]. The deconvoluted Gaussian peaks centred at 416, 440 and 467 nm ('figure 7') are the respective emissions from the $^1T_1$, $^3T_2$ and $^3T_1$ bands to the $^1A_1$ ground state of tetrahedral tungstate group [20].

The PL intensity is decreased at 2% of La doping and further enhancement in PL intensity is occurred with increase in Lanthanum concentrations. For the 6% lanthanum doping, much enhanced PL emission is observed. The reason for the enhancement in the PL intensity can be explained based on the defects and oxygen vacancies created. At high doping La concentration, the more replacement of Ce^{4+} ions along with Ce^{3+} substitution happens. When Ce^{4+} ions are substituted with La^{3+} ions, the formation of more oxygen vacancies occurs creating additional energy levels within the band gap and which in turn creates radiative centres in PL emission and hence the PL intensity is increased at high lanthanum doping.

4. Conclusion

The chemical precipitation method has been effectively used for the synthesis of lanthanum doped $Ce_2(WO_4)_3$ nanocrystals. XRD studies conclude that lanthanum doping results decrease in crystallite size, but with an increase in microstrain values without affecting the crystal structure. Characteristic vibrations of the WO_4^{2-} group are not affected by the doping with lanthanum ions. The optical band gap of the lanthanum doped sample shows variation in values that is associated with the various defects. The photoluminescent intensity is greater for the heavily doped sample. This enhancement in the PL intensity is associated with the oxygen vacancies created during the substitution of Ce^{3+} and Ce^{4+} ions by the La^{3+} ions. Hence, lanthanum doped cerium tungstate nanocrystals with suitable concentrations is a potential candidate for photoluminescent applications.

Acknowledgements

The authors acknowledge thanks to Nanoscience Research Centre (NSRC), Dept. of Physics, Nirmala College, Muvattupuzha for providing fundamental facility to undertake the experimental study. The authors also thank DST-FIST support, SAIF Cochin and Department of Physics, Mahatma Gandhi University, Kottayam for providing necessary facilities for characterization.

References

[1] Li W, Xiqi F and Yanlin Huang 2004 *Journal of physics: Condensed Matter* **16** 1325

[2] Kobayashi M, Usuki Y, Ishii M, Yazawa T, Hara K, Tanaka M, Nikl M, Baccaro S, Cecilia A, Diemoz M and Dafinei I 1998 *Nucl. Instr. and Methods in Phys. Res. Sect. A: Acc., Spect., Deectors and Associated Equipment* **404**149

[3] Kobayashi M, Usuki Y, Ishii M, Senguttuvan N, Tanji K, Chiba M, Hara K, Takano H, Nikl M,

Bohacek P and Baccaro S 2011 *Nucl. Instr. and Methods in Phys. Res. Sect. A: Acc., Spect., Detector. and Associated Equipment* **465** 428

[4] Abraham Y B, Holzwarth N A, Williams R T, Matthews G E and Tackett A R 2001 *Physical Review B* **64** 245109

[5] Gressling T and Müller-Buschbaum H 1995 Zeitschrift für Naturforschung B 50 1513-6

[6] Priyanka K P, Revathy V R, Rosmin P, Thrivedu B, Elsa K M, Nimmymol J, Balakrishna K M and Varghese T 2016 *Materials Characterization* **113** 144

[7] Chen T, Liu T Y, Zhang Q R, Li F F, Tian D S and Zhang X Y 2007 Physics Letters A 363 477-81

[8] Nikl M, Bohäcek P, Nitsch K, Mihokova E, Martini M, Vedda A, Croci S, Pazzi GP, Fabeni P, Baccaro S and Borgia B 1997 *Applied physics letters* **71** 3755.

[9] Lei N, Han B, Feng X, Hu G, Zhang Y and Yin Z 1998 *physica status solidi (a)* **170** 37

[10] Zhu W L, Feng X Q, Huang Y L, Lin Q S and Wu ZH 2002 *physica status solidi (a)***193** 211-

[11] Holzwarth U and Gibson N 2011 Nature nanotechnology **6** 534

[12] Vijayaprasath G, Murugan R, Mahalingam T and Ravi G *2015 Journal of Materials Science: Materials in Electronics* **26** 7205-13.

[13] Prabhu Y T, Rao K V, Kumar V S and Kumari B S 2014 *World Journal of Nano Science and Engineering* **27**

[14] Čejka J 2018 *Uranium: Mineralogy, Geochemistry and the Environment* Peter C. Burns and Robert J. Finch Berlin Boston De Gruyter 521-622.

[15] Burcham L J, Wachs I E 1998 *Spectrochimica Acta Part A: Molecular and Biomolecular Spectroscopy* **54** 1355-68.

[16] Babitha K K, Sreedevi A, Priyanka K P, Sabu B and Varghese T 2015 *Indian Journal of Pure & Applied Physics (IJPAP)* **53** 596-603.

[17] Treadaway M J and Powell R C 1974 *The Journal of Chemical Physics* **61** 4003-11.

[18] Tauc J 1974 *Amorphous and liquid semiconductors* 159

[19] Mott N F and Davis E A 2012 *Electronic processes in non-crystalline materials* (Oxford university press)

[20] Jose A, John M, Hitha H, Kuriakose S, Baiju K P and Varghese T 2021 *Results in Surfaces and Interfaces* **4** 100020

[21] Nikl M, Bohacek P, Mihokova E, Kobayashi M, Ishii M, Usuki Y, Babin V, Stolovich A, Zazubovich S and Bacci M 2000 *Journal of luminescence* **87** 1136-9

IOP Conf. Series: Materials Science and Engineering 1221 (2022) 012039 doi:10.1088/1757-899X/1221/1/012039

Study of Dielectric Constant of Glycine Doped Ammonium Dihydrogen Phosphate (ADP) Crystal

A Z Khan[1], Z S Khan[2] and A Pathade[3]
[1]Department of Applied Physics, YeshwantraoChavan College of Engineering, Nagpur, India
[2]Department of Applied Physics, Anjuman College of Engineering & Tech., Nagpur, India
[3]Jawaharlal Nehru Medical College, DattaMeghe Institute of Medical Sciences, Wardha, India
E-mail: arsalazamirkhan@gmail.com

Abstract. To enhance the properties of Ammonium dihydrogen phosphate (ADP) crystals, it has been doped with simplest amino acid, Glycine. ADP with Glycine dopant and Ammonium dihydrogen phosphate (ADP) crystals have been grown by slow conventional, rotation and SR methods and subjected to dielectric characteristics. The dielectric constant has been measured which increases with rise in temperature for pure ADP crystals and conventional, rotation and SR method grown Glycine doped ADP crystals. It is beingnoted that doped crystals have higher dielectric constant than pure ADP crystal and found to agree with the reported values. SR method grown crystals were found to possess highest dielectric constant among all the grown crystals. Higher dielectric constant directs that the grown crystalspossess good crystalline quality with lesser defects.

1. Introduction

Method of growing crystals differs broadly; it is mainly depictedfrom the physical characteristics and size of the material [1-2].The request for nonlinear optical crystals with higher quality is increasing because of the quantum jump in the design of nonlinear optical devices with superior performance. For this reason, materials which are having striking nonlinear characteristicshave been discovered at a faster rate[3-5]. Slow evaporation procedure has been extensively used to grow crystals. Crystals oriented along certain directions are very much useful for reducing the cost and wastage of materials during the fabrication of nonlinear optical devices. Researchers named Sankaranarayanan-Ramasamy invented the unidirectional methodof crystal growth which have been effectively used to grow the benzophenone single crystal with various orientations [6].Dielectric properties of doped and pure ADP crystals with the outcome of swift heavy ions the has beenverified by various researchers [7]. For the modification of ADP crystals and to improve the dielectric properties of ADP, it has been thought to dopeADP with amino acid Glycine.

2. Synthesis of the Crystals

Ammonium dihydrogen phosphate, ADP and glycine of Merck were starting materials. Measured amount of both the salts were dissolved in 1000 ml water. Stirring the solution for few hours with magnetic stirrer,saturated solution of ADP and Glycine doped ADP was prepared. Prepared saturated solution was filtered by means of filtration pump. Beaker containing saturated solution was then covered with lid having holes and permitted to evaporate at normal temperature. This procedure was recurrent for variousGlycine mole % dissolved in ADP solution. Conventionally grown ADP and ADP crystals with Glycine dopant have been gathered in few days. The crystals grown by slow evaporation process was utilised as seed crystals for rotation and SR methods[8].

3.Dielectric Studies

Crystals grown by all the three methods mentioned above have been subjected to study dielectric propertiesvia precision impedance analyser.The cut and polished plate of 3 mm thickness was utilized to measure dielectric constant. The crystal was then silvered on both the sides so that it acts as a parallel plate capacitor. Electrodes with silver pastedon boththe faces confirmappropriate electrical contacts. Crystal holder was mounted with the silver paste crystalwhich was placed inside a digitalized furnaceconnected with the precision impedance analyser. The output from this analyser is obtained in the excel sheet which comprisesof the tabularised values of capacitance, impedance, phase angle, dissipative factor, quality factor, etc.

The permittivity is a property of a substance that gives the force amongst2-point charges in the materials. Also, it is afeaturethrough which the electric field amongst the charges comparative to vacuum is decreasing. This relative permittivity is known as dielectric constant. For studying the lattice dynamics in the crystal, dielectric property is a very indigenous parameter.As single crystal possesses dielectric property, it holds electrostatic charges.Depending on the charge polarization displacement inside the crystalmagnitude of dielectric constant changes.In the low frequency areaparticles are exiled from regularlocations due to an applied electric field. Dielectric response of diatomic crystal has been illustrated by Phillips-Van Vechten (PV) theory.In the theory of Shengjun Zhu, with the notion of the effective, ionic charge and the isotropic displacement, the bond-chargemeasurement of bond non-linearity was illustrated [9].The interaction amongthe electrons in a solid and an externallyapplied electric field is the cause for its refractiveindex. The dielectric constant at larger wavelength could be measuredby means ofapproximately free electron model in the semiconductor.

4. Results and Discussion

Dielectric constant of crystals of undoped ADP and ADP doped with Glycine,preparedthroughconventional slow evaporation, rotation and SR methods with various concentrationshave been calculated.

The equationfor computing the dielectric constant or relative permittivityε' of the crystal is:

$$(\varepsilon') = \frac{C\ d}{\varepsilon_0 A}$$

where'A' is the specimen area;'C' is the capacitance;'d' is the distance amongst the parallel plates and 'ε_0' is the absolute permittivity of free space.Temperature dependence of dielectric constant of undoped and doped ADP (G-ADP) crystals with 1, 3 and 5 mole % concentration are shown in the figure 1.

The capacitanceis augmented due totheexistence of a dielectric sandwiched between the condenser plates. Basically, dielectric constant is the measure that the substance is easily polarized in an externallyapplied electric field. It is seen from figure 1, that dielectric constant increases with increase in temperature and it is also found to increase with increasing concentration of glycine dopant in pure ADP which is a typical dielectric performance of an ADP crystal. In this research module, it has been observed that doped ADP crystals doped with the Glycine have greaterdielectric constant values than the pure ADP crystal.

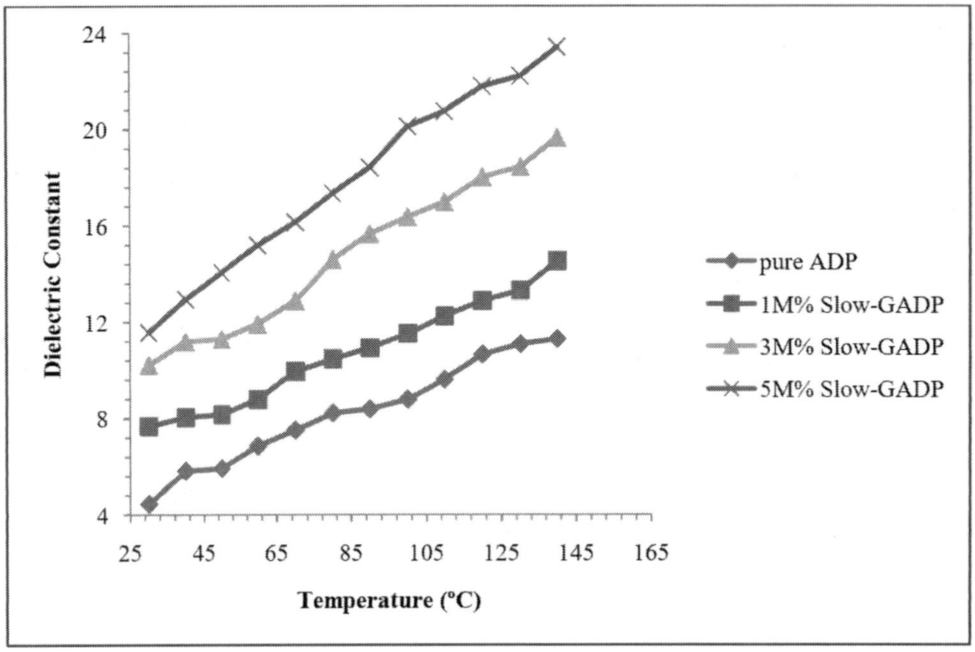

Figure 1. Variation of Dielectric Constant Vs Temperature with increase in concentration of Glycine in ADP.

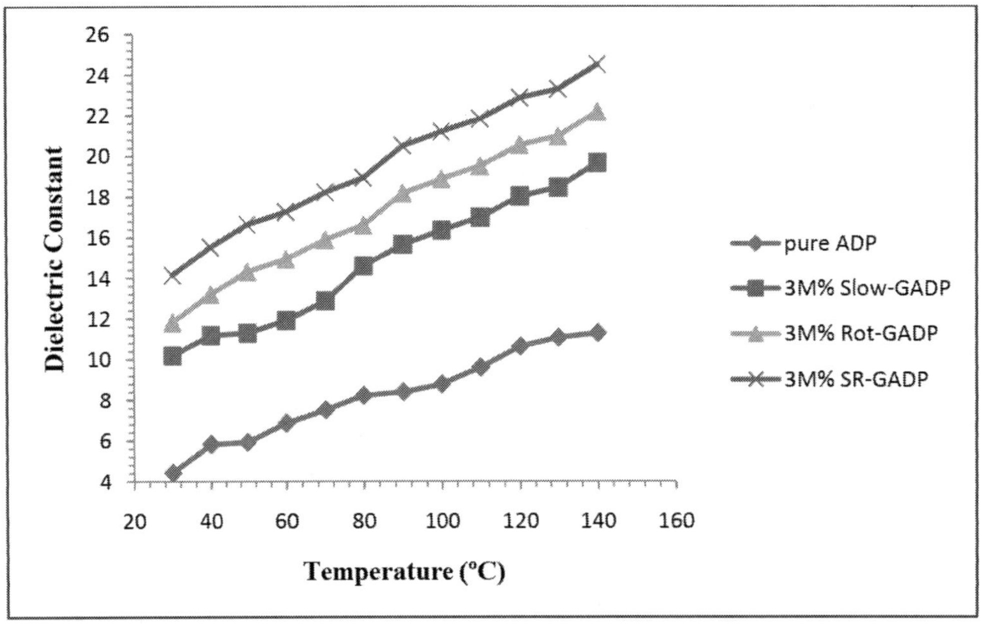

Figure 2. Variation of Dielectric Constant Vs Temperature for Slow, Rotation and SR grown (3M%) Glycine Doped ADP Crystals at 1 KHz Frequency

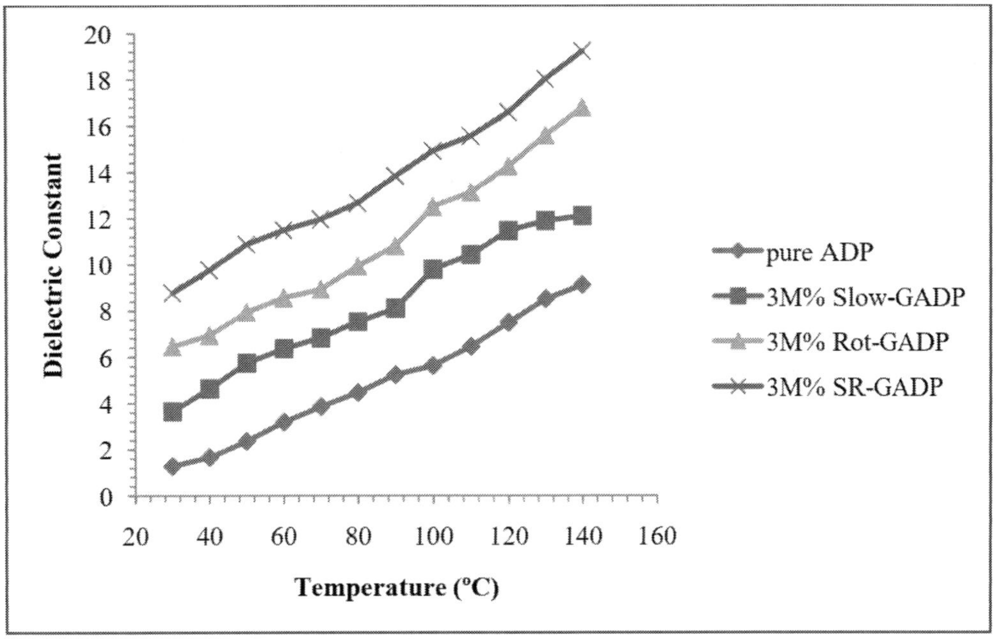

Figure 3. Variation of Dielectric Constant Vs Temperature for Slow, Rotation and SR grown (3M%) Glycine Doped ADP Crystals at 10 KHz Frequency

Figure 2 shows the comparative graph for dielectric constant of crystal of ADP doped with Glycine grown by conventional method, rotation method, SR method and conventionally grown undoped ADP crystal as a function of temp at constant 1 kHz frequency. It has been observed from the figure that as the temperature increases, dielectric constant also gets increasedfor all the crystals. Asimilartrend has been observed for 10 kHz frequency shown in figure 3.

The contributions from all the 4 identified causes of polarization that depends on the frequencies are space charge,electric, dipolar and ionic polarizations whichresults in the dielectric constant of a material.All these polarizations are active at very low frequencies[10]. The space charge polarization is usually active at high temperature andsmall frequencies. If the frequency is increased, the space charge polarisation could not withstand externally applied field and therefore the polarization reduces which in turnreduces dielectric constant values.

In the high frequency region,the value of dielectric constant decreases for higher frequencies due to the contributions of electric polarization,as shown in the figure 3. It has been observedfrom thefigure 2 and 3, that dielectric constant is morefor SR technique grown Glycine doped ADP crystals than Rotation and conventionally grown Glycine doped ADP crystals. It is also noted that doped crystals have higher dielectric constant than pure ADP crystal which agrees with the reported values. Similar results have been obtained for L-Lycine Monohydrochloride Dihydrate and DL-Malic acid doped Ammonium dihydrogen phosphate crystals[11-12].

This consequencecould be ascribed to the charge distribution effect through carrier hopping on defects. At lower frequencies, defects nearer to the positive end of the external applied field acquires negative charge andthatnearer to the negative end of the externalapplied field acquires positive chargebecause of the rapid redistribution of the charges on the defects. Thatultimately screens the field and hence there is anoverall decrease in the electric field. Asfield is inversely proportional capacitance, thelessening in the field for a certain voltage result in the increase in the value of capacitance observed for the low frequency. At higher frequencies, the defects could not have adequateperiod to reorder in response to the voltageapplied, hereafterdecreasing the value of the capacitance [13].

5. Conclusion

Dielectric characteristics are associated with the redistribution of electric field inside the solid substances. Study of dielectric characteristics affords with the qualitative evidence of the substance. The dielectric constant of the grown crystals was computed as a function of temp at a frequency of 1 kHz and 10 kHz. It was observed that as the temperature increases, the dielectric constant of the crystalof ADP dopedwith glycine dopant also increases. The same tendency is observed in case of increase in dopant concentration at different temperatures. It is noted that as the frequency is decreased, dielectric constant also gets decreased. Hence addition of Glycine dopant enhances thedielectric constant value. The present study alsospecifies that high dielectric constant is observed for the GADP crystal preparedthrough SR method equated to the crystalsprepared by other two methods. This finally indicates that the crystalsprepared by SR method havemore crystalline precisionas compared to the crystal prepared by rotation and conventional methods.The dislocation is lower in SR method grown crystals is the indication of crystalline precision.

References

[1] Buckley H E 1951 *Crystal Growth* (John Wiley, NY)
[2] Mullin J W 1976*Industrial Crystallization*(Plenum Press, New York)
[3] Baumert C et al 1987 *Appl. Phys. Lett.*, **51** 2192-2194
[4] Chemla S and Zyss J 1987 Nonlinear Optical Properties of Organic Molecules and Crystals (Academic Press, Orlando)
[5] Newman P R et al 1990 *Materials Research Society Symposium Proceedings***173** 557-561
[6] Sankarnarayanan K and Ramasamy P 2005 *Journal of Crystal Growth***280** 467-473
[7] Bhat S G and Dharmaprakash S M 1998 *Materials Research Bulletin***33** 833
[8] Sheikh A and Khan Z 2017 *Int. J. of Eng. Tech.Sci. and Research* **4** (9) 772-776
[9] Shengjun Zhu S W 2014 *Journal of Crystal Growth***388** 98-102
[10] Meena M et al 2008 *Cryst. Res. Technol.* **43** 166-172
[11] Rajesh P et al 2008 *Journal of Crystal Growth* doi:10.1016/j.jcrysgro.2008.11.091
[12] Rajesh P and Ramasamy P 2009 *Journal of Crystal Growth***311** 3491–3497
[13] Zukowski et al 1989 *Physica Status Solid (A)***112** 695-698

IOP Conf. Series: Materials Science and Engineering 1221 (2022) 012040 doi:10.1088/1757-899X/1221/1/012040

Highly Efficient Degradation of Organic Compounds via β-Bi_2O_3Semiconductor Under Visible Illumination

Khaled Alkanad[1], Othman Ali[1], Sujay Shekar G. C.[1], Abdo Hezam[2], Mohammed Abdullah Bajiri[3] and Lokanath N. K.[1*]

[1]*Department of Studies in Physics, University of Mysore, Manasagangotri, Mysuru 570 006, India.*

[2]*Department of Physics, Faculty of Science, Ibb University, Ibb, Yemen.*

[3]*Department of Studies and Research in Industrial Chemistry, School of Chemical Sciences, Kuvempu University, Shankaraghatta-577 451, India.*

*Email:lokanath@physics.uni-mysore.ac.in

Abstract. β-Bi_2O_3semiconductor nanoparticles synthesized by solution combustion method. The as-synthesized sample was systematically characterized by several techniques, including PXRD, DLS, Raman spectrometry, SEM, and DRS-UV–vis spectroscopy. PXRD and Raman analyses confirmed high crystallinity and phase purity of the prepared β-Bi_2O_3. SEM and DLS showed the morphology and the agglomeration of β-Bi_2O_3nanoparticles.DRS-UV-Vis showed the enhanced visible light absorption toward the visible light range with bandgap 2.43 eV. Significantly, β-Bi_2O_3 nanoparticles demonstrated outstanding visible-light photocatalytic properties, revealed enhanced methylene blue (MB) dye degradation. 91% of the dye content was degraded by 30 minutes under 50W visible LED light irradiation, which could be attributed to enhanced visible light absorption and the smaller size of the synthesized nanoparticles.

1. Introduction

Nanostructured materials are excellent photocatalysts due to their interest characteristics, including the large surface area and the redox capabilitycompared to their bulk material[3-6].TiO_2 semiconductor is considered the most efficient and the standard photocatalyst [6];The wide bandgap (Eg) of TiO_2 (3.0-3.2 eV) restricts light absorption only the UV reign, which is 4% of the solar spectrum[7, 8]. As a result of the reduced use ofsolar energy, the application of TiO_2 in photocatalytic systems is limited to some extent. Therefore, semiconductor materials with visible light responses have prompted growing interest in photocatalysis research. Bi_2O_3, Fe_2O_3, and g-C_3N_4arethe most famous semiconductors[1, 9-12], which are photocatalytically efficient due to their unique properties, including nontoxicity, powerful oxidization of valence holesnarrow bandgap in the range of 2.1 to 2.8 eV [1, 10, 13, 14]. These properties enhanced the photocatalytic system to use artificial light with low intensities as an alternative photo source in the absence of sun due to the climate condition[6, 11].In this work, we explored a fast, efficient, and cost-effective technique for producing β-Bi_2O_3semiconductor nanoparticles with porosity morphologies, rich surface oxygen vacancies and a narrow band gap, which facilitates absorbing a broad spectrum of solar irradiation and visible light photoactivation, resulting in increased photocatalytic performance. Under visible light, the photocatalytic degradation of methylene blue dye was performed. PXRD, DLS, SEM, and DRS-UV–visible methods analyzed the produced β-Bi_2O_3 nanoparticles.

2. Experimental work

2.1. Precursors

Bismuth nitrate pentahydrate [$Bi(NO_3)_3.5H_2O$], citric acid [$C_6 H_8 O_7$], Nitric acid [HNO_3], deionized water [H_2O].

2.2. Synthesis of β-Bi_2O_3 nanoparticles

β-Bi_2O_3 nanoparticles were prepared by solution combustion method. With the fuel oxidizer ratio equal to one, $Bi(NO_3)_3.5H_2O$ and $C_6H_8O_7$ were used as precursors. Practically three mmol of $Bi(NO_3)_3$was dispersed in 1 mmol of HNO_3 with 20 ml of deionized H_2O. A suitable quantity of $C_6H_8O_7$was added to the solution and stirred for 30 minutes at room temperature. The prepared solution was heated using a hot plate at 250°C. The solution starts boiling and evaporating. After a certain period of time, the solution starts to generate foam and then ignite. After the abrupt ignition, the dark yellow powder was produced. Later,the bulk was calcinated in a muffle furnace for two hours at 350°C. The final powder was collected and kept in a dry container for further characterization.

2.3. Analytical studies

The prepared sample was studied through various characterization techniques. Powder X-ray diffraction (PXRD) (Rigaku Smart Lab-II, CuKα radiation, λ = 1.540598) was utilized to confirm the crystallinity and phase of the prepared sample. The data collection is set on 0.02° step size and 2θ between 25° and 60°. The morphological structure was examined using a scanning electron microscope (SEM) (Hitachi (S-3400)). Raman microscope was employed in the resolution of 100 to 1000 cm^{-1} to detect the Raman spectrum of the studied sample. UV-vis differential reflectance spectroscopy (DRS) was used to determine the absorption spectrum and the calculated bandgap of the photocatalyst.

2.4. Photocatalytic activity assessment

Methylene blue (MB) was degraded photocatalytically using pure β-Bi_2O_3 nanoparticles with the influence of visible LED irradiation 50 (W). In 100 ml of MB (10 mg L^{-1}, PH = 3), 5 mg of pure β-Bi_2O_3 was dispersed, sonicated, and stirred in the dark to approach the adsorption-desorption equilibrium. After different time intervals of exposing the solution to visible LED irradiation and continuous stirring, the degradation efficiency was estimated using a Shimadzu UV-2450 spectrometer.

3. Results and discussion

The XRD pattern of the prepared nanoparticlesis shown in Figure 1(a). The diffraction peaks of the synthesizedBi_2O_3 are found at 28.1°, 30.3°, 31.5°, 32.7°, 46.9°, 47.7°, 54.0°, 56.8°, and 58.1° which are indexed to (201), (221), (002), (220), (222), (400), (203), (421) and (402) respectively(JCPDS-PDF No. 27-0050).The intensive and sharp peak at 28.1° (which is correlated to (201) plane) indicates the availability of tetragonal-Bi_2O_3.The diffraction pattern reports that Bi_2O_3 crystallized as β-phase with space group P-421c and lattice parameters a = b = 7.742 Å and c = 5.631 Å. As per the XRD pattern, the strong peaks demonstrate the high crystallinity of β-Bi2O3, which indicates the phase purity of the sample. The Debye-Scherrer equation is used to estimate the crystal size,

$$D = 0.9\lambda/\beta\ cos\theta \qquad (1)$$

Where λ =0.154059 nm (for CuKα) is the wavelength, β is the full width at the half maximum, and θ is the Bragg's angle. From the equation 1, the measured particle size was found to be 9 nm.

Dynamic liquid spectroscopy (DLS) was performed to calculate the size of the nanoparticle by dissolving 0.01g of the material in 20 ml of ethanol for half-hour. The average value of the particle size distribution curve is 10 nm, as shown in figure 1(b). The particle size is found to agree with the average size determined by XRD.
The morphological structure of the synthesized material has been visualized using a scanning electron microscope. Figure 1(c) shows the SEM image of the synthesized sample, which indicates irregular agglomeration of the particles with porous morphology. The porous structure gives rise to a large surface area, which enhances the absorption of the visible light and provides better interaction between the photocatalyst and the pigment [1, 2].

Raman spectroscopy was performed to investigate the microstructure of the porous β-Bi_2O_3 nanomaterial Figure 1(d). Four significant peaks, 93, 132, 226 and 467, were detected to be equivalent to the literature data [15]. The broadness of the peaks could be ascribed to the excellent mobility of surface oxygen vacancies [1, 15].

Uv-vis analysis was conducted to estimate the absorption characteristics of theβ-Bi_2O_3nanoparticles. Figure 1(e) demonstrate a broad visible light absorption exceeding the UV spectrum range toward the visible region with an absorption edge around 510 nm. The energy bandgap (E_g) of β-Bi_2O_3 nanoparticles was calculated using the equation $E_g = 1240/\lambda$, where λ is the absorption edges obtained from the DRS spectra [6]. The optical E_g of the β-Bi_2O_3 nanoparticles is found to be 2.43 eV.

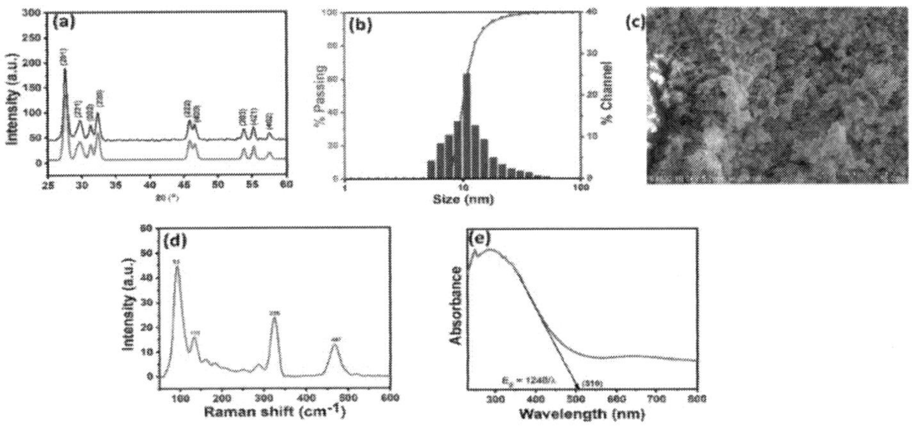

Figure 1. (a) Powder XRD pattern, (b) particle size distribution curve (c) SEM image, (d) Raman spectra, and (e) DRS UV–vis spectra of β-Bi_2O_3 nanoparticles.

4. Photocatalytic studies

The photocatalytic activity of the synthesized β-Bi_2O_3 nanostructure was examined by MB (10 mg L^{-1}) degradation tests under visible light irradiation. Figure 2(a) presents the MB degradation curves of the sample at different time intervals. The synthesized β-Bi_2O_3 showed a degradation rate of around 90% within 30 minutes of exposure to visible light. This efficiency gives the β-Bi_2O_3 photocatalyst a benign role in environmental treatments.

The schematic mechanism of photocatalytic degradation is demonstrated in figure 2(b). The photoreaction of the incident light radiation with theβ-Bi_2O_3 nanoparticles gives rise to the absorption of the photon energy,which leads to exciting the electrons from the VB towards the CB. This process leads to creating positive changes in the VB called holes (h^+). The generated h^+ in the VB interact with the H_2O molecules producing hydroxyl radicals ($^{\bullet}OH$). The interaction between the hydroxyl radicals and MB dye leads to the creation of CO_2 and H_2O, as disclosed in the following equations (2-4)

$$\beta\text{-}Bi_2O_3 + h\nu\beta\text{-}Bi_2O_3 + (e_{CB}^- + h_{VB}^+) \quad (2)$$

IOP Conf. Series: Materials Science and Engineering 1221 (2022) 012040 doi:10.1088/1757-899X/1221/1/012040

$$h^+_{VB} + H_2O \rightarrow {}^{\bullet}OH \quad (3)$$

$$^{\bullet}OH + \text{MBIntermediates} \rightarrow CO_2 + H_2O \quad (4)$$

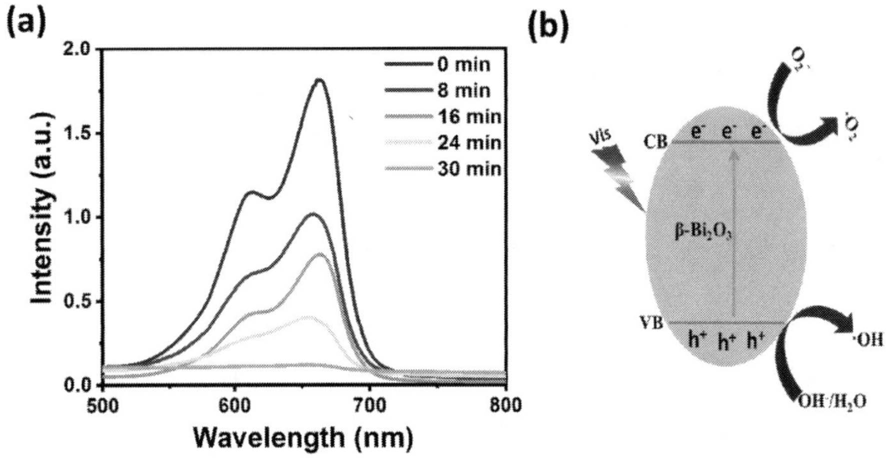

Figure 2.(a) The absorption spectrum of MB in β-Bi_2O_3nanoparticles at different time intervals, (b) schematic mechanism of the β-Bi_2O_3nanoparticles of photocatalytic degradation of MB dye.

5. Conclusion

The β-Bi_2O_3nanoparticle was synthesized usingthe solution combustion route and characterized with various techniques. The nanostructure of the prepared sample was emphasized using PXRD and Raman analyses. PXRD confirmed the excellent crystallinity of the sample and phase purity of β-Bi_2O_3.Raman studied reveal a presence of oxygen vacancies on the sample surface. The size of β-Bi_2O_3 nanoparticles was computed using DLS technique and found to be 10 nm correlated with the PXRD result. SEM image demonstrated the morphology of the nanoparticles as irregular shape particles with porous structures. The porosity of β-Bi_2O_3 nanoparticles was attributed to the oxygen vacancies and the evaporated gases due to the abrupt ignition during the combustion process. UV-vis analysis disclosed a wide visible light range absorption, which shows the evolved photodegradation activity of MBusing β-Bi_2O_3 nanoparticles. Accordingly, the prepared β-Bi_2O_3 nanoparticles are efficient to be used in wastewater treatments.

6. Acknowledgment

The authors are thankful to the National Single Crystal Diffractometer Facility, DoS in Physics, CPEPA and DST-Purse, Vijnana Bhavana, University of Mysore, Mysuru.

7. References

[1] Alkanad K., Hezam A., Shekar G S, Drmosh Q, Kala A.A., AL-Gunaid M Q and Lokanath N K 2021 Magnetic recyclable α-Fe_2O_3–Fe_3O_4/Co_3O_4–CoO nanocomposite with a dual z-scheme charge transfer pathway for quick photo-fenton degradation of organic pollutants *Catal. Sci.Technol.* **11** 3084-3097.

[2] Shekar G C S, Alkanad K, Hezam A and Lokanath N K 2020 Enhanced photocatalyticdegradation of Rhodamine B using porous tricobalt tetraoxide

nanoparticles*IRJET***7** 194-196.

[3] Petronella F, Pagliarulo A, Striccoli M, Calia A, Lettieri M, Colangiuli D, Curri M L and Comparelli R 2017 Colloidal nanocrystalline semiconductor materials as photocatalysts for environmental protection of architectural stone *Crystals***7** 30.

[4] Chen S, Thind S S and Chen A 2016 Nanostructured materials for water splitting-state of the art and future needs: a mini-review *Electrochem. commun.***63** 10-17.

[5] Hameed A, Montini T, Gombac V and Fornasiero P 2008 Surface phases and photocatalyticactivity correlation of Bi_2O_3/Bi_2O_4-X nanocomposite *J. Am. Chem. Soc.***130** 9658-9659.

[6] Alkanad K, Hezam A, Drmosh Q A, Ganganakatte Chandrashekar S S, AlObaid A A, Warad I, Bajiri M A and Neratur Krishnappagowda L 2021 Construction of Bi_2S_3/TiO_2/MoS_2 S- scheme heterostructure with a switchable charge migration pathway for selective CO_2 reduction *Sol. RRL* 2100501.

[7] Low J, Cheng B and Yu J 2017 Surface modification and enhanced photocatalytic CO_2 reduction performance of TiO_2: A Review *Appl. Surf. Sci.***392** 658-686.

[8] Nguyen T P, Nguyen D L T, Nguyen V -H, Le T -H, Vo D -V N, Trinh Q T, Bae S -R, Chae S Y, Kim S Y and Le Q V 2020 Recent Advances in TiO_2-based photocatalysts for reduction of CO_2 to fuels *J. Nanomater.***10** 337.

[9] Hezam A, Namratha K, Drmosh Q, Yamani Z and Byrappa K 2017 Synthesis of heterostructured Bi_2O_3–CeO_2–ZnO photocatalyst with enhanced sunlight photocatalytic activity *Ceram.***43** 5292-5301.

[10] Bajiri M A, Hezam A, Namratha K, Viswanath R, Drmosh Q, Naik H B and Byrappa K 2019 CuO/ZnO/g-C_3N_4 Heterostructures as efficient visible light-driven photocatalysts *J. Environ. Chem. Eng.***7** 103412.

[11] Bajiri M A, Naik H B, Hezam A, Viswanath R, Al-Zaqri N, Alsalme A, Alharthi F A and Alasmari R 2021 2D G-C_3N_4as a bifunctional photocatalyst for co-catalyst andsacrificial agent-free photocatalytic N_2 fixation and dye photodegradation *New J Chem***45** 7174-7184.

[12] Zahid A H and Han Q 2021 A review on the preparation, microstructure, and photocatalyticperformance of Bi_2O_3 in polymorphs *Nanoscale*.

[13] Bajiri M A, Hezam A, Keerthiraj N, Al-Maswari B M, Naik H S B, Byrappa K, Zaqri N A,Alsalme A and Alasmari R 2021 Non-noble metallic Cu with three different roles in Cu doped ZnO/Cu/G-C_3N_4 heterostructure for enhanced Z-scheme photocatalytic

activity *New J Chem.*

[14] Shekar GC, Alkanad K, Hezam A, Alsalme A, Al-Zaqri N and Lokanath N 2021 Enhanced photo-fenton activity over a sunlight-driven ignition synthesized A-Fe_2O_3-Fe_3O_4/CeO_2 heterojunction catalyst enriched with oxygen vacancies *J. Mol. Liq.* **335** 116186.

[15] Shinde N, Xia Q X, Yun J M, Singh S, Mane R S and Kim K -H 2017 A Binder-free wet chemical synthesis approach to decorate nanoflowers of bismuth oxide on Ni-foam for fabricating laboratory scale potential pencil-type asymmetric supercapacitor device *Dalton Trans.* **46** 6601-6611.

IOP Conf. Series: Materials Science and Engineering 1221 (2022) 012041 doi:10.1088/1757-899X/1221/1/012041

Growth of manganite nanoparticles with narrow size distribution using reverse micelle method

Pratheek[1,2], S V Bhat[2] and B G Hegde[1]
[1] Department of Physics, Rani Chennamma University, Belagavi-591156
[2] Department of Physics, Indian Institute of Science, Bengaluru-560012

E-mail: bghegde@rcub.ac.in; svbhat@iisc.ac.in

Abstract. Charge ordering is one of the interesting phenomena exhibited by doped rare earth manganites. Along with the application of a high magnetic field and variation of composition and temperature, the intrinsic method of size reduction to nanoscale is a much investigated method of 'melting' the charge order. However, the question of complete melting of charge order versus the melting of only the long range charge order is not yet fully settled. Investigating small enough and monodisperse nanoparticles should help clarify the point. We have prepared nanoparticles of $La_{0.375}Ca_{0.625}MnO_3$, $Nd_{0.375}Ca_{0.625}MnO_3$, and $La_{0.7}Sr_{0.3}MnO_3$ following the two-microemulsion reverse micelle method and characterized them using powder x-ray diffraction and transmission electron microscopy techniques. We find that nanoparticles of narrow size distribution are obtained and conclude that with further tuning of the preparation parameters it is possible to achieve monodisperse nanoparticles of the manganites.

1. Introduction

Doped rare earth manganites of the general formula $RE_xA_{1-x}Mn^{3+}_xMn^{4+}_{1-x}O_3$ where RE is a trivalent rare earth ion such as La^{3+}, Nd^{3+}, Pr^{3+} or bismuth ion and A is a divalent alkaline earth ion such as Ca^{2+}, Sr^{2+}, or Ba^{2+} have occupied the centre stage in condensed matter physics research for the last three decades [1-3]. As a function of the composition x and temperature T they exhibit exotic phenomena like colossal magnetoresistance and a multitude of phase transitions such as metal to insulator, paramagnetic to ferromagnetic or antiferromagnetic and charge disordered to charge ordered. The different phases are delicately balanced and are separated by fragile phase boundaries due to the intimate coupling present between the spin, charge, and structural degrees of freedom.

A particularly interesting type of phase transition in these materials is the one from charge ordered (CO), antiferromagnetic, insulating to charge disordered, ferromagnetic, metallic phase. In the CO phase the Mn^{3+} and Mn^{4+} ions are spatially ordered below a certain temperature T_{co}. The CO state can be 'melted' following a number of different routes such as applying a high magnetic field or changing the composition or temperature. It was found [4,5] that one particularly interesting and intrinsic way of doing it is by reducing the size of the particles to nanoscale. However, the exact nature of the disappearance of CO is still an unsettled question in the literature [6]. Zou et al [7] conclude that in nanomanganites of diameter of ~ 40 nm while the long range CO disappears the short range CO is still present. However, Anuradha et al., [8,9] find that in ~ 30 nm diameter nanoparticles of $Pr_{0.57}Ca_{0.41}Ba_{0.02}MnO_3$ and also in nanowires of the same material with the nominal diameter of ~ 80 nm the CO has completely disappeared but there may be evidence for CO fluctuations. This raises the question whether the short-range CO concluded by some workers is of dynamic nature. Part of the reason for this uncertainty can be traced to the polydisperse nature of the prepared nanoparticle systems, leaving out the possibility of small enough, monodisperse nanoparticles exhibiting complete disappearance of CO. In this work we present preliminary results of our attempts aimed at preparing monodisperse nanoparticles to bring clarity on this problem by preparing nanoparticles of certain chosen manganites. Our main focus is on the much-studied $La_{0.375}Ca_{0.625}MnO_3$, (LCMO), the composition corresponding to the most robust charge order. However, we also show that the technique can be extended to prepare

IOP Conf. Series: Materials Science and Engineering 1221 (2022) 012041 doi:10.1088/1757-899X/1221/1/012041

nanoparticles of other manganites as well such as $Nd_{0.375}Ca_{0.625}MnO_3$, (NCMO) and $La_{0.7}Sr_{0.3}MnO_3$, (LSMO). We follow the two-microemulsions reverse micelle method known [10-18] to be useful for the preparation of nanoparticles with good size control and narrow size distribution.

2. Experimental details

The preparation method for the nanoparticles of the manganite $La_{0.375}Ca_{0.625}MnO_3$ consisted of mixing of two microemulsions [15]. One of the microemulsions was formed by using $MnCl_2$, $La(NO_3)_3$ and $Ca(NO_3)_2$ in stoichiometric proportions as precursor salts, which were dissolved in water. To the solution the surfactant consisting of cetyl trimethylammonium bromide(CTAB), 1-hexanol and 1-butanol in stoichiometric proportions was added. The second microemulsion was formed by using aqueous solution of $(CH_3)_4NOH$ as a precipitating agent to the surfactant solution of CTAB+1-hexanol+1-butanol [16]. Before mixing the two microemulsions each of these microemulsions was stirred well until a clear solution is formed. Then the microemulsion containing the precipitating agent is added drop wise to the microemulsion containing the metal salts while continuously stirring. The mixture of the two microemulsions is aged for an hour where the exchange of reactants within the micelles takes place [15] and the final product is precipitated in powder form. After aging, the powder is collected by centrifugation and repeatedly washed with the aqueous solution of $(CH_3)_4NOH$ to remove any surfactant and oil phase molecules present. Finally, the powders were calcinated at different temperatures and durations with a view to obtaining particles of different sizes. It was found that calcination at 800 ^{0} C for 2 hours gave the best results. The process is schematically shown in figure 1. The ratio of water to the surfactant was varied with a view to studying the effect on the size and the polydispersity and the results are summarized in Table 1. The powders were analysed using a Rigaku Miniflex X-ray powder diffractometer and a FEI Titan Themis transmission electron microscope. Nanoparticles of $Nd_{0.375}Ca_{0.625}MnO_3$, and $La_{0.7}Sr_{0.3}MnO_3$ also were prepared following similar procedure.

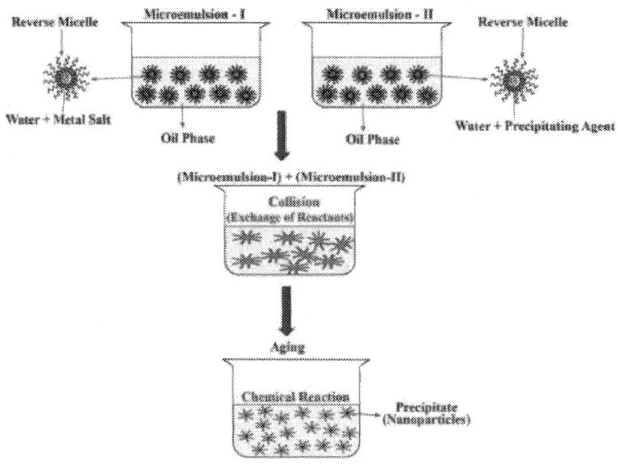

Figure 1. Schematic diagram of two-microemulsion reverse micelle method for the preparation of manganite nanoparticles.

3.Resultsanddiscussion

The powder XRD patterns of the three nanoparticle manganite samples are presented in figures 2-4. The patterns are consistent with the respective patterns published in literature [17]. Figure 2 presents the XRD patterns of the three $La_{0.375}Ca_{0.625}MnO_3$ samples prepared with different water to surfactant ratios. Following the recent work by Rabiei et al., [18] the sizes and the corresponding stress values of the nanoparticles of the three $La_{0.375}Ca_{0.625}MnO_3$ samples were estimated using UDM plots [19]. The results are presented in Table 1. It is seen that as the surfactant amount increases, size of the particle

IOP Conf. Series: Materials Science and Engineering 1221 (2022) 012041 doi:10.1088/1757-899X/1221/1/012041

decreases. TEM energy-dispersive X-ray spectroscopy (EDS) (fig. 6) confirms the chemical composition. The particle sizes were also estimated from the TEM images confirming the dependence of the size on the amount of surfactant. The accompanying histograms show the reasonably small distribution of the particle sizes obtained by the reverse micelle preparation technique (fig. 5). Some NCMO (fig. 3) and LSMO (fig. 4) particles are also prepared which shows that reverse micelle is a promising method to synthesize and take control over the size of the manganite nanoparticles.

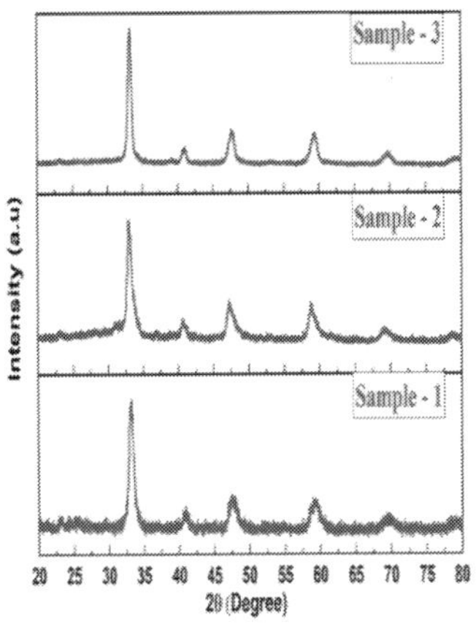

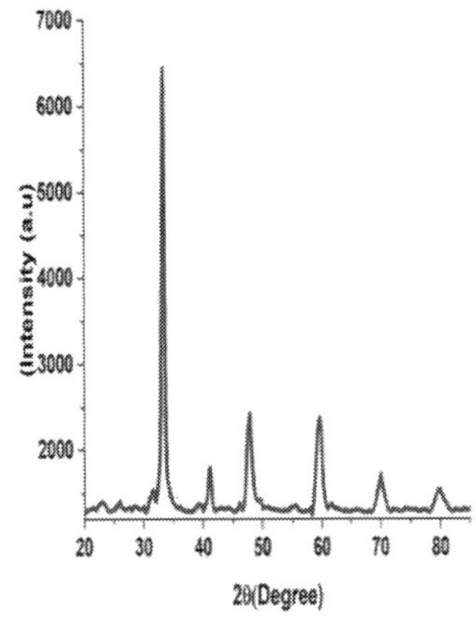

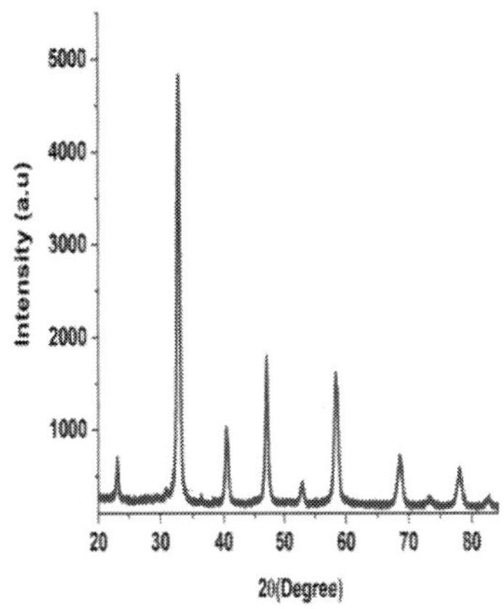

Figure 2. XRD patterns of 3 LCMO samples **Figure 3.** XRD pattern of NCMO **Figure 4.** XRD pattern of LSMO

Table 1. Dependence of XRD estimated crystallite sizes and TEM measured particle sizes on the surfactant content for $La_{0.375}Ca_{0.625}MnO_3$ nanoparticles calcinated at 800 ^{0}C for 2 hours

Sample	CTAB (g)	Water (ml)	Crystallite size from XRD (nm)	STRESS	Particle size from TEM (nm)
1	16.15	15.15	7.85	0.01144 ± 0.00206	15-40
2	15.15	16.15	8.72	0.00739 ± 0.00189	20-43
3	14.15	17.15	10.52	0.00792 ± 9.77831E-4	25-50

IOP Conf. Series: Materials Science and Engineering 1221 (2022) 012041 doi:10.1088/1757-899X/1221/1/012041

<table>
<tr><th>Sample</th><th>TEM image</th><th>Size distribution</th></tr>
<tr><td>1</td><td>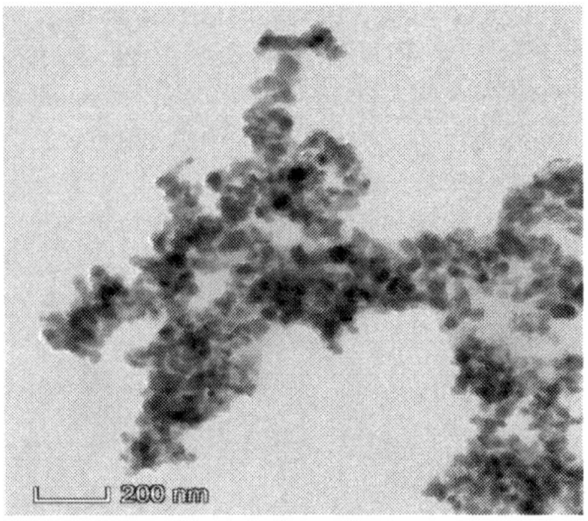
</td><td>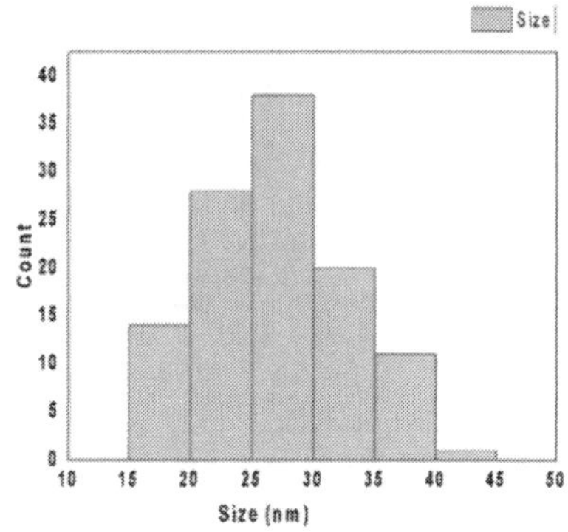
</td></tr>
<tr><td>2</td><td>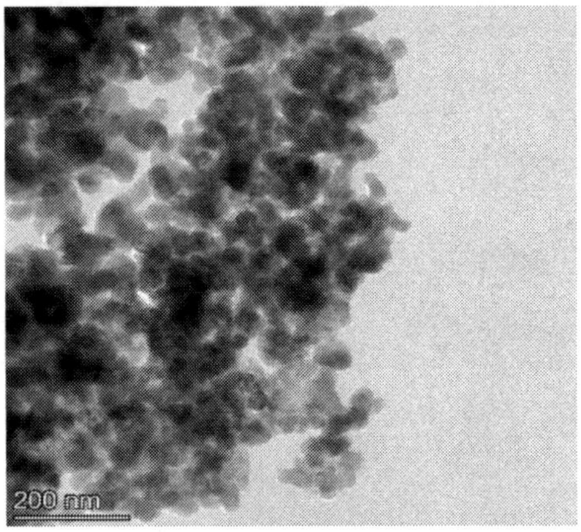
</td><td>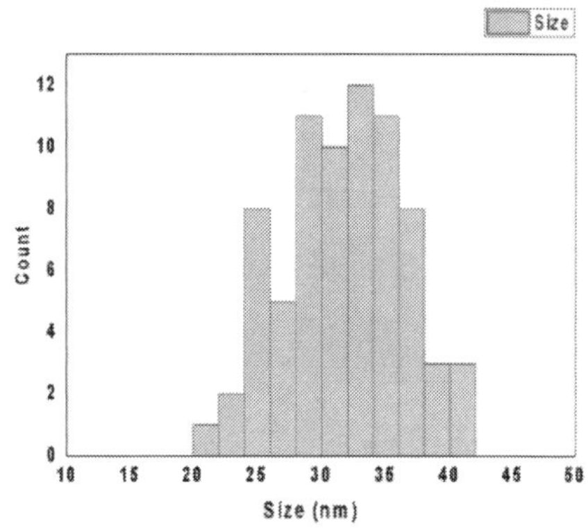
</td></tr>
<tr><td>3</td><td>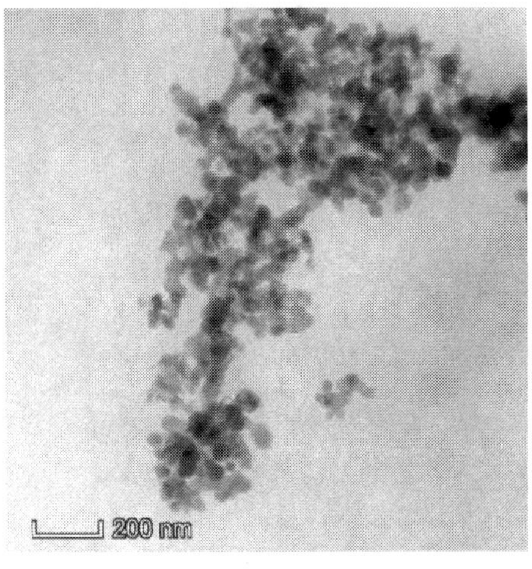
</td><td>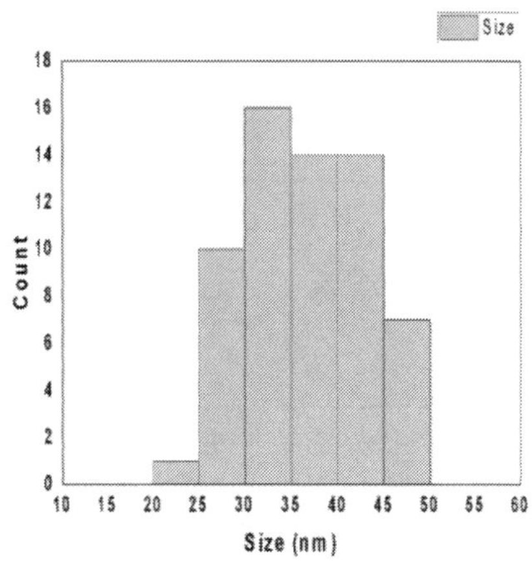
</td></tr>
</table>

Figure 5. TEM images and size distribution of the $La_{0.375}Ca_{0.625}MnO_3$ nanoparticles

IOP Conf. Series: Materials Science and Engineering 1221 (2022) 012041 doi:10.1088/1757-899X/1221/1/012041

Sample - 1

Sample - 2

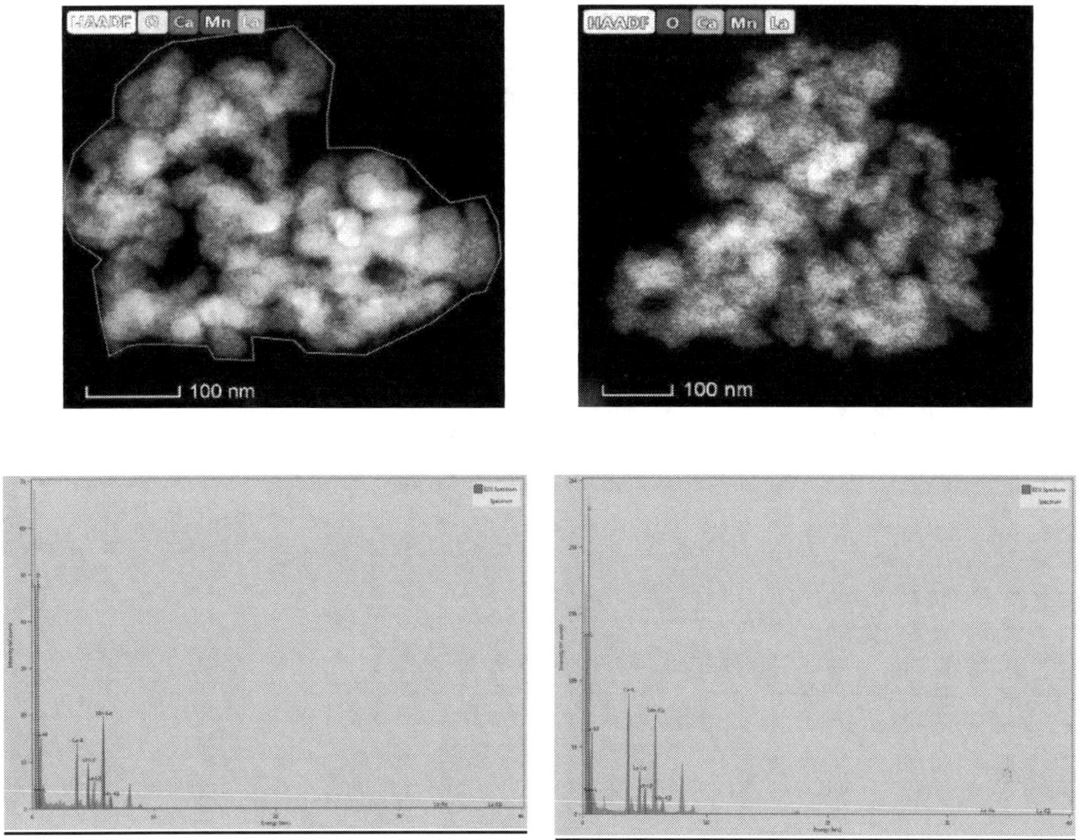

Figure 6. TEM EDS images of sample 1 and 2 of $La_{0.375}Ca_{0.625}MnO_3$

4. Conclusion

Nanoparticles of doped rare earth manganites were prepared following the two-microemulsion reverse micelle method and were characterized using powder XRD and TEM techniques. It was found that particles with narrow size distribution can be prepared using this method. Further optimization of the parameters is likely to help prepare monodisperse particles.Studies on these particles should help clarify the nature of charge order melting.

Acknowledgements

Pratheek and SVB thank the National Academy of Sciences, Prayagraj, for support.

References

[1] Tokura Y 2006 *Rep. Prog. Phys.* **69** 797

[2] Rao C N R, Raveau B 1998 *Colossal* Magnetoresistance, *Charge Ordering and Related Properties of Manganese Oxides* (Singapore: World Scientific)

[3] Xia W, Pei Z, Leng K and Zhu X 2020 *Nanoscale Research Letters* **15** 1-55
[4] Rao S S, Anuradha K N, Sarangi S, and Bhat S V 2005 *Appl. Phys. Lett.* **87**, 182503
[5] Rao S S, Tripathi S, Pandey D and Bhat S V 2006 *Phys. Rev. B* **74**, 144416
[6] Lora Rita Goveas, Bhagyashree K S, Anuradha K N and Bhat S V, AIP Advances 11, 025313 (2021)
[7] Zhou S, Guo Y, Wang C, He L, Zhao J and Shi L 2012 *Dalton Trans* **41**, 7109
[8] Anuradha K N, Rao S S and Bhat S V 2008 *Applied Magnetic Resonance* **33,** 127-136
[9] Anuradha K N, Rao S S and Bhat S V 2007 Journal of Nanoscience and Nanotechnology **7,** 1775-1778.
[10] Eastoe J, Hollamby M J and Hudson L 2006 *Advances in Colloid and Interface Science*, Volumes **128–130**, Pages 5-15
[11] Ahmad T, Ramanujachary K V, Lofl S E and Ganguli A K 2006 *J, Chem. Sci.*, Vol. **118**, No. 6, pp. 513–518
[12] Ganguli A K, Vaidya S and Ahmad T 2008 *Bull. Mater. Sci.*, Vol. **31**, No. 3, pp. 415–419
[13] Uskokovic V, Drofenik M 2007 *Materials and Design*, **28** 667
[14] Muzzio M, Li J, Yin Z, Delahunty I M, Xie J and Sun S 2019 *Nanoscale*, **11**, 18946
[15] Zielińska-Jurek A, Reszczyńska J, Grabowska E, and Zaleska A 2012 "Nanoparticles preparation using microemulsion systems."Microemulsions-an introduction to properties and applications, **229-250**
[16] Uskokovic V, Makovec D and Drofenik M 2005 *Materials Science Forum* Vol. **494**, pp-155-160
[17] Martinelli A, Ferretti M and Ritter C 2016 *Journal of Solid State Chemistry*, Vol. **239**, Pages 99-105
[18] Rabiei M, Palevicius A, Monshi A, Nasiri S, Vilkauskas A and Janusas G 2020 *Nanomaterials*, **10**, 1627
[19] Nath D, Singh F and Das R 2020 *Materials Chemistry and Physics*, Vol.**239**,122021

IOP Conf. Series: Materials Science and Engineering 1221 (2022) 012042 doi:10.1088/1757-899X/1221/1/012042

Structure assigned optical transitions from single walled carbon nanotube based high absorber composite thin film coatings

Sonia Saini[1,2], S Reshmi[3],Girish M Gouda[2], and Kuntala Bhattacharjee[1,3,*]

[1]Indian Institution of Space Science and Technology (IIST), Thiruvanthapuram, 695 547, India
[2]Laboratory for Electro-Optics Systems (LEOS), ISRO, Bengaluru, 560 058, India
[3]Institute of Physics, Sachivalaya Marg, Bhubaneswar 751005, Odisha, India

Corresponding author: kbhattacharjee@iopb.res.in , kuntala.iopb@gmail.com

Abstract. Single walled carbon nanotubes (SWCNTs) exhibit unique electronic and optical properties owing to their one-dimensional (1D) structures that are described by a set of chiral indices (*n, m*), diameter (*d*) and chiral angle (θ). Absorption and optical spectroscopies are powerful techniques to determine the structure and diameter distribution of the SWCNTs through bulk measurements. In this work, we correlate structure assigned absorption features with the corresponding optical transitions observed from the SWCNT based high absorber composite thin film coatings. Various optical signatures demonstrated here are direct manifestation of electronic transitions happening within the sharp van Hove (vH) states of valence band, v_2 to conduction band, c_2 from SWCNTs of particular chiral indices and diameter. These transitions arising within the vH singularities due to unique 1D structures of SWCNTs, correspond to the energy band E_{22}. The presentwork based on empirical calculations according to tight binding model (TB) approximation, represents unique way of correlating structure assigned absorption features with the optical transitions that can have direct impact in tuning the bandgap of SWCNTs of specific diameter and chirality for nanoelectronics and nanophotonic applications.

1. Introduction

Single walled carbon nanotubes (SWCNTs), first reported in 1993 [1] have shown unique electronic and optical properties due to their distinctive one-dimensional (1D) structure. These nanotubes, as a rolled-up graphene sheet are described by a set of chiral indices (*n, m*) which correlate their diameter (*d*) and chiral angle (θ) [2]. Structure-dependent optical properties of SWCNTs make them a strong candidate for research as well as for various applications. Optical absorption and photoluminescence (PL) characterizations are two non-destructive techniques to extract structure dependent spectroscopic properties of specific SWCNTs through bulk measurements. Optical absorption of SWCNTs can reveal a set of diameter-dependent absorption features corresponding to discrete electronic states for a particular spectral range of interest, whereas, PL studies allow for experimental verification of each nanotube's optical fingerprint and has been an essential tool in the qualitative determination of the (*n, m*) distribution of SWCNTs [3]. Optical absorption features are distinctive without much overlap if the diameter distribution of the SWCNT ensemble is relatively narrow. Strong PL transitions can be observed from individual SWCNT aqueous suspensions stabilized by the addition of surfactants or oxidants [1]. However, PL yield is usually quenched for the CNT bundles. Sharp transitions in the discrete density of states (DOS) for 1D materials in the optical region are recognized as van Hove singularities (vHs).Optical absorptions and emissions of SWCNTs are very sensitive to the (*n, m*) values which are expressed by sharp vH peaks. Bachilo et al. had demonstrated a spectrofluorimetric measurement procedure on SWCNTs isolated in aqueous surfactant suspensions that revealed distinct electronic absorptions and emissions for different nanotube species [3], [4]. They had shown the first model-independent prediction

IOP Conf. Series: Materials Science and Engineering 1221 (2022) 012042 doi:10.1088/1757-899X/1221/1/012042

of structure dependent first and second vHoptical transitions for a wide range of semiconducting SWCNTs [4].

The distinct transitions occurring between valance band, v_2 to conduction band, c_2 comprise the energy band, $E_{22}$of SWCNTs of specific *(n, m)* structures [5]. Tight binding (TB) approximation defines inverse correlation between the vHs to the *d*of nanotubes and for the energy band E_{22}, the relation is given by,

$$E_{22} = \frac{4\gamma_0 a_{C-C}}{d} \quad (a)$$

where $\gamma_0$is C-C tight-binding overlap energy (~2.9 eV), a_{C-C} is the nearest neighbor C-C distance (~0.142 nm).

In our previous work, high absorptance of the order of 97-98 % was obtained in the visible spectral range from the SWCNT based composite film coatings [6].The motivation to carry out the present work is to develop further understanding about the basic physical processes responsible for the observed high absorptance of the SWCNT based composite films.We try to correlate various structure specific transitions within the absorption band in the energy range 1.8 to 2.3 eV with the observed PL transitions in the similar visible spectral range. Analyses carried out on the absorption and PL data reveal signature of SWCNTs with different chiral indices and diameters in the coated film that are responsible for exhibiting transitions in the energy band E_{22}. This is mostly one of the reasons of these films to be a good candidate for stray light control space applications [6].

2. Experimental section

SWCNTs, produced by high-pressure carbon monoxide (HiPCO) methodwere obtained from NoPo Nanotechnologies India Pvt. Ltd. Bengaluru which have thediameter and length in the range between 0.6 to 1.1 nm and ~ 400 to 1000 nm respectively. Post-synthesis chemical purification is done to eliminate the contaminants present in the as-prepared CNTs. Purification process involves wet oxidation at 300°C followed by an acid wash using concentrated hydrochloric acid (HCl) and then annealing at 900°C in an inert environment. Purified SWCNTs were then functionalized using concentrated H_2SO_4 and HNO_3 with a volume ration of 3:1 and stirring at 70°C for 8 hrs followed by cooling down of the mixture at room temperature (RT) and neutralizing by rinsing with deionized water. For coated sample, acid treated CNTs mixed with polymer binder and solvent was hand sprayed over Al substrate [6]. Absorption properties of the spray-coated samples were analyzed using a UV-VIS-NIR spectrometer, PerkinElmer Lambda 950 in the wavelength range of 500 to 750 nm. Optical transitions from the coated sample were obtained by subjecting them to photoluminescence (PL) spectroscopy analysis using Lab Ram HR 800 instrument.

3. Results and discussion

Choice of correct line shape is important to extract information of individual *(n, m)* SWCNT species and various theoretical studies suggest symmetric line shapes as a better fit for the optical absorption measurements [7]. Voigt function is proposed to be best suited to fully capture the underlying physical processes that give rise to SWCNT absorption spectrum. Analysis of the optical properties of carbon nanotubes has revealed excitonic sidebands that are assigned to resonances emerging from the absorption of light by a bound exciton-phonon state [8], [9] .Figure 1 shows the absorption features obtained from the densified film of modified SWCNT based composite structures within the energy band, E_{22}. The fitting of the absorption data was performed using Voigt line profiles [7]. From the fitting, we obtained 7 contributions at different energies which are used to calculate *d* of the SWCNTs using equation (a). Once we know the *d* of the tubes, we can assign (*n, m*) according to the empirical formalism proposed by Bachilo et al. [4]. The calculated *d*values obtained from the transitions within the absorption band is in the range between 0.7 to 1 nm [Table 1]. Table 1 shows the energy - structure [*d* (nm), (*n, m*)] correlation according to the TB approximation discussed by Bachilo et al [3].

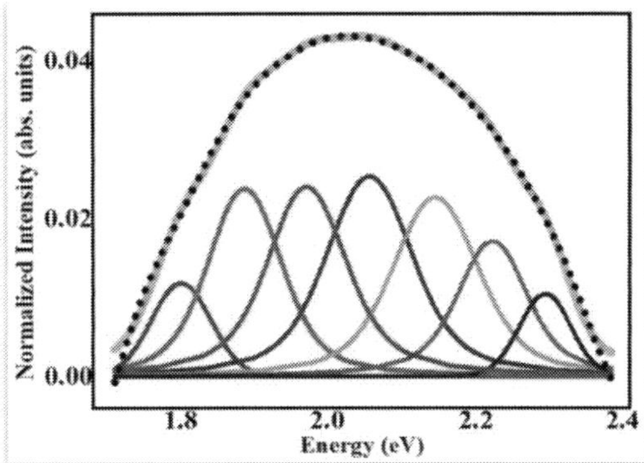

Fig. 1. Voigt line profile fitting of the absorption spectrum obtained for the coated film in the energy range 1.8-2.4 eV which corresponds to E_{22} band.

Table 1 Diameter assignment to the corresponding absorption energy transitions.

Energy (eV)	**Diameter, *d* (nm) & Chirality (*n, m*)**
1.80	1.04 (12,2)
1.89	0.89 (7,6)
1.97	0.94 (10,3)
2.06	0.84 (8,4)
2.14	0.69 (6,4)
2.22	0.81 (9,2)
2.29	0.79 (10,0)

In Figure 2, we show the fitted PL spectrum for the energy range between 1.8 - 2.3 eV of the SWCNT based coated film by using similar Voigt line profiles. We observe various optical transitions from the 1D vH states of the specific SWCNTs in the PL spectrum as a result of the corresponding absorptions shown in Figure 1.

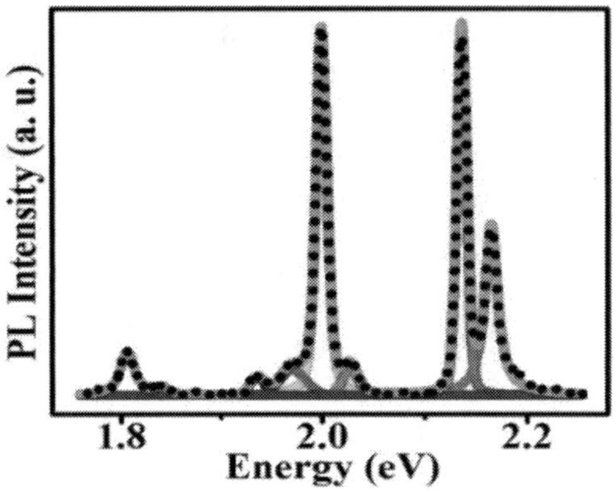

Fig. 2. PL transitions in the energy range 1.8-2.3 eV after performing Voigt function fitting of the coated film.

Each energy transition is attributed to a particular set of SWCNTs with specific (*n, m*)and *d* [7]. These optical signatures cover the diameter range from 1.8 - 2.2 nm. Table 2 depicts the energy - diameter correlation for the coated film along with assigning their chiral index.

Table 2 Diameter assignment to the corresponding Voight line fit energy transitions of PL data.

Energy (eV)	**Diameter, *d* (nm) & Chirality (*n, m*)**
1.79	0.76 (9,1)
1.84	0.98 (9,5)
1.94	0.83 (7,5)
1.97	0.65 (7,2)
2.01	0.92 (11,1)
2.03	0.92 (11,1)
2.14	0.69 (6,4)
2.16	0.69 (6,4)

4. Conclusion

In this study, an attempt to assign structural indices (*n, m*) and diameters (*d*) to the SWCNTs of the coated film is demonstrated effectively according to the absorption and optical transition energies. We report here absorption features with corresponding optical transitions in the energy range 1.8 to 2.3 eV for SWCNTs of specific structures. Absorption spectrum obtained from the coated film reveals discrete features within the absorption band coming from the SWCNTs with *d* between 0.8 to 1 nm. Similarly, PL spectrum acquired from the coated film exhibits optical transitions within the same energy window from the SWCNTs of analogous *d* ranging from 0.8 to 1 nm. Once, we know the *d* of the tubes, we can assign the chirality according to the empirical formalism proposed by Bachilo et al. [4]. Hence, in a unique way, we are able to correlate the structure dependent optical transitions from the SWCNTs of specific (*n, m*) and *d* with the discrete absorption features in the energy band, E_{22}. This can be considered as an important step forward to tune the optical and electronic properties of these coated films via selective choices of SWCNTs for applications related to fabrication of photodetectors, black absorber coatings and optoelectronic devices. This understanding may open up a new era in the field of nanotechnology related research based on SWCNTs.

5. Acknowledgement

K. B. sincerely thanks DST Mobility Scheme for providing with the necessary professional and personal stability and the funding for overall support.

References

[1] Bhushan B 2007 *Springer Handbook of Nanotechnology* (Berlin, Heidelberg, Heidelberg: Springer Berlin Heidelberg)
[2] Luo Z, Pfefferle L D, Haller G L and Papadimitrakopoulos F 2006 *Journal of the American Chemical Society* **128** 15511–6
[3] Bachilo S M, Strano M S, Kittrell C, Hauge R H, Smalley R E and Weisman R B 2002 *Science* **298** 2361–6
[4] Weisman R B and Bachilo S M 2003 *Nano Letters* **3** 1235–8
[5] Wang B, Poa C H P, Wei L, Li L J, Yang Y and Chen Y 2007 *Journal of the American Chemical Society* **129** 9014–9
[6] Saini S, Reshmi S, Gouda G M, Kumar S. A, K. V. S and Bhattacharjee K 2021 *Nanoscale Advances* **3** 3184–98
[7] Sanchez S R, Bachilo S M, Kadria-Vili Y, Lin C W and Weisman R B 2016 *Nano Letters* **16** 6903–9
[8] Dresselhaus M S, Dresselhaus G, Saito R and Jorio A 2007 *Annual Review of Physical Chemistry* **58** 719–47
[9] Perebeinos V, Tersoff J and Avouris P 2005 *Physical Review Letters* **94** 3–6

IOP Conf. Series: Materials Science and Engineering 1221 (2022) 012043 doi:10.1088/1757-899X/1221/1/012043

Photocatalytic performance of g-C_3N_4/Bi_2S_3/MWCNT ternary nanocomposite

A C Swathi and Maneesh Chandran*

Department of Physics, National Institute of Technology Calicut, Kerala, 673601 India

*Corresponding author email: maneesh@nitc.ac.in

Abstract. Graphite-like carbon nitride (g-C_3N_4), has gained significant attention in the field of photocatalytic dye degradation during the past decade, owing to its simple preparation method and visible light response. In this study, g-C_3N_4/Bi_2S_3/MWCNT ternary heterostructure was prepared via a simple hydrothermal method, and its photocatalytic performance was compared with g-C_3N_4, Bi_2S_3, and g-C_3N_4/Bi_2S_3 heterostructures, using malachite green dye as a test pollutant. The result showed that g-C_3N_4/Bi_2S_3/MWCNT heterostructure photocatalyst exhibit improved photocatalytic activity, compared to other catalysts, owing to the improved electron-hole separation efficiency and strong visible light absorption. The study offers a simple way to develop g-C_3N_4/Bi_2S_3/MWCNT visible-light-driven ternary heterostructure for dye wastewater treatment.

1. Introduction

Fabrication of micro and nanocomposite with functional properties has gained considerable attention in recent years because of their unique composition-dependent physical properties. With the rise of environmental issues in recent times, photocatalytic degradation of organic pollutants has emerged as one of the most efficient solutions. So, there is an urgent need to find environmentally friendly photocatalytic degradation materials. Therefore, the studies on the controlled synthesis of low-cost carbon nano-composite materials are of great interest and are actively being pursued. Among carbon composites, the graphitic carbon nitride (g-C_3N_4), a metal-free semiconductor, has been gained more focus due to its simple synthesis procedure, outstanding chemical stability, and photocatalytic activity, non-toxicity, visible light-responsive bandgap (E_g= 2.7eV), and high reduction ability[1]. However, photocatalytic efficiencies of g-C_3N_4 remain poor due to the fast charge carrier recombination. To get better photocatalytic efficiency, g-C_3N_4 has been commonly studied in conjunction with various semiconductors, such as ZnO, Fe_2O_3, SnO_2, WO_3, Ag_3VO_4, and so on [2–6].

Interestingly, g-C_3N_4 and bismuth sulfide (Bi_2S_3) have well-matched overlapping band structures, implying that they are suitable for developing heterostructures (g-C_3N4/Bi_2S_3) with high photocatalytic performance. The charge carrier mobility and the light absorption of Bi_2S_3, g-C_3N_4/Bi_2S_3composites still need to be improved for better photocatalytic performance under visible light. On the other hand, incorporation of Bi_2S_3 with carbonaceous materials, such as multiwalled carbon nanotubes (MWCNTs) has been widely considered for enhancing the photocatalytic activities of Bi_2S_3 [7]. Due to the one-dimensional structure, exceptional electrical characteristics, rapid electron transport, and large surface area of CNT; the combination of CNT with Bi_2S_3 could substantially improve the charge transport performance of the composites and decrease the charge carrier recombination rate [8]. Therefore, the addition of MWCNTs is expected to increase the photocatalytic activity of g-C_3N_4/Bi_2S_3 nanocomposite. So, the synthesis of a multicomponent photocatalyst emerges as a feasible method to achieve high photocatalytic activity by increasing light absorption and reducing photogenerated electron-hole recombination. In this paper, we present the synthesis of g-C_3N_4/Bi_2S_3/MWCNT ternary nanocomposites, with different weight percentages of MWCNT, using a simple hydrothermal method. The effect of MWCNT content on the photocatalytic efficiency of the ternary nanocomposites has been investigated and discussed in detail.

2. Materials and Methods

2.1. Sample Synthesis

g-C_3N_4 sample was prepared by thermal decomposition of nitrogen-rich melamine precursor. 5 g melamine was completely dissolved in 50 ml ethanol and 50 ml deionized water. The dissolved melamine was dried at 150 °C to obtain a dried melamine precursor. Then the g–C_3N_4 was synthesized by calcination of dried melamine powder at 530 °C at a heating rate of 5 °C/min, in a furnace for 2 h. Calcination temperature is set based on thermogravimetric analysis of precursors. After cooled down to ambient temperature, the product was collected and ground into a fine powder.

The g-C_3N_4/Bi_2S_3/MWCNT ternary nanocomposite was prepared using a simple hydrothermal method. Initially, 250.09 mg g-C_3N_4, 250.09 mg thiourea (99%), and 610.09 mg bismuth nitrate pentahydrate ($Bi(NO_3)_3.5H_2O$) (99.99%), and a suitable fraction of MWCNT was added to 60 ml deionized water and stirred to form a homogeneous solution at ambient temperature. This solution was poured into a teflon-lined autoclave and kept in an oven for 7 h at 140 °C. Afterward, the resultant solution was centrifuged, rinsed with deionized water and ethanol, and dried at 90 °C for 8 h to get g-C_3N_4/Bi_2S_3/MWCNT ternary hybrid. A series of g-C_3N_4/Bi_2S_3/MWCNT nanocomposites were prepared with varying MWCNT (0.4, 0.8, and 1.2 wt%) quantities using a similar method. The same process was used to synthesize the Bi_2S_3 and g-C_3N_4/Bi_2S_3 composites for comparison.

2.2. Characterization

The crystal structure and phase analysis of the produced catalysts were analyzed using X-ray diffraction (XRD, Rigaku-600). A field emission scanning electron microscope (FESEM; JEOL JSM-7610F PLUS), equipped with an energy dispersive x-ray spectrometer (EDXS) was used to examine the microstructure, morphology, and composition of the nanocomposites. The Fourier transform infrared spectra (FTIR) were analyzed using JASCO 4700 spectrometer. The optical properties of the samples were studied using a UV–visible diffuse reflectance spectrophotometer (DRS, Shimadzu UV-2600).

The photocatalytic performance of all the catalysts was examined for the decolouration of malachite green dye under 100 W LED irradiation at ambient temperature. A malachite green dye stock solution was prepared by mixing 0.0365 g of malachite green dye powder (C.F. = $C_{23}H_{25}ClN_2$) with 100 ml of deionized water, and a 10 ppm dye solution was prepared. In this procedure, 50 mg of the prepared catalyst was mixed with 50 ml malachite green dye aqueous solution. Before photocatalysis, the solution was stirred in the dark for 30 min, to achieve adsorption-desorption equilibrium. Subsequently, the rate of decolouration of malachite green dye was recorded, using a UV–Vis spectrophotometer. Under a similar procedure, the decolouration of malachite green dye was studied in the presence of g-C_3N_4, Bi_2S_3, g-C_3N_4/Bi_2S_3, and g-C_3N_4/Bi_2S_3/MWCNT-X (X = 0.4, 0.8, and 1.2 wt% MWCNT). The formula used for photocatalytic efficiency is as follows[9],

$$Degradation\,(\%) = \frac{C_0 - C}{C_0} \times 100 \quad (1)$$

where C_o and C are initial malachite green concentration and concentration of malachite green after LED light irradiation at time t. Characterization of photocatalysis process was done using a pseudo-first-order reaction, based on the Langmuir-Hinshelwood kinetics model [10],

$$\ln(\frac{C_0}{C}) = kt \quad (2)$$

where k is the first-order rate constant.

3. Results and Discussion

3.1. *XRD analysis*

The XRD patterns of g-C_3N_4 samples showed in figure 1(a). The intense diffraction peak at 27.5°, corresponding to the (002) plane of g-C_3N_4, which denoting a graphite-like interlayer stacking. The peak located at 13° corresponds to the (100) plane and can be attributed to the periodic arrangement of tri-s-triazine units [11]. Figure 1(b) shows the XRD pattern of Bi_2S_3, g-C_3N_4/Bi_2S_3, and g-C_3N_4/Bi_2S_3/MWCNT hybrid at different wt% of MWCNT. The XRD pattern of g-C_3N_4/Bi_2S_3 and g-C_3N_4/Bi_2S_3/MWCNT hybrid showed peaks of both g-C_3N_4 and Bi_2S_3 (JCPDS NO.17-0320), which confirmed the formation of the composites. The diffraction peak of MWCNT does not appear in the XRD pattern, which may be due to the low amount of MWCNT in the nanocomposite.

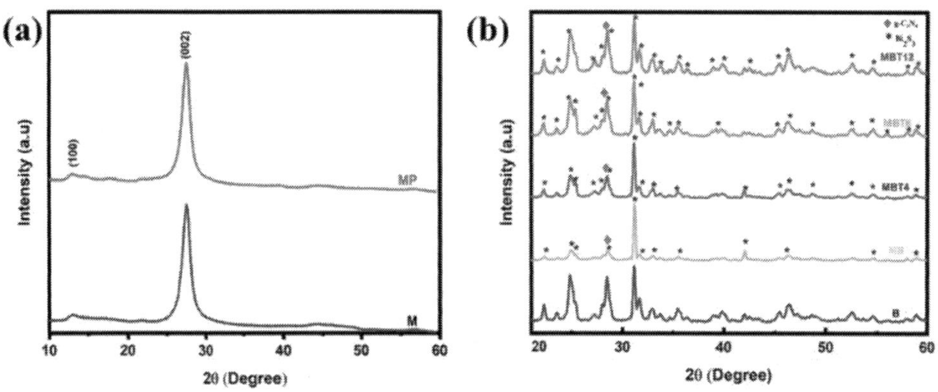

Figure 1. XRD patterns of (a) g-C_3N_4 prepared by calcination of pretreated melamine (MP) and pure melamine (M), and (b) g-C_3N_4/Bi_2S_3 (MB), and g-C_3N_4/Bi_2S_3/MWCNT-X (MBTX)nano-composites.

3.2 Surface morphology characterization

The morphological analysis of g-C_3N_4, Bi_2S_3, g-C_3N_4/Bi_2S_3, and g-C_3N_4/Bi_2S_3/MWCNT-1.2 wt% was done using FESEM. The g-C_3N_4 has sheet-like morphology with a fluffy structure (figure 2(a)) and Bi_2S_3 has a flower-like morphology (figure 2(b)). Figure 2(c) shows g-C_3N_4/Bi_2S_3 morphology. Figure 2(d) shows irregular deposition of g-C_3N_4 and MWCNT onto the surface of Bi_2S_3, confirming the successful formation of the nanocomposite. EDXS was used to determine the elemental distribution and composition of the g-C_3N_4/Bi_2S_3/MWCNT1.2 nanocomposite sample, which identified N, C, Bi, and S elements in the g-C_3N_4/Bi_2S_3/MWCNT1.2 sample.

3.3. Optical properties

The optical properties of the prepared samples were studied using DRS. Figure 3(a) and (b) shows the bandgap energy estimation of g-C_3N_4, Bi_2S_3, g-C_3N_4/Bi_2S_3, and g-C_3N_4/Bi_2S_3/MWCNT nanocomposite samples using Tauc's relation (equation 3), where absorption coefficient (α) estimated using Kubelka-Munk function (F(R)) [12].

$$A\left(h\nu - E_g\right) = (\alpha h\nu)^{1/n} \quad (3)$$

The bandgap values are tabulated (Table 1). The optical bandgap of g-C_3N_4/Bi_2S_3/MWCNT composites was found to be reduced with an increase in MWCNT content. This is attributed to the presence of MWCNT, which improved visible light absorption.

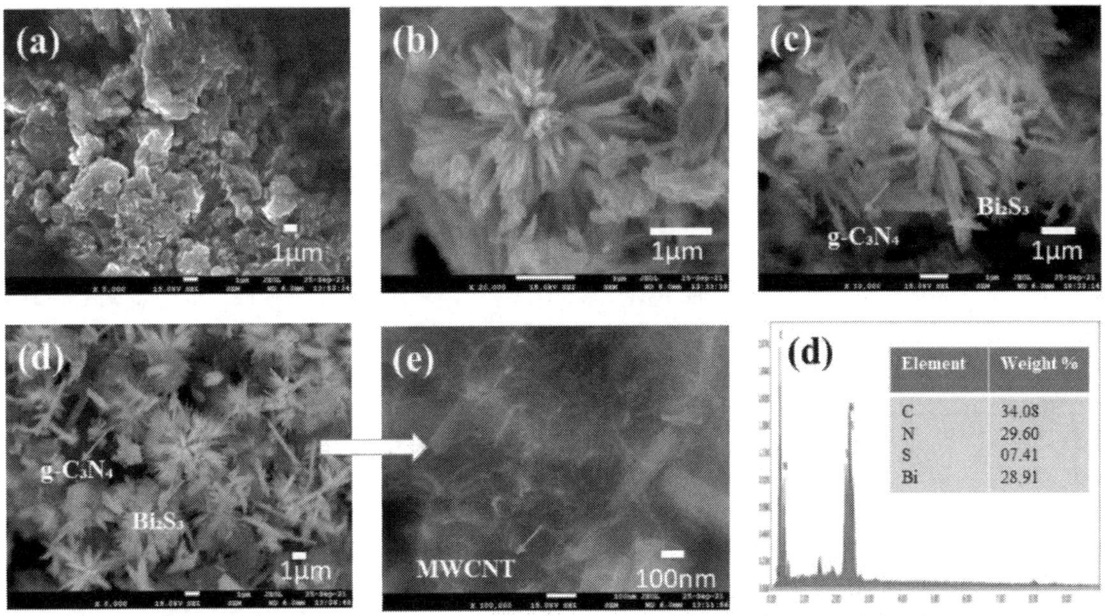

Figure 2. FESEM images of (a) g-C_3N_4, (b) Bi_2S_3, (c) g-C_3N_4/Bi_2S_3, (d) g-C_3N_4/Bi_2S_3/MWCNT-1.2, (e) FESEM images of g-C_3N_4/Bi_2S_3/MWCNT-1.2 sample at a larger magnification, and (f) EDX spectra of g-C_3N_4/Bi_2S_3/MWCNT-1.2 nanocomposite.

Table 1. The bandgap of prepared samples.

Samples	Bandgap
g-C_3N_4	2.82
Bi_2S_3	1.59
g-C_3N_4/Bi_2S_3	1.64
g-C_3N_4/Bi_2S_3/MWCNT-0.4 wt%	1.64
g-C_3N_4/Bi_2S_3/MWCNT-0.8 wt%	1.64
g-C_3N_4/Bi_2S_3/MWCNT-1.2 wt%	1.62

3.4. *FTIR analysis*

The FTIR spectra of the samples are presented in figure 3(c). In the case of g-C_3N_4 sample, the peak at 808 cm^{-1} is the characteristic breathing mode of triazine units. The peak at 1234 cm^{-1}, 1315 cm^{-1}, 1403 cm^{-1}, 1551 cm^{-1}, and 1632 cm^{-1} originated from C-N heterocycles stretching mode [13]. A band around

3000-3500 cm^{-1} indicating the stretching of the N-H bond [14]. In the case of Bi_2S_3, the peak appearing at 804 cm^{-1}, 1019 cm`1, 573 cm^{-1}, and 1423 cm^{-1} corresponding to the C–H and C-O bond stretching and C-H and C-S bond bending vibrations, representing the Bi^{3+} formation [15]. The occurrence of main characteristic peaks of Bi_2S_3 and g-C_3N_4 in the FTIR spectrum, indicating the g-C_3N_4/Bi_2S_3/MWCNT ternary hybrid formation. However, due to the low content of g-C_3N_4, some peaks between 1000 and 1700 $cm^{-1}$are difficult to identify on the spectra. Furthermore, peaks of MWCNT might also be present in the same region as the C-N triazine heterocycles [16]. The FTIR analysis confirmed the formation of the heterostructure.

3.5. *Photocatalytic activity*

Figure 3(d) showed the photocatalytic efficiency of all the samples. From that figure, it is clear that g-C_3N_4/Bi_2S_3/MWCNT-1.2 ternary nanocomposite has the highest photocatalytic performance, compared to the g-C_3N_4/Bi_2S_3, g-C_3N_4 and, Bi_2S_3. The g-C_3N_4/Bi_2S_3/MWCNT-1.2 decolorizes 78.8% dye in 50 min irradiation of light. The weak photocatalytic efficiency of the binary and the single-component system is attributed to the rapid recombination of charge carriers. Interestingly, the addition of 0.4 wt% MWCNT on the binary component system shows an increase in photocatalytic efficiency. When the amount of MWCNT increases, the photocatalytic activity changes from 44.4 % to 78.8 %. This indicates MWCNT serves as an electron reservoir and decreased photogenerated electron-hole recombination.

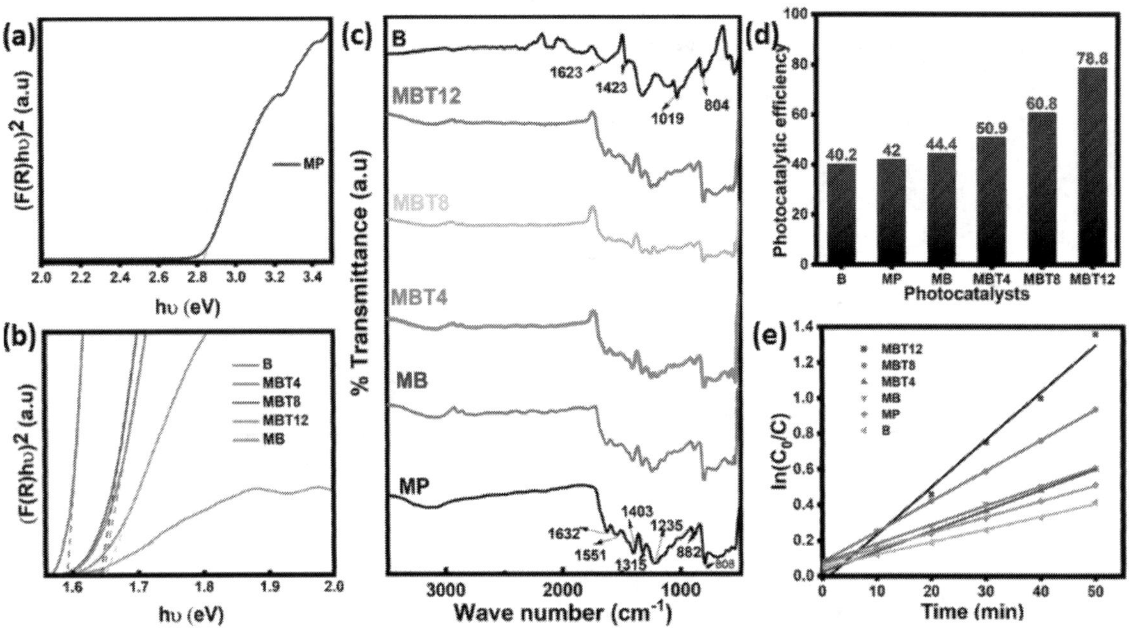

Figure 3. (a) The bandgap of g-C_3N_4 prepared by calcination of pretreated melamine (MP), (b) bandgap of Bi_2S_3 (B), g-C_3N_4/Bi_2S_3 (MB), and g-C_3N_4/Bi_2S_3/MWCNT (0.4, 0.8, and 1.2 wt%) (MBT4, MBT2, MBT6) nanocomposites, (c) FTIR spectra and (d) Photocatalytic efficiency of g-C_3N_4 (MP), Bi_2S_3 (B), g-C_3N_4/Bi_2S_3 (MB), and g-C_3N_4/Bi_2S_3/MWCNT (0.4, 0.8, and 1.2 wt%) (MBT4, MBT2, MBT6) nanocomposites, and (e) Graph of ln(C_0/C) against time representing the kinetics of the photocatalytic reaction.

Using the Langmuir-Hinshelwood model (equation (2)), the rate constant values can be estimated from figure 3(e). The g-C_3N_4/Bi_2S_3/MWCNT-1.2 ternary hybrid has the highest rate-constant value of 0.0265 min^{-1} as compared to g-C_3N_4/Bi_2S_3 (0.0107 min^{-1}), Bi_2S_3 (0.0071 min^{-1}), and g-C_3N_4 (0.0088 min^{-1}), under LED light exposure. The g-C_3N_4/Bi_2S_3/MWCNT-0.4 and g-C_3N_4/Bi_2S_3/MWCNT-0.8 samples have the rate constant values of 0.0171 and 0.0115 min^{-1}, respectively. The value of the rate constant increases with MWCNT content, which indicates that the amount of MWCNT plays a significant role in photocatalytic dye degradation.

4. Conclusion

In summary, g-C_3N_4/Bi_2S_3/MWCNT ternary nanocomposite photocatalyst was synthesized using a simple hydrothermal technique. The improved photocatalytic efficiency of g-C_3N_4/Bi_2S_3/MWCNT hybrid is attributed to improved charge carrier separation efficiency and strong visible light absorption, originated from the heterojunction formation between g-C_3N_4, Bi_2S_3, and MWCNTs. The g-C_3N_4/Bi_2S_3/MWCNT-1.2 hybrid exhibited the highest photocatalytic efficiency (78.8% in 50 min irradiation of light), which confirms that the amount of MWCNT plays a crucial role in the photocatalytic dye degradation, as an electron acceptor. The g-C_3N_4/Bi_2S_3/MWCNT ternary nanocomposite is expected to be a promising visible-light-driven photocatalyst for dye wastewater treatment.

Acknowledgments

The author, ACS would like to thank University Grants Commission (UGC), New Delhi for the research fellowship. The authors also acknowledge the Department of Physics, University of Calicut, for XRD characterization.

REFERENCES

[1] Perumal K, Shanavas S, Karthigeyan A, Ahamad T, Alshehri S M, and Murugakoothan P 2020 *Diamond and Related Materials* **110**
[2] Sun J X, Yuan Y P, Qiu L G, Jiang X, Xie A J, Shen Y H and Zhu J F 2012 *Dalton Transactions* **41** 6756–63
[3] Qingbo Y, Yang K, Li H, and Li X 2021 *Physicochemical and Engineering Aspects* **616**
[4] Zang Y, Li L, Li X, Lin R, and Li G 2014 *Chemical Engineering Journal* **246** 277–86
[5] Zang Y, Li L, Zuo Y, Lin H, Li G and Guan X 2013 *RSC Advances* **3** 13646–50
[6] Mei F, Dai K, Zhang J, Li W and Liang C 2019 *Applied Surface Science* **488** 151–60
[7] Memon Z M, Yilmaz E and Soylak M 2017 *Talanta* **174** 645–51
[8] Jung J Y, Lee D and Lee Y S 2015 *Journal of Alloys and Compounds* **622** 651–6
[9] Shakir M, Faraz M, Shoeb Khan M and Al-Resayes S I 2015 *Comptes Rendus Chimie* **18** 966–78
[10] Shet A and Vidya S K 2016 *Solar Energy* **11** 67–78
[11] Zhang Y, Liu J, Wu G, and Chen W 2012 *Nanoscale* **4** 5300–3
[12] Zhao G, Zhang D, Yu J, Xie Y, Hu W and Jiao F 2017 *Ceramics International* **43** 15080–8
[13] Wang G, Kuang S, Zhang J, Hou S and Nian S *Electrochimica Acta* **187** 243–8
[14] Zhao Y, Liu Z, Chu W, Song L, Zhang Z, Yu D, Tian Y, Xie S and Sun L 2008 *Advanced Materials* **20** 1777–81
[15] Arumugam J, Dhayal Raj A and Albert Irudayaraj A 2017 *Journal of Materials Science: Materials in Electronics* **28** 3487–94
[16] Saleh T A and Gupta V K 2012 *Journal of Colloid and Interface Science* **371** 101–6

IOP Conf. Series: Materials Science and Engineering 1221 (2022) 012044 doi:10.1088/1757-899X/1221/1/012044

Sol-Gel Synthesis of Multifunctional Bismuth Ferrite Nanostructures: Effect of Heating Conditions on their Photocatalytic Activity

Manjusha Eledath[1], Shalini Vishwanathan[2], and Maneesh Chandran[1*]
[1]Department of Physics, National Institute of Technology Calicut, Kerala, India 673601
[2] Department of Chemical Engineering, National Institute of Technology Calicut, Kerala, India 673601
*Email: maneesh@nitc.ac.in

Abstract: Bismuth ferrite ($BiFeO_3$) is a multifunctional perovskite material well-known for its lead-free chemistry, multiferroic properties at room temperature, photovoltaic properties, and photocatalytic activity. $BiFeO_3$ nanostructures were synthesized using sol-gel method by varying annealing temperature and annealing rate. The formation of phase-pure $BiFeO_3$ was confirmed using X-ray diffraction and Raman spectroscopy.Using scanning electron microscopy, the size of $BiFeO_3$nanostructures was evaluated to be around 100 – 500 nm depending on the annealing parameters. The bandgap of the synthesized nanostructures was estimated to be 2.18 – 2.26 eV, with UV diffuse reflectance spectroscopy. The photocatalytic activity was assessed by the degradation of malachite green dye under visible light irradiation and their correlation with heating parameters is discussed in detail.

1. Introduction

Among the metal oxide perovskite materials, bismuth ferrite ($BiFeO_3$,BFO) has received wide attention as a multifunctional material. $BiFeO_3$ possesses multiferroic nature at room temperature, with a ferroelectric Curie temperature of 1103 K and an antiferromagnetic Neel temperature of 643 K[1]. The bandgap of $BiFeO_3$ lies in the range of 2 – 2.8 eV, which leads to interesting applications in the fields of photocatalysis and photovoltaics [2]. The lead–free chemistry of the material adds to the advantages. As a result, the material in different forms, such as film, nanostructure, and bulk, has been investigated for various applications over decades [3–5].
Wet chemical methods are commonly used to synthesize BFO nanostructures of various sizes and morphology. However, the presence of impurity phases, such as $Bi_{36}Fe_{24}O_{57}$, $Bi_2Fe_4O_9$, and $Bi_{25}FeO_4$,often deteriorates the properties of the material. The photocatalytic activity of phase pure and doped BFO nanostructures with dopants such as lanthanum (La), selenium (Se), zirconium (Zr), etc., have been investigated by several groups[6,7]. Enhanced photocatalytic activity of $BiFeO_3$ based binary and ternary composites such as $BiFeO_3$/rGO, $BiFeO_3/ZrO_2$, andg-$C_3N_4/BiFeO_3$/carbon nanotube has also been reported[8–10]. Different process parameters have a significant influence in determining the phase formation, size, and morphology of BFO nanostructures, which can affect their photocatalytic applications. Hence, exploring these parameters and their effects on the aforementioned properties are crucial in optimizing their use in photocatalysis. Although the role of parameters like the molar concentration and annealing temperature has been investigated, the effect of annealing rate on the properties and photocatalytic activity of BFO nanostructures remains unattended.In this work, we present the synthesis of BFO nanostructures using sol-gel technique at various annealing conditions.

The effect of annealing temperature and the rate of annealing on the crystallinity, size, and morphology of the prepared structures are studied. The photocatalytic activity of the BFO nanostructures was explored by the decolorization of malachite green dye under visible light irradiation, and their correlation with the annealing parametersis discussed in detail.

2. Materials and Methods

2.1 Synthesis of bismuth ferrite nanostructures

BFO nanostructures were synthesized by a facile sol-gel method.Bismuth nitrate pentahydrate [$Bi(NO_3)_3$. $5H_2O$, Sigma Aldrich] and iron nitrate nonahydrate [$Fe(NO_3)_3$. $9H_2O$, Sigma Aldrich] were chosen as the precursor materials, and a mixture of ethylene glycol and acetic acid was used as the solvent. The precursors were dissolved in the mixture of ethylene glycol and acetic acid (4:1) by stirring at room temperature to obtain a 0.2 M sol. The sol was kept at 100 °C until the formation of dried gel. Prior to annealing, the dried gel was ground into fine powders. The annealing was carried out at 500 °C and 600 °C, at different heating rates of 2, 5, and 10 °C/min.

2.2 Characterization techniques

The phase formation in the BFO nanostructures was identified using X-ray diffraction (Rigaku) with CuKα irradiation. Raman spectra were recorded with Horiba Scientific Xplora Plus Raman microscope at an excitation wavelength of 638 nm. Scanning electron microscopy (SEM, Carl Zeiss Sigma) was used for the analysis of their size and morphology. The ultraviolet-visible diffuse reflectance spectra (UVDRS) were recorded with Shimadzu UV-2600 UV-vis spectrophotometer.The photocatalytic activity of the prepared samples was studied by the degradation of malachite green dye under visible light exposure.The catalyst was added to 50 mL of the prepared aqueous solution of malachite green (5 ppm) at a concentration of 1 mg/mL.Before the visible light irradiation, the solution loaded with the catalyst was stirred in the dark for 1 h to attain a homogeneous dispersion of powders and to establish the adsorption-desorption equilibrium. Following, it was irradiated with visible light using a 100 W LED bulb for 240 min. A 3 mL aliquot was collected at 30 min intervals and subsequently centrifuged at 9000 rpm for 10 min. The supernatant was collected to determine the absorbance in the UV-visible range. The photocatalytic degradation efficiency (η) was evaluated using the following equation.

$$\text{Photocatalytic degradation efficiency, } \eta\ (\%) = \frac{C_0 - C}{C_0} \times 100 \qquad (1)$$

where C_0represents the dye concentration at the beginning of light irradiation after attaining adsorption-desorption equilibrium in the dark, and C represents the final dye concentration after light irradiation.

3. Results and Discussion

3.1 Structural Analysis

The x-ray diffractogram of BFO nanostructures is given in figure 1(a). As indicated in the figure, all diffraction peaks from the samples correspond to the reflections from various planes of rhombohedral BFO, which are consistent with the standard values. This confirms the formation of phase pure perovskite BFO with rhombohedral crystal structure belonging to the space group R_3C. It was observed that the crystallinity improves with annealing temperature. The variation of FWHM of the diffraction peak corresponding to (012) plane is shown in figure 1(b). The FWHM decreases with an increase in annealing temperature, which indicates an increase in crystallite size according to the Scherrer formula[11].

Figure 1(c) shows the Raman spectra of BFO nanostructures prepared at various annealing conditions. According to group theory, there are $4A_1 + 5A_2 + 9E$ vibrational modes for BFO, out of which $4A_1 +$ 9E modes are Raman active while the A_2 modes are Raman inactive [12]. We have observed 12 Raman modes for each sample, which further confirms the phase formation. The E-1, A(TO1), and A(TO2) modes are of pronounced intensity, while all other modes appear weak. Bi centers participate in the three lowest-frequency vibrations, whereas Fe or O atoms are involved in the higher frequency modes[13]. The observed Raman modes are tabulated in Table S1 (See supplementary information).

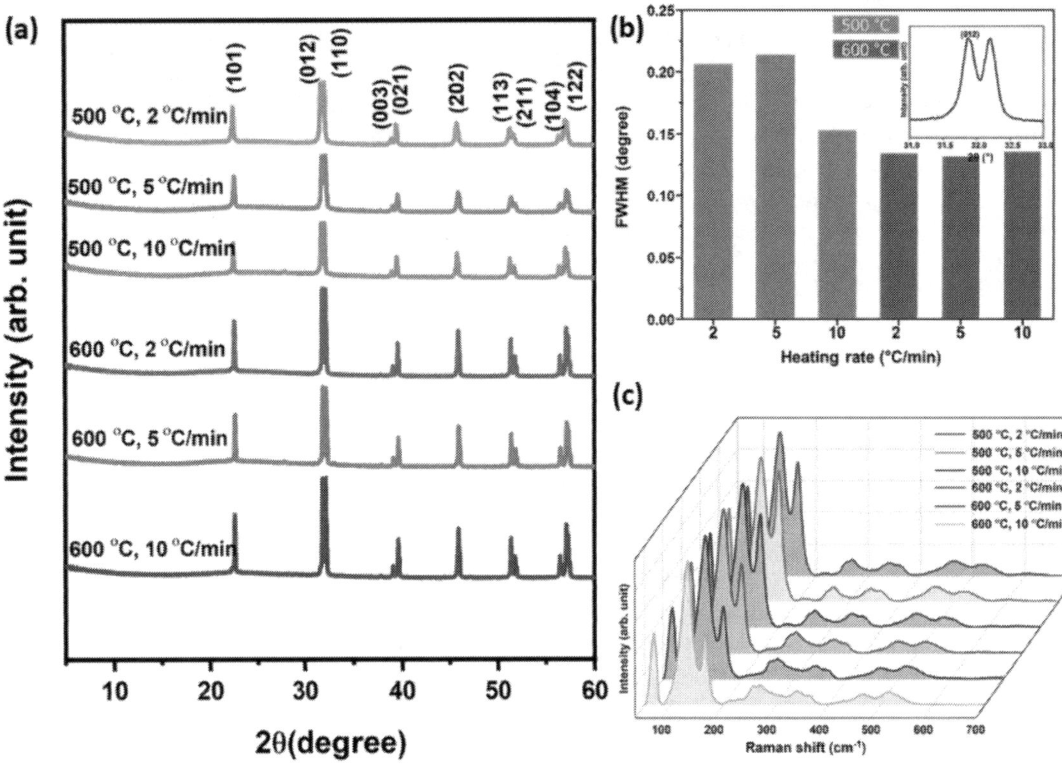

Figure 1. (a) XRD pattern of BFO prepared at different annealing conditions, (b) variation of FWHM of x-ray diffraction peak corresponding to (012) plane with respect to annealing conditions, and (c) Raman spectra of BFO nanostructures, recorded at an excitation wavelength of 638 nm.

3.2 SEM analysis

The SEM images of BFO nanostructures synthesized at various annealing conditions are given in figure 2.It was observed that the size and its uniformity depend on the annealing conditions. The particle size increases with the increase in annealing temperature. Further, more homogeneity can be observed with a lower heating rate. With the highest annealing temperature and heating rate (600 °C and 10 °C/min), there is a tendency to form grains of micrometer size, and the standard deviation appears nearly half of the average size. These observations may be explained as a result of the dominance of crystal growth over the nucleation process with the increase in annealing temperature and rate of heating. The sizes of the nanostructures obtained from the SEM images are tabulated in Table 2.

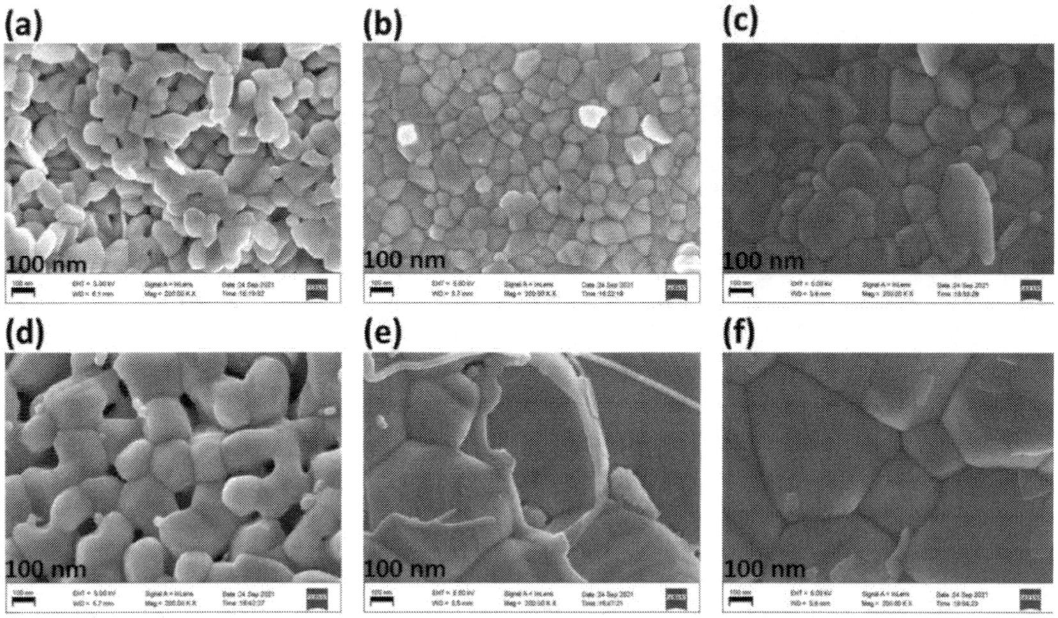

Figure 2. SEM images of BFO NPs annealed at 500 °C with heating rates (a) 2 °C/min (b) 5 °C/min, and (c) 10 °C/min, and at 600 °C with heating rates (d) 2 °C/min, (e) 5 °C/min, and (f) 10 °C/min.

Table 1. Sizes of the BFO NPs obtained from the SEM images

Annealing temperature (°C)	500			600		
Annealing rate (°C/min)	2	5	10	2	5	10
Size (nm)	100 ± 17	114 ± 36	153 ± 77	194 ± 51	495 ± 188	523 ± 275

3.3 Optical Properties

Figure 3(a) shows the UVDR spectra of BFO NPs prepared by annealing at different temperatures and heating rates. The UVDR spectra showed a strong transition in the wavelength range 500 to 600 nm, which corresponds to the electronic transition from the O 2p states of the valence band to the Fe 3d states of conduction band[14]. The bandgap of each sample was evaluated from the diffuse reflectance spectra using the Kubelka-Munk function[15]. The bandgap lies in the range 2.18 to 2.26 eV, which is suitable for visible light photocatalysis (figure 3(b)). The variation in the bandgap may be correlated with the crystallite size. The bandgap shows a slight increase with an increase in FWHM, i.e., as the crystallite size decreases(see, figure 1(b)), which is in agreement with the previous reports[16]. The obtained values are comparable with the previously reported values of BFO nanostructures.

3.4 Photocatalytic activity

The study of the photocatalytic activity of BFO nanostructures evaluated using malachite green dye is summarized in figure 3(c) and 3(d). From the C/C_0 versus time graph (Figure S1(a)), it can be seen that all the catalysts exhibit considerable adsorption in the dark region. The BFO nanostructures annealed at 500 °C possess better catalytic properties compared to those annealed at 600 °C. The catalytic activity also depends on the annealing rate. From figure S1(a), BFO nanomaterial annealed at 500 °C at a rate of 2 °C/min is found to exhibit the highest catalytic activity. This may be explained as a combination of several factors: (i) smaller particle size and uniform size distribution, which increases

IOP Conf. Series: Materials Science and Engineering 1221 (2022) 012044 doi:10.1088/1757-899X/1221/1/012044

the surface area and thus the active sites available for adsorption and catalysis, (ii) higher absorbance in the visible region of the electromagnetic spectrum (figure 3(a)), and (iii) small crystallite size which could reduce the diffusion length between the sites where charge carriers are photogenerated, and the surface sites where they react[17]. The photocatalytic rate constant (κ) for malachite green degradation is evaluated from the first order plot (figure3(c)) using equation 2.

$$\ln \frac{C_0}{C} = \kappa t \tag{2}$$

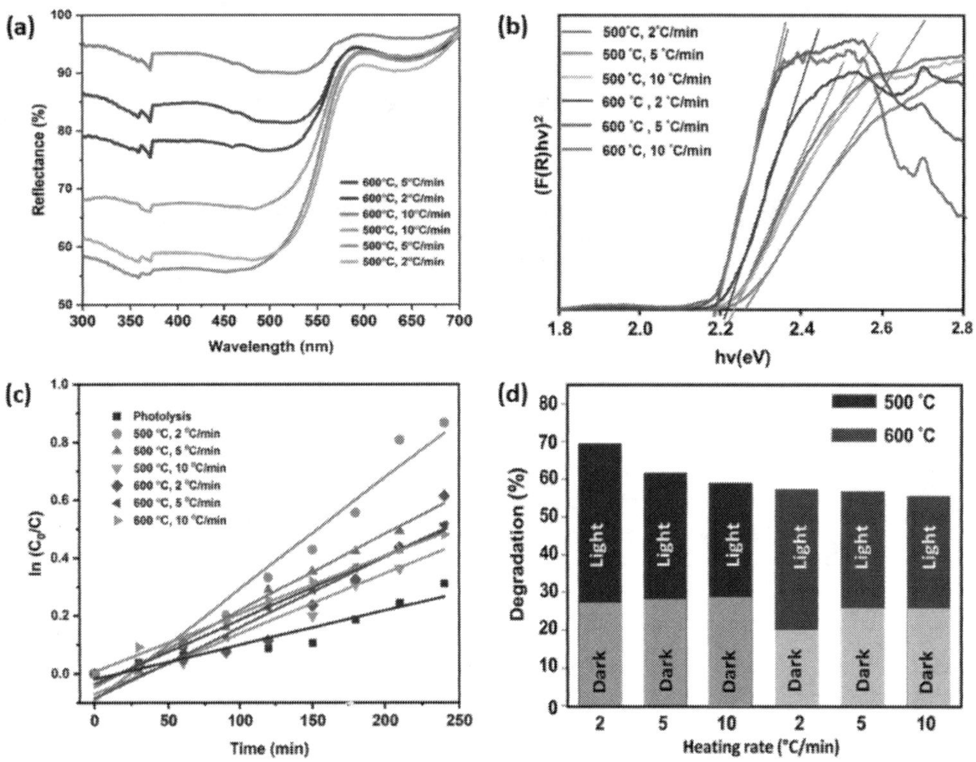

Figure 3. (a) Diffuse reflectance spectra and (b) bandgap of BFO samples prepared at different annealing conditions; (c) photo degradation kinetics and (d) percentage degradation of malachite green dye in the presence of BFO nanostructures

For the catalyst with the highest photodegradation, the rate constant was evaluated as 0.0038 min^{-1}. Figure 3(d) illustrates the percentage degradation of malachite green in the presence of different BFO samples. In the presence of BFO annealed at 500 °C at a rate of 2 °C/min, the degradation of ~ 70 % was observed. The degradation reduced to 56 % when the annealing temperature and rate were 600 °C and 10 °C/min, respectively. As shown in figure 3(d), the highest degradation is observed when the annealing rate is 2 °C/min for both annealing temperatures. With an increase in the annealing rate, the % degradation decreases. The higher degradation of malachite green dye in the presence of catalyst annealed at 500 °C and 2 °C/min may be inferred as a combined effect of both annealing temperature and rate of annealing. However, the role of heating rate needs to be thoroughly studied in a wider span, along with the variation of annealing temperature, to distinguish the effect of annealing temperature and rate of annealing more evidently.

IOP Conf. Series: Materials Science and Engineering 1221 (2022) 012044 doi:10.1088/1757-899X/1221/1/012044

4. Conclusion

Phase pure BFO nanostructures were synthesized using sol-gel technique by varying the annealing parameters. The prepared materials were characterized using XRD, SEM, Raman spectroscopy, and UVDR spectroscopy, and the crystallinity, size, and optical properties of the samples were determined. The photocatalytic activity of the BFO nanostructures was evaluated by the degradation of malachite green dye in the presence of visible light. It was observed that nanostructures prepared at annealing temperature 500°C and annealing rate 2 °C/min displayed the highest photocatalytic activity with a percentage degradation of 70.

Acknowledgment

The authors would like to acknowledge Dr. Ratheesh K Vijayaraghavan, IISER Kolkata, for the XRD characterization.

References

[1] Yin L and Mi W 2020 *Nanoscale* **12** 477–523
[2] Gao T, Chen Z, Huang Q, Niu F, Huang X, Qin L and Huang Y 2015 *Rev. Adv. Mater. Sci.* **40** 97–109
[3] Chatterjee S, Bera A and Pal A J 2014 *ACS Appl. Mater. Interfaces* **6** 20479–86
[4] Jayababu S, Inbasekaran M and Narayanasamy S 2021 *Inorg. Chem. Commun.* **123** 108306
[5] Tahir M, Riaz S, Sajjad Hussain S, Awan A, Xu Y B and Naseem S 2020 *J. Magn. Magn. Mater.* **503**
[6] Wang F, Chen D, Zhang N, Wang S, Qin L, Sun X and Huang Y 2017 *J. Colloid Interface Sci.* **508** 237–47
[7] Irfan S, Li L, Saleemi A S and Nan C W 2017 *J. Mater. Chem. A* **5** 11143–51
[8] Soltani T and Lee B K 2016 A *Chem. Eng. J.* **306** 204–13
[9] Huo H, Hu X, Wang H, Li J, Xie G, Tan X, Jin Q, Zhou D, Li C, Qiu G and Liu Y 2019 *Int. J. Environ. Res. Public Health* **16** 3219
[10] Hu L, Hu H, Lu W, Lu Y and Wang S 2019 *Mater. Res. Bull.* **120** 110605
[11] Holder C F and Schaak R E 2019 *ACS Nano* **13** 7359–65
[12] Snežzana Miljanić, Leo Frkanec, Tomislav Biljan 3 Zlatko Meić Mladen Žini ć 2007 *J. Raman Spectrosc.* **38** 1538–53
[13] Hong Y, Li J, Bai H, Song Z, Li G, Wang M and Zhou Z 2020 *Appl. Phys. Lett.* **116**
[14] Mocherla P S V, Karthik C, Ubic R, Ramachandra Rao M S and Sudakar C 2013 *Appl. Phys. Lett.* **103**
[15] Das S and Srivasatava V C 2016 *Smart Sci.* **4** 190–5
[16] Wang L, Xu J B, Gao B, Chang A M, Chen J, Bian L and Song C Y *Mater. Res. Bull.* **48** 383–8
[17] Devi L G, Murthy B N and Kumar S G 2010 *Mater. Sci. Eng. B Solid-State Mater. Adv. Technol.* **166** 1–6

Investigation on the impacts of thickness on thermal stress on Thermoelectric materials $MoSi_2$ and Mo_5SiB_2

Sarabjeet Singh[1] and Yogesh Chandra Sharma[2]

[1] Research Scholar, Department of Electronics & Communication, Vivekananda Global University, Jaipur

[2] Innovation, Research and Development, DR CBS Cyber Security Services LLP, Jaipur- 302006 (Rajasthan) India

Email: ece_sarabjeet@yahoo.co.in

Abstract. Thermoelectric cooler employs Peltier effect for dissipating heat in an electronic casing structure. It shows exceptional rewards over conservative cooling skill via quiet process, extended life span, and effortless integration. Nevertheless, Joule heating results in the accumulation of internal heat thereby exposes thermoelectric cooler towards the risk of thermo-mechanical breakdown all through continuous operations in pragmatic thermal surroundings. A relative analysis of the effect of thickness size on thermal stress on MoSi2 and Mo5SiB2 by the COMSOL-Multiphysics platform is offered. Mo5SiB2 in comparison to MoSi2 has lower anisotropic single crystal elastic moduli, along with lower shear modulus. Mo5SiB2 has a slightly higher bulk, shear and Young's than MoSi2. RT Vickers hardness of Mo5SiB2 is much larger than those of MoSi2. Fracture toughness is comparable to those of MoSi2. In this paper, a 3D module of thermoelectric materials MoSi2 and Mo5SiB2 is designed on the way to examine the effect of thermal stress with increasing thickness of the material taking into consideration the temperature reliant TE material traits. One side of the module is kept at 300K with fixed constraints while the other side is kept at 1200K. It has been observed that the thermal stress induced in MoSi2 and Mo5SiB2 decrease exponentially with increase in thickness of the material. Beyond thickness of 500 nm, the incremental difference in thermal stress is not large although a slight rise in stress level is observed at thickness 700 nm. It was found that the induced thermal stress for a particular thickness in Mo5SiB2 is lower than MoSi2. For MoSi2, the voltage swings across the length is from -2.42mV to 1.09 mV whereas for Mo5SiB2, the voltage swings across the length is from -1.87 mV to 1.64 mV. It was found that excessive elevated levels of thermal strain may source the dislocations as well as cracks in the layers of the material.

Keywords: Thermoelectric effect, peltier effect, COMSOL simulation, thermoelectric cooler, thermal stress.

1. Introduction

Analyzing and formatting stress concentrations in components forms an extremely significant division of the design life cycle of a component. Structural components not often form perfect convex geometries and have to be tailored to aesthetic in addition to design constraints having irregular cross sectional areas. The irregular geometry from the pure convex shape results in notches, where stress concentration arises, i.e., areas where stress is much higher than in the neighborhood. These areas of stress concentrations are placed where the component starts to fail, i.e., cracks begin to appear. Thermoelectric cooler (TEC) employs Peltier effect for dissipating heat in an electronic casing structure, which shows noteworthy dominance into thermal supervision along with engrossing interest both in academic as well as industrial community [1-4]. Past few years, noteworthy advancement in nano structured material enhanced thermoelectric traits through conventional size as well as interface effect [2]. Though, the means of transportation of charge carriers in such intricate resources is entirely unstated [5], capabilities of such nano structured materials are not easy in synthesis in addition to fall short in meeting industrial wishes. Currently, researchers have focused in optimizing the performance of device, like Thomson-effect [6,7], contact resistance [8-10], heat exchanger [11,12], leg numbers [13] and geometry [14,15]. Those most favorable design techniques [16-19] have technically grown with full-fledge. The assemblage of TE devices requires stacking of several layers The most critical malfunction mechanism for TE devices for prolonged procedure [20-22] is the higher thermo-mechanical stress at the boundary.

2. Model depiction

2.1 Model Geometry

The 3D geometric models shown in Fig.1 are taken in account which consists of thin films of thermoelectric materials Molybdenum Silicide (MoSi2) and Mo5SiB2 with dimensions of 5 x 2x 0.1 μm^3 each. It is well-known that the occurrence of thermal stress is owing to the gradients in temperature in addition to mismatching of thermal expansions in thermoelectric material. Occurrence of dislocations and cracks eventually direct in the malfunction of TE module. In the study, levels of thermal stress were obtained with respect to equivalent stress (also called von Mises stress), competent to provide an assessment of resolved shear stress which is capable in resulting plastic deformation.

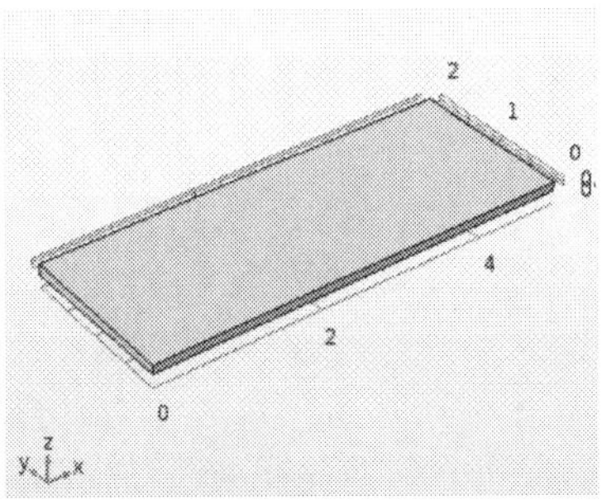

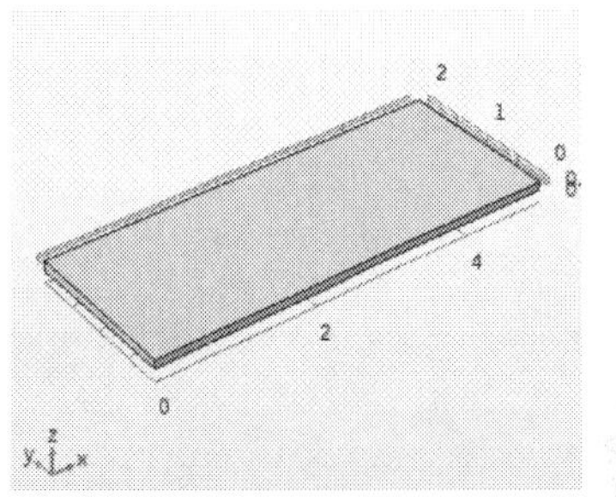

Figure. 1(a) The 3D geometric model of MoSi2 **Figure. 1(b) The 3D geometric model of Mo5SiB2**

2.2 Material properties

The material properties of MoSi2 and Mo5SiB2 are listed in table 1 which has been used as the input parameters of simulation. Properties are chosen from the COMSOL Multi-physics material library.

Table1. Material properties of MoSi2 and Mo5SiB2.

Property	**MoSi2**	**Mo5SiB2**
Thermal Conductivity (k)	66.2 W / (m.K)	27 W / (m.K)
Density (□)	6240 Kg / m^3	8200 Kg / m^3
Heat capacity at constant pressure (Cp)	430 J / (kg.K)	330 J / (kg.K)
Relative Permittivity (□)	1	1
Young's modulus	430 Pa	383 Pa

2.3 Experimental model

Failure of thermal stress due to too much rise in temperature has turn out to be a procedural block hindering the performance along with consistency of electronic equipments. Efficient control of temperature along with thermal drawings is vital assurance in the natural function of these equipments. Two 3D geometric models are taken in account which consists of thin films of thermoelectric materials Molybdenum Silicide (MoSi2) and Mo5SiB2 with dimensions of 5 x 2x 0.1 μm^3 each. One side each of both models are kept at 300K while the other side is kept at 1200K. Force of -100 N and +100 N is applied on each horizontal plane of the model as shown in fig.1.

IOP Conf. Series: Materials Science and Engineering 1221 (2022) 012045 doi:10.1088/1757-899X/1221/1/012045

3. Results and analysis

In this segment, we have presented the results which are figured from the simulation studies of the materials using the thermoelectric module of COMSOL. We look at thermoelectric parameters as well as thermo elastic parameters such as thermal stress distribution for various values of thickness, total displacement and surface temperature distribution across the length of the module, isothermal contours across the span of the material. Then, on analyzing the impact of parameters in designing, the unit optimized design to attain the least thermal stress was conceded. The results are presented in the figures (Fig.2 to Fig.6).

3.1 Surface temperature Distribution

Fig. 2 corresponds to the surface temperature and variation in temperature to the length of the TE materials for $MoSi_2$ as well as Mo_5SiB_2. Variation in the temperature throughout the length of the module is figured at 300 K to 1200 K.

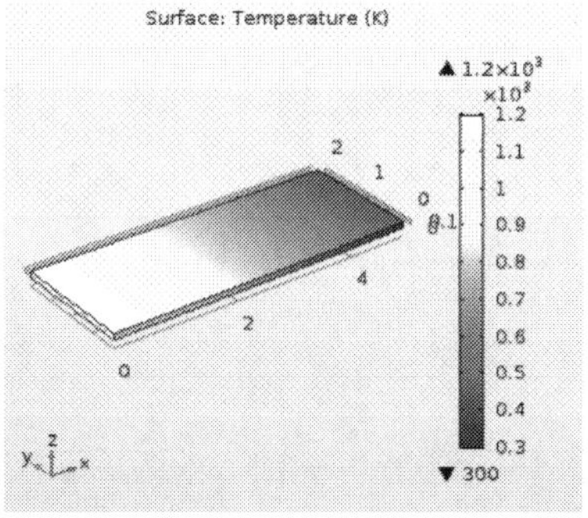

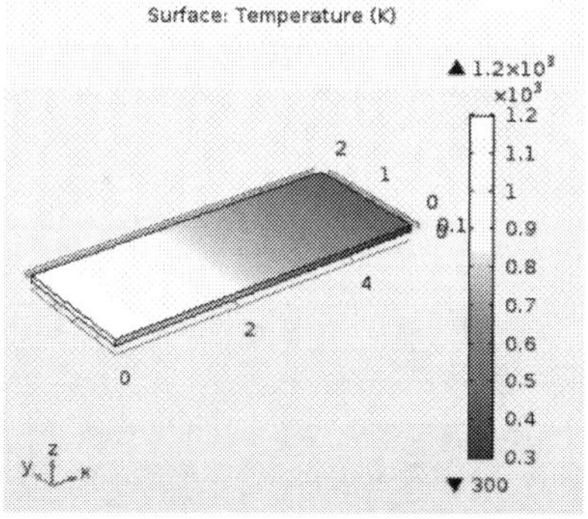

Figure. 2(a) Temperature distribution across MoSi2 **Figure. 2(b) Temperature distribution across Mo5SiB2**

3.2 Thermal stress analysis

Thermal stress results due to the occurrence of gradients in temperature along with not matching in thermal expansions of the material. In common, the degree of thermal stress is principally associated to displacement, strain as well as temperature field. Table 2 shows the values of thermal stress occurred in 3D unit due to varying thickness of material from 100 nm to 1000 nm. The outcomes are depicted in figures 3 and 4. It was found that the induced thermal stress for a particular thickness in Mo5SiB2 is lower than MoSi2.

Table 2. Thermal stress occurred on varied thickness.

Thickness (nm)	Thermal Stress in (N/μm²) (MoSi2)	Thermal Stress in (N/μm²) (Mo5SiB2)
100	655	619
200	251	238
300	136	130
400	132	123
500	91	84
600	79	74
700	83	76
800	71	65
900	58	54
1000	48	46

IOP Conf. Series: Materials Science and Engineering 1221 (2022) 012045 doi:10.1088/1757-899X/1221/1/012045

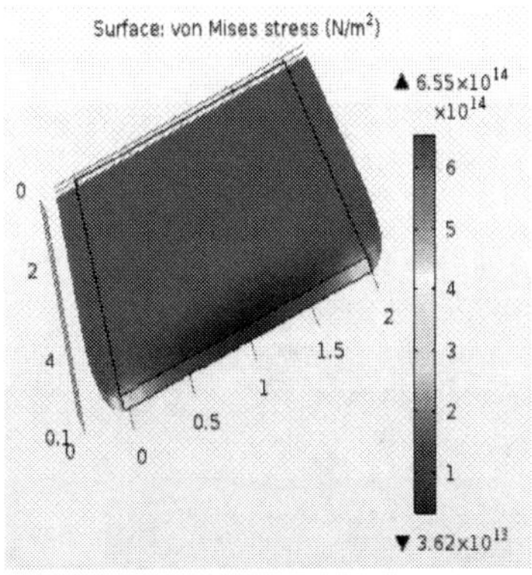

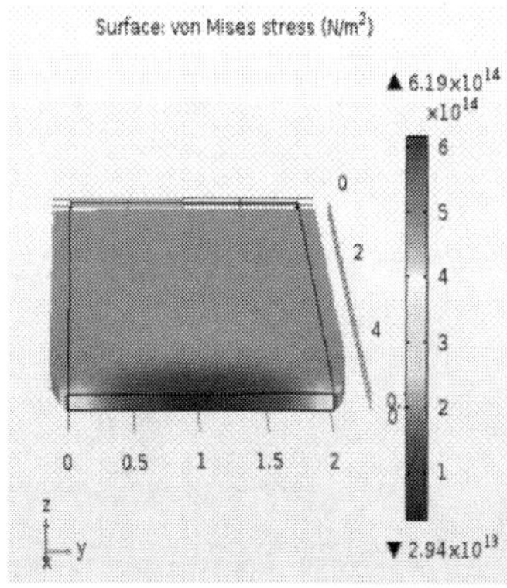

Figure. 3(a) Thermal stress across MoSi2 at 1200K **Figure. 3(b) Thermal stress across Mo5SiB2 at 1200K**

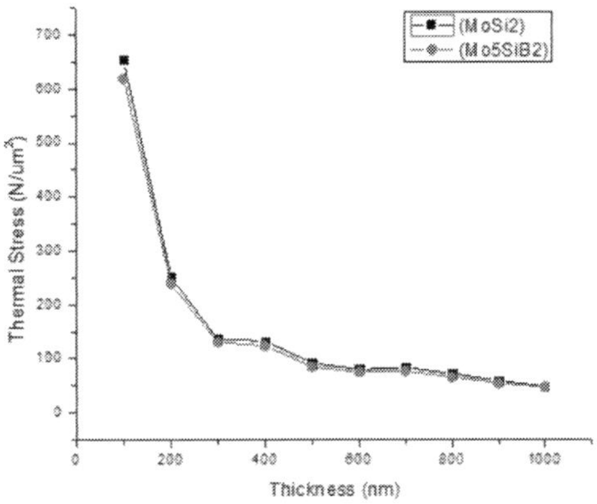

Figure. 4 Thermal stress v/s Thickness of the model

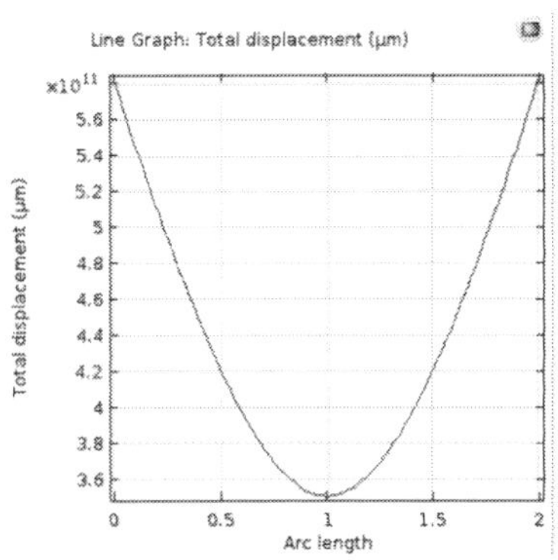

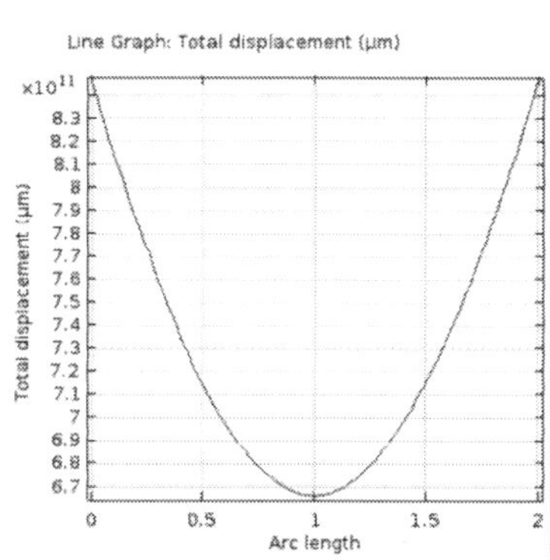

Figure. 5(a) Total displacement across MoSi2 at 1200K **Figure. 5(b) Total displacement across Mo5SiB2 at 1200K**

3.3 Electric Potential

Figure 6 illustrate the potential distribution across the materials. For $MoSi_2$, the voltage swings across the length is from -2.42mV to 1.09 mV whereas for Mo_5SiB_2, the voltage swings is from -1.87 mV to 1.64 mV.

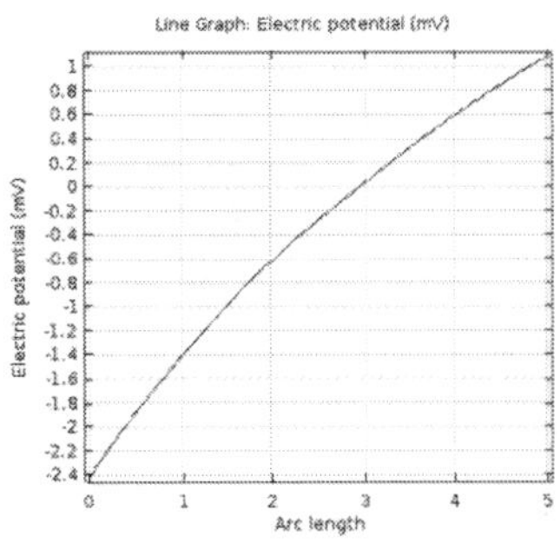

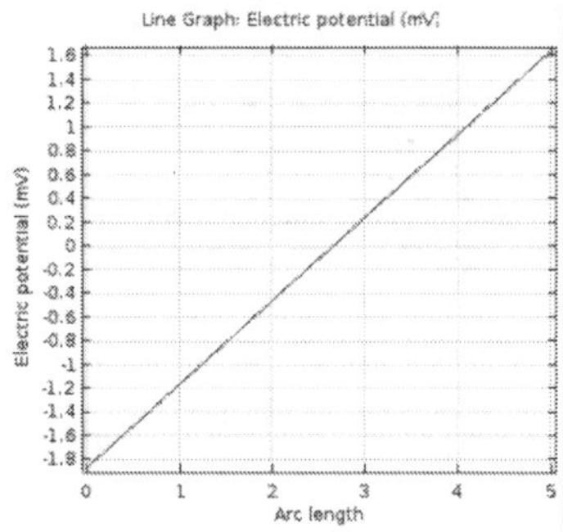

Figure. 6(a) Potential distribution across $MoSi_2$

Figure. 6(b) Potential distribution across Mo_5SiB_2

4. Conclusion

The study build up a 3D geometrical analysis representation on the basis of thermoelectric along with thermal stress taking in report the properties of temperature reliant TE material. The influence of temperature on the levels of thermal stress were examined. The difference in temperature due to the Peltier effect can too show the way in amazing thermal stress levels at the hot sight of the unit. COMSOL Multiphysics clearly demonstrates the model and displays the thermoelectric behavior of the materials taken. It has been observed that the thermal stress induced in MoSi2 and Mo5SiB2 decrease exponentially with increase in thickness of the material. Beyond thickness of 500 nm, the incremental difference in thermal stress is not large although a slight rise in stress level is observed at thickness 700 nm. It was found that the induced thermal stress for a particular thickness in Mo5SiB2 is lower than MoSi2. For MoSi2, the voltage swings across the length is from -2.42mV to 1.09 mV whereas for Mo5SiB2, the voltage swings across the length is from -1.87 mV to 1.64 mV. Due to high- melting-point and outstanding resistance towards oxidation, MoSi2 is worn in soaring-temperature furnaces as it can endure extended exposure in air. The study demonstrates the service for such applications but also display the probable behavior of the chosen materials apart from aiding in determining optimum parameters and doing a comparative analysis.

References

[1] Majumdar A2 009 Thermoelectric devices: helping chips to keep their cool. Nat Nanotechnol; 4:214.

[2] Minnich A J, Dresselhaus M S, Ren Z F and Chen G 2009 Bulk nanostructured thermo- electric materials: current research and future prospects. Energy Environ Sci 2:466e79.

[3] Liu D, Zhao F Y, Yang H X and Tang G F 2015 Thermoelectric mini cooler coupled with micro thermosiphon for CPU cooling system. Energy 83:29e36.

[4] Martinez A, Astrain D and Aranguren P 2016 Thermoelectric self-cooling for power electronics: increasing the cooling power. Energy; 112:1e7.

[5] Zebarjadi M, Esfarjani K, Dresselhaus M S, Ren Z F and Chen G 2012 Perspectives on thermoelectrics: from fundamentals to device applications. Energy Environ Sci ; 5:5147e62.

[6] Snyder G J, Toberer E S, Khanna R, Seifert W 2012 Improved thermoelectric cooling based on the Thomson effect. Phys Rev B 86:41e8.

[7] Lee H S 2013 The Thomson effect and the ideal equation on thermoelectric coolers. Energy 56:61e9.

[8] Fang E, Wu X, Yu Y, Xiu J 2017 Numerical modeling of the thermoelectric cooler with a complementary equation for heat circulation in air gaps. Open Phys 15:27e34.

[9] Bjork R 2016 An analytical model for the influence of contact resistance on ther- moelectric efficiency. J Electron Mater 45:1301e8.

[10] Sullivan O, Alexandrov B, Mukhopadhyay S and Kumar S 2013 3D compact model of packaged thermoelectric coolers. J Electron Packag 135 031006.

[11] Lin Z, and Tan H, Yu J 2013 Analysis on optimal heat exchanger size of thermoelectric cooler for electronic cooling applications. Energy Convers Manag 76: 685e90.

[12] Liu D, Cai Y and Zhao F Y. 2017 Optimal design of thermoelectric cooling system inte- grated heat pipes for electric devices. Energy 128:403e13.

[13] Wang N, Li X-C and Mao J F 2015 Improvement of thermal environment by thermo- electric coolers and numerical optimization of thermal performance. IEEE Trans Electron Dev 62:2579e86.

[14] Kishore R A, Kumar P, Sanghadasa M and Priya S 2017 Taguchi optimization of bismuth-telluride based thermoelectric cooler. J Appl Phys; 122 337e46.

[15] Manikandan S and Kaushik S C 2015 Energy and exergy analysis of an annular ther- moelectric cooler. Energy Convers Manag 106:804e14.

[16] Jeong E S 2014 A new approach to optimize thermoelectric cooling modules. Cryogenics 59:38e43.

[17] Soprani S, Haertel J H K, Lazarov B S, Sigmund O and Engelbrecht K 2016 A design approach for integrating thermoelectric devices using topology optimization. Appl Energy 176:49e64.

[18] Shen L, Chen H, Xiao F and Wang S 2015 The practical performance forecast and analysis of thermoelectric module from macro to micro. Energy Convers Manag 100:23e9.

[19] Huang Y X, Wang X D, Cheng C H and Lin T W. 2013 Geometry optimization of thermo- electric coolers using simplified conjugate-gradient method. Energy 59: 689e97.

[20] Ming T, Yang W, Wu Y, Xiang Y, Huang X, Cheng J 2017 Numerical analysis on the thermal behavior of a segmented thermoelectric generator. Int J Hydrog Energy 42:3521e35.

[21] Ming T, Wang Q, Peng K, Cai Z, Yang W and Wu Y 2015 The influence of non- uniform high heat flux on thermal stress of thermoelectric power generator. Energies 8 12584e602.

[22] Pan T, Gong T, Yang W and Wu Y 2018 Numerical study on the thermal stress and its formation mechanism of a thermoelectric device. J Therm Sci 27: 249e58.

IOP Conf. Series: Materials Science and Engineering 1221 (2022) 012046 doi:10.1088/1757-899X/1221/1/012046

In-situ STS studies and first principles calculations on bare and Sn adsorbed UHV exfoliated WS_2 layers

Manu Mohan[1], Vipin Kumar Singh[2], Mihir Ranjan Sahoo[3], Reshmi S[4], Sudipta Roy Barman[2] and Kuntala Bhattacharjee[1,4*]

[1] Department of Physics, Indian Institute of Space Science and Technology, Valiyamala, Trivandrum, 695547

[2] UGC-DAE Consortium for Scientific Research, University campus, Khandwa Road, Indore,452001, M.P, India

[3] Harishchandra Research Institute, Prayagraj, 211019, Uttarpradesh, India

[4] Institute of Physics, Sachivalaya Marg, Sainik School (PO), Bhubaneswar, 751005, Odisha, India

*Corresponding email: kbhattacharjee@iopb.res.in; kuntala.iopb@gmail.com

Abstarct. Two dimensional (2D) derivatives of tin (Sn) have obtained special deliberations recently due to practical realization of planar, as well as, buckled hexagonal lattice of Sn called stanene. However, it has been observed that proper choice of substrate is very important for growth of stanene like films owing to large core size of Sn that prefers sp^3 hybridization over sp^2. Transition metal dichalcogenides (TMDs) like MoS_2 or WS_2 with honey comb lattice structure seem to be promising substrate candidates for 2D growth of Sn. In the present work, we report mechanical exfoliation of few layers of WS_2 under ultra-high vacuum (UHV) conditions and investigations of growth and local electronic structure by in-situ scanning tunneling microscopy (STM) and spectroscopy (STS) studies. Flat WS_2 surface with honeycomb lattice structure in the atomic scale with a lattice constant of 0.34 nm is evident in the STM investigations, whereas, STS measurements reveal local density of states (LDOS) of WS_2 with a bandgap of approximately 1.34 eV. Density functional theory (DFT) calculations performed by considering bulk WS_2 reveal conduction and valence band states comprised of S *p* and W *d* at both sides of the Fermi energy (E_F) and an indirect bandgap of 1.38 eV. Experimental observations upon Sn adsorption, reveal commensurate growth of Sn atoms on the sulfur 'S' sites with a buckling height of 40 $\pm$10 pm. STS measurements exhibit local electronic structure of the Sn adsorbed surface with clear evidence of in-gap states. DFT calculations quantify the experimental results demonstrating 'S' sites as the most stable sites for the atomic adsorption of Sn with a buckling height of around 80 pm and reveal signature of in-gap hybridized states comprised of Sn *p* and W *d* orbitals.

1. Introduction

With isolation of graphene from graphite [1, 2], a new domain of two dimensional (2D) materials comprising a diversified list has emerged. Present materials research is based on transitionmetal dichalcogenides (TMDs) [3-15], stanene and other 2D group-IV elements of X-enes [15-23],hexagonal boron nitride (hBN) [24-26], borophenes [27], phosphorene [28, 29], aluminene [23] etc. that show superior properties and potential applications. Stanene, a monolayer of Sn with hexagonal lattice structure, has attracted special interest owing to its outstanding electronic properties governed by strong spin-orbit coupling (SOC). It is a large gap (0.3 eV) quantum spin Hall insulator with emergence of robust 2D topological insulating states that are considered to be ideal for quantum transport [30]. Epitaxial growth of stanene and 2D Sn superstructures in buckled or planar forms on the substrates such as Cu(111) [30], Bi_2Te_3(0001) [31], Ag(111) [32], Au(111) [33-35] have been successfully demonstrated. Buckling in stanene happens as a result of the substrate as well as due to preference of sp^3 hybridization of Sn over sp^2 which lowers the bond angles from 120^0 and increases the bond lengths. Strain induced growth of epitaxial stanene or allotropes of 2D Sn on a substrate is driven by lattice mismatch and substrate-induced charge transfer that affects the electronic structure. As a result, conductivity of stanene can reach very high at room temperature (RT) with tunable electronic properties making it as one of the most important members of atomically thin group-IV family. MoS_2 or WS_2[36, 37] in the TMDs family with honeycomb lattice structure and thickness dependent tunable electronic properties have received enormous research attention in the recent times. Effect of TMD substrate on the structural and electronic properties of stanene was investigated by first-principles calculations[36]. It was predicted that MoS_2 as a substrate helps to open an energy gap at the K point of the Brillouin zone of the isolated stanene which has a Dirac cone at the K point, whereas, there is no evidence of additional transport channels near the Fermi level [36].

Our previous scanning tunneling microscopy (STM) investigations reveal a 2D like commensurate growth of Sn on the `top' `S' sites of the WS_2 surface [38]. Further investigations related to local electronic structure of the Sn/WS_2 system are important to study hybridized states and electronic stability. Here, we present our experimental scanning tunneling spectroscopy (STS) measurements and first principles theoretical calculations carried out on bare WS_2, as well as, on Sn adsorbed WS_2 surfaces. STS measurements conducted on the cleaved WS_2 surface reveal a bulk bandgap of approximately 1.34 eV, whereas, after Sn deposition, emergenceof additional states is evident within the bandgap. Density functional theory (DFT) calculations performed on bulk WS_2 crystal exhibit a band gap of around 1.38 eV which is very close to our observed bandgap value of 1.34 eV. In this paper, experimentally acquired local density of states (LDOS) extracted from STS measurements on WS_2, as well as, on Sn/WS_2 surfaces are discussed, compared and quantified with the results obtained from the theoretical calculations.

2. Experimental and Calculation Details

WS_2 crystal from 2D Semiconductors, USA with 99.99% purity was mechanically exfoliated under ultra-high vacuum (UHV) conditions in a load-lock chamber of pressure around low 10^{-9} mbar. The load lock chamber is attached with a sample preparation and growth chamber which has in-situ low energy electron diffraction (LEED) and Auger electron spectroscopy (AES) characterization facilities. Variable temperature scanning tunneling microscopy and spectroscopy (VT-STM/STS) system is mounted with the growth chamber via a gate valve for in-situ transfer and characterizations of the sample surface and local electronic properties. The whole system was procured from Omicron Nanotechnology, Germany. WS_2 crystal after mechanically exfoliated in the load lock chamber under UHV conditions was transferred in-situ to the STM chamber via magnetic transfer rods to perform RT STM and STS measurements [39]. The base pressure of the VT-STM chamber during measurements was of the order of low 10^{-10} to high 10^{-11} mbar. After performingSTM and STS measurements on the bare WS_2 surface, the crystal was transferred to the growth chamber where Sn was evaporated from a water-cooled Knudsen cell[40]at a temperature of 800 °C (1073 K) for 10 min with the WS_2 substrate at RT. Post grown Sn/WS_2 surface was then examined by in-situ STM and STS measurements.DFT calculations were carried out using Vienna Ab-initio Simulation Package (VASP) [41]within the projector augmented wave (PAW) approach to describe the ion-electron interactions and the ground state of the electronic structure is described using the generalized gradient approximation (GGA) with PBE exchange correlation functional. The cut off energy for expansion of the wave functions is 500 eV.The Monkhorst- Pack scheme of k-points sampling method is used for the integration in the first Brillouin zone and 4x4x4 k mesh is employed. The atomic positions are fully relaxed until the force on each atom is less than 0.02 eVÅ^{-1}.

3. Results and Discussion

STM micrographs acquired at RT in constant current mode on the bare WS_2, as well as, on the Sn/WS_2 surfaces are shown in Fig.1. Fig.1(a) shows a large scale STM image of the UHV cleaved WS_2 surface, whereas, Fig.1(b) shows an atomically resolved surface of the WS_2 crystal. Flat WS_2surface with dark spots, a typical signature on the TMD crystal surfaces [42, 43] is evident in Fig.1(a).We do not find evidence of contaminants or foreign adsorbed species in the STM investigations [38] as shown in Fig.1. Bare WS_2 in the atomic scale exhibits clear honeycomb lattice structure with lattice constant of approximately 0.34 nm [38]as can be seen in Fig.1(b). We assign the bright protrusions in the STM micrograph [Fig.1(b)] as the `S' atoms [38]. STM measurements carried out on the Sn grown sample at RT are shown in Fig.1(c, d). Fig.1(c) is a large area scan which does not reveal any island or layered growth of Sn on the WS_2 surface. However, atomic scale investigations clearly disclose adsorption of Sn atoms as bright protrusions at the `S' sites of the WS_2 [Fig.1(d)] surface. Distance between two nearest Sn atoms is ~0.34 nm confirming a commensurate growth of Sn on the WS_2 crystal. The buckling height of the adsorbed Sn is 40 (±10) pm [38].

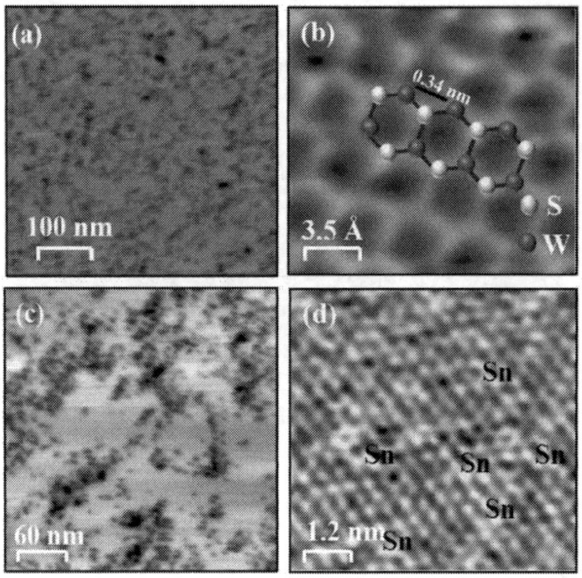

Figue 1. STM images of bare WS_2. (a) Large scale and (b) atomically resolved WS_2 surface with a lattice constant of 0.34 nm. Schematic representation of hexagonal surface motif of WS_2 is superimposed for better clarity. (c) and (d) shows the wide area and atomically resolved STM images after Sn deposition.

STS measurements carried out on bare WS_2 and Sn/WS_2 surfaces are shown in Fig.2. dI/dV spectra [Fig.2] are numerically extracted from the I-V data acquired at different positions on the surface with the feedback loop off and by ramping the bias voltage from -1.5 V to +1.5 V. Fig.2(a) shows such a dI/dV spectrum obtained from bare WS_2 revealing prominent signature of conduction band (CB) and valence band (VB) states [shown by arrows in Fig.2(a)]. dI/dV measurements also exhibit a bulk bandgap of WS_2 crystal which is around 1.34 eV [Fig.2(a)]. Similar STS measurements carried out on the Sn deposited surface reveal emergence of in-gap electronic states around Fermi energy (E_F) [marked by oval in Fig.2(b)], as well as, clear evidence of CB and VB states [shown by arrows in Fig.2(b)].

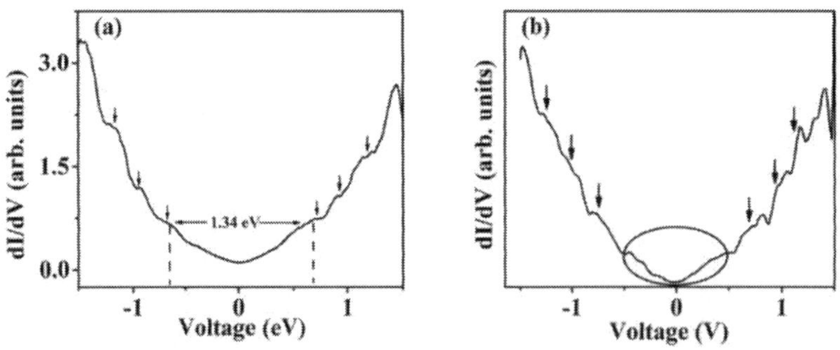

Figure 2. STS measurements carried out on bare (a) and Sn deposited (b) WS_2 surfaces.

DFT calculations were carried out for bare WS_2 considering bulk unit cell. The relaxed WS_2 (lattice parameter a=3.18 Å) structure is shown in Fig.3(a, b). Fig.3(a) & (b) are the side and top views of the relaxed structure respectively. Fig.3(c) & (d) depict the band structure and partial density of states (PDOS) of the bulk WS_2 crystal. From the calculations, we get an indirect bandgap of ~1.38 eV with VB maxima at the Γ point and CB minima between the K-Γ line [Fig 3(c)].PDOS calculations [Fig. 3(d)]reveal mainly the S *p* and W *d* states which spread over in the VB and CB.

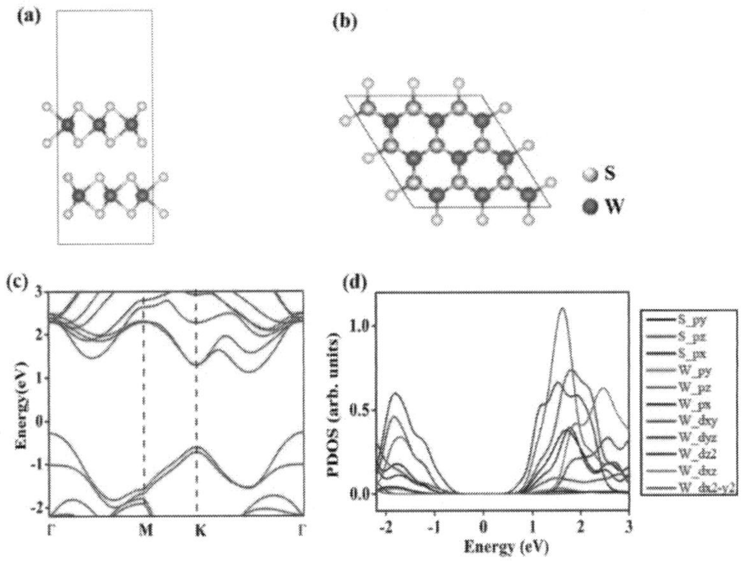

Figure 3. (a) side view and (b) top view of the relaxed structure of bulk WS_2. (c) Band structure and (d) PDOS vs. energy of the bulk WS_2 crystal. Right panel in (d) shows the color coding of different S *p* and W *d* orbitals.

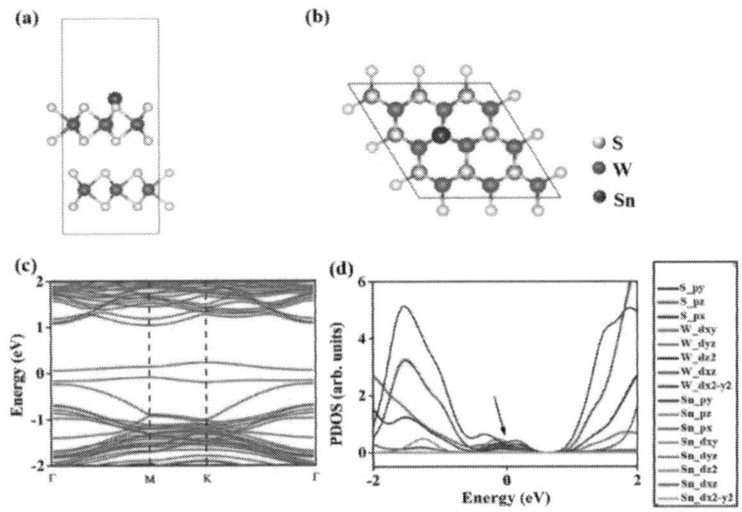

Figure 4. Relaxed structure of Sn/WS_2. (a) Side view and (b) top view. (c) Band structure and (d) PDOS of Sn/WS_2. Evidence of in-gap states are shown by arrow in (d).Right panel in (d) shows the color coding of different Sn *p*, *d*; S *p* and W *d* orbitals.

First principlescalculations on the Sn adsorbed WS_2 layers reveal `S' sites as the most stable sites for the adsorption of Sn atoms with a buckling height of ~80 pm [Fig.4(a, b)]. These theoretical results are in direct correlation with the experimentally observed STM investigations where Sn atoms getting adsorbed at the `S' sites with a buckling height of 40 (±10) pm [38] on the WS_2 lattice is clearly administered in the STM micrographs [Fig.1(d)]. Calculated band structure and PDOS of the Sn/WS_2 system are shown in Fig.4(c, d). Calculations reveal emergence of in-gap hybridized states comprised of Sn *p* and W *d* around E_F [Fig.4(c, d)] corroborating our observed STS results.

4. Conclusion

In-situ STS studies and first principles theoretical calculations were carried out on the mechanically exfoliated WS_2 and Sn/WS_2 surfaces. WS_2 surface exhibits a bandgap which is 1.34 eV according to the STS measurements and 1.38 eV as per DFT calculations. The calculations reveal a bulk indirect bandgap of the WS_2 with VB maxima at the Γ point and CB minima between the K-Γ points. Calculated PDOS exhibit various S *p* and W *d* states within the VB and CB of the WS_2 that can be directly correlated with the emergence of local electronic states in the observed STS data. The Sn/WS_2 system reveal signature of hybridized in-gap states. These electronic states near E_F within the bandgap are mainly attributed to Sn *p* and W *d* hybridized states.

Acknowledgement

M. M. and K. B. sincerely thank the funding and infrastructure provided by UGC-DAE CSR Indore through the project grant: Ref.: CSR-IC/CRS-152/2015-16/20 to carry out this work. K. B. also thanks DST Mobility Fellowship to provide her with the personal and professional stability and the funding for overall scientific support. The simulation works related to DFT calculations presented here were supported in part by the SAMKHYA: high performance computing facility provided by Institute of Physics, Bhubaneswar.

References

[1] Novoselov K S, Geim A K, Morozov S V, Jiang D, Katsnelson M I, Grigorieva I V, Dubonos S V and Firsov A A, 2005 *Nature* **438** 197–200
[2] Geim A K and K S Novoselov 2007 *Nat. Mater.* **6** 183–191
[3] Eda G, Yamaguchi H, Voiry D, Fujita T, Chen M, and Chhowalla M 2011 *Nano Lett.* **11** 5111–5116
[4] Ganatra R, and Zhang Q, 2014 *ACS Nano* **8** 4074–4099
[5] Feng C, Huang L, Guo Z, and Liu H, 2007 *Electrochem. Commun.* **9** 119–122
[6] Kibsgaard J, Chen Z, Reinecke B N, and Jaramillo T F, 2012 *Nat. Mater.* **11** 963–969
[7] Tenne R, Margulis L, Genut M and Hodes G, 1992 *Nature* **360** 444–446
[8] Chuang H-J, Chamlagain B, Koehler M, Perera M M, Yan J, Mandrus D, Tománek D, and Zhou Z, 2016 *Nano Lett.* **16**
[9] Li X, and Zhu H, 2015 *J. Mater.* **1** 33–44
[10] Singh A K, Kumar P, Late D J, Kumar A, Patel S and Singh J, 2018 *Appl. Mater. Today* **13** 242–270
[11] Shi H, Yan R, Bertolazzi S, Brivio J, Gao B, Kis A, Jena D, Xing H G and Huang L. 2013 *ACS Nano* **7** 1072–1080
[12] Li S, Wang S, Salamone M M, Robertson A W, Nayak S, Kim H, Tsang S C E, Pasta M and Warner J H, 2017 *ACS Catal.* **7** 877–886
[13] Venkata Subbaiah Y P, Saji K J and Tiwari A, 2016 *Adv. Funct. Mater.* **26** 2046–2069
[14] Xu K, Wang Z, Du X, Safdar M, Jiang C and He J, 2013 *Nanotechnology* **24** 465705
[15] Miró P, Audiffred M and Heine T, 2014 *Chem. Soc. Rev.* **43** 6537–6554
[16] Luo B, Wang B, Li X, Jia Y, Liang M and Zhi L, 2012 *Adv. Mater.* **24** 3538–3543
[17] Garcia J C, de Lima D B, Assali LV C and Justo J F, 2011 *J. Phys. Chem.* C **115** 13242–13246
[18] Cahangirov S, Topsakal M, Aktürk E, Şahin H and Ciraci S, 2009 *Phys. Rev. Lett.* **102** 236804
[19] Şahin H, Cahangirov S, Topsakal M, Bekaroglu E, Akturk E, Senger R T and Ciraci S, 2009 *Phys. Rev. B* **80** 155453
[20] Roome N J and Carey J D, 2014 *ACS Appl. Mater. Interfaces* **6** 7743–7750
[21] Liu C C, Jiang H, Yao Y, 2011 *Phys. Rev. B* **84** 195430

[22] Vogt P, De Padova P, Quaresima C, Avila J, Frantzeskakis E, Asensio M C, Resta A, Ealet B and Le Lay G, 2012 *Phys. Rev. Lett.* **108** 155501
[23] Kamal C, Chakrabarti A and Ezawa M, 2015 *New J. Phys.* **17** 083014
[24] Zhang K, Feng Y, Wang F, Yang Z and Wang J, 2017 *J. Mater. Chem. C* **5** 11992–12022
[25] Khan A F, Brownson D A C, Foster C W, Smith G C and Banks C E, 2017 *Analyst* **142** 1756–1764
[26] Khan A F, Brownson D A C, Randviir E P, Smith G C and Banks C.E. 2016 *Anal. Chem.* **88** 9729–9737
[27] Mannix A J, Zhou X F, Kiraly B, Wood J D, Alducin D, Myers B D, Liu X, Fisher B L, Santiago U, Guest J R, Yacaman M J, Ponce A, Oganov A R, Hersam M C, and Guisinger N P, 2015 *Science* **350** 1513–1516
[28] Li L, Yu Y, Ye G J, Ge Q, Ou X, Wu H, Feng D, Chen X H and Zhang Y 2014 *Nat. Nanotechnol.* **9** 372–377
[29] Qin G, Yan Q -B, Qin Z, Yue SY, Cui H J., Zheng Q R and Su G, 2015 *Sci. Rep.* **4** 6946
[30] Deng J, Xia B, Ma X, Chen H, Shan H, Zhai X, Li B, Zhao A, Xu Y, Duan W, Zhang S C, Wang B and Hou J G, 2018 *Nat. Mater.* **17** 1081–1086
[31] Zhu F, Chen W, Xu Y, Gao C, Guan D, Liu C, Qian D, Zhang S C and Jia J, 2015 *Nat. Mater.* **14** 1020–1025
[32] Yuhara J, Fujii Y, Nishino K, Isobe N, Nakatake M, Xian L, Rubio A and Le Lay G, 2018 *2D Mater.* **5** 025002
[33] Maniraj M, Stadtmüller B, Jungkenn D, Düvel M, Emmerich S, Shi W, Stöckl J, Lyu L, Kollamana J, Wei Z, Jurenkow A, Jakobs S, Yan B, Steil S, Cinchetti M, Mathias S, and Aeschlimann M, 2019 *Commun. Phys.* **2** 12
[34] Sadhukhan P, Barman S, Roy T, Singh V K, Sarkar S, Chakrabarti A and Barman S R, 2019 *Phys. Rev. B* **100** 235404
[35] Sadhukhan P, Pandey D, Singh V K, Sarkar S, Rai A, Bhattacharjee K, Chakrabarti A, And Roy Barman S, 2020 *Appl. Surf. Sci.* **506** 144606
[36] Ren C-C, Feng Y, Zhang S-F, Zhang C-W, and P-J Wang, 2017 *RSC Adv.* 7 9176
[37] C-X Zhao and J-F Jia 2020 Front. Phys. **15** 53201
[38] Manu Mohan, Vipin Kumar Singh, Reshmi S, Sudipta Roy Barman, Bhattacharjee K, 2020 Surface Science **701** 121685
[39] Rai A, Nayak J, Roy S, Barman 2014 *Surf. Sci.* **625** 97–103

[40] Shukla A K, Banik S, Dhaka R S, Biswas C, Barman S R, Haak H, 2004 *Rev. Sci. Instrum.* **75** 4467–4470
[41] Kresse G and Furthmüller J, 1996 *Phys. Rev. B* **54** 11169
[42] Sengoku N and Ogawa K 1995 *Jpn. J. Appl. Phys.* **34** 3363–3367
[43] Yuhara J, Isobe N, Nishino K, Fujii Y, Chan L H, Araidai M and Nakatake M, 2019 *J. Phys. Chem. C* **123** 22293–22298

Ag_2O Decorated ZnO Nanorods Demonstrating Two-Step Visible-Light Photocatalytic Dye-Degradation Phenomena

Payel Sahu and Debajyoti Das*

Energy Research Unit, School of Materials Sciences,
Indian Association for the Cultivation of Science, Jadavpur, Kolkata – 700 032, India.
*Corresponding author: erdd@iacs.res.in

Abstract. ZnO/Ag_2O heterostructures, as the functional materials for photodegradation of MB dye under visible light irradiation, have been grown using inexpensive hydrothermal and ultrasonic-assisted synthetic routes. The heterostructures have revealed wurtzite ZnO peaks that demonstrate its highly crystalline and hexagonal structure, as well as the cubic Ag_2O peaks. The ZnO/Ag_2O photocatalyst has demonstrated two-step dye degradation slopes, with the higher-slope arising later in the process when the metallic-Ag traps the valence electrons of Ag_2O nanoparticles and advances e^-/h^+ separation across the ZnO/Ag^0/Ag_2O heterojunction structures, making them readily accessible for superior dye degradation.

1. Introduction

Zinc oxide (ZnO), an *n*-type semiconductor with a direct band gap (3.37 eV), has attracted a lot of research interest for a variety of potential applications due to its biocompatible design, ease of crystallization, high photosensitivity, structural flexibility, and anisotropic growth [1]. Complete mineralization of organic contaminants from wastewater via photocatalysis is one such viable application in the environmental treatment field [2]. Owing to its wide optical band gap, ZnO hardly utilizes the visible light component of the sunlight and results in poor quantum yield along with rapid recombination of the photogenerated e^-/h^+ pairs. Ag_2O, a narrow band-gap (~1.2 eV) material with high photo-oxidative capability, is planned to be strategically coupled to ZnO and be used as an efficient visible light absorber [3,4] as well as an electron trapping agent to reduce rapid recombination of photogenerated e^-/h^+ pairs and effectively utilize those for photocatalytic degradation of the pollutants [5].

2. Synthesis of ZnO/Ag_2O heterostructures

Initially, the ZnO nanorods were synthesized from zinc acetate dihydrate ($Zn(C_2H_4O_2.2H_2O)$), sodium hydroxide (NaOH) and hydrazine hydrate (N_2H_4, H_2O) pursuing hydrothermal method at 150 °C for 6 h. After washing the product with deionized (DI) water and ethanol (EtOH) it was dried in oven at 60 °C for 12 h.

ZnO/Ag_2O heterostructures, as the functional materials for photodegradation of MB dye under visible light irradiation, have been achieved using an inexpensive co-precipitation method. The mixture of certain amount of ZnO nanorods and $AgNO_3$ solution was magnetically stirred following by dropwise addition of NaOH solution till the complete precipitation of Ag_2O NPs from $AgNO_3$. The final product was obtained by centrifugation, washing with DI water and EtOH several times and drying in oven at 60 °C for 6 h.

3. Characterization

An X-ray powder diffraction (XRD) investigation employing a Cu-K radiation source (1.5418) revealed the crystal structure of the heterostructures. A JEOL JSM-6700F field-electron SEM was used to assess the structural morphology.

3.1. Photocatalytic Study

The photocatalytic activity of pure ZnO, Ag_2O, and ZnO/Ag_2O composite catalysts was investigated individually with MB dye and varied periods of light soaking in the photo-reactor cell under visible light illumination by a 400 W metal halide lamp, after 30 minutes of dark degradation. The quantitative determination of MB was performed monitoring their absorbance contribution, using a UV-vis spectrophotometer (Lambda 25, PerkinElmer, USA, and UV-3101PC, Shimadzu). The degradation % of the dye was estimated using the equation [6]:

$$\text{Degradation (\%)} = (1 - \frac{C}{C_0}) \times 100$$

where, C_0 is the initial dye-concentration when the adsorption-desorption equilibrium is attained and that changes to C on photo-irradiation.

4. Results and discussions

Figure 1 represents the X-ray diffraction spectrum of the prepared photocatalyst materials. In particular, peaks appeared at 2θ = 32.05°, 34.67°, 36.50°, 47.78°, 56.8°, 63.09°, 66.58°, 68.14° and 69.30° were identified as the (100), (002), (101), (102), (110), (103), (200), (112) and (201) planes, respectively, of wurtzite ZnO (JCPDS card no. 36-1451) [6], and the peaks at 33.12°, 38.36°, 55.27° and 65.8° were indexed as the (111), (200), (220), and (311) planes of cubic Ag_2O phases (JCPDS card no. 41-1104), respectively [7]. Furthermore, the absence of any Ag (111) peak at ~38.1° and Ag (200) peak at ~44.3° confirms the nonexistence of any metallic-Ag as any impurity in the synthesized ZnO/ Ag_2O composite material.

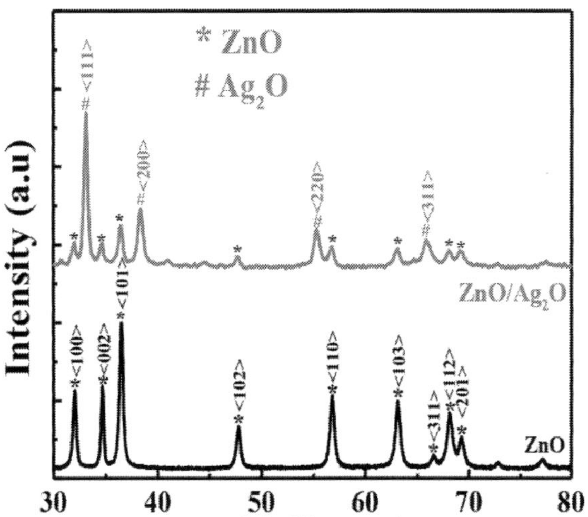

Figure 1. XRD patterns of bare ZnO and ZnO/Ag_2O heterostructures.

Figure 2(a) and 2(b) show the low and high-magnification FE-SEM images of the ZnO/Ag_2O heterostructures revealing that the spherical Ag_2O nanoparticles are decorated on to the clustered rod like structures of ZnO. The one-dimensional rod-like structures are highly facilitating for the rapid transfer of the charge carriers for their effective participation in the dye degradation process [8]. Furthermore, surface of the ZnO nanostructures is generally highly porous and that helps in the efficient adsorption of the dye molecules and their onward degradation during the photocatalytic process [9].

Figure 2. (a) Low and (b) high-magnification SEM images of the ZnO/Ag_2O composite sample.

Photocatalytic Degradation Study

The absorption spectra of MB dye, after photodegradation at different intervals of time in presence of the ZnO/Ag_2O photocatalyst and measured using UV-vis spectrophotometer, are shown in Figure 3, exhibiting an absorption maximum at 664 nm which is typical for the dye.

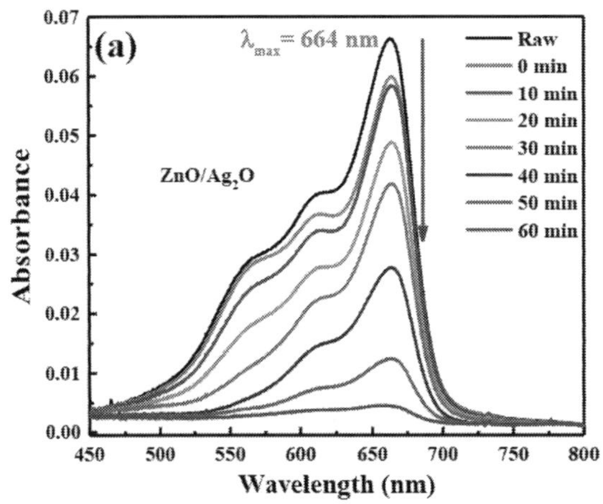

Figure 3. Absorbance spectra of MB dye after different periods of visible light irradiation on the ZnO/Ag_2O photocatalyst

From the slope of $\ln(C_0/C)$ vs. t (min) plot, shown in Figure 4, the rate constant of photodegradation (k) was estimated following: $\ln(C/C_0) = k$ t [10,11]. The degradation profile is almost similar for both pure ZnO and Ag_2O photocatalysts upto a certain period of visible light irradiation, however, on extended time of exposure Ag_2O photocatalyst shows a relative faster degradation characteristic. Interestingly, in ZnO/Ag_2O heterostructure photocatalyst the degradation rates are amplified at both the shorter and longer time zones, following a modified two-step photocatalytic degradation kinetics: $\ln(C/C_0) = k_1 t + k_2 (t - t_C)$, where k_1 and k_2 are the lower and higher rate constants of photodegradation occurring across a certain crossover time (t_C). The t_C is virtually the point of intercept between two slopes over the extended exposure time. The ZnO/Ag_2O

heterostructure has demonstrated a rate constant, k_2 ~0.10064 min^{-1}, which is 10.83 and 3.83 times higher than that of pure ZnO and Ag_2O, respectively.

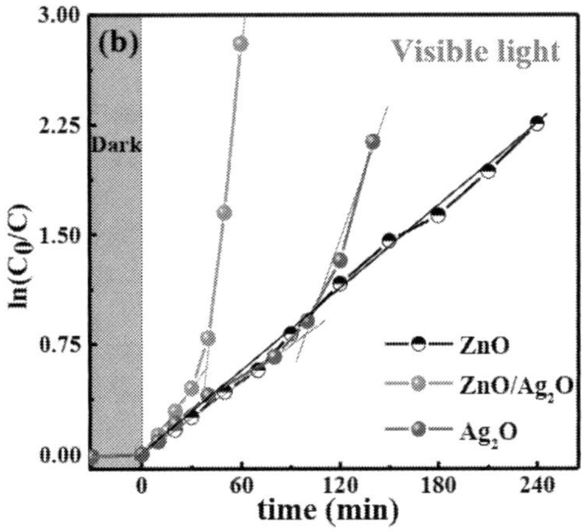

Figure 4. Two step photocatalytic degradation of MB dye in ZnO/Ag_2O catalyst.

5. Discussions:

The observed XRD patterns reveal the well-defined high crystalline nature of the synthesized heterostructures without any trace of impurity while the morphological and microstructural studies using FE-SEM analysis confirms the growth of Ag_2O nanoparticles on the surface of the clustered ZnO nanorods creating *p*-Ag_2O/*n*-ZnO nano-junctions. The synergistic interaction of ZnO and Ag_2O, which broadens visible-light absorption and allows extremely effective separation of photogenerated e^-/h^+, is responsible for the heterostructure's improved visible-light photo-catalytic activity. However, the production of metallic Ag causes a similar, but more pronounced, impact via the formation of ZnO/Ag0/Ag2O heterojunction structures later in the process. The metallic Ag serves as an electron trapping agent after being exposed to light. It scavenges the valence electrons of Ag_2O nanoparticles, increases the e^-/h^+ separation, and therefore accelerates dye degradation. [12,13].

6. Conclusions

In the ZnO/Ag_2O heterostructures grown by hydrothermal and co-precipitation methods, highly crystalline and hexagonal wurtzite ZnO and cubic Ag_2O peaks were found individually from the XRD spectra, without any trace of impurities. Furthermore, the FE-SEM images reveals that the Ag_2O nanoparticles were nicely adorned on the surface of porous ZnO nanorods, creating *p*-Ag_2O/*n*-ZnO nano-junctions that accelerate photocatalytic dye degradation on the surface of the synthesized heterostructures. During an extended photocatalysis process, the growth of Ag^0 helps forming the $ZnO/Ag^0/Ag_2O$ heterojunction structures which facilitates in scavenging the valence electrons of Ag_2O nanoparticles, delaying e^-/h^+ recombination and making them available for superior photocatalytic degradation.

Acknowledgement
This work was done under Projects funded by the DST and CSIR, Govt. of India. P.S. acknowledges DST for providing an INSPIRE Fellowship.

References

[1] Qiu Y, Chen W and Yang S 2010 Facile hydrothermal preparation of hierarchically assembled, porous single-crystalline ZnO nanoplates and their application in dye-sensitized solar cells. *J. Mater. Chem.* **20** 1001–1006

[2] Das D and Nandi P Ternary 2020 ZnCdSO composite photocatalyst for efficient dye degradation under visible light retaining Z-scheme of migration pathways for the photogenerated charge carriers *Sol. Energy Mater. Sol. Cells* **217** 110674

[3] Ma S, Xue J, Zhou Y and Zhang Z 2014 Photochemical synthesis of ZnO/Ag_2O heterostructures with enhanced ultraviolet and visible photocatalytic activity. *J. Mater. Chem.* A **2** 7272–7280

[4] Mohamed R M, Ismail A A, Kadi M W, Alresheedi A S and Mkhalid I A 2020 Facile synthesis of mesoporous Ag_2O–ZnO heterojunctions for efficient promotion of visible light photodegradation of Tetracycline. *ACS Omega* **5** 33269–33279

[5] Liang Y C, Liu Y C and Hung C S 2019 Sputtering control of Ag_2O decoration configurations on ZnO nanorods and their surface arrangement effects on photodegradation ability toward methyl orange. *Nanotechnology* **30** 495701–495712

[6] Nandi P and Das D 2019 Photocatalytic degradation of Rhodamine-B dye by stable ZnO nanostructures with different calcination temperature induced defects *Appl. Surf. Sci.* **465** 546–556

[7] Pham Q T, Huya B T and Lee Y I 2015 New Highly efficient electrochemical synthesis of dispersed Ag_2O particles in the vicinity of the cathode with controllable size and shape. *J. Mater. Chem. C* **3**, 7720–7726

[8] Makal P and Das D 2019 Superior photocatalytic dye degradation under visible light by reduced graphene oxide laminated TiO_2-B nanowire composite. *J. Env. Chem. Eng.* **7** 103358

[9] Mondal P and Das D 2013 Transparent and conducting intrinsic ZnO thin films prepared at high growth-rate with *c*-axis orientation and pyramidal surface texture. *Appl. Surf. Sci.* **286** 397–404

[10] Das D and Nandi P 2021 Synthesis of CdS/GO modified ZnO heterostructure for visible light dye degradation applications *Appl. Surf. Sci.* **570** 151260

[11] Makal P and Das D 2018 Self-doped TiO_2 nanowires in TiO_2-B single phase, TiO_2-B/anatase and TiO_2-anatase/rutile heterojunctions demonstrating individual superiority in photocatalytic activity under visible and UV light *Appl. Surf. Sci.* **455** 1106–1115

[12] Lamba R, Umar A, Mehta S K and Kansal S K 2017 Enhanced visible light driven photocatalytic application of Ag_2O decorated ZnO nanorods heterostructures. *Sep. Purif. Technol.* **183** 341–349

[13] Kadam A, Dhabbe R, Gophane A, Sathe T and Garadkar K 2016 Template free synthesis of ZnO/Ag_2O nanocomposites as a highly efficient visible active photocatalyst for detoxification of methyl orange. *J. Photochem. Photobiol. B.* **154** 24–33

Preparation of CdS and its Photocalytic behaviour for Dye Degradation

Bharati N Patil[1] and Smita A Acharya[2]

[1] Department of Applied Physics, Yeshwantrao Chavan College of Engineering, Nagpur-441110, M.S. India
[2] Department of physics Rashtrasant Tukdoji Maharaj Nagpur University, Nagpur- 440033, M.S. India
E-mail:bharatinpatil@gmail.com

Abstract. A short time ago, transition metal sulphides like CdS has been arduously studied because it has unique catalytic functions. This disclosed that CdS nanocrystals are good photocatalysts because it can quickly generate electron-hole pairs by photo-excitation. In the present work, CdS is prepared by hydrothermal process. Prepared samples were characterized by X-ray diffractions (XRD), SEM, Optical study, and for evaluation of photophysical properties, UV/VIS spectra were recorded. By UV irradiation and photocatalytic decay of dyes, the photoactivity of synthesised CdS was accessed. The CdS shows higher catalytic activity.

Keywords: CdS, Congo Red, dyes, Photocatyalatic activity.

1. Introduction:

In the visible region, CdS is recognized as an effectual photocatalyst but due to Cd^{2+} ion leaching, it shows low stability. To enhance its photocatalytic stability, taking into account its moderately high photoactivity, many endeavours have been taken; like (i) changing surface structure of CdS nanoparticles [1-2] (ii) doping with transition metal ions into CdS. [3] Depositing CdS to Nafion membranes, graphene sheets, or carbon sheets to get uniform, homogeneously distributed CdS QDs [4-9].

Due to its small bandgap (2.4 eV) corresponding to the visible light spectrum, cadmium sulfide (CdS), a very interesting photocatalyst has been anormously studied, which makes it photo-excited effortlessly under visible light irradiation. However, this results in photo corrosion due to the low separation efficiency of electron-hole pairs, which restricts the global applications of CdS. [10] And also, CdS becomes very unbalanced and badly recyclable because the photo corrosion problem that the sulphide ion can easily get oxidised in photocatalytic activity. To solve these problems, several viewpoints have been developed.

2. Experimental:

2.1 Materials: cadmium nitrate Cd $(NO3)_2$, thiourea (NH_2 (SNH_2), All the predecessor were GR grade and acquired from Merck (India) with 98% purity.

2.2 Synthesis of CdS :

CdS, obtained using synthesis process, 50 ml of D.I. water added to 10.6612g cadmium nitrate $Cd(NO3)_2$ mixed up to 30 min. 3.0448 g. thiourea was put into the mixture, once again mixed for about 15 minutes. Then using autoclave the combination was warmed upto 170°c for 24 hrs. The combination was cooled, strained, and washed a few times with D.I water, then the product was dehydrated at room temperature to get CdS.

3. Results and Discussion:

3.1 Structural confirmation by X-Ray Diffraction:

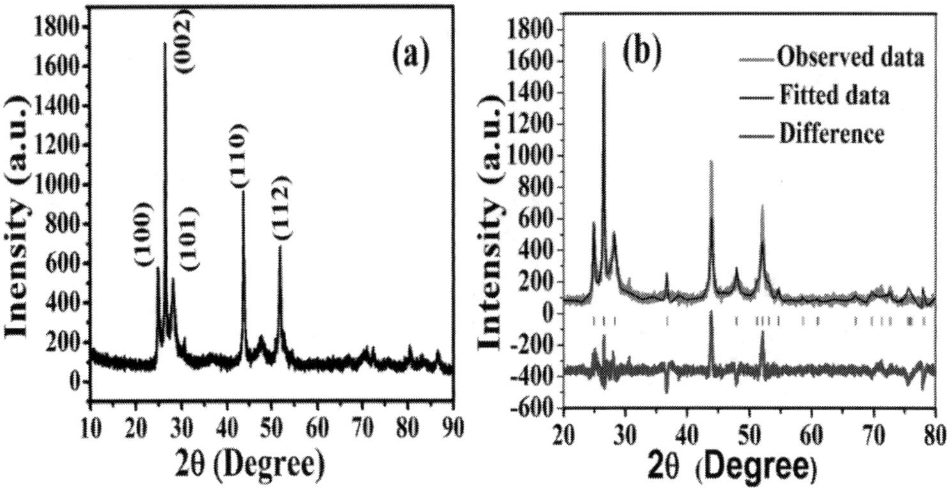

Figure 1 **(a)** XRD of CdS **(b)** Rietveld refinement of XRD pattern of CdS

Figure 1 (a and b) shows the XRD patterns of as-synthesized CdS, the phase identification was carried out with help of standard JCPDS file no 10-0454 and indexed as hexagonal quartzite structure. The peaks at 24.93^0, 26.42^0 28.23^0, 43.71^0, and 51.88^0 are corresponded to (100), (002), (101), (110), and (112) planes of qrtzite crystalline CdS. For more systematic explanation, XRD data are purified by the CdS quartzite structure with a space group of P 6_3 mc. From Reitveld Refinement the constants acquired and are shown in **Table 1.** From lattice constants '*a*', '*b*', and '*c*' quartzite phase can be observed, and found very similar with observed lattice constant from XRD and standard JCPDS no. 10-0454. Rietveld fitted XRD of CdS is shown in **Figure 1 (b)** with the goodness of fit is found to be 4.76.

Lattice Parameter	a = b = 4.1243 c = 6.7210
Goodness of fit	χ^2 = 4.76
FWHM	0.230
Particle size by Scherrer formula	40 nm
R_P	43.3
R_{WP}	41.6
R_{EXP}	19.09

3.2 Optical studies:

UV absorption spectrum of as-synthesized CdS prepared by the hydrothermal method is shown in Figure 2 (a). A strong absorption peak is observed around the 488 nm region. The optical band gap is determined by using Tauc equation. The intercept obtained from extrapolation of the linear portion of the curve is energy bandgap (Eg). An employed linear fit to absorption edge in the Tauc plot is shown in Figure 2 (b).

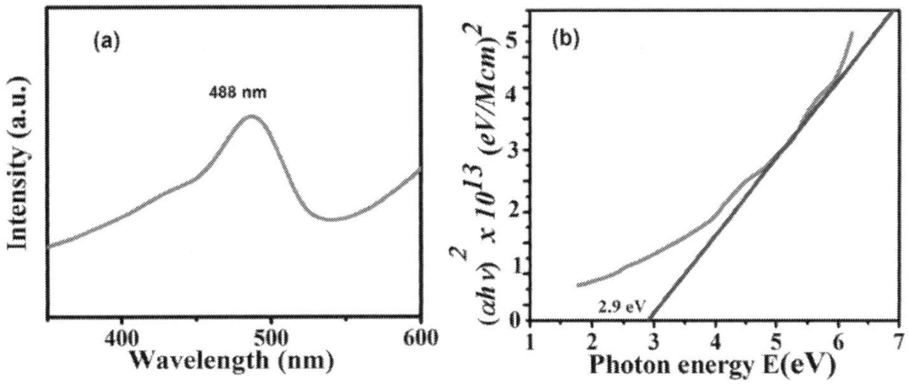

Figure 2 (a) UV Visible spectra CdS NPs and **(b)** $(\alpha h\nu)^2$ Vs photon energy (hν)

3.3 Scanning Electron Microscopy:

Figure 3 is a classic SEM image of the CdS nanostructures. The SEM image indicates the dendritic-like geometrical morphology of CdS grain. This nanostructure is composed of the main branch and

Figure 3. SEM images of hydrothermally prepared CdS at different

IOP Conf. Series: Materials Science and Engineering 1221 (2022) 012048 doi:10.1088/1757-899X/1221/1/012048

some small equal branches with about hundreds of nanometers wide and a few micrometers in length. SEM image in Figure 3 depicts that such dendritic structures are about entirely in the product with a mean length equal to 1-2 μm along the trunk, and this specifies the good consistency and high yield attained hydrothermally by synthesized CdS. Some of the dendrites show flower-like morphology with symmetrical trunks. Thus microstructural study discloses a clear and well-defined dendritic structure with a noticeable trunk and extremely ordered branches.

3.4 Photodegradation ability to single-phase CdS :

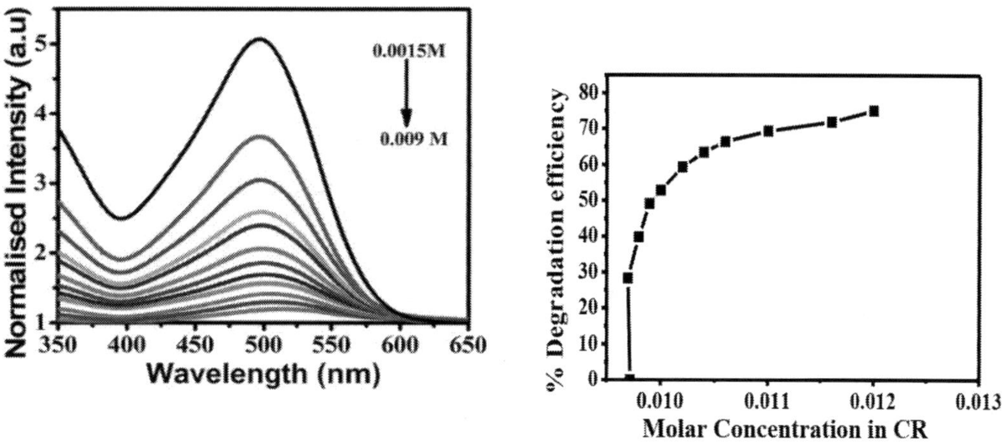

Figure 4 shows a graph of photodegradation of CR, with different amounts of doses of CdS. After particular doses scattering of light also occurs. These entire graphs clearly show a decrease in the intensity of CR with an increase in doses of photo-degrading agents. Doses concentrations varying in between 0.0015 - 0.009 M to get maximum decoloration of given dye. Around 70% degradation is obtained for dose level 9 mM. CdS are found to be an effectual photocatalyst in the visible region but because of Cd^{2+} ion leaching CdS shows low quantum efficiency and low stability in solution. To enhance its photocatalytic stability, considering its comparatively high photo-activity, many attempts are being made. To enhance the photocatalytic efficiency of CdS, various attempts have been made, like graphene sheets, changing the surface structure of CdS nanoparticles by controlling morphology, carbon nanotubes to get uniform, depositing CdS to Nafion membranes, doping of transition metal ions into CdS, uniformly distributed CdS quantum dots and combination of CdS with other semiconductor. Among these, the productive way to control the potential of conduction and valence bands by successive changes in the combination is combination of distinct bandgap semiconductor forming nanocomposites. The combination of these two phases with the formation of nanocomposite may lead to the tailoring of energy bandgap across the interface thus manipulate electron-hole formation and recombination rate in presence of visible light. It helps to tune the photodegradation rate.

4. Conclusions :

In summary, widely studied II-VI group semiconductor CdS nanostructured systems are prepared by hydrothermal synthesis routes. XRD confirms high degree structural purity with single phase formation of as-synthesized CdS without any post-processing techniques after synthesis. A microstructural study by SEM confirms nanoscale dimensions of the samples. The dendritic shape grains are detected for CdS. Optical spectra of CdS exhibit a blue shift in UV-VIS absorption edge. The energy band gap was estimated by Tauc plot and exhibited a rise in energy bandgap in comparison to their respective bulk system. The photodegradation of water dissolved dye CR in presence of visible light by CdS NPS is revealed and found CdS is more efficient. The system exhibited the potential to be used as a photocatalyst for the disintegration of organic dyes in presence of visible light.

References

[1] Rajeshwar K, Tacconi N R, and Chenthamarakshan C R, *Chem. Mater.*, 13 (2001) 2765.

[2] Gandhi N, Sharma V, Khant A, Khandelwal R C, *Bulletin of the Catalysis Society of India*,9 (2010) 25.

[3] Huating H, Xianbao W, Fangming L, Jingchao W, Chunhui X, *Synthetic metals,* 161 (2011) 410.

[4] Chavan R, Kumar A, Chaudhary R P. *Journal of Luminescence,* 145 (2014) 6.

[5] Mohammed M A, Shitu A, Ibrahim A, *Res.J.Chem.Sci.*, 4 (2014) 91.

[6] Neelakandeswari N, Sangami G, Dharamaraj N, Taek N K, Kim H Y, *Spectrochimica Acta.,* 78 (2011) 1592.

[7] Soltani N, Saion E, Mahmood W, Frafani M, Rezaee K. *Applied Surface Science.,* 290 (2014) 440.

[8] Mehata M R, Goyal N, Salvi K N, Prajapati V P, *Acta Chin.Pharam Indica*, 295 (2012) 100.

[9] Chu J, Li X, Qi J, *Cryst Eng Comm.,* 14 (2012) 1881.

[10] Davis A P and Huang C, Water Res., 25 (1991) 1273.

IOP Conf. Series: Materials Science and Engineering 1221 (2022) 012049 doi:10.1088/1757-899X/1221/1/012049

Antibacterial Performance of Synthesised Superparamagnetic $Fe_3O_4@TiO_2$ Nanoparticles

[1*]Sachin A Kulkarni, [2]P S Sawadh, [1]Anuruddha R. Chabukswar, [1]Swati Jagdale

[1] Dr. Vishwanath Karad MIT World Peace University, Pune-411038, Maharashtra, India

[2] B.D. College of Engineering, Sevagram- 442102, Maharashtra, India

*Corresponding Author's mail Id: kulkarnisachin42@gmail.com

Abstract. Co-precipitation method is used to prepare Fe_3O_4 nanoparticles, and its surface was coated with titania by sonication of ammonium sulphate. These nanoparticles were analyzed using SEM, UV-Vis, FTIR and antibacterial activity. The mono dispersion and increment in size with TiO_2 concentration confirmed through SEM and good antibacterial activity is observed against gram positive and negative bacteria.

1. Introduction

The interest level on magnetite particles has increased recently as nanomaterials and their potential impact on technology advancement is being realized. It was reported that super paramagnetic particles are compatible for bio-applications, for example as MRI contrast agents, medium for targeted drug delivery, protein-specific taggers, ferrofluids and MR fluids [1-5]. However, not modified Fe_3O_4 nanoparticles forms clusters and exhibited poor anti-oxidation and thermal stability. These applications typically use magnetite in the form of particles.

Several theoretical and applied studies have shown that by combining titanium dioxide with iron ions (Fe^{2+}, Fe^{3+}) the photo catalytic activity increases significantly [4-8]. Therefore, driven by these reports, in the present work TiO_2 has been selected to coat on the magnetite nanoparticles. This work presents an ease chemical synthesis approach to coat TiO_2 layer on Fe_3O_4 nanoparticles. These nanocomposites were used in magnetic photocatalysts, even though their antibacterial properties were less reported. Here is a report on antibacterial activity of Fe_3O_4-TiO_2 nanocomposites against gram positive and gram-negative bacteria.

2. Experimental Section

The synthesis procedure and experimental details were followed as in our earlier report [9]. The surface details of the nanocomposites were observed by JEOL JSM 6480 LV scanning electron microscope (SEM). The chemical details of the nanoparticles were analyzed with Nicolet FT-IR Avatar 360 (Nicolet, USA) by standard KBr pellet technique. The UV-absorption analysis done by UV CARY 100 Scan UV-Visible Spectrophotometer.

2.1. Measurement of Antibacterial activity

Antibacterial activity of the prepared core shells was tested using diffusion method against four bacteria, two Gram positive b. subtilus, S. aureus, two Gram negative E. coli and P. aeuriginosa. The antibacterial property of Fe_3O_4/TiO_2 nanocomposite particles was investigated by using a zone inhibition method and calculated minimum inhibitory concentration (MIC) values. The MIC provides information regarding the required concentration for complete inhibition of the bacterial growth. the nanoparticles were placed in a four different glass plates and then mixed with nutrient agar. At last b. subtilus, S. aureus, E. coli and P. aeuriginosa was added into above mixtures separately and shaken well to mix the materials as shown in fig. 4. These inoculated samples were incubated at 37 °C for 24 h [10].

3. Results and Discussion

Fig.1 SEM micrographs of (a) pure Fe_3O_4 nanoparticles and TiO_2 coated Fe_3O_4 nanoparticles with Fe/Ti weight content (b)1:1.25 (c) 1:1.50 (d) 1:1.75.

Fig.1 displays the SEM micrographs of pure and TiO_2 coated magnetite nanocomposites. It could be observed from SEM that with increase in the concentration of titania, the titanium dioxide covered magnetite particles have a uniform size and are uniformly distributed throughout the surface. It has been observed from SEM micrographs that there is an increment in size of the particle with increment in TiO_2 concentration. From SEM it is evident that magnetite composites getting darker to brighter with TiO_2 content. The dark region is due to presence of Fe_3O_4 and bright region is due to presence of TiO_2. It may be due to higher electron binding ability of Fe_3O_4 than TiO_2. In this regard, the cores were identified as the darker compared to the shell [11].

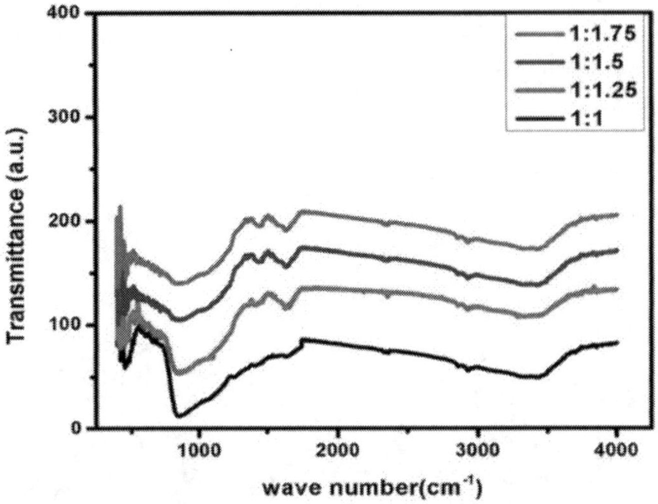

Fig.2 FTIR Spectra of TiO_2 modified Fe_3O_4 nanoparticles

Fig. 2 shows FT-IR spectra of different Fe_3O_4/TiO_2 nanocomposites in ratios. The presence of OH stretching vibrations confirmed by a band at 3412cm^{-1}. The band observed at 1636cm^{-1} is due to the H-O-H bending. The C-H bending is observed at 1448 cm^{-1}. The bands at 1087 cm^{-1} and 950cm^{-1} are characteristic Ti-O-C vibrations. The observed absorption band at 638 cm^{-1} represents stretching vibrations of the Fe-O and OH groups at Fe_3O_4. Same types of characteristics stretching vibrations were observed from literature supporting our result [12].

The synthesized materials were characterized using UV-Vis Spectroscopy and presented in Fig. 3. From the plots, seen that increase in TiO_2 concentration the absorption peak is red shifted. It is due to increase in particle size reduces the band gap, leads to have a red shift in absorption peak. Shift in discrete energy levels is observed as per to the quantum size effect. It is because when particle size increases the energy levels loses its discreteness due which band gap decreases. This study reports that increase in TiO_2 concentration, the particle size increases, band gap decreases [14,15].

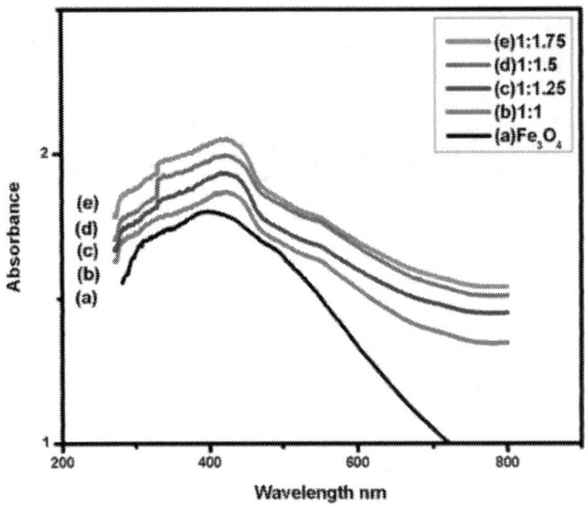

Fig. 3 UV-Vis. Spectra of TiO_2 modified Fe_3O_4 nanoparticles.

Table 1. Band Gap Energy of Fe_3O_4-TiO_2 Nanocomposite

Sr. No.	Fe_3O_4-TiO_2 (g) Nanocomposite	λ max (nm)	Energy Band Gap (ev)
1	1:1	434	2.85
2	1:1.25	445	2.78
3	1:1.5	451	2.74
4	1:1.75	454	2.73

According to Table 1, the band gap of TiO_2 modified Fe_3O_4 nanocomposites can be calculated by using the relation, $\lambda_g = 1240/E_g$. it is observed that the rise in absorption wavelength threshold of lower titania content to higher titania content. It concludes that with increase in TiO_2 content to Fe_3O_4 nanocomposites, the band gap reduced and the absorption edge of Fe_3O_4-TiO_2 core-shell red shifted, which support our SEM result of increase in particle size of nanocomposites.

Table 2. Zone of inhibition of Fe_3O_4-TiO_2 Nanocomposite.

Sr. No.	Name of Composite	Zone of Inhibition (In mm)			
		b. subtilus	*S. aureus*	*E. coli*	*P. aeuriginosa*
1	Fe_3O_4-TiO_2	26	--	28	21
2	Standard (Ciprofloxacin)	27	28	29	27

Max inhibition > 20mm, Moderate inhibition >15mm, Poor inhibition < 15 mm

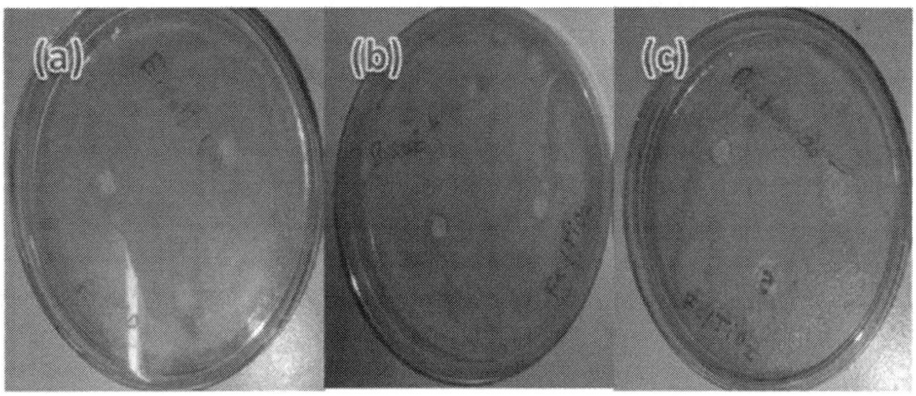

Fig.4. Antibacterial activity Photographic images of (a) E. coli, (b) b. subtilus and (c) P. aeuriginosa of Fe_3O_4-TiO_2 nanocomposite.

various pathogenic bacteria cause many infectious diseases. Therefore, development of antibiotic resistance, antibacterial agents are increasingly required. The agglomeration behavior of TiO_2 nanoparticles leads to decrease the antibacterial activity of pathogenic bacteria. To solve the problem in this study, synthesized a magnetic core modified with titania nanoparticles. As a result, the protected magnetic core can be used to control bacterial activity.

Table 2 shows ZOI values of Fe_3O_4-TiO_2 nanocomposite for tested micro-organisms. The titania surface on magnetite nanoparticle is responsible for the antimicrobial action as observed from Fig.4. In the test system ciprofloxacin at 10 μg/mL was used as a positive control. The average for the ZOI obtained from three replicas of the Fe_3O_4-TiO_2 nanocomposites. The MIC values were determined for those microorganisms showed ZOI greater than 25 mm. E. coli and b.subtilus shows the highest ZOI value (28 mm and 26 mm) while P. aeuriginosa was observed to be the less sensitive (21mm). MIC value of Fe_3O_4-TiO_2 nanocomposite was found to be 50 μg/mL. Fe_3O_4-TiO_2 core/shell composites shown the capacity to target pathogenic bacteria as well as effectively inhibited the cell growth of the bacteria (>99.9%) which is targeted by the magnetic nanoparticles under UV light irradiation of maximum wavelength 306 nm and power 0.412 mW/cm^2 for 20 min [16]. From the observation it is evident that, synthesized Fe_3O_4-TiO_2 nanocomposite can effectively inhibit gram positive and gram-negative bacteria without UV-irradiation

4. Conclusions:

The interaction between Fe_3O_4 and TiO_2 were confirmed from FT-IR spectra. Fe_3O_4-TiO_2 nanocomposites shows prominent antibacterial agent against gram positive and gram-negative bacteria.

References

[1] YS Kim, YH Kim, J. Magn. Mater. 267 (2003) 105-110.
[2] K. Raj, R . Moskowitz, IEEE Trans. Magn. 16 (2002) 358-363.
[3] D. Beydoun, R. Amal, G. K. C. Low, S. McEvoy, J. Phys. Chem. B 104 (2000) 4387-4396.
[4] R. D. McMichael, R. D. Shull, L. J. Swartzendruber, L. H. Bennett, R. E. Watson, J. Magn. Magn. Mater. 111 (1992) 29-33.
[5] Wei Y, Han B, Hu X, Lin Y, Wang X, Deng X., Energy Procedia 27 (2012) 632–637.
[6] C. Teh, A. Mohamed, J. Alloys Compd. 50 (2011) 1648–1660.
[7] W. Hsieh, J. Pan, C. Huang, Y.Su, Y.Juang, , Sci. Total Environ. 408 (2010) 672–679.
[8] C. A.Castro-López, A. Centeno, S. A. Giraldo, Catal. Today 157 (2010) 119–124.
[9] S. A. Kulkarni, P. S. Sawadh, 2265, 030142 (2020).
[10] M. Shuanglong, S. Zhan, J. Yanan, Q. Zhou, ACS Appl. Mater. Interfaces 2015, 7, 39, 21875–21883
[11] Hua Li, Xianzhe Shi, Lizhen Qiao, Xin Lu, Guowang Xu, J. Chromatogr.,A 1275 (2013) 9– 16.
[12] Tiejun Xin, Mingliang Ma, Hepeng Zhang, Junwei Gu, Shuangjie Wang, Mengjiao.Liu, Qiuyu Zhang, Appl. Surf. Sci. 288 (2014) 51– 59.
[13] N. Yeh , Y.C. Lee , C.Y. Chang , T.C. Cheng, Thin Solid Film, 549 (2013) 93–97.
[14] Qinghang He , Zhenxi Zhang , Jianwen Xiong , Yuying Xiong , Hua Xiao, Opt Mater 31 (2008) 380–384.
[15] Chen WJ, Chen YC, Nanomedicine-UK, 5 (2010) 1585-93.

IOP Conf. Series: Materials Science and Engineering 1221 (2022) 012050 doi:10.1088/1757-899X/1221/1/012050

Estimation of the nano-pores diameter by conductometric measurements

M Chander[1,2], S Kumar[1*] and S Kumar[3]

[1]*Materials Science Lab., Department of Physics, Chaudhary Devi Lal University, Sirsa-125055, India*
[2]*Govt. College Israna, Panipat, Haryana-132103*
[3]*Govt. Senior. Sec. School Ugalan, Hisar, Haryana-136027*

E-mail: mukesh.chander86@gmail.com

Abstracts. Ionic transport properties and diameter of asymmetric conical nano-pores in polyethylene terephthalate membranes were calculated. These nano-pores were synthesized by chemical etching method in irradiated polyethylene terephthalate (PET) films. After etching, the transport characteristics weredocumented under stepping voltage with 2M ionic concentrations of Potassium halides i.e. KCl, KBr and KI. The symmetric bathing condition (same temperature and same pH value solution in both side of chamber) in electrochemical cell were accomplished. It is investigated that diameter of nano-pores is also evaluate by measuring the current rectification properties of track etched nano-pores.

Keywords: PET, Etching, Nano-pores.

1. Introduction

The radiationphysics and chemistry of polymer membranes has also been emergingin recent days. These membranes are classically formed by irradiation on polymer foils having thickness of micron[1]. The trail of charged heavy particles, has been intentional to disclose the relevancyof radiation damage on the linear energy transfer (LET) of the ion. These trails of a heavy energetic ions in polymer membrane were used for etching with different chemicals. PET, polycarbonate (PC), polyimide(PI), etc. were used in this technology[2]. The chemical etching method permits the construction of protracted uniform and non uniforms geometry of poresin nanometer (nm) range diameter[1]. Chemical etching techniques, in combination with conductometric measurements of nano-pores was functional to investigate the geometry of the trails of numerous charged ions with atomic numbers from 18 to 92[3, 4-6].

The recent research work were motivated on the production of different designednano-pores in polymer film by ion track etching, the study of their transport and the estimation of diameter of conical nanopores by current-voltage (I-V) characteristics. Conical shaped nano-pores with charged surfaces are cations selective and show superiorcations flow (i.e. rectification) from the thin opening entrance to the wide opening exit of the cone [7-8]. The current rectification is a suitable method for evaluating the diameter of nanopores and can easily be applicable on cylindrical pores also.Thesenanoporeshave been engaged in biosensing [9], nanofludic devices [10], production of nanowires, nanotubes and nanospheres [11] etc..

2. Experimental methods

Polyethyleneterephthalate(PET) foils were irradiated by heavy charged ions, and by succeeding one-side etching for producing conical nano-pores. The diameter of the conical pores in PET varied between 4-20 nm at the narrow opening and 50-100nm at the wide opening. Ions passing through the foils wererecorded by a particle detector. Chemical track etching of PET foilstrajectories was performed with 9M NaOH etchantat room temperature[12, 13].For etchingtrajectories on one side, the irradiated foil was placed between the two assembly room of anelectrochemical cell, connected to a current/ voltage converter of nA sensitivity.

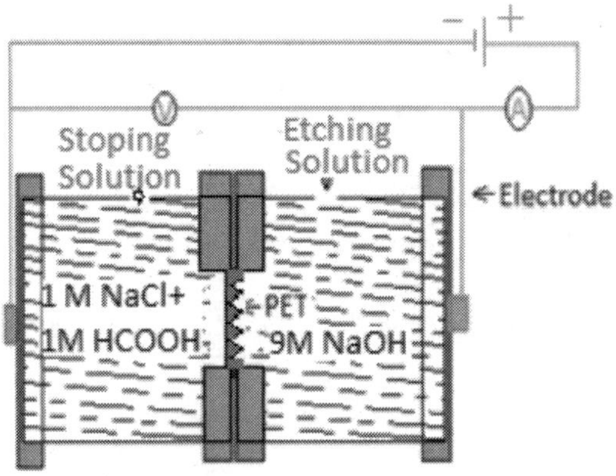

Fig. 1 Schematic of an electrochemical etching cell

One assembly room of the electromagnetic cell was occupied with the etchant (9M NaOH in this case)and another were filled with stopping (1M NaCl + 1M HCOOH)solutions, shown in Fig. 1 [14].When the etchant touches the rear surface of the sample, an acidic stopping solution deactivates the NaOH etchant.The conical channel created in the PET membrane has condensed wallbe made up of carboxylic end groups positioned at the surface designed during track etching process. The Positive charged particle of these groups detach above a specific pH value leaving behind their fixed negatively charged COO− part. Finaly, a mobile counter ion coveradjacent to the surface is designed in the solution compensating the external charge.

3. Measurement of cone diameter and discussion of result

In most of the precedingresearch it was presumed that asymmetric nanopores were conical in nature. The diameter of the cone thin entrance was predictable from the conductance 'G' of a conical nano-pore, using the formula [15,16].

$$G = \frac{k\pi dD}{4L} \qquad --(1)$$

As
d –narrow entrance tip diameter,
D - Base diameter of nano-pores,
k - Specific conductivity of the electrolyte,
L -Nanoporeseffective length.

According to this model, circular cones were formedin the foil. The bulk etching of the material is characterized by the bulk etching speed $V_B$and the track etching speed V_T. After breakthrough ($t > t_b$), the pore has the geometry of cones characterized by a cone base diameter
D and tip diameter d.

IOP Conf. Series: Materials Science and Engineering 1221 (2022) 012050 doi:10.1088/1757-899X/1221/1/012050

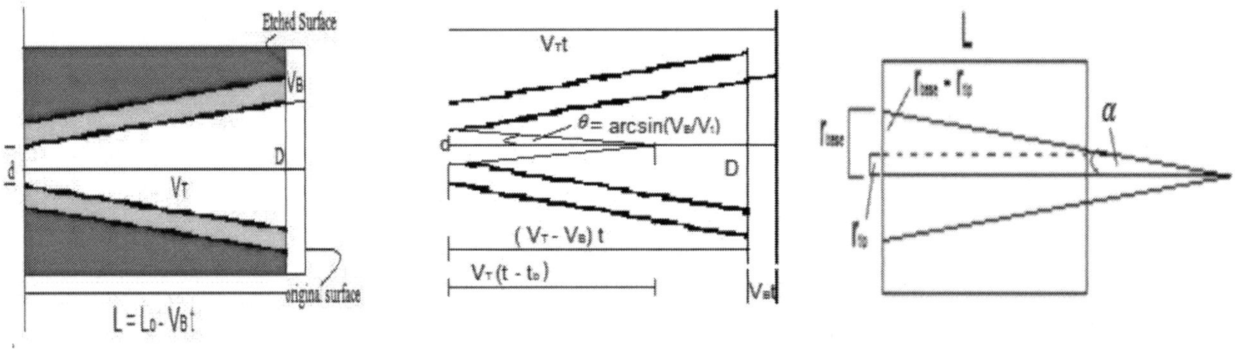

(a) (b) (c)

Fig. 2(a) Cone base/wide diameter D, tip diameter d and Foil thickness/pore length L. The cone angle corresponds to the track etch ratio V_T/V_B, (b) Geometry for driving equations (3), (c) Cone base geometry.

We used the result of double cone model geometry as [15]

$$G = \frac{k\pi V_B^2 t(t - t_b)}{(L_o - 2V_B t)} \quad --- (2)$$

If we apply for single side cone then eq. 2 convert in this form

$$G = \frac{k\pi V_B^2 t(t - t_b)}{(L_o - V_B t)} --- (3)$$

As we compare the eqs. (1) and (3), Then

$$\frac{dD}{4L} = \frac{V_B^2 t(t - t_b)}{(L_o - V_B t)} \quad --- (4)$$

where L is the effective length of the pore,therefore

$$\frac{dD}{4} = V_B^2 t(t - t_b) \quad --(5)$$

For calculating D using the geometry as shown in Fig. 4 (c)

$D = d + 2L\, tan\alpha$ --- (6)

where α is the cone half angle

$r_{eff} = \frac{1}{2}(dD)^{1/2}$corresponds to the geometrical mean of the cone diameter i.e. D is cone base diameter and d istip diameter. Theeffective radius r_{eff} is also nearly equal to the radius of a cylindrical pores of length $(L_o - V_B t)$ with the electrical conductance G[13].

If we have N pores per cm^2 than the conductance is

$$\frac{1}{R} = N.G = N.\frac{k\pi dD}{4L} \qquad ---(7)$$

where'R' is the resistance of the pores which can be calculated by I-V characteristics as shown in figure 3. Typical pores density of sample is approximate 10^4 to 6×10^8 pores/cm^2, effective pore length is 10μm, k is conductivity of solution than the diameter of the pore is calculated.If we change the pores density then the resistance of the nanopores changes. As we take the case of cylindrical pores in the membrane then the no. of pores in both side of membrane is same but if we take the conical pores then the pores density is different on both sides of membrane so the resistance of pores in both sides is different as shown in Fig. 3. It shows that the flow of current from base to tip side of the membrane show high resistance due to the no. of pores is less as compared to tip side. By this method we can also distinguish whether the membrane is conical or cylindrical.

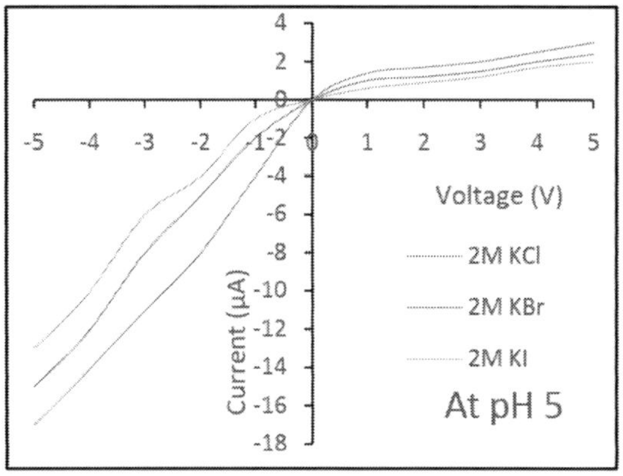

Fig. 3: Rectification ofKCl, KBr and KI electrolyte through conical nanopores at pH~5 and atconcentrations 2M.

The ohmic resistance for conical nonporous can be measured by I-V characteristics of polymer nonporous membranes .

$$R = \frac{V}{I} \qquad ----(9)$$

By comparing eq. (6) and (9) we would easily calculated diameter of nonporous if knowing the pores density and the conductivity of solution. The nonporous tip radii is nearly equal to size of proteins, DNA molecules etc. so is selected touse sensing or separation application. These applications include to detect/analyse DNA molecules, genomics, pH sensitive sensor as well as used as template of production for nanowires, nanotubes and nanospheres. The transport properties/rectification of conical pores differ from cylindrical pores[16].

4. Conclusion

The ionic current from end to endinnanoporesfluctuatesnoticeably, and the changeability depends on the applied voltage as well as concentrations of the solution. This unlikeperformance has been recognizedwith the chemical structure of polymer manipulating surface characteristics of the nano-pores in membrane.I-V characteristics are strongly nonlinear in nature. It is dependent of solutionconductivity, pore geometry/diameter and density of pores in membrane. In this method we easily calculate the radius of pores.

References

[1] Siwy Z, Apel P, Dobrev D, Neumann R, Spohr R, Trautmann C and Voss K2003*Nucl. Instr. and Meth.B***208**143.
[2] Kabanov V, Feldman V, Ershov B, Polikarpov A, Kiryukhin D and Apel P 2009*Radiation Chemistry ofPolymers, High Energy Chemistry***43** **1**1.
[3] Apel P and Khim V 1991 *Energ.***252**132.
[4] Apel P, Schulz A, Spohr R, Trautmann C andVutsadakis V1998 *Nucl. Instrum.Meth.Phys. Res. B***146**468.
[5] Apel P, Spohr R, Trautmann C andVutsadakis V 1999*Radiat. Meas.* **31** 51.
[6] Apel P, Blonskaya I V, Oganessian V, Orelovitch O and Trautmann C2001 *Nucl. Instrum.Meth.Phys. Res. B***185**216.
[7] Kumar S, Kumar S andChakarvarti S K 2003*Radiat. Meas.* **36** 757.
[8] Schiedt B, Healy K, Morrison A P, Neumann R and Siwy Z 2005*Nucl. Instrum.Meth.Phys. Res. B* **236** 109.
[9] Siwy Z, Trofin L, Kohli P, Baker L A, Trautmann C and Martin C R 2005 *J. Am. Chem. Soc.***127**5000.
[10] Siwy Z 2006*Adv. Funct.Mater.***16**735.
[11] Kravets L I, Dmitriev S N, Satulu V, Mitu B and Dinescu G 2014*Romanian Reports in Physics* **664** 1165, 2014.
[12] Apel P, Korchev Y E, Siwy Z, Spohr R and Yoshida M 2001*Nucl. Instr. and Meth.B* **184** 337.
[13] Chander M, Kumar R, Kumar S, Kumar N and Chakarvarti S K 2018*NANO: Brief Reports and Reviews***131**1850011.
[14] Apel P, Blonskaya I V, Orelovitch O L, Sartowska B A andSpohr R 2012*Nanotechnology***23** 225503.
[15] Apel P, Ramirez P, Blonskaya I V, Orelovitch O L andSartowska B A 2014*Phys. Chem. Chem. Phys.***16**15214.
[16] Chander M, Kumar R, Kumar S, Kumar N and Chakarvarti S K 2018 *Digest J. Nanomat. Biostruct.***13**113.

IOP Conf. Series: Materials Science and Engineering 1221 (2022) 012051 doi:10.1088/1757-899X/1221/1/012051

Mosquito Larvicidal Activity of DNA Capped Colloidal Silver Nanoparticles

Reena V N [1], Shanasree M[2], Subin Kumar K[3], Bhagyasree G S[1] and Nithyaja B[1]

[1]Department of Physics, Government College Madappally, Vadakara, Kerala, India
[2]Department of Zoology, Government College Madappally, Vadakara, Kerala, India
[3]Department of Chemistry, Government College Madappally, Vadakara, Kerala, India
Email: rinanarayan@gmail.com

Abstract. Mosquitoes spread terrible diseases like chikungunya, yellow fever, malaria, dengue, Japan encephalitis and lymphatic filariasis in human beings even causes many deaths in every year. Chemical insecticides cause adverse effect in human beings, also to the environment. In this work, we sought a credible, eco friendly tool to control the vector mosquitoes at their larva stage. It reveals an eco-friendly and effective larvicidal activity of the Deoxyribonucleic acid (DNA) capped colloidal Silver nanoparticles. Characterizations of the Silver nanoparticles were done using absorption spectra and SEM. Spectra shows a predominant peak at 425nm and the SEM image shows the particles size is about 45nm. The third instars of *Culex* larvae were treated using these colloidal Silver nanoparticles, also the mortality rate for 24 hours was observed with 100% mortality rate at 10 ppm. The LC_{50}, LC_{90} and confidence interval at 95% were investigated. Colour analysis and the SEM image of the treated larva were performed and confirm that the colloidal Silver nanoparticles were accumulated all over the skin surface of the larva which causes death.

1. Introduction

Mosquitoes are the major insecticide causative of transmitted diseases like yellow fever, zika, chikungunia, dengue and malaria [1]. There are over 3,000 species of mosquitoes, about 100 variety species of mosquitoes are vectors of human diseases. Rapid spreading of highly invasive mosquitoes and the diseases spread by them are responsible for killing more people than all the wars in history.[2-3] Nowadays several insecticides namely, DDT, organophosphorus, dieldrin, fenithothion and propoxur were widely used for mosquito prevention and control. Constant use of chemical insecticides leads to undesirable consequences including ecological imbalance and extinction of non-target organisms in the environment [4-5]. In this scenario, mosquito larvae are usually targeted to control growth of mosquitoes. Some eco-friendly methods are needed to manage and control mosquitoes at larvae stage. Nanotechnology is a novel method with comprehensible potential applications in the field of insect control [6-7]. There are a number of works has been experimented to seek the larvicidal activity of metal nanoparticles like Cd, Au, Cu and Ag [8-11] However, biological or green synthesis of nanomaterials has received substantial interest due to its eco-friendly and non-hazardous nature. Particularly, the synthesis of silver nanoparticles (AgNPs) incorporating extracts of plants displaying insecticidal activity has been shown to produce larvicidal compounds which are efficient against mosquitoes at very low concentrations. Moreover, the synthesis of such plant-derived AgNPs does not require high pressure, energy, temperature or the use of highly toxic chemicals; therefore, it is considered cost-effective, environmentally friendly and time-efficient. A number of reports are published in the larvicidal activity of green synthesized AgNPs. The common procedure they used that taking the extract of specified plants parts and used as a capping agent [12]. Extraction itself a long procedure as Devan Elumalai et al. reports, it takes almost one day to separates out the extract of L. aspera plant. De-oxyribo nucleic acid (DNA), the biopolymer, has been used widely as a capping agent to synthesize nanoparticles because of its physicochemical stability, well defined sequence of DNA base and a diversity of super helix structures. Commonly, synthesis of nano particles based on DNA template has been done by incubating metal ion coated DNA in a reduction agent solution. The ion DNA complexes controlled the release of metal ions, slowed down their reduction and effectively inhibited metal ions from growing into big clusters. One of the author of this paper has been reported previously the synthesis of highly stable silver nanoparticles in aqueous solution at room temperature by standard reduction method using DNA as capping agent [13-14].

Work has been done on the linear and non-linear optical properties of AgNPs prepared in DNA biotemplate. However, no work has been done in the research field on the antilarval properties of colloidal AgNPs prepared on DNA biotemplate. Comparing most of the green synthesized AgNPs the DNA capped AgNPs can be synthesized easily and since DNA is a biomaterial, DNA capped Silver nanoparticles can be more a biocompatible and non-toxic particle. Rigo C. et al. discussed the biocompatibility of AgNPs and their wound therapy applications. DNA templated AgNPs are colloidal in nature and hence it would uniformly be distributed and cause higher activity [15].

The present work aimed to expose larvicidal activity of AgNPs synthesized using DNA biotemplate. Characterizations of nanoparticles were done using UV-Vis spectroscopy and SEM. Images of the treated and non treated larvae were investigated using light microscope. SEM image of the treated larva was also observed to compare with the microscopic image of the treated larva. Study on the microscopic image of treated dead larva using RGB and HSV colour analysis was also carried out here; which can reveal the variation in skin colour of treated and non- treated larvae.

2. Materials and Methods

All chemicals and biopolymers used in this work were obtained from Sigma-Aldrich.

2.1. Preparation of Silver Nanoparticles

Silver nano particles were synthesized in aqueous solution of DNA by standard reduction technique [14]. A 20mL solution with a concentration of 0.25mM $AgNO_3$ and DNA with concentration (5 wt %) in water was prepared. While stirring vigorously 0.6ml of 10mM NaBH4 was added at once. Stirring was stopped after 30s. DNA capped AgNPs were formed with deep golden yellow colour.

The prepared AgNPs were characterized using UV-Visible Spectroscopy. Surface topology was examined using SEM analysis.

2.2. Antilarval studies of Silver nanoparticles

Larvae were collected from Kannur district, Kerala (11054'30.2"N 75019'3.0"E) in November 2020. Mosquito larvae were successfully cultured in tap water and the third instars healthy larvae were collected for the present work [16]. The temperature during this assignment is $30^0C \pm 2^0C$. The antilarval studies were carried out using standard rules described by World Health Organization (WHO) [17]. Different concentrations of AgNPs (1ppm, 2ppm, 5ppm and 10ppm) were taken in beakers containing 200 ml distilled water. For performing the antilarval studies, about 25 larvae were introduced into the beaker containing water and AgNPs. One beaker containing 200ml water was taken as control. For each concentration under observation, five trials were constituted. The larval mortality rates were analyzed and found the mosquito larvae's acute toxicity levels after 24 hours. The mortality rates of larvae were reported as the average of five replicates performed under observation. The mortality rate was calculated using the Abbott formula [18].

2.3. Statistical Analysis

The mortality rate in accordance with concentration of AgNPs was calculated by using Abbott's formula. The mortality data were subjected to probit analysis for finding LC_{50}, LC_{90}.

2.4Colour analysis of the microscopic image of Larva

The AgNPs treated and non treated larvae were viewed under Olympus CX43 bio-microscope and taken photographs were analyzed with RGB and HSV colour analysis by Python Programming (Version-Python 3.8) to visualize the image in another colour space. By doing this, one can visibly recognize the changes in a treated and non-treated larva.

The method of representing the contained colors in digitally is generally referred to as colour space models. Each colour spaces describes the colour information in special ways which makes certain computations more beneficial and also contribute a way to find colours that is more perspective. The colour spaces in this study used are RGB and HSV. A brief description of each is shown below.

2.4.1RGB colour spaces

RGB colour space is the base model which is derived from three orthogonal additives red, green and blue chromaticity commonly known as primary colours. All the other colour spaces are obtained by mathematical transformation of RGB colour space.

2.4.2 HSV colour space

HSV colour space is a cylindrical coordinate representation of RGB colour space formulated through non-linear transformation. HSV colour space is closer to the way human perceive a colour. Hue (H) represents the dominant wavelength present in a mixture of colour. Saturation (S) gives the measure of pure colour diluted by white light. The Value (V) component provides the gray value from negative value (blue colour) to positive value (yellow). It encodes the perceptual difference in colours when viewed by human [19]

3. Result and Discussions

3.1 Absorption Spectra

1 ml prepared AgNPs were diluted with 5 ml of distilled water and then measured by the UV-Vis spectroscopy using a quartz cuvette with distilled water as a reference. The UV-Vis spectroscopy analysis of AgNPs was carried out on Schimadzu spectrophotometer at a resolution of 2.5 nm. The absorption spectrometric readings were recorded at a scanning speed of 300-900 nm. Fig.1. represents the UV-Visible spectrum of AgNPs and the absorption peak was observed around 425 nm. The SPR peak of silver nanoparticles mentioned in many other previous reports is matches with this value [14].

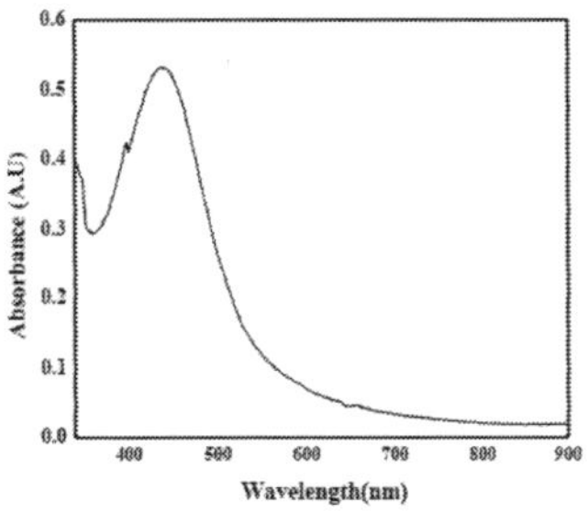

Figure1. Absorption spectra of AgNPs

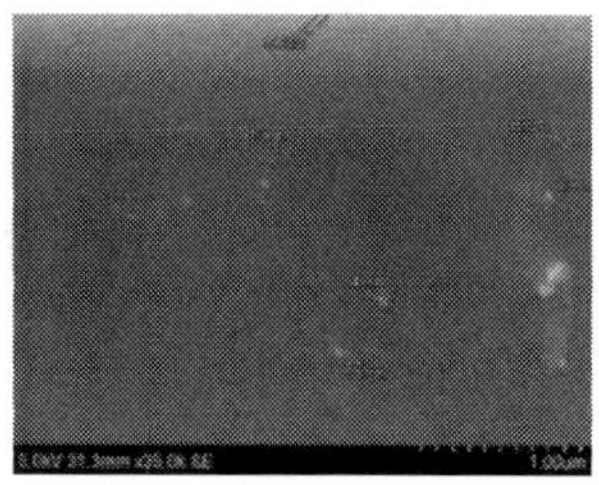

Figure2. SEM image of AgNPs

3.2 Scanning Electron Microscope Studies

SEM images shows the AgNPs are of spherical in shape and having an average size of 45nm.

4. Larvicidal Activity of AgNPs

Percentage of dead larvae for all bioassays in 4 hours interval is shown in Fig.3 and Table1 represents the mortality value at 24 hours. AgNPs shows 100% mortality at 10ppm, at the same time control group and the capping agent, DNA seldom showed any mortality. Fig.4. Shows the probit analysis of

AgNPs; the LC_{50} and LC_{90}values of AgNPs are 1.1089 and 5.661 respectively; their confidence interval at 95% is also found out and marked in the table.

To determine the toxicity of AgNPs, a non-target organism P.*recticulata* (Guppy) of 1g±0.25g were collected and 30 healthy P.*recticulata* were placed in 400 ml water solution with three replicates. Each group was exposed to AgNPs (10ppm) with one control group where 30 P.*recticulata* were placed in unchlorinated water and observed for duration of 3 days with 24 hours periods. It is found that no P.*recticulata* were died in the presence of AgNPs. This indicates that larvicidal activity of DNA capped colloidal AgNPs are targeted to larva only.

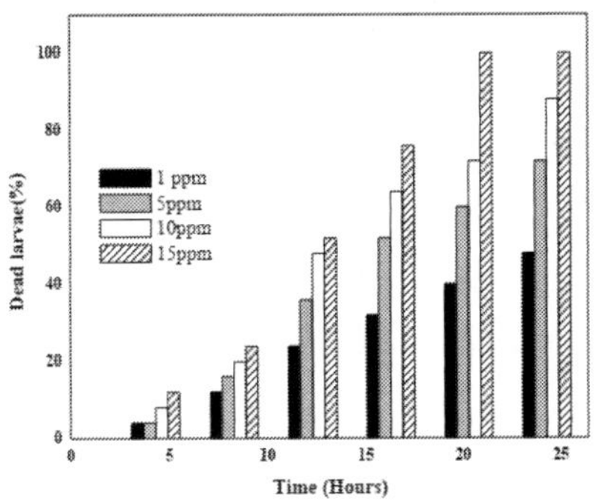

Figure3. Percentage of Larvae time of exposure

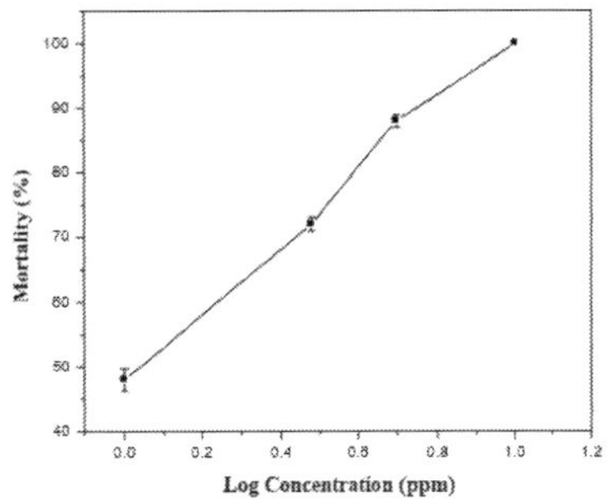

Figure4. Probit Analysis

Table1. Larvicidal activity of AgNPs against Culex third instars larvae

Mosquito larva	Concentration of AgNPs (ppm)	24 hour mortality (%)± SD	LC_{50}(ppm) (LCL-UCL)	LC_{90}(ppm) (LCL-UCL)
Culex	Control (Water)	0.0±0.0	1.10891 (0.671 - 1.55)	5.661 (5.22 - 6.1)
	DNA (10 ppm)	0.0±0.0		
	1	48±1.79		
	2	72±1.06		
	5	88±0.97		
	10	100±0.0		

5. Colour Analysis of Treated and Non-treated larva

Fig.5. represents different colour models of dead larva treated in 10ppm, in which the mortality is 100% in 24 hours. The light microscopic image of this larva shows yellowish shade as seen in Fig. 5(a). It indicates that the AgNPs were covered all over the body of larva. The microscopic image of the treated larva in RGB colour space was converted into HSV colour, which is shown in Fig. 5(b). The fluorescent green colour indicates the dark shades of 5(a). It could not clearly differentiate between fluorescent green and yellow colour from the HSV colour image, masking is done for 5(b) for all the colours other than the colour of AgNPs and it is shown in Fig. 5(c). Since head, thorax and tail already have dark shade, it is difficult to differentiate AgNPs colour from the microscopic image. In Fig. 5(d), to reveal the AgNPs colour on the head, thorax and tail of the larva, all other sections were masked. From all these analysis it is clearly observed that the AgNPs were deposited on the surface of the body, the skin, of each larva. The size of pores in the skin may be comparable to the size of the AgNPs. The pores on the skin of larva may be blocked. It leads to the death of the larva very fast. The number of AgNPs increases it takes less time to block the pores so that the mortality rate is directly proportional to the number of AgNPs and time.

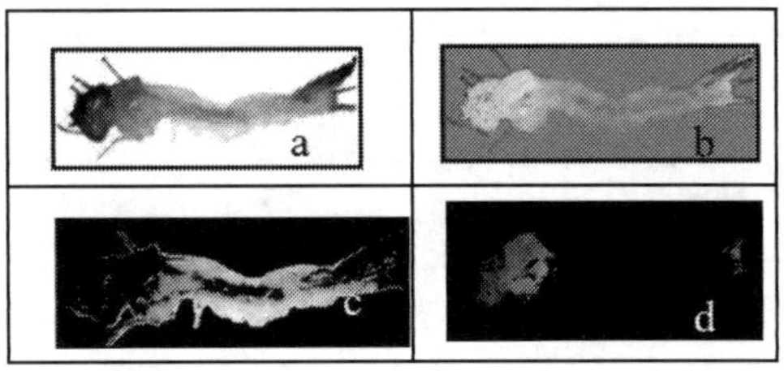

Figure5. Colour analysis of treated larva

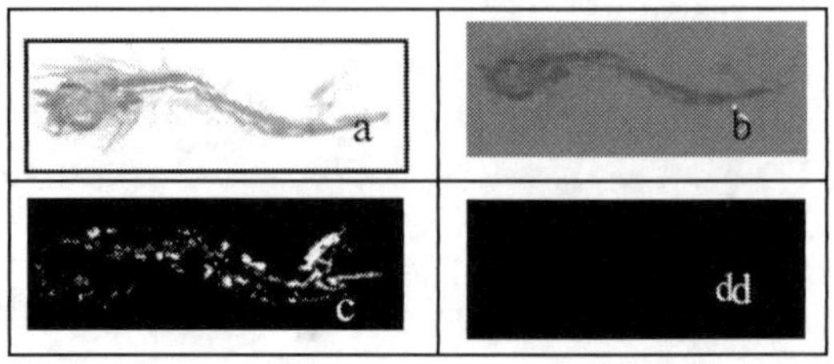

Figure6. Colour analysis of non-treated larva

Fig.6. shows the colour analysis of the larva immersed in distilled water. All the colour analysis done for the treated larva was again done for this non treated larva. In Fig.6(b), 6(c) and 6(d) it could not observed any colour other than the original.

The colour analysis reveals that AgNPs is deposited all over the body. Even if the mosquito swallows the particles, the main reason for death is the deposition of AgNPs on the surface of the body. From RGB colour analysis, it is found that the nanoparticles were deposited wholly on the skin of the mosquito.

6. SEM analysis of treated larva

The sample, treated larva is prepared for SEM analysis as per protocol. Initially treated larvae were fixed with formaldehyde followed by dehydration using acetone after that the sample were dried in air. Fig.7. is the SEM image of treated larva, it could be clearly seen the AgNPs are deposited all over the skin as primarily confirmed in RGB colour analysis.

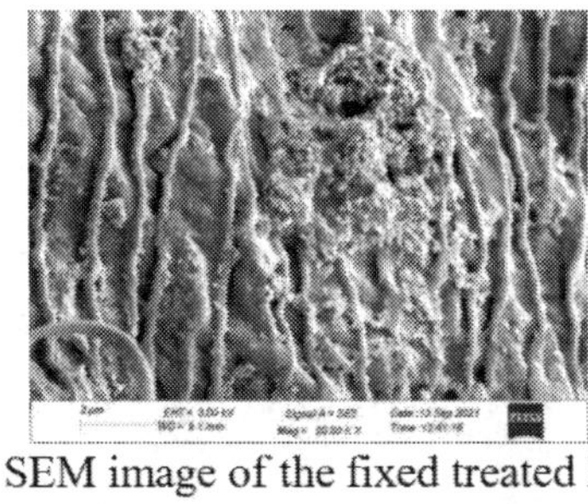

Figure7. SEM image of the fixed treated larva

7. Conclusion

In the present report, larvicidal activity of DNA capped colloidal AgNPs were studied and found that it is targeted larvicidal nanoparticle. Colour analysis of microscopic image of the treated larva and SEM analysis of the treated larva confirmed the cause of fast death. Thus this biologically synthesized AgNPs could be a potential and eco-friendly material to control mosquitoes.

Acknowledgement
The authors wish to thank CSIF, Calicut University for providing research facilities.

References

[1] Lee H, Halverson S and Ezinwa N 2018, Mosquito-borne diseases *Primary Care: Clinics in Office Practice* **45(3)** pp 393-407.

[2] Tolle and M A 2009, Mosquito-borne diseases *Current problems in pediatric and adolescent health care* **39(4)** pp 97-140

[3] Nazareth T, Craveiro I, Moutinho A, Seixas G, Gonçalves C, Gonçalves L, Teodósio R and Sousa C A 2020, What happens when we modify mosquitoes for disease prevention? A systematic review *Emerging microbes & infections* **9(1)** pp348-365

[4] Rahman M M 2013, Insecticide substitutes for DDT to control mosquitoes may be causes of several diseases *Environmental Science and Pollution Research* 20(4) pp 2064-2069

[5] Aktar M W, Sengupta D and Chowdhury A 2009 Impact of pesticides use in agriculture: their benefits and hazards *Interdisciplinary toxicology* **2(1)** p1

[6] El Wakeil N, Alkahtani S and Gaafar N 2017, Is nanotechnology a promising field for insect pest control in IPM programs? *In New Pesticides and Soil Sensors* Academic Press pp 273-309

[7] Ragaei M and Sabry A K H 2014, Nanotechnology for insect pest *control Int J Sci Environ Technol* **3(2)** pp 528-545

[8] Hajra A, Dutta S and Mondal N K 2016, Mosquito larvicidal activity of cadmium nanoparticles synthesized from petal extracts of marigold (Tagetes sp) and rose (Rosa sp) flower *Journal of parasitic diseases* **40(4)** pp 1519-1527

[9] Cheng S S, Liu J Y, Tsai K H, Chen W J and Chang S T 2004, Chemical composition and mosquito larvicidal activity of essential oils from leaves of different Cinnamomum osmophloeum provenances *Journal of agricultural and food chemistry* **52(14)** pp 4395-4400

[10] Rajagopal G, Nivetha A , Sundar M , Panneerselvam T, Murugesan S, Parasuraman P, Kumar S, Ilango S and Kunjiappan S 2021, Mixed phytochemicals mediated synthesis of copper nanoparticles for anticancer and larvicidal applications *Heliyon* pe07360

[11] Morejón B, Pilaquinga F, Domenech F Ganchala D, Debut A and Neira M 2018, Larvicidal activity of silver nanoparticles synthesized using extracts of Ambrosia arborescens (Asteraceae) to control Aedes aegypti L.(Diptera: Culicidae) *Journal of Nanotechnology* **2018**

[12] Amarasinghe L D, Wickramarachchi P A S R., Aberathna A A A U, Sithara W S and De Silva C R 2020, Comparative study on larvicidal activity of green synthesized silver nanoparticles and Annona glabra (Annonaceae) aqueous extract to control Aedes aegypti and Aedes albopictus (Diptera: Culicidae) *Heliyon* **6(6)** p e04322

[13] Slesiona N, Thamm S, Stolle H L K S, Weißenborn V, Müller P, Csáki A and Fritzsche W 2020 DNA-Biofunctionalization of CTAC-capped gold nanocubes *Nanomaterials* **10(6)** p 1119

[14] Nithyaja B, Misha H and Nampoori V P N 2012, Synthesis of silver nanoparticles in DNA template and its influence on nonlinear optical properties *Nanosci Nanotechnol* **2(4)** pp 99-103

[15] Rigo C, Ferroni L, Tocco I, Roman M, Munivrana I, Gardin C, Cairns W R, Vindigni V Azzena B, Barbante C and Zavan B 2013, Active silver nanoparticles for wound healing *International journal of molecular sciences* **14(3)** pp 4817-4840

[16] Kumar K S, Reena V N and Aravindakshan K K 2021, Synthesis, anticancer and larvicidal activities of a novel Schiff base ligand, 3-((2-((1-(4-hydroxyphenyl) ethylidene) amino) ethyl) imino)-N-(p-tolyl) butanamide and its Mn (II), Fe (III), Co (II), Ni (II) and Zn (II) complexes. *Results in Chemistry* **3** p100166

[17] WHO 2006, Guidelines for testing mosquito adulticides intended for Indoor Residual Spraying (IRS) and Insecticide Treated Nets (ITNs) Document WHO/CDS/NTD/WHOPES/GCDDP/2006 3 Geneva: *World Health Organization*

[18] Abbott W S 1987, A method of computing the effectiveness of an insecticide *Journal of the American Mosquito Control Association* **3(2)** pp 302-303

[19] Menesatti P, Angelini C, Pallottino F, Antonucci F, Aguzzi J and Costa C 2012, RGB color calibration for quantitative image analysis: The "3D Thin-Plate Spline" warping approach *Sensors* **12(6)** pp 7063-7079

IOP Conf. Series: Materials Science and Engineering 1221 (2022) 012052 doi:10.1088/1757-899X/1221/1/012052

Synthesis and Stabilization of Palladium Nanoparticles in Various Ionic Liquids

Chanchal Singh
Assistant Professor, Department of Chemistry, D.S.N. (PG) College, Unnao,
Chhatrapati Shahuji Maharaj University, Kanpur
Email: premichanchal@gmail.com

Abstract. Synthesis of Palladium nanoparticles (PdNPs) was carried out in various Ionic Liquids under variable temperature and with different palladium precursors. Further study showed that nitrogen containing functional group are better for generation of PdNPs. The size of nanoparticles was determined by TEM which showed the high variation in shape and size, and presence of a built-in stabilizing agent enhanced dispersibility of PdNP in different organic solvents. PdNPs formed was found to be catalytic active under standard carbon-carbon bond coupling reaction.PdNPs was synthesized by the thermal reduction in nitrile functionalized ionic liquid which act as in-situ catalyst C-C coupling reaction.

1. Introduction

Palladium catalyzed reactions holds a very significant place in transition metal catalysis for synthesis of complex structure in organic chemistry.[1]Generally these reactions require the use of supporting ligands such as $Pd(PPh_3)_4$ or some others electron-rich phosphines.[2]Majority of these ligands are susceptible to moisture, lengthy synthetic procedure, and costly. ThusligandlessPd catalysis approachis the greenest and economical approach to cross-coupling reactions. Since last decade substantial amount of research has been gone into Palladium nanoparticle (PdNP) assisted catalysis. Nanoparticles brings selectivity and efficiency to transition metal catalysis. The products can be easily separated from the reaction media, and the catalyst can be reused and recycled which offers the economical approach to overall process.[3] One major disadvantage of NPs is that they are thermodynamically and kinetically unstable hence, some kind of stabilizing agent is required to avoid amalgamation of the NPs. Recently ionic liquids (ILs) emerged as stabilizing media for transition-metal NPs. They are usually liquid at room temperature.[4]They possess distinctive properties such as no vapour pressure, good thermal stability.They comprise of organic cations such as quaternary ammonium, imidazolium and pyridinium salts, and variety of anions such as tetrafluoroborate (BF_4), hexafluorophosphate (PF_6) etc. Imidazolium ringis frequently used nucleus in room-temperature ionic liquids (RTILs).[5]

2. Synthetic procedurefor ionic liquids

3-(3-cyanopropyl)-1-methyl-1H-imidazol-3-ium bromide CN[bmim]Br (IL1)(Scheme S1): Reaction mixture was prepared by adding N-methyl-1H-imidazole (3 mL, 1eq.) and 4-bromobutanenitrile (3.5 mL, 1.2 eq.), solution were stirred for 24 hours at 70°C. The reaction was supervised by TLC, crude product was washed with diethyl ether and ethyl acetate,viscous liquids was obtained in 92% yield.

3-(3-cyanopropyl)-1-methyl-1H-imidazol-3-ium hexafluoro phosphate CN[bmim]PF_6 (IL 2)(Scheme S1): CN[bmim]Br was subjected to anion exchange withpotassium hexafluorophosphate (KPF_6). Reaction mixture was prepared by adding CN[bmim]Br (2 mL, 1eq.), KPF_6 (1.6 g, 1 eq.)in water (5 mL), and solution was stirred for 24 h at room temperature. The reaction was supervised by TLC, water was evaporated on rotary evaporater. Crude liquid was dissolved in DCM, dehydrated completely onmoleculer sieves (4A), and finally DCM was evaporated on rotary evaporater to obtain viscous product in 80% yield.

3-(3-cyanopropyl)-1-methyl-1H-imidazol-3-ium flouride CN[bmim]F (IL 3) (Scheme S1): Reaction mixture was prepared by adding CN[bmim]Br (2 mL, 1eq.), water (3 mL, 1 eq.) solution of AgF (2.4 g in 4 mL water, 1.15 eq.) in water was gradually addedand solution was stirred for 2 h. The solution sieved over celite to separate precipitated silver bromide (AgBr). The clear solution obtained was rested overnight to separate out any remaining AgBr. The water was evaporatedon rotary evaporaterand the viscous liquid was dissolved inACN:CH_3OH (7:3), dehydrated completely overmoleculer sieves (4A), and filtrate obtained was evaporatedon rotary evaporaterto obtain viscous product in 80% yield.

3-butyl-1-methyl-1H-imidazol-3-ium bromide [bmim]Br (IL 4) (Scheme S2): Reaction mixture was prepared by adding N-methyl-1H-imidazole (4 mL,1eq.), 4-bromobutane (8 mL,1.2 eq.) and the solution was refluxed for 24 h at 70°C. The reaction was supervised by TLC, the crude product was washed with diethyl ether and ethyl acetate and viscous liquid was obtained in 98% yield.

3-butyl-1-methyl-1H-imidazol-3-ium fluoride [bmim]F (IL 5)) (Scheme S2): [bmim]Br was subjected to anion exchange withsilver fluoride (AgF). Reaction mixture was prepared by adding [bmim]Br (2 mL, 1eq.), water (4mL) and solution of AgF (2.4 g in 4 mL water, 1.15 eq.) in water was added gradually and solution was stirred for 2 h. The solution was sieved over celite to separate precipitated silver bromide. The clear solution obtained was rested overnight toseparate out any remaining silver bromide. The water was evaporated on rotary evaporaterand the crude product was dissolved inACN:CH_3OH (7:3), dehydrated completely overmoleculer sieves (4A), and concentrated on rotary evaporaterto obtain viscous product in 80% yield.

3-butyl-1-methyl-1H-imidazol-3-ium acetate [bmim]OAc (IL 6)) (Scheme S2):Reaction mixture was prepared by adding [bmim]Br, sodium acetate (1.12 g, 1.5 eq.) and water (2 mL, 1 eq.) and solution was stirred for 24 h. The water was evaporated on rotary evaporaterand the crude product was dissolved in acetonitrile, filtered overcelite to separateunreacted sodium acetate, and concentrated to obtain viscous product in 84% yield.

Experimental procedure for the synthesis of Palladium nanoparticles (Pd NPs): Palladium precursor [$Pd(OAc)_2$/$PdCl_2$] (4 mg), IL 2/IL 3 (2 mL) and acetonitrile as cosolvent (5 mL) was taken in round bottom flask and contents was stirred at 80°C. After 2h acetonitrile was evaporated on rotary evaporater, and reaction mixture were stirredvigrously for 6 h at 140°C and brownish-blackprecitipate was obtained. Pd NPs of size 2-5 nm/25-30 nm were obtained. The catalyst embedded in IL 2 was used for Hiyamacross-coupling.

Procedure for Hiyama coupling: Preformed PdNPs in CN[bmim]PF_6 were utilized as catalyst in Hiyama coupling reaction. Reaction mixture was prepared by adding Bromobenzene (7) (1 mmol), trimethoxyphenylsilane (8) (2 mmol), TBAF (2.5 mmol) and acetonitrile (2.5 mL) as solvent, reaction was carried out at 70°C for 8h. Product was extracted with ethylacetate. Purification was done by column chromatography. Biphenyl (9) obtained were confirmed by 1H NMR, and 13C NMR.
Preparation of sample for UV-Vis spectroscopy: Sample (2 µL) was taken out in eppendorf with help of capillary, diluted with acetonitrile (3 mL), sonicated for 20 minutes and Uv-Vis spectra was recorded.Same process was repeated after an interval of every one hour. It showed the typical pattern of palladium nanoparticles.

IOP Conf. Series: Materials Science and Engineering 1221 (2022) 012052 doi:10.1088/1757-899X/1221/1/012052

2.1. Preparation of ionic liquids

Herein we report the synthesis of PdNPs starting from a Pd (II) precursor in various ionic liquids. Study mainly includes the four different type of ILs, two of which are nitrogen functionalized and remaining two are unfunctionalized (figure 1).

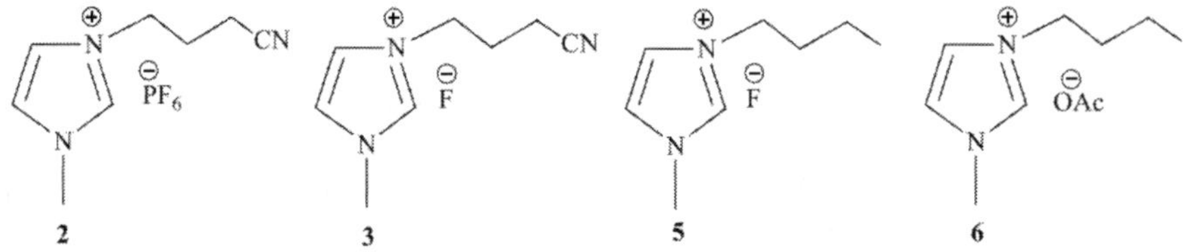

Figure 1. Ionic liquids used for generation of Palladium nanoparticles

The synthesisof ionic liquidsCN[bmim]PF_6 (IL **2**), CN[bmim]F (IL**3**) [bmim]F (IL**5**), and [bmim]OAc (IL**6**) was carried out byliteratureprocedures.[6]ILs **2**and **3**were synthesized as shown in (Scheme S1).Refluxing of 4-bromobutanenitrile and N-methylimidazoleat 60 °C resulted in the formation of CN[bmim]Br (IL**1**) which on anion exchange with potassium hexafluorophosphate and silver fluoride gaveIL**2**and IL**3**as transparent viscous liquids respectively. Anion exchange of [bmim]Br (IL**4**) with silver fluoride and sodium acetate yielded IL**5**and IL **6** respectively (Scheme S2).

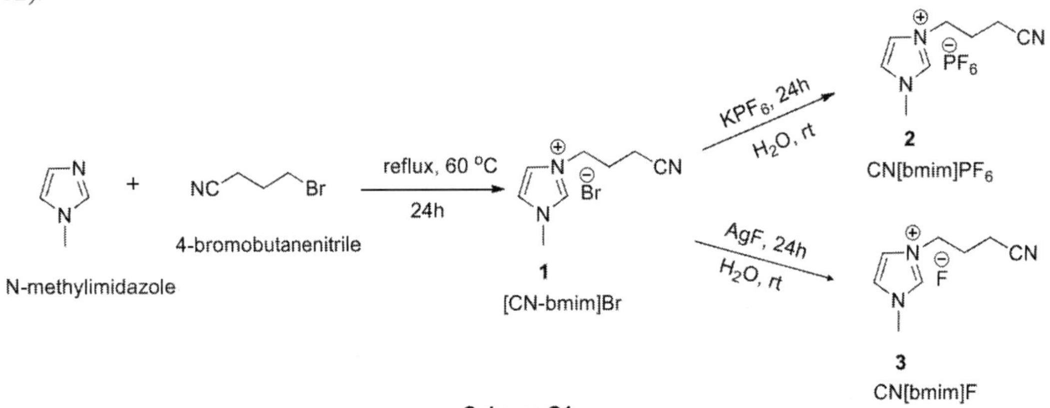

Scheme S1

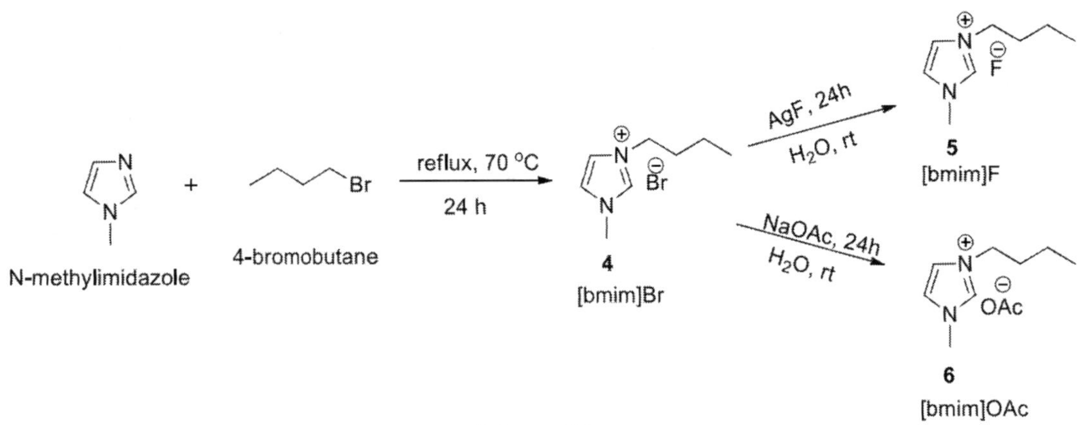

Scheme S2

Figure 2. Synthesis of Various ionic liquids

To determine the function of ionic liquidsin generation and sustainationof NPs, extensive work were carried out with CN[bmim]PF_6 (**2**), CN[bmim]F (**3**), [bmim]F (**5**) and [bmim]OAc (**6**) and$Pd(OAc)_2$ and $PdCl_2$ as the catalytic salt.

Table 1

S.No.	Palladium precursor	Ionic liquid	Temp. (°C)	Pd NPs (nm)
1.	$PdCl_2$	2	140	25-30
2.	$PdCl_2$	3	140	50-80
3.	$Pd(OAc)_2$	2	140	2-5
4.	$Pd(OAc)_2$	3	140	10-20
5.	Pd(OAc)2	2,3	100	-
6.	$Pd(OAc)_2$	5,6	100	-
7.	$Pd(OAc)_2$	5,6	140	-

Best Reaction condition:4 mg of $Pd(OAc)_2$ in 2 mL of IL**2**

$PdCl_2$as catalyst precursor gives NPs of size 25-30 nm in IL **2** and while in IL **3** size increased upto 50-80 nm (Table 1, entries 1, 2). While $Pd(OAc)_2$ in ILs **2** gives PdNPs of range 2-5 nm and in IL **3** of range 10-25 nm (entries 3, 4). When reaction was carried out at 100°C no black-brownprecitipate was found in ILs **2** and **3** (entry 5).Similarly in ILs **5**and **6**no Pd^0 precipitate was foundat temperatures of 100 and 140 °C (entry 6, 7). This showed that transformation of Pd^{II} into Pd^0 was not only the thermal decomposition of palladium precursor but it was the nitrile side chain on N-methylimidazole that played the important role in generation of PdNPs. The best reaction conditions were found to be 4 mg of $Pd(OAc)_2$ in 2 mL of IL **2**at 140 °C for 8 hours (entry 3).

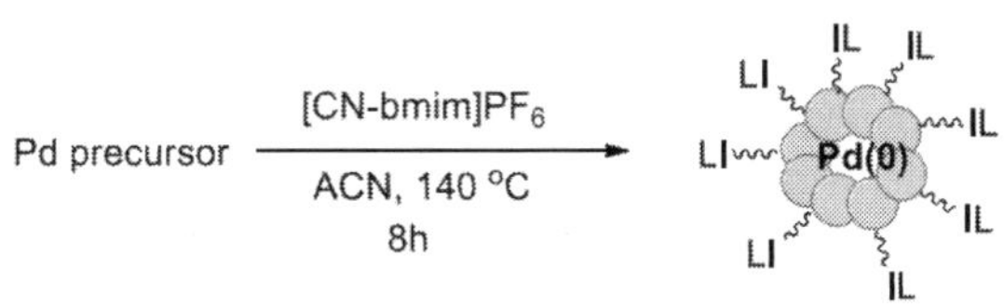

Figure 3. Scheme for generation of PdNPs

3. Characterization of Pd NPs:

The size of PdNPsgoverned by TEM and SEM which shows the distribution of nanoparticles in IL **2** and **3** in accords with table1. The 0.5 mL of reaction mixture was centrifuged, washed carefully with ethanol and nanoparticles was redispersedin hexane, ultrasonicated for 20 minutes and TEM was recorded.
The $PdCl_2$ yielded PdNPs of size 25-30 nm in IL **2**while in IL **3** size 50-80 nm was obtained (figure 4 a,b). $Pd(OAc)_2$ in IL **2** yieldedevenly distributed PdNPs of much smaller size 2-5 nm as shown in figure 5 (c) while in IL **3** size increased to 10-20 nm (figure 5d). Similarly figure 6 (a, b) and figure 7 (a, b) show the SEM images of PdNPs prepared by $PdCl_2$ and $Pd(OAc)_2$respectively.Higher size of NPs by $PdCl_2$ could be attributed to lower solubility of $PdCl_2$ in ILs, which impacts the heterogeneous distribution of reaction mixture resulting in larger sized particles.

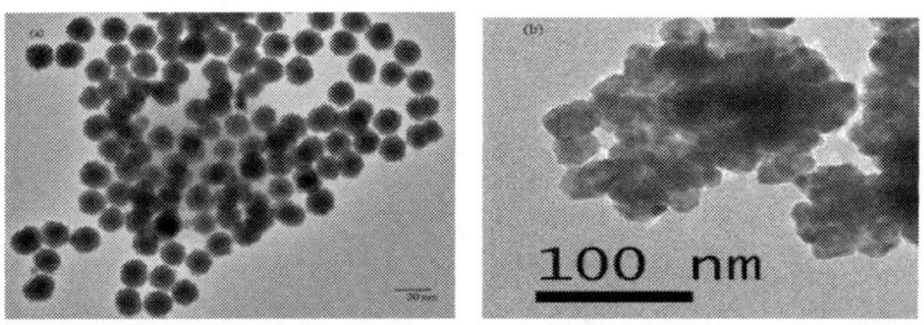

Figure 4.TEM images of PdNPs by $PdCl_2$ in (a) IL 2 (b) in IL 3

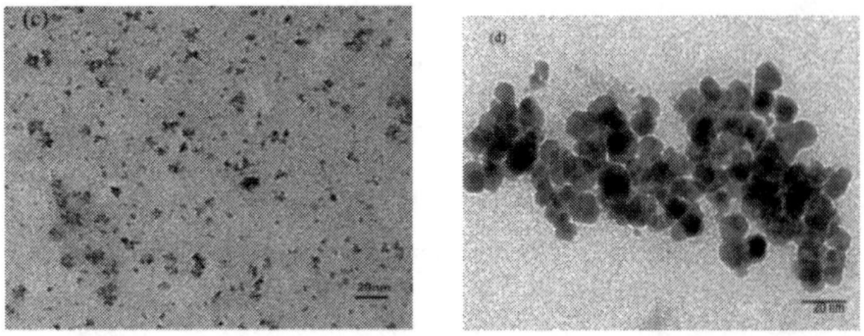

Figure 5.TEM images of PdNPs by $Pd(OAc)_2$ in (c) IL 2 (d) in IL 3

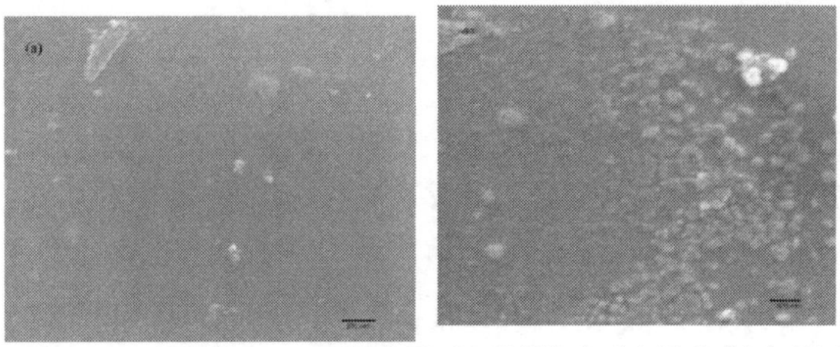

Figure 6. SEM images of PdNPs by $PdCl_2$ in (a) IL 2 (b) in IL 3

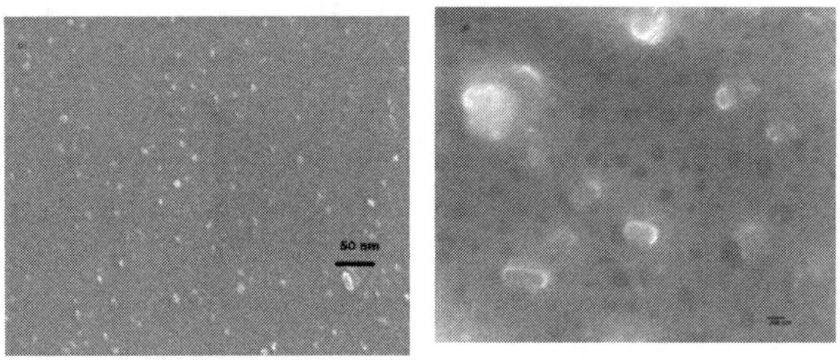

Figure 7. SEM images of PdNPs by $Pd(OAc)_2$ in (c) IL 2 (d) in IL 3

The characterisation study was further carried out by UV-Vis and PXRD with 2-5 nm nanoparticles.UV-Vis (Figure 8 a) showed the progress of reaction, an aliquot of reaction mixture was taken after 1 hour interval.$Pd(OAc)_2$ was dissolved inIL **2** and acetonitrile as cosolvent, the colourof solution was brownish orange and exhibited a peak at 240 nm. After 1h, the colour of the reaction mixture was switched to yellow and anabsorption peak appeared at 260 nm. Finally when the colour of the solution was changed to black-brown and a typical UV-Vis pattern suggested the formation of Pd(0) NPs.[7]No change was observed in colour of reaction mixture and UV-Vis pattern even after 24 h.

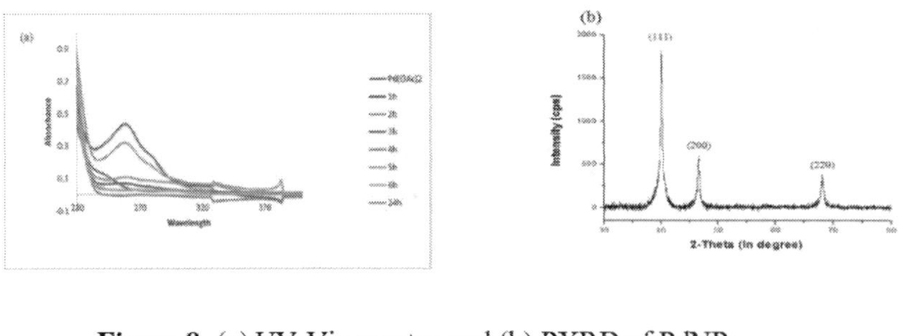

Figure 8. (a) UV-Vis spectra and (b) PXRD of PdNPs

The synthesizedPd(0) was further confirmed by and PXRD. Figure 8 (b) represent the characteristicpeaks corresponding to three planes (111), (200), and, (220) with lattice constant, a = 3.871 A°,which can be indexed to crystalline Pd both in peak position and relative intensity and coincide with JCPDS file 87-0638.[8] (figure 8 b)

To study catalytic activity of PdNPs,Hiyama coupling was carried was literature method[9], only the catalyst and ligand was replaced with preformed PdNPs in CN[bmim]PF_6. The reaction went smoothly and biphenyl was obtained which was characterized by ^{1}H-NMR and ^{13}C spectroscopy.

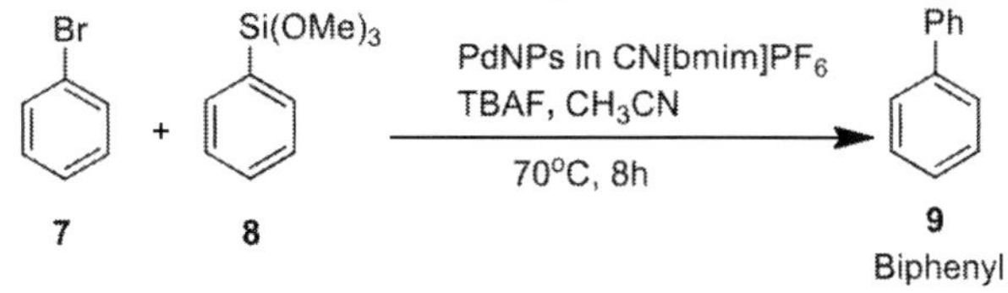

Figure 9. Scheme for Hiyama coupling

4. Conclusion

New protocol was developed for synthesis of PdNPs in nitrile functionalized ionic liquid. Nitrile functionalized IL offer the finest medium for synthesis and stabilization of Pd NPs. The characterization of PdNPs wascarried out by UV, PXRD and size was calculated by SEM and TEM.For sample preparation for SEM,PdNPs were centrifuged at 5000 rpm, washed with ethanol and PdNPswas re-dispersed in hexane and sonicated for about 45minutes. For SEM PdNPs was mounted on a sample holder with spin coating technique to get evenly dispersed nanoparticles. The sample was then coated in a sputter coater with a gold layer in vacuum. For TEM, 2μL of PdNPs in CN[bmim]PF_6were taken ineppendrof and diluted with hexane (1 mL) which was coated on the carbon-coated copper grid and solvent was subjected to evaporate at room temperature.

Reference

[1] (a) Beletskaya I P and Cheprakov A V., *Chem. Rev.* 2000, **100**, 3009-3066 b) Wu X-F,
(b) AnbarasanP, Neumann H and Beller M, *Angew. Chem. Int. Ed.* 2010, **49**, 9047-9050c) Leadbeater
(c) E, *Nature Chemistry*, 2010, **2**, 1007-1009.

[2] (a) Rosner T, Bars J L, Pfaltz A and Blackmond D G, *J. Am. Chem. Soc.* 2001, **123**, 1848-1855
(b) Dupont J, Pfeffer M and Spencer J, *Eur. J. Inorg. Chem.* 2001, 1917-1927c) Bedford R B, *Chem. Commun.* 2003, 1787-1796.

[3] (a) Shenhar R and Rotello V M, *Acc. Chem. Res.* 2003, **36**, 549-561 b) *Reetz* M T , Westermann
(b) E, *Angew. Chem. 2000,* **112**, 170c) Astruc D, Lu F, Aranzaes J R, *Angew. Chem.* 2005, **117**, 8062

[4] (a) Crooks R M, Zhao M, Sun L, Chechi V and Yeung L K, *Acc. Chem. Res.* 2001, **34**, 181-190
(b) Roucoux A, Schulz J and Patin H, *Chem. Rev.*, 2002, **102**, 10, 3757-3778.

[5] Dupont J and Scholten J D, *Chem. Soc. Rev.* 2010, **39**, 1780-1804 (b) Dupont J, *Acc. Chem. Res.* 2011, **44**, 1223-1230.

[6] (a) Swatloski R P, Holbrey J D and Rogers R D, *Green Chem.* 2003, **5**, 361-363;
b) Maiti APagoria P F, Gash A E, Han T Y, Orme C A, Gee R H and Fried L E *Phys.Chem.Chem.Phys.* 2008, **10**, 5050–5056

[7] Nemamcha A, Rehspringer J-L, and Khatmi D*The Journal of Physical Chemistry B* 2006 110 (1), 383-387 DOI: 10.1021/jp0535801

[8] Navaladian S, Viswanathan B, Varadarajan T K and Viswanath R P , *Nanoscale Res. Lett.* 2009, **4**,181-186

[9] Premi C and Jain N,*Eur. J. Org. Chem.* 2013, 5493–5499

IOP Conf. Series: Materials Science and Engineering 1221 (2022) 012053 doi:10.1088/1757-899X/1221/1/012053

Electron Transport in N-functionalised Armchair Graphene Nanoribbons: Computational Insight

Sonal Agrawal[1*], Anurag Srivastava[1] and Gaurav Kaushal[2]

[1]Advanced Material Research Group, CNT lab, ABV-Indian Institute of Information Technology and Management Gwalior 474015, India
[2]VLSI design lab, ABV-Indian Institute of Information Technology and Management Gwalior 474015, India
Email: sonalagrawal711@gmail.com

Abstract: Density functional theory (DFT) theoryhas been used to explore the electronic and transport properties of (3,0) pristine armchair graphene nanoribbons (AGNR) and its Nitrogen (N) doped counterparts. The analysis has been performed for four different Nitrogen doping concentrations i.e., 6.25%, 12.5%, 18.75% and 25% by replacing the carbon atoms of the pristine AGNR. The computed formation energy (E_{form}) confirms that the stability of N-doped AGNR is comparable to pristine AGNR. Band structure and density of states profile shows that pristine AGNRs havea band gap of 1.06 eV and introduction of N doping changes its behavior to n-type semiconductor. Interestingly, in N doped AGNR, its metallic character remains intact with increasing concentration of N and shows superior current-voltage characteristics compared to linear atomic chains of metals, with 25% doping concentration. This interesting behavior of N doped AGNR can be exploited for its application as electrodes or interconnects in electronic devices.

1. Introduction:

At nanoscale, the carbon materials show unique properties, depending on their morphologies.Single layer graphenehas honeycomb lattice structure ofcarbon atoms, is an allotrope of carbon and havesp^2-hybridization. Graphene in the recent years,has attracted the scientific community in big way, due to its remarkable properties [1].Graphene have Dirac cone in the bandstructure profile [2], high charge carrier mobility,ballistic conduction, astonishing quantum properties andmassless carrier behavior [3]. In spite of these properties, the graphene limits its use in the electronics industry due to its zero bandgap.To optimize the performance of semiconductor devices, the bandgap of graphene can be tuned by using the one-dimensionalform of graphene[4], [5]. Sufficient reports are available on doping of Boron (B) and Nitrogen (N) on Graphene Nano Ribbons (GNRs). In [6] Singh at al analyzed the effect of site dependent doping of boron and nitrogenon the bandgap ofzigzag nanoribbons. In another study, doping of zigzag graphene nanoribbons with Be, B, and N used as a technique to make them semiconductor by openingthe band gap[6].In [7] themost stable site for the B and N doping have been determined and effect of same on the electronic and transport properties of zigzag GNRs have been analyzed.In one of the report[8], B and N has been introduced as doping in eight atom wide armchair graphene nanoribbon and observed that impurity level in doped graphene is independent on the dopant site. In another report, effect of boron doping on the electron transport properties of AGNR has been investigated with different doping concentration for interconnect application [9]. The present work reports the effect of nitrogen doping on the electronic properties of armchair GNRs with four different doping concentrations has been discussed for its possible application in transistor and oscillators. The stability, electronic properties (Bandstructure, density of states), electron density and quantum transport properties (I-V Characteristics, transmission characteristics) of the pristine and doped AGNRs have been analyzed.

2. Computational details:

To understand the impact of nitrogen doping on the AGNRs, the computation has been performed using Density Functional Theory (DFT) based Quantum Wise- Atomistix Toolkit (ATK)[10] package. The Monkhorst-pack scheme has been used in the Brillouin zone with $1 \times 1 \times 50$kpoint sampling.Energy cut off of 75 Hartree with double zeta double polarization basis set have been used in the calculations.Generalized gradient approximation has been used as exchange–correlation functional in the calculation[11]. The geometries of pristine and doped AGNRs are optimized until the maximum stress and maximum force on each atom reaches to 0.05eV/Å^3 and 0.02eV/Å, respectively. For the stability analysis the formation energies of these structures have been calculated using the following equation [12].

$$\mathbf{E_{form}=E_{total}-\Sigma_P E_{total}(N) -\Sigma_Q E_{total}(C) -\Sigma_R E_{total}(H)} \qquad \mathbf{(1)}$$

Where E_{form} represents the formation energy of the system and E_{total} is the total energy or ground state energy of the system, $E_{total(N)}$, $E_{total(C)}$ and $E_{total(H)}$ are the total energy of thenitrogen,carbon atom and hydrogen atom respectively. P, Q and R represent the number of nitrogen, carbon and hydrogen atoms respectively.

The transport properties of these AGNRshave been computed using two probe modelthroughLandauer–Büttiker formula, reported elsewhere [13]

$$I = \frac{2e}{h}\int_{\mu L}^{\mu R}(T(E)[\mathrm{F_L(E)} - \mathrm{F_R(E)}])\mathrm{dE} \qquad (1)$$

$$\mathrm{T(E,V)} = \mathrm{Tr}(\Gamma_L(\mathrm{E,V})G^R(\mathrm{E,V})\Gamma_R(\mathrm{E,V})G^A(\mathrm{E,V})) \qquad (2)$$

Here μ_L, $F_L(E)$ and μ_R, $F_R(E)$ represents the chemical potential and Fermi distribution functions of the left andright electrodes, respectively. T(E) is the total transmission probability of all the channels at applied voltage V and energy E. $G^R$and G^A are the retardedand advanced Green's function for left and right electrodes respectively. $\Gamma_L$and Γ_R are the coupling function for the left electrode and right electrode respectively.

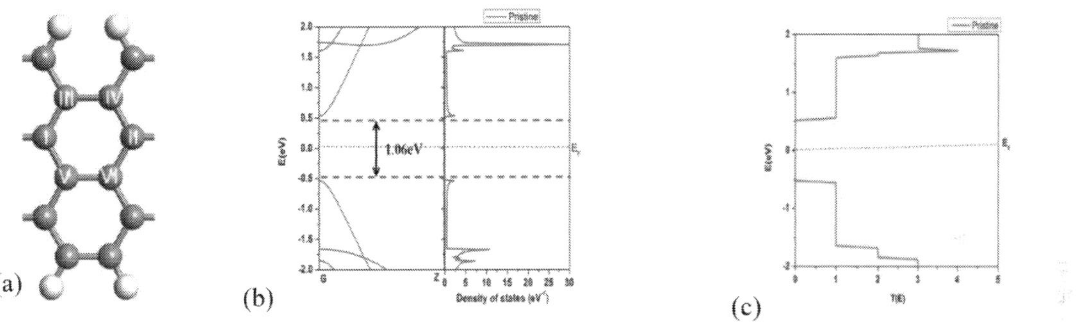

Figure 1. (a) (3, 0) AGNR labeled with doping site I-VIII (b) Bandstructure and Density of states of Pristine AGNR (c)Transmission spectrum of pristine AGNR

3. Results and Discussions:

The nitrogen doped AGNRs have been analysed using their optimized geometries shown in Fig. 1. The lattice constant, Carbon-Carbon and Carbon-Hydrogen bond lengths in optimized geometry of pristine AGNR are found to be 4.31 Å, 1.43 and 1.10 Å, respectively. The unit cell of pristine AGNR composed of one fourth of hydrogen atoms and three fourth of carbon atoms and four different doping concentration 6.25%, 12.5%, 18.75% and 25% has been performedon AGNR.The carbon atoms labelled as lattice sites I-VIII have been chosen for introducing the dopant and the most stable sites out of theall attempted sitesselected, where, for 6.25% doping concentration lattice position labelled as III is the most stable position. Similarly, for 12.5% doping concentration, lattice position labelled as IV and V, for 18.5% doping concentration I, II and IV are the most stable position and for 25% doping concentration, III, IV, V, VI are the stable positions.The bond length between neighbouring carbon and dopant atoms in optimized geometries for N doped AGNR at 6.25%, 12.5%, 18.75% and 25% doping concentration are found to be 1.40Å, 1.40Å, 1.39Å and 1.42Å, respectively, are relatively reduced from that of pristine AGNR, because of smaller covalent radius of nitrogen incomparison to that of carbon.The stability of AGNRs have been analysed, in terms of formation energy (E_{form})

IOP Conf. Series: Materials Science and Engineering 1221 (2022) 012053 doi:10.1088/1757-899X/1221/1/012053

calculated using equation1 and reported in table1, confirms that the formation energy of nitrogen doped AGNRs are slightly less very close to that of pristine AGNR.

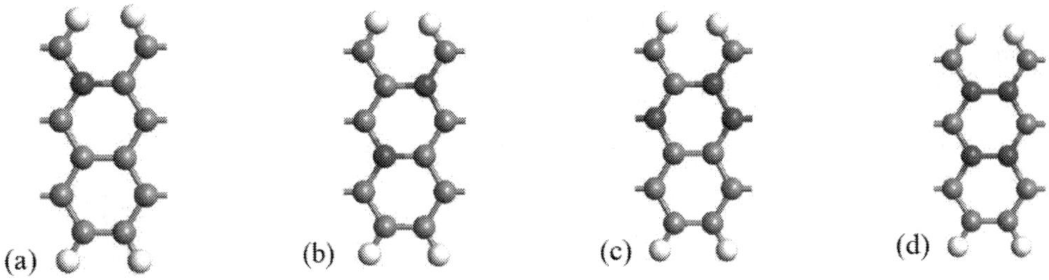

Figure 2. Most stable positions of N doping with different doping concentration (a) 1N doped AGNR (b) 2N doped AGNR (c) 3N doped AGNR (d) 4N doped AGNR, respectively

Table 1 Formation energies of Nitrogen doped AGNR

Morphologies	Formation Energy (E_{form}) (eV)
Pristine	-8.145
1N doped AGNR	-8.021
2N doped AGNR	-7.848
3N doped AGNR	-7.660
4N doped AGNR	-7.411

3.1 Electronic properties analysis:

To investigate the effect of N doping on the electronic properties of AGNRs, the bandstructure and Density of States (DOS) profiles have also been computed. From the Fig. 1(b),it has been observed that the pristine (3, 0) AGNR show bandgap of 1.06 eV and the Density of states (DOS)comprehend thebandstructure profile,as shown in Fig.1(b). It can clearly observe thatintroduction of the N type impurity, shifts theFermi level towards the conduction band (CB) in AGNRs. From Fig. 2(a-d), it has been observed that, N-doped AGNR shows metallic behaviour, and thisbehaviour of AGNR enhances with increasing N doping concentration. Hence, it can be concluded that nitrogen doping is responsible for a phase transformation in AGNRs from its original semiconducting to metallic nature.

(a)  (b)

(c) (d)

Figure 3: Bandstructure and density of states of N doped AGNRs (a) 1N doped AGNR (b) 2N doped AGNR (c) 3N doped AGNR (d) 4N doped AGNR, respectively

3.2 Transmission spectrum analysis:

To further analyse the electron transport in these AGNRs, the transmission spectrum has been computed using transmissioncoefficient,in case of pristine AGNR the magnitude of transmission peak at the Fermi level is found to be zero due toband gap,as shown in Fig 1(c).In N-doped AGNR with 6.25% doping concentration (Fig. 4(a)) the magnitude of transmission coefficient is 1 andincreasesup to magnitude of 2 with 12.5% doping, as shown in Fig. 4(b).Whereas, at 18.75% doping, the transmission coefficientagain decreases to magnitude of1 (shown in Fig. 4(c)) and further increasing the doping concentration leads to increase in the magnitude of transmission peak to2.5as shown in Fig. 4(d), hence the nitrogen doping of AGNR with 25% doping concentration showsrelatively better conductancethan the other counterparts in this work.

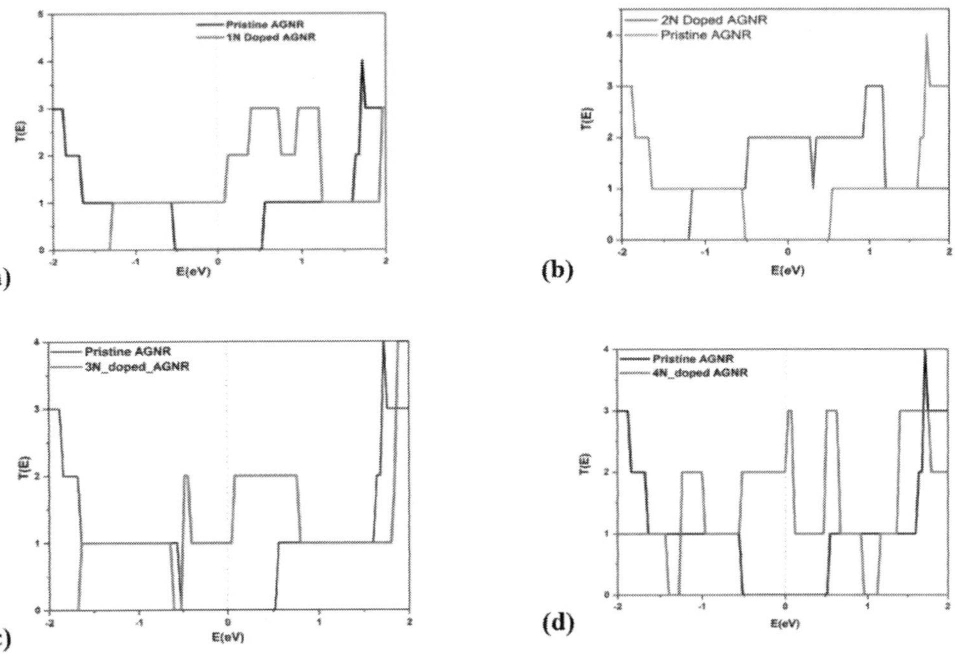

(a) (b) (c) (d)

Figure 4. Transmission Spectrum of N doped AGNRs (a) 1N doped AGNR (b) 2N doped AGNR (c) 3N doped AGNR (d) 4N doped AGNR, respectively

3.3 Electron densityand Transconductance analysis

In the electron density plot for N doped AGNR at different concentrations i.e.,6.25%, 12.5%, 18.75%, 25%, as shown in Fig. 5, confirms that electron density on nitrogen atom is between 1.8 -3.6 as compared to carbon 0 to 1.7due to high electro negativity of nitrogen, then the carbon. The transconductanceof N doped AGNR has been calculated at four different doping concentrations and reportedin table 2.From the table 2, it can be observed that the N doped AGNR with 25% concentration shows superior conductance than other counterparts.

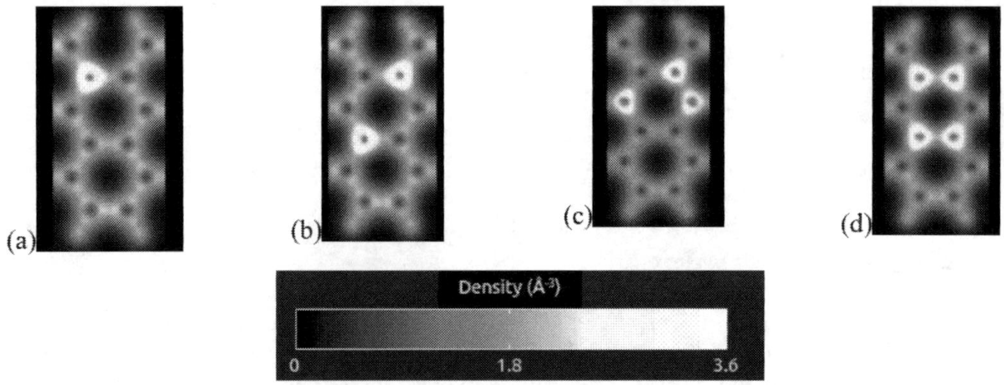

Figure 5. Electron density plot for N doped AGNR at different concentration (a) 1N doped AGNR (b) 2N doped AGNR (c) 3N doped AGNR (d) 4N doped AGNR, respectively

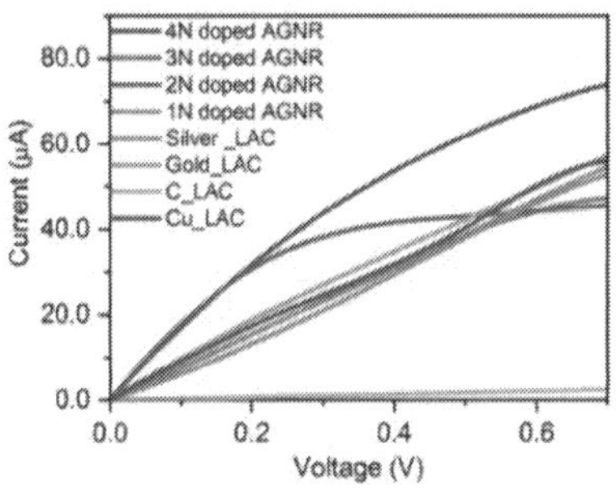

Figure 6. I-V characteristics of linear atomic chains and N doped AGNRs

Table 2. Transconductance of doped AGNRs at different concentrations at bias voltage of 0.7 V

Environment	Transconductance (μS)
1N doped AGNR	82.250
2N doped AGNR	66.007
3N doped AGNR	67.832
4N doped AGNR	107.634

3.4 Current-Voltage characteristics

For the application point of the I-V characteristics of reported AGNRs have also been computed and ithas been observed that when AGNRs are dopedwith nitrogen metallicity increases, and the I-V characteristics showsmonotonically increasing nature, with maximum value of current reaches to 57575.37nA, 46205.05nA, 47482.77nA, 75343.81nA at 0.7Vfor 6.25%, 12.5%, 18.75% and 25% doping concentrations, respectively.With 6.25% and 12.5% doping concentrations, the current is almost same up to bias voltage of 0.55V, that implies that the doping does not have any significant impact on the I-V characteristicsup tobias voltage of 0.55V. With 6.25% and 25% doping concentrations, the AGNR shows monotonically increasing trend in I-V characteristics. Oncomparing the I-V characteristics of N doped AGNRs with the Linear Atomic Chains (LACs) of metals, it is observed that the N- doped AGNR at 25% doping concentration have relatively better current,and that can be exploited these GNRs for interconnect application, where, maximum conductivity can be observed with 25% doping concentration.

4. Conclusion

The present work reportsthe effect of varied concentration of nitrogen doping, on the stability, electronic and transport properties of AGNR, computed using Density Functional Theory. The report includes the stability analysis, with the most stable position for the N doping. Further, to investigate the electronic properties, band structure and density of states has been computed and shows that metallicity in AGNR enhances with N doping, and prompted us to compute and confirm through the electron density and transport properties for its potential interconnect application of N doped AGNRs. The conductance and I-V characteristics analysis with increasing doping concentration, confirms that the current in the AGNR increases with nitrogen concentration, and gives relatively better current in comparison to the popular linear atomic chains (Silver, gold, copper),with maximum conductance at 25% doping concentration. In the view of these observations, it can be concluded that the proposed doping in AGNRs defends their possible applications in the electrode of transistors or interconnect application.

References

[1] Novoselov K S, 2004, Electric field effect in atomically thin carbon films," *Science (80-.).*, vol. 306, no. 5696, pp. 666–669,.

[2] Novoselov K S 2005, Two-dimensional atomic crystals," *Proc. Natl. Acad. Sci.*, vol. 102, no. 30, pp. 10451–10453,.

[3] Novoselov K S, 2005. "Two -dimensional gas of massless Dirac fermions in graphene," *Nature*, vol. 438, no. 7065, p. 197,

[4] Marmolejo-Tejada J M and Velasco-Medina J 2016, Review on graphene nanoribbon devices for logic applications," *Microelectronics J.*, vol. 48, pp. 18–38,

[5] Nakada K, Fujita M, Dresselhaus G, and Dresselhaus M S, Edge state in graphene ribbons: Nanometer size effect and edge shape dependence," *Phys. Rev. B*, vol. 54, no. 24, p. 17954, 1996.

[6] Chowdhury R, Adhikari S, Rees P, Wilks S P, and Scarpa F, 2011, Graphene-based biosensor using transport properties," *Phys. Rev. B - Condens. Matter Mater. Phys.*, vol. 83, no. 4, ,.

[7] Chauhan S S, Srivastava P, and Shrivastava A K, 2014, Electronic and transport properties of boron and nitrogen doped graphene nanoribbons: an ab initio approach," *Appl. Nanosci.*, vol. 4, no. 4, pp. 461–467.

[8] Singh S, De Sarkar A, and Kaur I, 2017, A comparative and a systematic study on the effects of B, N doping and C-atom vacancies on the band gap in narrow zig-zag graphene nanoribbons via quantum transport calculations," *Mater. Res. Bull.*, vol. 87, pp. 167–170

[9] Agrawal S, Srivastava A, and Kaushal G, 2021 Electron transport in boron functionalised armchair graphene nanoribbons : Potential interconnects," *Solid State Commun.*, vol. 327, no. May 2020, p. 114209,

[10] Soler J M 1996, The SIESTA method for ab initio order-N materials simulation," *J. Phys. Condens. Matter*, vol. 14, no. 11, pp. 2745–2779

[11] Perdew J P, Burke K, Ernzerhof M, Phys. Rev. Lett. 77 3865.

[12] Ma D. 2018, C 3 N monolayers as promising candidates for NO 2 sensors, *Sensors Actuators, B Chem.*, vol. 266, no. 2, pp. 664–673.

[13] Lake R and Datta S, 1992, Nonequilibrium Green's-function method applied to double-barrierresonant-tunneling diodes," *Phys. Rev. B*, vol. 45, no. 12, pp. 6670–6685, Mar.

The structural and optical properties of combustion synthesized Ag doped ZnO nanoparticles

A Mahesha[1], M Nagaraja[1, $], S Satyanarayana Reddy[2], A Madhu[3], N Srinatha[4, *]
[1]Department of Physics, SSIT, Sri Siddhartha Academy of Higher Education, Tumakuru-572 105, India
[2]Department of Physics, Bengaluru City University, Dr. Ambedkar Veedhi, Bengaluru 560001, India
[3]Department of Physics, Bangalore university, Bengaluru-560 056, India.

[4]Department of Physics, RV Institute of Technology and Management, Bengaluru, 560 076, India
*Corresponding Author Email: *srinatha007@gmail.com; [$]madihally.nagaraja@gmail.com.*

Abstract
We report the combustion synthesis of Ag (1 to 5 mol%) doped ZnO nanoparticles and their structural and optical characteristics as a function of dopant concentration in this work. The phase, purity, structure, and optical characteristics of the produced samples were thoroughly investigated using XRD and UV-Visible spectroscopy methods. The production of polycrystalline nanoparticles in their as-synthesized condition is confirmed by XRD findings. To estimate the different structural parameters and examine the structure of the produced samples, Rietveld refinement was used. The optical properties were evaluated using Tauc's plots (band gap energy). With increasing dopant concentration, the band gap was determined to be increasing.

1. Introduction

Organic dyes are becoming more widely used in everyday life as the world's population, commercialization, and civilizations advance [1]. Food, leather, medicines, cosmetics, paints, plastics, paper, and textiles are just a few of the sectors that use synthetic organic colours. The management of organic dyes, which migrate to waterbodies via waste water, is the most difficult problem in the post-production of these sectors' goods [2,3]. These manufactured colours (organic dyes) are harmful to both human health and the environment [1–3]. To remove colour, coagulation, filtration, adsorption, and reverse osmosis have all been employed in the past. However, because of their aromatic structural stability, these colours are difficult to remove from water. Nano-catalysts [4] have been identified as one of the most promising agents for reducing synthetic dyes.

ZnO is a versatile semiconductor with a very large band gap (3.27 eV) among all transition metal oxides [5]. It also has a high exciton binding energy (60 meV), making it a unique semiconducting material with benefits such as great abundance, low cost, high catalytic efficiency, chemical and thermal stability [6,7]. Furthermore, the band structure of ZnO may be altered with the addition of an appropriate dopant for specific purposes. Many dopants have been employed to alter the band structure of ZnO to date, including Co, Fe, Al, Ga, In, Sb, and others [8–11]. Because of their appropriate band gap and enhanced photocatalytic properties (Ag decreases electron hole recombination), Ag ions have become a primary dopant for many research groups. Silver is also nontoxic, non-hazardous, and ecologically benign [12,13]. Furthermore, due to their diverse optical, thermal, electrical, and biological properties, silver nanoparticles are critical in the current context. Until yet, photocatalytic degradation of organic pigments mediated by silver nanoparticles has been described [14–18]. Furthermore, various factors such as particle size, structure, morphology, and silver nanoparticle concentration influence the degradation rate. Particle size is one of them, and it plays an important role in dye degradation. To put it another way, the smaller the particle, the higher the surface-to-volume ratio, the stronger the adsorption to the reactants (organic dye), and therefore

IOP Conf. Series: Materials Science and Engineering 1221 (2022) 012054 doi:10.1088/1757-899X/1221/1/012054

the faster the reaction (dye degradation). AgNPs function as electron transfer mediators during the dye degradation process, according to studies [14]. As a result, it's essential to consider their structural and optical characteristics, which are crucial to the photocatalyst's efficiency. As a result, in this study, we used a solution combustion synthesis approach to synthesise Ag doped ZnO nanoparticles, and we examined their structural and optical characteristics before testing their photocatalytic activity.

2. Experimental and Characterization

2.1 Synthesis of $ZnAl_2O_4$ nanoparticles

Zinc nitrate hexahydrate (Zn(NO3)2.6H2O), Silver nitrate (Ag(NO3)3), and Leucine (C6H13NO2) were utilised as a source, dopant, and fuel, respectively, in the combustion process to manufacture 1 to 5 mol percent Ag doped ZnO nanoparticles. For full combustion, stoichiometric amounts of oxidizer, dopant, and fuel were employed, with the oxidizer to fuel ratio set to unity. In de-ionized water, the stoichiometric amounts of oxidizers (source and dopant) and fuel were fully dissolved. The solution was then maintained at 450°C in a muffle furnace that had been preheated. The solution is burned, which involves an igniting flame and the release of large amounts of combustion gases from the final product. The resultant substance was scraped and mashed using a pestle and mortar before being collected. The as-synthesized final powder samples were also calcinated for 2 hours at 600 °C to remove any leftover carbon in the sample after burning.

2.2 Characterization Techniques

The synthesized samples were evaluated using a Rigaku ULTIMA IV instrument with a Cu K wavelength of 1.5418 for phase purity and structural information. The diffused reflectance of the prepared samples was measured using a UV-Visible Spectrometer (USB4000+ES, Ocean Optics Inc. equipment) in the wavelength range of 250 to 800 nm.

3. Results and Discussions

3.1 Structural studies by PXRD

Figure 1 shows the powder x-ray diffraction (PXRD) patterns of Ag doped ZnO nanoparticles as synthesised. All of the patterns in Fig. 1 have well-defined multiple peaks, indicating that the sample is polycrystalline. All of the peaks (except those marked with an asterisk) correspond to JCPDS card no. 36-1451, which is a hexagonal ZnO wurtzite structure. Furthermore, the presence of a few asterisk-marked peaks, which correspond to metallic silver nanoparticles, shows that Ag has partially replaced Zn in the lattice. This might be related to the fact that Ag (1.72 Å) and Zn (0.74 Å) have different ionic radii. This implies that Ag's substitutional limit at the Zn position is less than 1 mol%. It's worth noting that when the Ag concentration rises, the strength of the ZnO peaks diminishes. Furthermore, as seen in the Right picture of Fig. 1, the strength of the impurity peak rises with increasing Ag concentration, particularly at 38.2°. That is, when the concentration of silver metal nanoparticles increases due to doping, the intensity of (111) peaks increases.

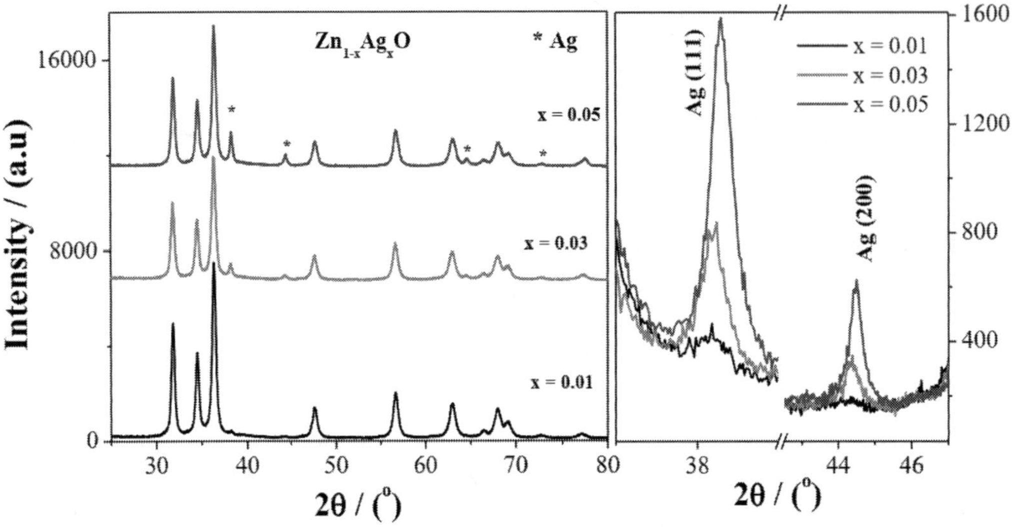

Fig. 1 XRD patterns of Ag doped ZnO nanoparticles (left) and enlarged view showing the peak intensity variation of Ag (111) and Ag (200) peaks (right).

Rietveld refinement was used employing FullProf software to determine the proportion of Zn1-xAgxO (Ad doped ZnO – single phase) and Ag phase (impurity) as well as structural information. The revised structural parameters are presented in the Table, and the Rietveld fitted XRD patterns are displayed in Fig. 2. 1. The impurity (Ag) phase is found to be very low, with 0.23 percent, 1.81 percent, and 4.25 percent for 1 mol percent, 3 mol percent, and 5 mol percent Ag doped ZnO nanoparticles, respectively. The revised lattice parameters and unit cell volume values drop until they reach 3 mol%, then increase to 5 mol%. The average crystallite size Ag doped ZnO nanoparticles were computed using Scherrer's equation; $D = \left({k\lambda}/{\beta Cos\theta} \right)$. The crystallite size of Ag doped ZnO was shown to increase with Ag concentration, owing to the replacement of bigger ionic sized Ag (1.72 Å) for smaller sized Zn (0.74 Å). In addition, the estimated crystallite size of Ag metal nanoparticles in the host matrix was determined to be in the range of 25 – 30 nm. The strain rises and the dislocation density reduces as a result of Ag doping.

3.2 Optical properties

Diffused reflectance spectra (Fig. 3(a)) were acquired using a UV-Visible spectroscope in the visible wavelength band from 300 nm to 800 nm to examine the optical characteristics of Ag doped ZnO nanoparticles, as shown in Fig. 3. Interestingly, surface plasmon resonance (SPR) peak around 410 nm wavelength due to silver is observed in the reflectance spectra. Further with increase in the Ag concentration, the SPR peak shifts towards the higher wavelength region. This could be due to the increase in the Ag nanoparticle concentration in the host matrix as observed in the XRD results. In addition, the absorption band edge shifts towards lower wavelength region indicates the increase in the band gap energy. The optical energy band gap was computed using the modified Tauc's relation; $[F(R)h\upsilon]^2 = B(h\upsilon - E_g)$ after converting diffused reflectance into Kubelka Munk function F(R). The optical bandgap energy was calculated by extrapolating the linear component of the curve to the y-axis, i.e. $[F(R)h\upsilon]^2 = 0$ from Fig. 3(b). The calculated bandgap energy of ZnO samples is lower than the optical band gap energy of bulk ZnO (E_g^{Bulk} = 3.27 eV) for low Ag doping concentration and with increases further for higher concentration. The increase in the band gap energy with Ag doping is attributed to the Moss-Burnstein effect [19].

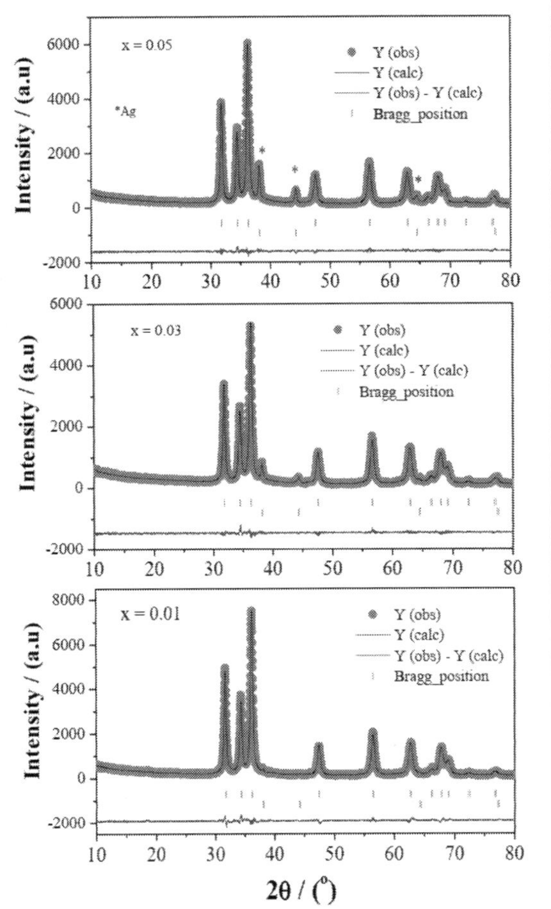

Fig. 2 Rietveld fitted XRD patterns of Ag doped ZnO nanoparticles. The experimental data is shown by red circles, the fitted line is represented by black lines, the anticipated Bragg locations of single phased ZnO wurtzite structure is represented by olive vertical lines, and the difference between the experimental and fitted lines is represented by blue lines.

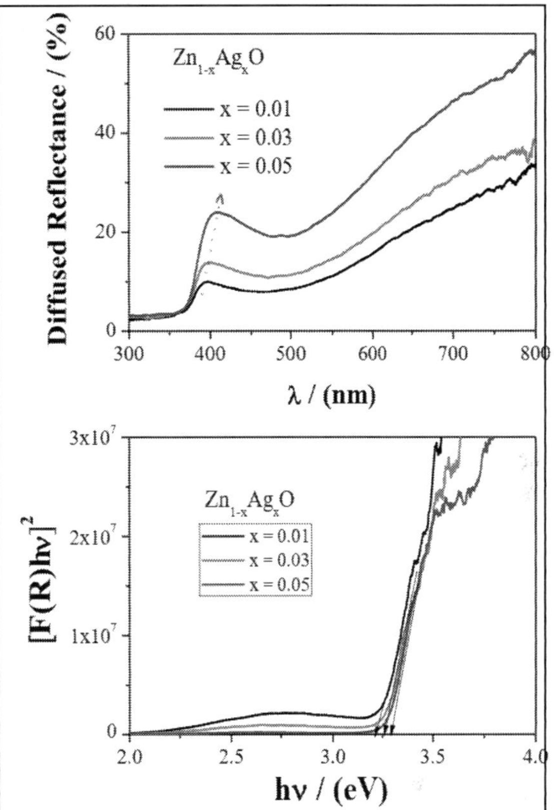

Fig. 3 **(a)** Diffused reflectance spectra and **(b)** Tauc's plots of Ag doped ZnO nanoparticles.

Table 1 Summary of structural parameters obtained from Rietveld refinement of XRD patterns of Ag doped ZnO nanoparticles.

Refined Parameters	$Zn_{1-x}Ag_xO$					
	x = 0.01		**x = 0.03**		**x = 0.05**	
	ZnO	**Ag**	**ZnO**	**Ag**	**ZnO**	**Ag**
Crystal Structure	*ZnO: Hexagonal*			*Ag: Cubic*		
Space group	*ZnO: P63mc*			*Ag: Fm-3m*		
Crystallite size, D / (nm)	29.3	30.1	43.1	25.3	46.97	26.7
Strain, ε / (%)	0.2707	--	0.3774	--	0.3738	--
Dislocations density, δ / (10^{14} lines m^{-2})	11.6755	--	5.3858	--	4.5333	--
Lattice parameters (a = b = c) / (Å)	$a = b =$ 3.2591 (2) c = 5.2206 (1)	$a = b = c$ = 4.0993 (1)	$a = b =$ 3.2501 (3) c = 5.2053 (5)	$a = b = c$ = 4.0871 (7)	$a = b =$ 3.2516 (2) c = 5.2073 (4)	$a = b = c$ = 4.0888 (4)
Unit cell volume / (Å^3)	48.022 (3)	68.886 (24)	47.618 (9)	68.274 (20)	47.680(6)	68.358 (13)
Percentage of Phase	99.77	0.23	98.19	1.81	95.75	4.25
Bragg R-factor	1.11	9.29	1.05	3.20	0.861	2.94
R_F / (%)	0.877	5.52	0.760	1.76	0.602	1.53
χ^2	1.93		1.63		1.70	
R_p / (%)	4.88		4.49		4.54	
R_{wp} / (%)	6.47		6.22		6.34	
R_{exp} / (%)	4.66		4.87		4.87	
GoF	1.4		1.3		1.3	
Zn (x,y,z)	(0.3333, 0.6667, 0)		(0.3333, 0.6667, 0)		(0.3333, 0.6667, 0)	
O (x,y,z)	(0.3333, 0.6667, 0.38596)		(0.3333, 0.6667, 0.38697)		(0.3333, 0.6667, 0.38800)	
Ag (x,y,z)	(0.3333, 0.6667, 0)	(0, 0, 0)	(0.3333, 0.6667, 0)	(0, 0, 0)	(0.3333, 0.6667, 0)	(0, 0, 0)

4. Conclusions

Using the combustion synthesis approach with Leucine as a fuel, we were able to effectively synthesise Ag doped ZnO nanoparticles. XRD results demonstrate the formation of polycrystalline nanoparticles in their as-synthesis state. Crystallite size and strain increase as Ag concentration rises, but dislocation density decreases. Furthermore, because the ionic radii of Ag and Zn ions differ, the proportion of impurities that belong to Ag metallic particles increases as the Ag concentration rises. Rietveld refinement was used to evaluate structural features such as lattice parameters, fitting parameters, and percentage of impurities. Presence of silver in the host matrix was confirmed by the presence of SPR peak in UV-Visible spectra. Further, the estimated band gap energy increases with Ag doping. The research showed that 1 mol % Ag doped ZnO is ideal for photocatalytic applications.

REFERENCES

[1] Tkaczyk A, Mitrowska K and Posyniak A, 2020 *Sci. Total Environ.* **717** 137222.
[2] M. Shahid, Shahid-Ul-Islam, F. Mohammad, 2013 *J. Clean. Prod.* **53** 310–331.
[3] D. Singhee, 2020 Review on Natural Dyes for Textiles from Wasters. In: Samanta, A.K., Awwad, N., Algarni, H.M. (Eds). *Chemistry and Technology of Natural and Synthetic Dyes and Pigments*, Intech Open.
[4] Somwanshi S B, Somvanshi S B and Kharat P B, 2020 *J. Phys.: Conf. Ser.* **1644** 012046
[5] Srinatha N, Dinesh Kumar V, Nair K G M, Angadi B, 2015 *Adv. Powder Technol.* **26** 1355.
[6] Gahlaut U P S, Kumar V and Goswami Y C, 2020 *Phys. E: Low-Dimen. Syst. Nanostru.* **117** 113792.
[7] Zhang Y B and Assadi M H N and Li S, 2011 *J. Phy.: Conden. Mat.* **23(6)** 066004.
[8] Srinatha N, Nair K G M and Basavaraj Angadi, 2017 *Adv. Powd. Techn.* **28** 1086.
[9] Srinatha N, K J R, M R S,Pattar K V,Madhu A and Angadi B, 2021 *Appl. Nanosci.* **11** 1197.
[10] Srinatha N, Angadi B, Nair K G M, Deshpande N G, Shao Y C and Pong W F, 2014 *J. Ele. Spect.Rel. Phen.* **195** 179.
[11] Elilarassi R and Chandrasekaran G, 2010 *Mat. Chem. Phys.* **123** 450.
[12] Adur A J, Nandini N, Shilpashree Mayachar K, Ramya R, Srinatha N, 2018 *J. Photochem.Photobio. B: Bio.* **183** 30.
[13] Liu W T, 2006 J. Biosci. Bioeng. **102** 1.
[14] Thakur D, Sharma A, Rana D S, Thakur N, Singh D, Tamulevicius T, Andrulevicius M, Tamulevicius T, Shukla S K and Thakur S, 2020 *Chemosens.* **8** 1.
[15] Moaddab A and Ghasemi S2021 *Microchem. J.* **166** 106222.
[16] Gajbhiye S, and Sakharwade S, 2016 *Derm. Sci. Appl.* **06** 48.
[17] Barabadi H, Mojab F, Vahidi H, Marashi B, Talank N, Hosseini O and Saravanan M, 2021 *Inorg. Chem. Commun.* **129** 108647.
[18] Knetsch M L W and Koole L H, 2011 *Polymers.* **3** 340.
[19] Srinatha N, Nair K G M, Basavaraj Angadi, 2015 *Phys. B: Conden. Matter.* **474** 97

IOP Conf. Series: Materials Science and Engineering 1221 (2022) 012055 doi:10.1088/1757-899X/1221/1/012055

Effect of plant extracts on structural & morphological features of CuO nano structured material

Rajeshwari.B.Rotti[a], M Ramya [b], K R Venkatesha Babu[c] and D.V Sunitha[a*]

[a] Department of Physics, School of Applied Sciences, REVA University , Bangalore 560 064, India
[b]Department of Biotechnology, School of Applied Science, REVA University,Bangalore 560 064,India
[c] Department of Physics, Nrupatunga University,Bangalore 560 009 , India

*Corresponding author: E-mail: sunithaprasad8@gmail.com;

Abstract. Cupper oxide (CuO) nano structured material has been synthesized by using eco-friendly simple solution method using plant leaf extracts(Mangifera indica, Carica papaya and Azadirachta indica) as a reducing agent. The obtained material is subjected to calcination at 500 °C for further studies. Powder X-ray diffraction (PXRD) and Fourier transmittance Infrared spectroscopy (FTIR) was performed for the structural confirmation of the calcined CuO samples.The CuO prepared by using Mangifera indica, Carica papaya and Azadirachta indica leaf extracts exhibited flake shape, irregular and spherical structures respectively in scanning electron microscope (SEM) micrographs.In addition, the UV-Visible absorption spectrum was recorded and the energy gap determined by using Tauc relation was 3.5, 3.75 and 3.9eV for Mangifera indica, Carica papaya and Azadirachta indica respectively.

1. Introduction

Nanotechnology is being widely used in many fields and branches such as material science, physics, chemistry, and biology [1]. Nanostructured materials are prepared by using various synthesis technique exhibit the unique properties [1]. The unique property are due to high surface to volume ratio, quantum confinement effect and they are chemically reactive. Various kinds of nanomaterials are synthesized such as sulphates, phosphate, oxides, silicates, and perovskites. Amongst metal oxides are widely used in many application such as capacitors, catalysts, display devices etc. Because metal oxides are having unique properties and exhibits different types of surface morphology such as nanoflakes, nanowire, nanorods, nanobelts etc [2]. Metal oxide nano structured materials can be prepared by various synthesis technique such as sol-gel method, co-precipitation, hydrothermal method, spray pyrolysis, sono chemical method etc [2]. But in these methods,the reducing agents used are highly chemical and toxic producing hazardous material. Whereas in green synthesis technique, safe and non-toxic nanostructured material can be synthesized. There are many metal oxides that can be synthesized such as ZnO, CuO, MgO, Al_2O_3, TiOare synthesized using simple solution technique using plant extracts (flower, leaves ,seeds, bark, juice and microoragnisms [2]. Amongst them CuO is one of the simple oxides possessing unique physical, chemical, optical and thermal properties. Used in wide variety of applications as a catalyst, gas sensor, for solar energy transformation and preparation of organic/inorganic nanocomposite, in the field of antimicrobial, anti-biotic and anti-fungal etc. Copper and copper-based products also used in pesticide and pharmaceutical companies.

In this paper, we have used three different plant leaves extract of Mangifera indica, Azadirachta indica and Carica papaya as a reducing agent during simple solution technique. These plant extracts contains phytochemicals such as tannis, flavonoids, phenolic and alkaloids that are used to reduce the precursor material. Also, green synthesis technique is cost effective and consumes a less energy while synthesis metal oxide. Therefore, we have used simple solution synthesis technique to obtain CuO using plant leaves extract as a fuel and its structural and morphological features are compared.

2. Material and methods

2.1. Synthesis of CuO

Copper sulphate (Fisher scientific, 99.5% purity), used as precursor material and the plant leaves Mangifera indica , Carica papaya, Azadirachta indica were collected from Yelahanka situated in Latitude N 13°6' 11.7756 and Longitude E 77°36'15.044 of Bangalore, Karnataka.Fresh plant leaves were collected and washed multiple times in tap water and kept for drying in the shade for 5 to 6 days. Then, the dried leaves ware grinded to make a fine powder. In order to extract the plant solution, 1:10 ratio of leaf powder (20gm) was dispersed in 200 ml double distilled water and was placed in the shaker incubator maintained 130 rpm at room temperature for 8 hours and obtained solution was filtered using whatmann filter paper in a beaker. Further, the 1:1 ratio of copper sulphate (0.018 gm) and plant extract solution was taken in a beaker and mixed thoroughly using magnetic stirrer maintained at 150 rpm for 15 min. Then, the reduced mixture was placed on hot water bath for 6 hr to evaporate the water content. Finally, a solid brown colored sample is settled at the bottom of the dish. The final product obtained was scratched and grinded into fine powder using agate and mortar, subjected to calcination for 500°C in a hot air oven. Further, the obtained sample was subjected to structural and morphological studies.

2.2. Instruments Used

CuO nano material diffraction pattern was recorded using XPERT-3, PXRD instrument made by Panalytical, using CuKα radition having a wavelength of 1.54nm, in the 2θ range of 30°- 80°. FTIR spectrawas recorded in the range of 500 – 4000cm^{-1} using alpha II spectrometer made by Bruker. Surface morphology was analysed through SEM microscope(VEGA3 LMU).

3. Result and discussion

3.1 PXRD

From fig.1 shows the PXRD pattern of CuOprepared using Mangifera indica, azadirchta indica and carica papayaand compared with standard JCPDS card no. 48-1548.The diffraction pattern exhibits CuO peaks along with impurity peaks of KCl and Aphthitalite. Also, it is observed that the impurity peaks are more in CuO prepared using mangifera indica, Azadirachta indica than carica papaya. This indicates phytochemicals cannot be decomposed at 500°C in CuO sample.

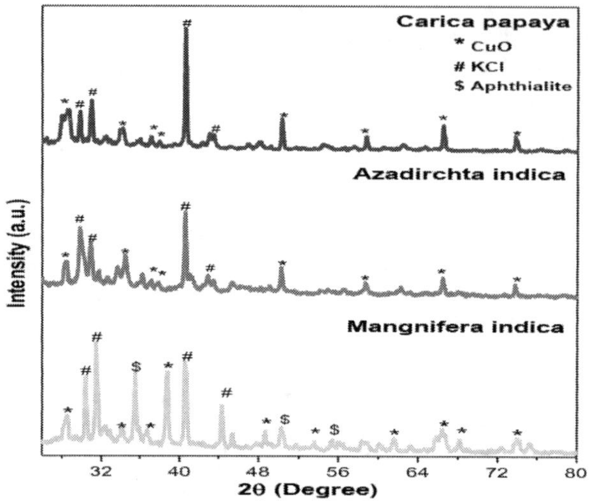

Fig.1 PXRD pattern of CuO

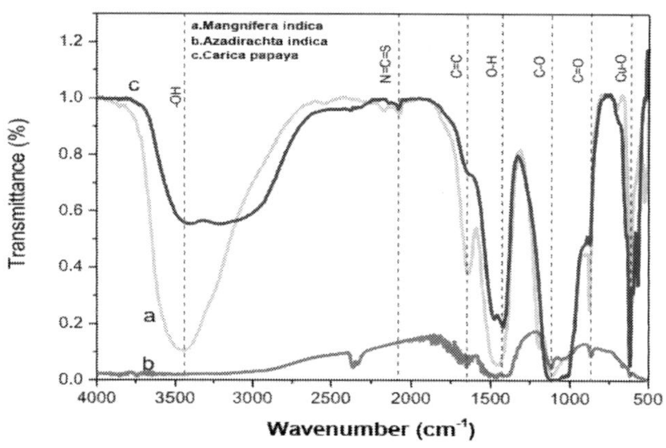

Fig.2.FTIR spectra of CuO

3.2 Fourier transform infrared radiation

The FTIR spectra recorded for CuO prepared by Mangifera indica, azadirachta indica and carica papaya respectivelyare shown in Fig 2. The strong absorption peaks recorded in the range613-618, 878-864, 1078-1111, 1422-1450, 1640-1650, 2077-2354 and 3444-3407 cm^{-1} corresponds to Cu-O , C=O (bending), C-O (stretching) , O-H (bending) ,(C-C stretching), N=C=S (stretching) and -OH (bending) modes of vibration [2]. Fig.2 indicates the absence of 2077-2354 and 3444-3407cm^{-1} in Azadirachta indica due to breaking of N=C=S and -OH bonds.

3.3 UV-Visible spectra

Fig.3(a) exhibits the UV-Visible absorption spectra of CuO prepared by mangifera indica, azadirachta indica and carica papaya. The absorption peak was recorded in the range 281-289nm. Further, the energy gap has determined using Tauc and Wood plot (Fig 3 (b)) and corresponding equation [3] is given below

$$(\alpha h\nu)^{1/r} = \beta\,(h\nu - E_g) \qquad \text{..............(1)}$$

Were, α is absorption coefficient, β is constant, h is Planck's constant, ν is absorption frequency, E_g is energy band gap, r=1/2 for direct band gap , r= 2 for indirect band gap [3]. The energy band gap obtained by mangifera indica, Azadirachta indica and carica papaya are 3.75, 3.9 and 3.5eV respectively. The presence of phytochemical present in the plant extract are dominating in CuO has greater impact on band gap.

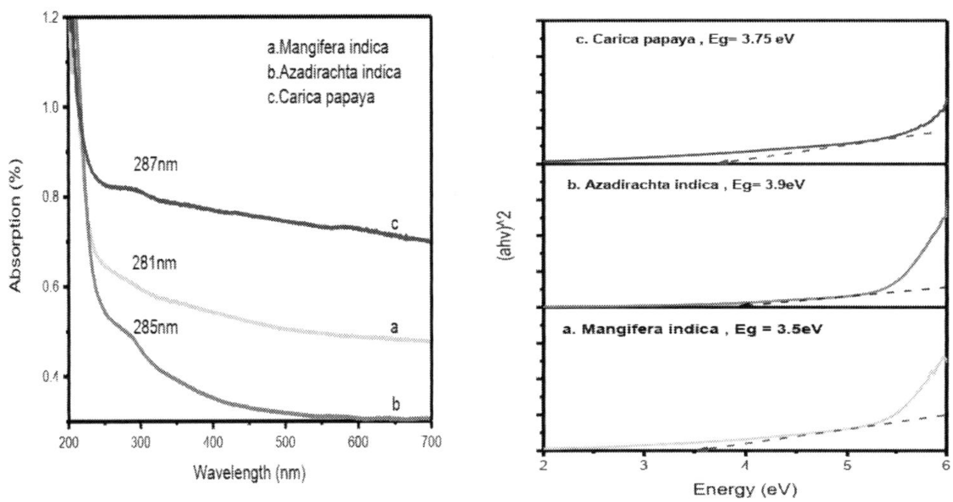

(a) (b)

Fig. 3(a) Absorption spectra and (b) Tauc and Wood plot for CuO nanostructure material

3.4 SEM

Surface morphology of CuO nanostructured material was analyzed by SEM images as shown in Fig4 .The variation in CuO nanostructure are due to chemical composition present in the reducing agent,these plant extract contains phytochemicals which are mainly responsible for change in surface morphology of CuO nanoparticles [4]. Fig 4 (a-c) indicates irregular shaped particles layered structure and spherical shaped particles corresponding to CuO prepared using the leaf extracts of Mangifera indica, Azadirachta indica andcarica papaya respectively.

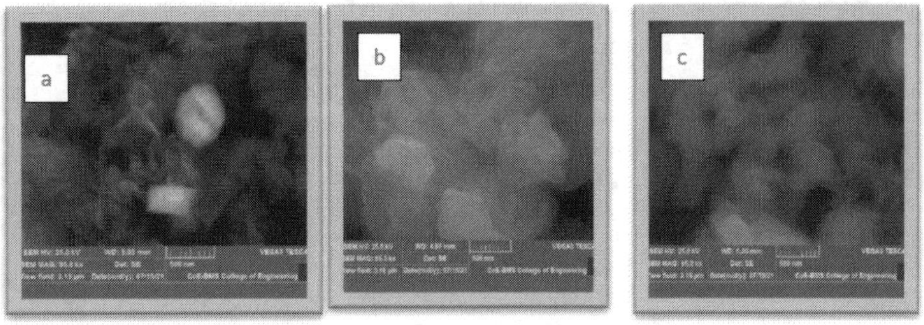

Fig.4 SEM images of CuO prepared using leaf extract 4(a) Mangifera indica, 4(b) Azadirachta indica, 4(c) Carica papaya

4. Conclusion

CuO was prepared by simple solution technique using plant extracts of mangifera indica, Azadirachta indica and carica papaya as a reducing agent. Obtained samples where calcinated at 500°C and subjected to structural and morphological studies. The PXRD pattern recorded for leaf extract indicates that the presence of impurity peak and FTIR spectrum was in accordance with the XRD result. The impurities KCl and Apithitalite present in the prepared sample had a greater impact on the optical band gap of the sample. Also, SEM micro graphs evident the impact of the phytochemical on the structure and morphological features of CuO nanomaterial.

Acknowledgement

The authors thank REVA university, Bangalore to provide seed money for carrying out the research work.

References

[1] shafey A. M. E., 2020, Green synthesis of metal and metal oxide nanoparticles from plant leaf extracts and their applications, J. of Green processing and synthesis, 9 304-339.
[2] Radhakrishnan R., Ali Khan F. L., Muthu A., Manokaran A., Savarenathan J. S., and Kasinathan K., 2021, Green synthesis of copper nanoparticles mediated by aqueous leaf extract of leucas aspera, and moringa tinctoria, letters in applied nano bio science, 10 2706 – 2714.

[3] Tauc J, 1968, Optical properties and electronic structure of amorphous Ge and Si., Materials Research Bulletin 3 37–46.

[4]. Ramalingam R., Hussain M., Fazil U. T., Verma N. K., Arunachalam K. D., 2019, Green synthesis, characterization and antibacterial evaluation of electrospun nickel oxide nanofibers, Material letters, 256 126616.

[5] Sharma J. K., Akhtar M. S., Ameen S., Pratibha, Srivastava, GurdipSingh, 2015, Green synthesis of CuO nanoparticles with leaf extract of Calotropis gigantea and its dye- sensitized solar cells applications, J of Alloys and compds., 25 321-325.

[6] Sunday Adewale, Akintelu Aderonke, Similoluwa Folorunso, Femi Adekunle FolorunsoAbel, Kolawole Oyebamiji, 2020, Green synthesis of copper oxide nanoparticles for biomedical application and environmental remediation, Bulletin, Heliyon, 6 e04508

[7] Suresh Chand Mali, Anita Dhaka, Chanda Kumari Githala, RohiniTrivedi, 2020, Green synthesis of copper nanoparticles using Celastrus paniculatus Wild. Leaf extract and their photocatalytic and antifungal properties –Biotechnology reports, 27 00518.

IOP Conf. Series: Materials Science and Engineering 1221 (2022) 012056 doi:10.1088/1757-899X/1221/1/012056

Chalcogenide Photonic Crystal: Channel Drop Filter

Rajpal Singh* **and Anami Bhargava**
Nanophysics Laboratory, Department of Physics, Government Dungar College, Bikaner, Rajasthan, India.
*rajubkn@gmail.com

Abstract. In the present paper, we have studied a two-dimensional photonic crystal structure of square lattice as channel drop filter application. Photonic crystal structure of square lattice was constructed with circular rods are embedded in air medium. The channel drop filter was designed with a point defect cavity nearby a waveguide, which constructed by removing a single row of rods near cavity. The cavity was designed by removing a number of rods, which characterized with dimension of the cavity. The transmission spectra were analyzed for the channel drop filter with the dimension of the cavity.

1. Introduction

Photonic crystals (PhC) have attracted much attention of research community because of their control over the motion of photon in it. PhC has periodic structure of dielectric with alternate high and low refractive index. Such periodicity of refractive index causes photonic bandgap (PBG) [1]. So, PhCs have various applications such as sensors, optical switch, DWDM, filters. Logic gate etc. [2-13]. Unwanted signal or noise has to be filtered from rest part of desired signal. For that, channel drop filter can be used to drop undesired noise signal from the set of signals. Such optical channel drop filter can be designed using two-dimensional (2-D) PhC structures.

Chalcogenide glasses can be formed using chalcogen elements (S, Se or Te), along with As, Si and Ge. Such kind of chalcogenides shows a good transmission in IR region. Chalcogenides can be fabricated in wide range of compositions. The refractive index of chalcogenides have high values between 2.4 to 3, over large wavelength range in IR region along with low absorption losses and a large $\chi^{(3)}$ nonlinearity (much higher than Si) [2]. Thus, Chalcogenide PhC could be a promising architecture to confine and guide light through it [14-16]. In this paper, we present channel drop filter (CDF) using 2-D PhC of square lattice structure. The transmission spectra of CDF based 2-D PhC is studied to show filter application.

2. Theoretical model and method

We have designed a Channel drop filter by placing a defect cavity near a line defect waveguide in 2-D photonic crystal. The drop channel is created because of capture of resonant mode by the defect cavity. Channel drop filter of 2-D square lattice of As_2S_3 is shown in Figure 1. The chalcogenide rods of As_2S_3 are embedded in air. The refractive index of As_2S_3chalcogenide rods is 2.405 and the radius of rods is taken r = 0.2a, where a is the lattice constant [17]. By removing one row of chalcogenide rods a wave guide is created. The selective coupling is achieved by introducing resonant cavity near the waveguide, obtained by removing chalcogenide rod with different dimensions of defect cavity as shown in Figure 1. We have taken horizontal and vertical dimensions of cavity as h and w as shown in Figure 1.

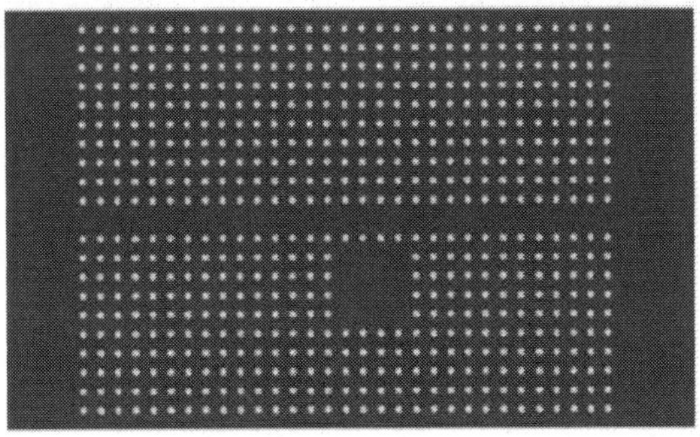

Figure 1. Geometrical Structure of a channel drop filter with defect cavity dimension *h x w*.

Here, the well-known finite difference time domain (FDTD) method is used for computation with computational domain 31x21 (total 651 unit cells) [18]. The waveguide is along the horizontal direction of 2-D PhC. The computation domain is surrounded with PML. The total number of time steps is 10,000 with each time step $\Delta t = 0.99/c(\Delta x^2+\Delta y^2)$, where c is the speed of light, Δx and Δy are space intervals [5]. The source is placed at the input port waveguide to insert signal. Moreover, one can easily normalize the transmission spectra, by comparing the energy flow (Poynting vector) through the output port with that without PhC waveguide in between.

3. Result and Discussion

Here, our aim is to study the effect of cavity's dimension on the transmission at output port for CDF. We have chosen the horizontal and vertical dimension are equal as $h = w = n$. So, the dimensions of cavity are n x n as shown in fig. 1. The geometrical structure with various dimension are shown in figure 2 (a-d).

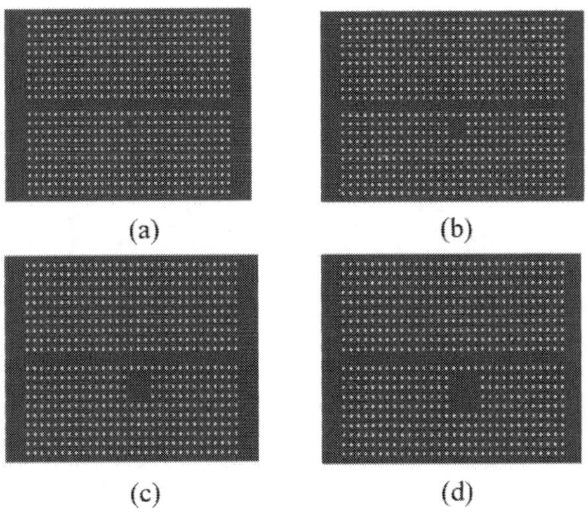

Figure 2. Geometrical structure of CDF based on 2-D PhC with different dimensions as (a) 1 x 1, (b) 2 x 2, (c) 3x3 and (d) 4x4.

The normalized transmission spectra are calculated using FDTD method for CDF devices as shown in fig. 2. The transmission spectra for various dimensions of cavities are plotted in fig. 3.

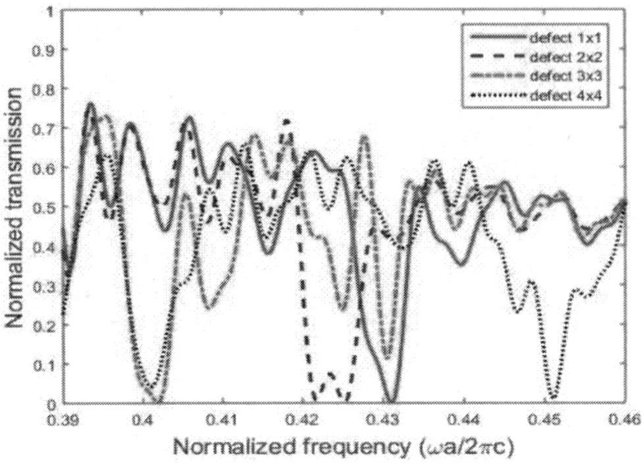

Figure 3. The transmission spectra for channel drop filter with varying vertical cavity dimensions 1 x 1, 2 x 2, 3 x 3 and 4 x 4.

In the transmission spectra, it is found that various channels/frequencies are dropped in cavity due to resonant mode. This resonant frequency depends on the dimension of the cavity. It is found that the drop is very sharp showing very good quality factor ($Q = \Delta\omega/\omega_0$) of 62 for 1 x 1 point defect cavity. It is shown by the fig. 3 that the number of drop channel increases with the dimension of cavity. Thus, such kind of geometrical structure can be used to design a CDF based on 2-D ChancogenidePhC of a desired frequency.

The drop frequency of input signal has dependence on the cavities dimensions. In this case, when the input signal has frequency equal to the resonant mode frequency of the cavity then such frequency signal couples to the cavity and the power at the output port becomes zero. The small dimension of the cavity causes a single drop channel can be achieved in CDF; whereas the large dimension of the cavity causes multi-drop channels can be achieved. The effect is shown here in the form of electric field distribution for CDF based on Chalcogenide PhC in Figs. 4 & 5.

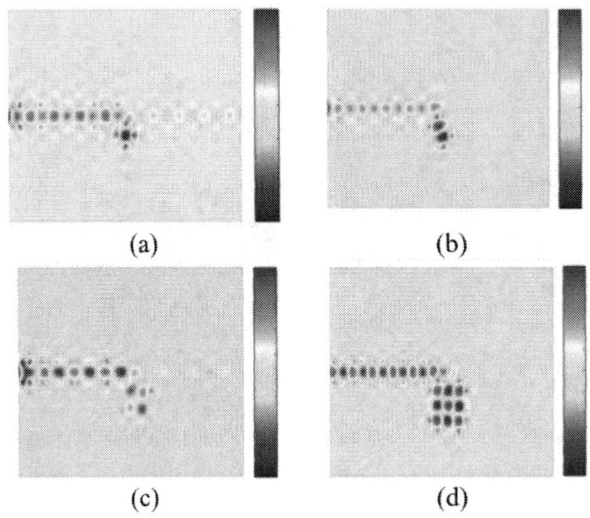

Figure 4. Electric Field distribution of channel drop filter (a) ON resonance at $\omega a/2\pi c = 0.4310$ for point defect cavity 1 x 1 (b) ON resonance at $\omega a/2\pi c = 0.4218$ for point defect cavity 2 x 2 (c) ON resonance at $\omega a/2\pi c = 0.4020$ for point defect cavity 3 x 3 (d) ON resonance at $\omega a/2\pi c = 0.4014$ for point defect cavity 4 x 4.

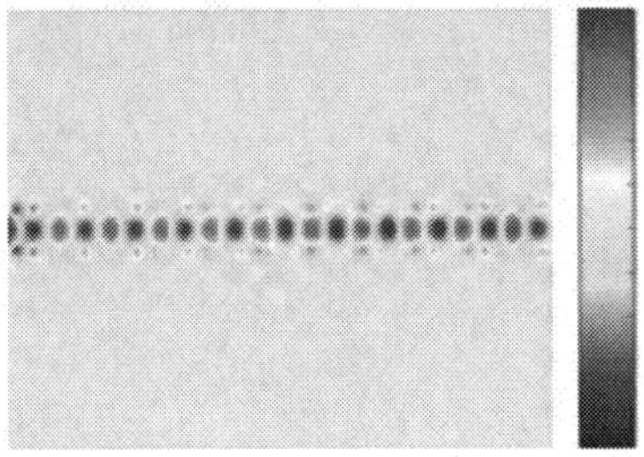

Figure 5. Electric Field distribution of channel drop filter OFF resonance at $\omega a/2\pi c = 0.41$.

From figs. 4 and 5, it can be seen that the resonant frequency is coupled with the cavity and rest of frequencies propagate in waveguide and reached at output port.

4. Conclusion

In this paper, the CDF is designed using 2-D Chalcogenide PhC of square lattice structure with a linear defect near a cavity. The transmission of CDF structure reflects multiple drop channels can be achieved with larger dimension of the cavity. The good quality factor is achieved for the present study. Such kind of CDF based on 2-D Chalcogenide PhC can be used to filter application and also used as optical switching application with external control.

References

[1] Joannopoulos J D, Meade R D, and Winn J N, 1995, Photonic Crystals: Molding of Flow of Light, Princeton University Press, Princeton.

[2] Bhargava A, and Suthar B 2013, Chalcogenide photonic crystals - Exploiting nonlinearity for nanophotonic applications, AIP Conference Proceedings 1536, 15-18.

[3] Ankita, Suthar B and Bhargava A 2021, Biosensor Application of One-Dimensional Photonic Crystal for Malaria Diagnosis, Plasmonics 16 59-63.

[4] Suthar B and Bhargava A 2021, Pressure Sensor Based on Quantum Well-structured Photonic Crystal, Silicon 13 1765-1768.

[5] Chhipa M K, Radhouene M, Robinson S and Suthar B 2017, Improved dropping efficiency in two-dimensional photonic crystal-based channel drop filter for coarse wavelength division multiplexing application, Optical Engineering 56 015107.

[6] Kumarm V, Suthar B, Kumar A, Singh Kh S, Bhargava A and Ojha S P 2014 Silicon based one-dimensional photonic crystal as a TM-mode filter, Silicon 6 73.

[7] Bhargava A and Suthar B 2009, Optical Switching in Kerr Nonlinear Chalcogenide Photonic Crystal, Journal of Ovonic Research 5 187-193.

[8] Singh R and Bhargava A 2018, Design of a Ring Resonator using 2-D chalcogenide based Directional Coupler Photonic Crystal, European Journal of Advances in Engineering and Technology 5 328-332.

[9] Suthar B, Nagar A K and Bhargava A 2009, Tuning the localized mode in point defect chalcogenide photonic crystal, Chalcogenide Letters 6 623-627.

[10] Zhu Z H, Ye W M, Ji J R, Yuan X D and Zen C 2008, Wavelength demultiplexers based on multimode interference effect in photonic crystals, Phys. Lett. A 372, 2534.

[11] Suthar B and Bhargava A, 2015, Optical properties of plasma photonic crystals, Silicon 7 433-435.

[12] Kumar V, Suthar B, Kumar A, Singh Kh S and Bhargava A 2013, Design of a wavelength division demultiplexer using Si-based one-dimensional photonic crystal with a defect, Optik 124 , 2527-2530.

[13] Chhipa, M K, Madhav, Robinson B T P, Janyani S, Suthar V B 2021, Realization of all-optical

logic gates using a single design of 2D photonic band gap structure by square ring resonator, Optical Engineering 60 075104.

[14] Suthar B, Nagar A K and Bhargava A 2010, Slow light transmission in chalcogenide photonic crystal waveguide, Journal of electronic science and technology 8 39-42.

[15] Suthar B 2015, Tuning of guided mode in two dimensional chalcogenide based photonic crystal waveguide, Optik 126 3429-3431.

[16] Singh R, Suthar B and Bhargava A 2018, Effect of Se Concentration on Photonic Bandgap of 1-D As-S-Se/Air multilayers, AIP Conference Proceedings 1953, 060027.

[17] Laniel J M, Menard J M, Turcotte K, Villeneuve A, Vallee R, Lopez C and Richardson K A 2003, J. Non-Cryst. Solids 328, 183-191.

[18] Taflove A 1995, Computational Electrodynamics: The Finite Difference Time-Domain Method, Artech House Inc., Norwood.

Approach toward Alpha Alumina Synthesis

Manisha Upasani[1*], Ashutosh Bagde[2]
[1]Department of Applied Physics,Yeshwantrao Chavan College of Engineering, Nagpur, Maharashtra, India
[2]Jawaharlal Nehru Medical College, Datta Meghe Institute of Medical Sciences, Sawangi (M), Wardha, Maharashtra, India
[*] *Corresponding author: E-mail address:* mupasani1@gmail.com

Abstract. The recent developments and trends towards the synthesis of alumina nanoparticles are discussed. Different modifications made to synthesis approaches for preparation of alumina nanoparticles are critically analyzed. Special attention is emphasized towards combustion synthesis of alpha alumina.

Keywords: nano-scale, alumina, combustion synthesis

1. Introduction

Nanotechnology has become a significant field in the science and engineering development. It includes the production or application of materials that have unit sizes of about 10–100 nm. Properties of nanoparticles depend strongly on their dimension as particle size, surface/volume ratios or specific surface area, crystallinity, morphology and many more.

During the last two decades, nanoscaled materials such as nanofibers, nanoparticles, nanorods, nanotubes, nanoribbons, nanodisks, nanowhiskers, etc. have been investigated widely. Properties like enhancement in strength, hardness, variations in electrical conduction, field-ion-emission through tunneling phenomenon, enhancement in magnetic,optical, and chemical properties etc are found when the nano size approaches. Nanostructures and materials, both metallic and non-metallic, have made incredible strides in their potential applications.

Ceramic materials are one such class of materials that have received considerable interest in materials science due to their potential applications in electronic, optics, catalytic, automotive and biomaterial industries. The ability to synthesize ceramic nanoparticles with controlled shapes and sizes opens new avenues in existing technologies. High surface area nanoparticles can serve as more efficient adsorbates in the refining industry and serve to improve the efficiency of catalytic converters. A proper awareness of these potentials of nanoceramics begins with the synthesis and processing of aggregate free nanophase materials. The ability to synthesize nanoparticles with precise size and morphology outcomes in the capacity to control the surface area and packing density of the particles. [1] In currentscenario, due to increasing demand of nanoparticles of oxides there is a interest in its preparation.

Alumina, with exclusive physical properties such as hardness, high resistivity with bases and acids, withstands very high temperature and exceptional wear resistance, have various applications in fields of engineering. Nano-alumina has many advantages as compared with micron-sized and nano-sized alumina particles.

2. Properties of Alumina

Alumina is a common name of aluminium oxide (Al_2O_3)is a chemical compound with strong ionic interatomic bonding between aluminium and oxygen. Alumina is the only oxide formed by aluminum in nature and occurs as the mineral corundum (Al_2O_3); diaspore (Al_2O_3•H_2O); gibbsite (Al_2O_3•$3H_2O$). and most commonly as bauxite which is an impure form of gibbsite.[3] Boehmite and diaspore are trihydroxides that occur as either crystalline or amorphous particles, respectively, while gibbsite, bayerite and nordstrandite form hydroxides with well-characterized crystal or amorphous forms. Alumina exists in the transition and stable forms. Of the different forms of alumina as (α, χ, η,δ, κ, θ, γ, ρ), all are in the transition forms except α-alumina. γ-alumina and α-alumina are the only two types of commercial alumina that are manufactured for various applications. Due to the hexagonal plate-like crystal structure of γ-alumina, this material is incorporated into many catalysis and absorbent applications. There are several types of crystal structures in α- alumina, most of which have rounded shapes or hexagonal shapes with a small surface areas. The mechanical, chemical, and electrical properties of Alumina (99.7% purity) often used in the ceramic industry are depicted in the table (1) & table (2):The wide-ranging applications of alumina are due to its abundance and its several forms and properties like purity, stability, refractoriness, and chemical inertia. The significant and extensive use of nano alumina in the ceramics field is

- as an insulating material,in the electronics industry for passive components, for hybrid circuits;
- used as a substitute material for several applications as abrasive and cutting tools;
- in biomedical fields to manufacture medical tubes and other scientific, medical products;
- in the manufacturing of refractories.
- As catalysts, fillers, paints, infrared- absorbing materials, fluorescent materials, waterish sensor.

Consequently, there is an increasing need to develop technologies that will produce nanoparticles of these oxides with superior properties for a lower price.

Table 1[2]

Mechanical properties	***Units of Measure***	***SI / Metric***	***Thermal properties***	***Units of Measure***	***SI / Metric***
Density	g/cm^3	3.89	Thermal Conductivity	W/m°K	35
Porosity	%	0	Coefficient of Thermal Expansion	10–6/^{0}C	8.4
Flexural Strength	MPa	379			
Elastic Modulus	GPa	375	Specific Heat	J/Kg ^{0}K	880
Shear Modulus	GPa	152			
Bulk Modulus	GPa	228	**Electrical properties**		
Poisson's Ratio	—	0.22	Dielectric Strength	kv/mm	16.9
Compressive Strength	MPa	2600	Dielectric Constant	@ 1 MHz	9.8
Hardness	Kg/mm^2	1440	Dissipation Factor	@ 1 kHz	0.0002
Maximum Use Temperature	^{0}C	1750	Volume Resistivity	ohm•cm	>1014

Table 2[2]

Physical and Chemical properties	***Aluminium oxide***
Solubility	Not soluble in water
Boiling Point	2980°C
Melting Point	2000°C
Stability	Stable under normal conditions
RiskyPolymerisation	Not occur
Incompatibilities	Chlorine trifluoride, Ethylene oxide
Circumstances to Avoid	Incompatibles
Absorption of water	Nil
Crystal structure	Cubic
Particle Size	10 nm - 50 nm
Specific Surface Area	35 m^2/g - 45 m^2/g
Odour	Odourless

3. Synthesis of Alumina

An important first step in studying nanoparticulated oxides (alumina) is synthesis of the material. An important challenge currently is to develop systematic studies to synthesize oxide nanoparticles with the needed size, shape, and morphology.Nano-alumina can be synthesized by both physical and chemical methods. Few methods; specifically for synthesis of alpha alumina; are only discussed in this paper. Chemical methods include co-precipitation, sol–gel processing, microemulsion technique, chemical vapor deposition, solutioncombustion decomposition. Physical methods include Pulsed wire discharge, laser ablation, Multiple-pulsed laser deposition, flame spray pyrolysis. [3]

3.1 Chemical Methods

3.2 Co-precipitation methods

To precipitate the oxo-hydroxide form, salt precursors (chloride, nitrate, etc.) are dissolved in water (or another solvent). In the case of mixed-metal oxides, chemical homogeneity and controlling size can be very challenging. As a result, surfactants, sonochemical methods, and high gravity reactive precipitation could be considered as new and potentially useful methods of optimizing solid morphological characteristics.J.Y. Park et al synthesis alpha alumina; after sintering as-synthesized prepared particles by precipitation of aqueous aluminum chloride solution, using ammonium acetate and ammonium hydroxide. [4,5,6,7]

3.3 Sol-gel processing

Sol-gel is a technique based on phase conversion of a solution obtained from a metallic alkoxide or organometallic precursor. The sol, is a suspension of particles in liquid, is polymerized at low temperature in order to form a wet gel. In the next step, the solvent is removed by drying the gel, followed by the correct heat treatment. Kumagai and Messing densified alumina ceramics to full extent while keeping grain size less than 0.5 mm by using boehmite sol–gels seeded with less than2.0 wt.% α-alumina powder in aboehmite hydrosol at pH-3. [8] Leilei Xu et al reported facile synthesis of nano-crystalline α-Al_2O_3at 400 °C by employing aluminum isopropoxide, absolute ethanol and nitric acid or hydrochloric acid as starting materials. However, the γ-Al_2O_3 phase also appeared at 800 °C which transform into alpha alumina at 1000^0C.[9] F.Mirjaliliaet al synthesis alumina nanoparticles(20-30 nm) after 48 hours stirring through an alkoxide route by using aqueous solutions of aluminum isopropoxide and 0.5M aluminum nitrate nanohydrate. Sodium dodecylbenzensulfonate as the surfactant stabilizing agent.[10] Sol-gel is an advantageous method as it allows for the synthesis of high purity materials, provides an easy way to introduce trace elements, and utilizes low processing temperatures to save energy.

3.4 Water/Oil Microemulsion technique

The technique was invented in the 1980s as a method for preparing nanoparticles. Water and surfactant molecules aggregate into reverse micelles in the oil phase in W/O microemulsions. Nanoscale reactors can be performed by these micelles. Once two micro emulsions, one of which contains the precursor and the other precipitating agent, are mixed consistently, the reaction will occur in nanometer-sized micelles in a controlled manner, resulting in the formation of nanoparticles with controlled characteristics. According to Huang Ke-long et al, they synthesized Al_2O_3 nanoparticles from aluminum nitrate /Triton X-100/ n-butyl alcohol/ ammonia /cyclohexane/water upon sintering at 1150^0C. This method has several advantages such as soft chemistry, easy handling, and no need for extreme pressures or temperatures.[11]

3.5 Solvothermal methods

Metal complexes can be decomposed thermically either by boiling in an inert atmosphere or using a pressure autoclave. Surfactants are usually added to the reaction medium to control particle size growth and limit agglomeration.A Young Kim and his associates have synthesized alpha alumina using gibbsite alumina precursor in 1, 4-butanediol solvent according to various pH conditions. They further proposed that for pH = 9 at 300 °C for 36 h a rhombohedral α-Al_2O_3 structure hexagonal shaped with about 1.5 to 2.0 μm of particle size have been synthesized.[12]

3.6 Synthesis using supercritical fluids

Supercritical fluids have been used to prepare nanoparticles. It allows the variation in crystallinity by modifying fluid nature. In this technique in a high temperature stirred vessel reactor the precursor and liquid solvents are added. Then the reactor is closed after filled with reaction media and reaches the operating conditions. The organic residue from precursor decomposition is removed at the end of the experiment. The powder as synthesized is collected in the solvents mixture, filtered and washed. After necessary sintering the morphology can be achieved. C. Bousquet et al reported that when metal precursor aluminium acetylacetonate, solubilized in ethanol with the reaction media mixture of H_2O/EtOH and CO_2/EtOH produce different kind of transition alumina; boehmite AlO(OH) or amorphous Al_2O_3; which after sintering led both to alpha-alumina. For this method special or expensive equipment are required and may synthesis particles different microstructures and densities.[13]

4. Physical Methods

4.1 Pulsed wire discharge

In this method for the synthesis of nanoparticles, thin metal wires are evaporated inside an oxygen atmosphere with a high electric current. It is however only possible to synthesize small numbers of powders by a single discharge. S Ishihara et al synthesis γ and α alumina powders from aluminium wire with different air flow speed for industrial viewpoints on large scale.[14]

4.2 Flame spray pyrolysis

In a novel flame spray pyrolysis method gas and aqueous precursors are used to synthesis nanoparticles. In comparison with other methods, this method has many advantages, such as the low cost, controlling particle size, ease of processing, high production yield, and ease of assembling in mass quantities.High temperature flames are used in this process to heat the feedstock material and spray it into a chamber, where it will condense into nanoparticles. A.I.Y. Tok et al. successfully produced nano-sized Al_2O_3 of about 10–30 nm (α-phase + γ-phase), and 80–100 nm (α-phase, calcined) from $AlCl_3$ vapour.[2] This method is very costly as it require specialized equipment and costly chemicals.

4.3 Laser Ablation Technique in Deionized Water

Nanoparticles can be synthesized using laser ablation in gas, vacuum, or liquid. This process involved making nanoparticles in liquid (deionized water) using a short pulse laser with a nanosecond pulse width. In addition, Al is used as a starting material for laser ablation, and this technique offers several advantages, such as rapid and high purity processes.The collection of nanoparticles formed by laser ablation in liquid is easier than that in a gaseous atmosphere.V. Piriyawong et al. reported that alumina nanoparticles were synthesized using pulsed laser ablation from Nd:YAG laser of wavelength of 1064 nm in deionized water with different laser energies. The result shows that the nanoparticles mainly consist of κ-Al_2O_3 and a small quantity of α-Al_2O_3. [15]

4.4 Solid state method

It is one of the most widely used method for the preparation of materials. In this process high temperature is used since some of the solids do not react at room temperature. The probability and rate of reaction depends on various factors such as properties of reactants, their reactivity, surface area, reaction conditions etc. P Kaur and coworkers synthesized alumina nanoparticles by thermal dehydration of Al(OH)3 (gibbsite). The heat treatment about 1150^{0}C produces alpha alumina (84%) and the remnant θ-phase (16%). [16]

5 . Combustion synthesis (CS)

The combustion synthesis process achieves alpha alumina at very low temperatures compared to most of the other processing routes and has a high degree of reproducibility and high potential for large-scale production.It has been found that several methods can be used to lower the phase transformation temperature so that nanocrystalline Al_2O_3 can be produced because lowering this temperature prevents the particles from agglomerating as well as reducing the energy input needed for the process.Combustion synthesis (CS) has established as asignificanttechnique for the synthesis as well as processing of catalysts,alloys, advanced ceramics, composites and nanomaterials.

In combustion or burning, a fuel reacts with an oxidant in a series of exothermic reactions. It results in the production of heat and the conversion into a chemical species.A flame or light can be produced as a

result of the heat. A majority of fuels is organic compounds, especially hydrocarbons, primarily found in liquid, solid, or gaseous form. The material used as fuel is intended for burning or modification in order to obtain energy, heat or to modify objects [17]. The theoretical definition of CS is self-propagating high temperature synthesis (SHS), which depends on the type of reactants; the type of elements (solid, liquid, or gas); and the exotherm (adiabatic temperature, T). With the auto combustion method, oxidizing agents such as metal nitrates react with a fuel such as urea, glycine, citric acid, or hydrazine to form metal oxides. [18,19,20]. When the reaction is initiated,it is self-sustainable and due to it's exothermic nature results in high internal temperature. This resultsin the crystallization and oxides formation. [18, 21] The redox mixture is made by mixing a stoichiometric amount of the metal nitrate or perchlorate (oxidiser) with urea or hydrazide derivative as carbohydrazide (CH), oxalyldihydrazide (ODH), malonic acid dihydrazide (MDH), tetra formal trisazine (TFTA) and igniting at temperature between 350°C and 500°C. The mass left after complete combustion is the oxide material.[20]
The redox reaction in the combustion synthesis of alumina can be written as:

$$2Al(NO_3)_{3(aq)} + 5NH_2CONH_2(aq)\ Al_2O_3(s) + 5CO_2(g) + 10H_2O(g)$$

Further, since there is no need for washing, filtering, drying, calcining, or any acid or base to hydrolyze the salts, this process saves both energy and time.The fuels are the source of C and H, which on combustion liberate CO_2 and H_2O gases with heat.As a result, the metal ions form complexes with the cations in solution, facilitating homogenous mixing.[22]Combustion differs depending on the type of fuel used, from flaming to non-flaming (smoldering). Solution combustion synthesis enables control of various properties of the final product, including size of crystallites, specific surface area, morphology, phases, and degree and nature of agglomeration.
The combustion process is characterized by the following features:
a)It is possible to prepare oxides and these composites at low temperatures.
b)They are homogeneous, crystalline, soft, fine, and have a high surface area.
c)The particle size distribution is narrow, with nanoscale particles.
d)A high level of purity is ensured in the preparation of materials.
e)A large quantity can be prepared inexpensively, and the particles show low agglomeration.
f)The product yields higher than 96 percent in a short period.
It was reported that application of fuels containing amino groups can successfully prepare nanoparticles of alumina by solution combustion synthesize. Norouzbeigi and Edrissi have used four amino acids (glutamine, histidine, lysine and arginine) as main fuels to improve the specific surface area of prepared alumina nanoparticles by solution combustion synthesis.[23,24] Recently F. Bustanafruza,& M. Jafar-Tafreshi have been used ammonium carbonate $(NH_4)_2CO_3$ and ammonium sulfate $((NH_4)_2SO_4$ as two new mineral fuels in combustion synthesis to produce alumina nanoparticles with higher specific surface area and finer particle sizes. They have synthesized γ alumina which on calcination converts to α alumina.
Most of the above synthesis techniques are capable of producing amorphous or γ phase nanoparticles.Though these materials found important catalyst related applications, they lose their nanocrystalline nature upon heat treatment due to γ- to α-phase transformation accompanied by a rapid grain growth.[25] As discussed above, the temperature required to transform into alpha phase and the presence of some of the intermediate phases depends not only on the size of the particles, but also on any impurities or surface stabilizers present.[26] To control phase transition temperature and particle size, seeding starting material with well-controlled nanosized alpha polymorphs may be an alternative to doping. The particle size has been improved by mixing fuels of different exothermicities.Using the urea process, nanosizedα-Al_2O_3can also be produced by microwave-initiated solid-phase combustion. [22]

6. Conclusions

Alpha alumina has proved to be difficult to manufacture on a large scale due to the extremely low yield rates obtained from chemical methods. The effectiveness of physical methods is limited since the nanoparticles' sizes cannot readily be controlled, and these methods can only be used for a limited number of materials. Sol–gel, laser ablation, and vapour deposition are also quite expensive methods, as they require specialized equipment such as vacuum systems, high-powered lasers, and expensive precursor chemicals.Using combustion synthesis, it is possible to create materials with desired composition, structure, and properties as demanded by Material Science Engineering. With the proper control of processing parameters, including microwave initiation, precursors, fuels, pressure, seeding, and redox mixture additives, it is possible to obtain narrow particle size distribution nanosized alpha alumina by combustion synthesis with the desired morphological features. Various parameters like fuels, fuel to oxidizer ratio, initiation temperatures, surfactants, dopants etc. are major key factors to achieve the results.

Reference

[1] Kumar R 2005,ThePennsylvania State University, PhD manuscript
[2] Frederick A 2008 Auckland University of Technology Auckland, New Zealand, PhD manuscript
[3] Tok A.I. Y., BoeyF.Y.C andZhao X.L 2006, Journal of Materials Processing Technology **178** 270
[4] ChangY F, NingG L, LIN Y, Chang Hou LIU and Yuan Bo LIN; 2003 Chinese Chemical Letters 104 –107
[5] Suslick, K.S. Choe, S.B.Cichowlas, A.A and Geenstaff, M.W1991,Nature, 353 - 414
[6] Chen, J.-F. Wang, Y.-H. Gou, F. Wang and X.-M. Zheng, C 2002, Ind. Eng. Chem. Res**39**948
[7]Ji-Young Park ,Seong-Geun Oh , Ungyu Paik and Sei-Ki Moon2002 Materials Letters **56** 429–434
[8] M. Kumagai, G.L and Messing1984J. Am. Ceram. Soc. **67** (11) C230–C231
[9] L Xu, H Song and L Chou 2012, Materials Chemistry and Physics,**132** 1071-1076
[10] MirjaliliaF, ChuahL andMohamad H2011 Journal of Ceramic Processing Research.**12**738 -741
[11] Ke-long H, Liang-guoY, Su-qin L andChao-jianL 2007, Transactions of Nonferrous metals, Society of China**17**633 – 637/
[12] Kim A-Y, Kim H S , Park N-K, Lee T J , Lee WG, Kim H D,Park J and Kang M2012 Journal of Nanomaterials, **2012-3**3
[13] BousquetC., ElissaldeC., AymonierC., MaglioneM., CansellF andHeintzJ.M., 2008, J. Eur. Ceram., **28** 223 -228
[14] IshiharaS, SuematsuH, NakayamaT, Suzuki T and NiiharaK 2012, International,**38** 4477-4484
[15]PiriyawongV, ThongpoolV, AsanithiP and LimsuwanP2012 Journal of Nanomaterials
[16] kaurP, KhannaA, KaurN, NayarP and Chen B, 2020, Phase Transitions, **93-3** 596-605
[17] Fuels. 2005 World Encyclopedia.Oxford University Press
[18]Suresh Kand PatilK C, 1992, J. Solid State Chem. **99** 12
[19] TonioloJ C, LimaM D, TakimiA Sand BergmannC P 2005, Mater. Res. Bull. **40** 561
[20] MimaniTand PatilK C, 2001, Mater. Phys. Mech. **4** 134 -137
[21]PedersonL R, MaupinG D, WeberW J, McReadyD Jand StephensRW,1990, Mater. Lett.**10**
[22] Sharma A, Modi O P and Gupta G K 2012, Advances in Applied Science Research, **3** 2151-2158
[23] Edrissi M and Norouzbeigi R 2011, J.A.M. Ceram, **94** 4052-4058
[24] EdrissiM andNorouzbeigiR, 2007, J. MaterialsScience-Poland **25** 1029-1040
[25] Ibram Ganesh, Paula M Cand Torres J.M.F. Ferreira , 2009, Ceramics International **35** 1173–1179
[26] Mather G.C, and Martinez-Arias; A. 2007, (Rodríguez, J.A. Fernández-García, M; Eds.) N.J.,**13**

Dye degradation study of Malachite green and Congo red by using $ZnIn_2S_4$

Mool Chand[a1], A.S. Rawat[a], Manika Khanuja[b], Seema Rawat[c] and Mitesh Ranwa[a]

[a]Department of Physics, H.N.B. Garhwal University Srinagar, Garhwal, Uttarakhand, 246174, India
[b]Center for Nanoscience and Nanotechnology, Jamia Millia Islamia, New Delhi 110025, India
[c]Department of Physics, Zakir Hussain Delhi College, Jawahar Lal Nehru Marg, New Delhi, 110002, India
[a)] Corresponding author: aashukhichar742@gmail.com

Abstract
In this manuscript, we demonstrate photocatalytic study of $ZnIn_2S_4$ powder sample under visible light irradiation on Malachite green and Congo red dye. The powder sample was synthesized usingIndium (III) sulphate,thioacetamide andZinc sulphate in aqueous solution. X-ray diffraction and FESEM were used to identify the hexagonal crystal structure and microspheres shape, while UV-Vis spectroscopy was used to acquire absorption and reflectance spectra under visible-light (λ= 420 nm) irradiation, the effects of process factors such as contact duration and adsorbent dose were studied on Malachite green and Congo red. Degradation efficiency of Malachite green and Congo red are 96.33%, 99.79% respectively.The results revealed that visible light photocatalytic degradation of dyes was highly efficient.

Key words: $ZnIn_2S_4$; Visible-light irradiation, Malachite green (MG), Congo red (CR)

1. Introduction

From last few years a great interest aroused for the investigation of photocatalytic dye degradation because lost about 15% of total production in worldwide. Peoples are also affected because due to cause of polluted water containing dye also creates a worldwide problem. Another various organic dyes methyl orange, malachite green, congo red are also very harmful and can be used as an antimicrobial and antifungal containing heavy water pollutant which are very toxic, affected for environment [1-3]. In order to remove these problems continuous search from last few decades to find the suitable and efficient catalysts to degrade of dye contaminants. To overcome this problem semiconductor photocatalytic and another nanostructured materials such as ZnO, TiO, andCds etc. because of their low cost and ecofriendly nature [4-7]. Out of which several are broadly investigated. These photocatalytic catalyst absorb photon in the range of UV due to containing a large band gap [8]. After tremendous effort we find a new micro/Nano catalytic $ZnIn_2S_4$ to remove pollutants of organic/inorganic pollutants, have a low cost, more efficient and synthesis by a simple root method in water solution. $ZnIn_2S_4$ have two polymorphous have been reported at hexagonal and cubic lattices [9]. $ZnIn_2S_4$ is degraded in the visible light range and widely use in photocatalytic applications [10]. In this work, Zinc Indium Sulphate powder sample was successfully synthesis by simple root method at low temperature. The crystal structure, morphology and photocatalytic activity of zinc indium sulfate were investigated. The effect of contact time and absorbent amount on Malachite green and Congo red in photocatalytic reaction was studied.

2. Synthesis method

$ZnIn_2S_4$ photocatalysts were made in aqueous solution at low temperatures using $ZnSO_4$, $In_2(SO_4)_3$, and thioacetamide (TAA). The quantities of Zinc Sulphate (4.0 mmol), Indium Sulphate (4.0mmol), and Thioacetamide (20.0 mmol) are dissolved in 250 ml flask of D.I water (80ml) which was sealed partially and placed at 80^0C in water bath for a specific duration of four hours. After final suspension, it was cool down at room temperature for four hours, and then final solution was filtered many times by DI-water and ethanol. After filtration sample dried for six hours at 60^0C in muffle furnace, and then collected sample was grinded by motor pestal. The final product essential for the study of dye degradation.

3. Characterization

IOP Conf. Series: Materials Science and Engineering 1221 (2022) 012058 doi:10.1088/1757-899X/1221/1/012058

The XRD pattern of ZnIn2S4 powder sample obtain from Rigaku smart lab with Cu Kα radiation (λ=0.1554 nm) with the scan rate 0.2 in the 2Θ range 0 to 90 degree. The morphologies of powder sample were observed using FESEMfrom Quanta 3D FEG instrument.The UV-Vis absorption spectra were obtained with a Cary 100 series, Agilent Technologiesspectrophotometer equipped.

4. Photocatalytic

The photocatalytic behaviour of $ZnIn_2S_4$ powder sample prepared under sun light irradiation was tested using Malachite green and Congo red dyes at room temperature. For the investigation of tests for photocatalysis, its different amount of photocatalyst in 100 ml aqueous solution of dye was suspended for two hours in the presence of sunlight. The effect of contact time and amount of absorbent dosage was investigated in photocatlysis process. The dye degradation efficiency of Malachite green and Congo red was calculated with the help of Beer-Lambert lawand recorded by using UV-Vis spectroscopy.

$$R= \frac{C-Co}{CO} X100$$

Where (C_0, C) are the concentration of Malachite green and Congo red at initial time 0 and t, respectively.

5. Results and Discussion

The XRD pattern of $ZnIn_2S_4$ powder samples as shown in Figure 1, result indicate that powder sample have hexagonal crystal structure [11]. The diffraction peaks corresponding to (0 0 2), (0 0 6), (1 0 2), (1 1 2) and (1 0 4) of the ZIS crystal structure with corresponding angle 15.0°, 20.0°, 27.41°, 47.33° and 52.15° respectively. In $ZnInS_4$ all the peaks of diffraction in hexagonal structure whose (JCPDS card no.65-2023), which shows that the high quality synthesization of $ZnIn_2S_4$ particles. The structure at higher magnification of $ZnInS_4$, investigated with the help of FSFM, as shown in Fig 2, and has a unique spherical superstructure of Marigold like, combination of various nanosheets [12].

Figure 3(a), 3(b), 3(c)and 3(d), 3(e), 3(f) shows UV-Vis absorbance plot of MG and CR aqueous solution to test the visible-light photocatalytic behaviour tests of $ZnIn_2S_4$products sample. The y-axis is defined as absorbance, while x-axis is defined as wavelength, the absorption intensity peak of dye at a wavelength of 614 nm for MG and 498nm for CR under each irradiation time interval with different amount of powder sample, and C_0 is absorption intensity of the as-prepared dye solution.In the visible light degradation of Congo red and Malachite green, resulting $ZnInS_4$ powder sample with different amount and irradiation times as shown in fig 3(a), 3(b), 3(c) for Malachite green and 3(d), 3(e), 3(f) for Congo red dye. The solution of Malachite green and Congo red are not properly degraded without the addition of $ZnInS_4$ products. After addition a significant degradation obtained. Their absorption peaks are around 614nm and 498nm respectively for Malachite green and Congo red. The effect of contact timing in photocatlysis mechanics clearly shown in figure and result indicate that with the increasing time 0 to 120 minute, concentration of dye in aqueous solution degrade under sun light and at 120 minute approx completely degrade.

The effect of absorbent dosage in photocatlysis mechanics clearly shown in figure 3(a), 3(b), 3(c) for Malachite green and 3(e), 3(f), 3(g) Congo red dye solution with different amount of $ZnIn_2S_4$ powder sample 10mg, 20mg, 50mg respectively. The result shows that higher amount sample have higher degradation efficiency [13].

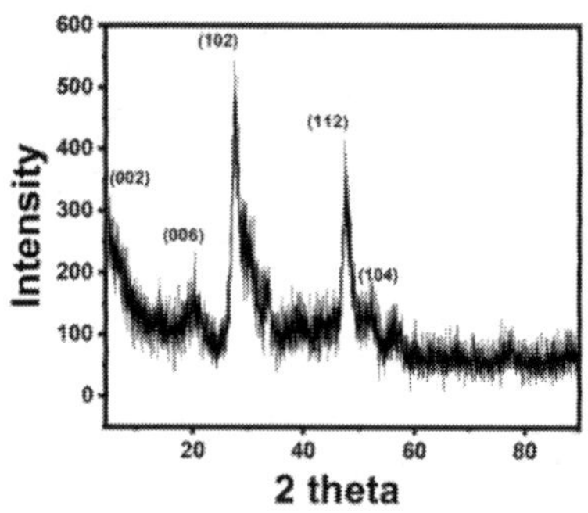

Fig. (1)- XRD pattern of $ZnIn_2S_4$ powder sample

Fig. (2)- FSEM images of $ZnIn_2S_4$ powder sample

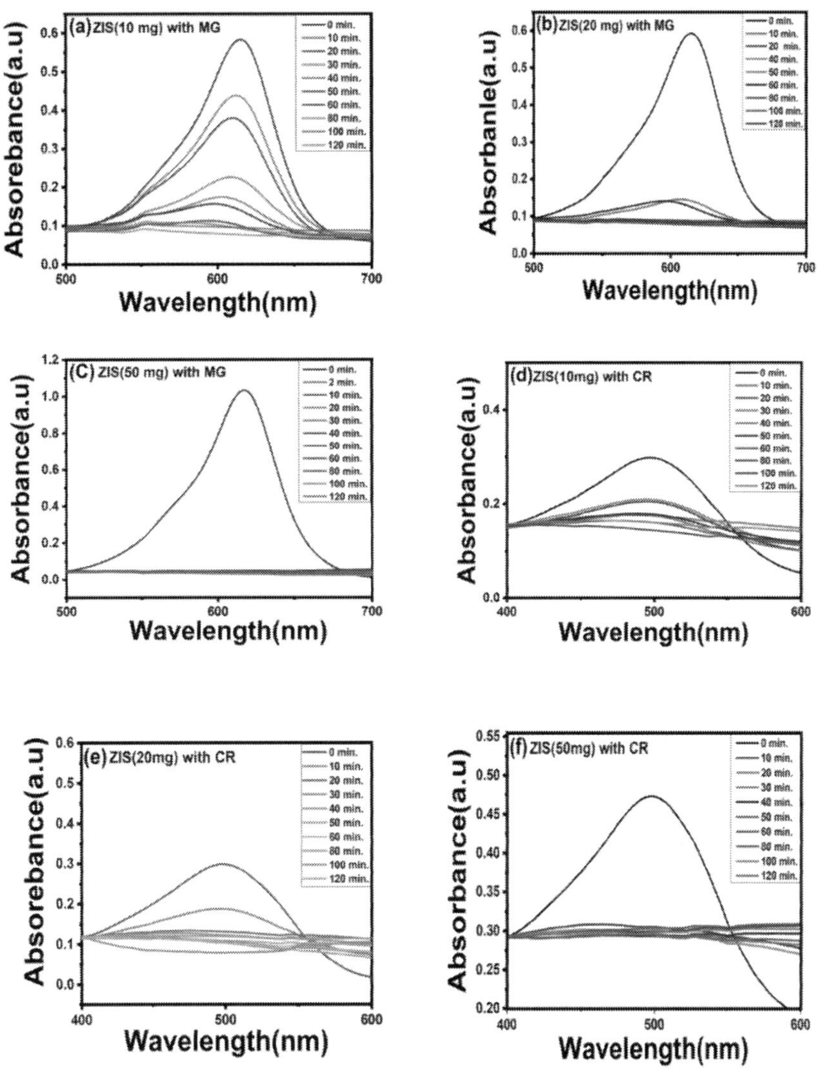

Fig.3 UV-Vis absorption spectra of (a), (b), (c) for MG dye and (d),(e),(f) for CR dye.

6. Conclusion

In this manuscript, Zinc Indium Sulphide microspheres have been successfully synthesis by a simple process which involves low temperature, low pressure or withought any solvent. The hexagonal crystal was identifying by XRD and FSEM image reveal that $ZnIn_2S_4$ have microsphere type superstructure composed of numerous nanosheets. The prepared $ZnIn_2S_4$ microsphere is the suitablephotocatalystfor use in degradation of Malachite green and Congo red dye. The visible light photocatalytic degradation of Malachite green and Congo red by $ZnIn_2S_4$ was investigated. Degradation efficiency of Malachite green and Congo red are 96.33%, 99.79% respectively.

Acknowledgements

Mr. Mool Chand, one of the authors, is grateful to the CSIR–New Delhi India for providing financial support in the form of a JRF-NET fellowship under award number 09/386(0060)/2018-EMR-1.

References

[1] Ashraf, W., Fatima, T., Srivastava, K. and Khanuja, M., 2019, Superior photocatalytic activity of tungsten disulfide nanostructures: role of morphology and defects. *Applied Nanoscience*, *9*(7), pp.1515-1529.

[2] Gautam, R.K., Rawat, V., Banerjee, S., Sanroman, M.A., Soni, S., Singh, S.K. and Chattopadhyaya, M.C., 2015, Synthesis of bimetallic Fe–Zn nanoparticles and its application towards adsorptive removal of carcinogenic dye malachite green and Congo red in water. *Journal of Molecular Liquids*, *212*, pp.227-236.

[3] Chaudhary, N., Raj, K., Harikumar, A., Mittal, H. and Khanuja, M., 2020, October. Comparative study of photocatalytic activity of hydrothermally synthesized ultra-thin MoS2 nanosheets with bulk MoS2. In *AIP Conference Proceedings* (Vol. 2276, No. 1, p. 020030). AIP Publishing LLC.

[4] Mittal, H. and Khanuja, M., 2020, Optimization of MoSe2 nanostructure by surface modification using conducting polymer for degradation of cationic and anionic dye: Photocatalysis mechanism, reaction kinetics and intermediate product study. *Dyes and Pigments*, *175*, p.108109.

[5] Li, F., Luo, J., Chen, G., Fan, Y., Huang, Q., Luo, Y., Li, D. and Meng, Q., 2014, Hydrothermal synthesis of zinc indium sulfide microspheres with Ag+ doping for enhanced H 2 production by photocatalytic water splitting under visible light. *Catalysis Science & Technology*, *4*(4), pp.1144-1150.

[6] Lei, Z., You, W., Liu, M., Zhou, G., Takata, T., Hara, M., Domen, K. and Li, C., 2003. Photocatalytic water reduction under visible light on a novel ZnIn 2 S 4 catalyst synthesized by hydrothermal method. *Chemical Communications*, (17), pp.2142-2143.

[7] Kumar, A., Singh, S. and Khanuja, M., 2020, October. Temperature based morphological study of graphitic carbon nitride for photocatalytic application. In *AIP Conference Proceedings* (Vol. 2276, No. 1, p. 020029). AIP Publishing LLC.

[8] Cai, W., Zhao, Y., Hu, J., Zhong, J. and Xiang, W., 2011, Solvothermal synthesis and characterization of zinc indium sulfide microspheres. *Journal of Materials Science & Technology*, *27*(6), pp.559-562.

[9] Kempken, B., Dzhagan, V., Zahn, D.R., Alcocer, M.J., Kriegel, I., Scotognella, F., Parisi, J. and Kolny-Olesiak, J., 2015, Synthesis, optical properties, and photochemical activity of zinc-indium-sulfide nanoplates. *RSC advances*, *5*(109), pp.89577-89585.

[10] Chai, B., Peng, T., Zeng, P., Zhang, X. and Liu, X., 2011, Template-free hydrothermal synthesis of ZnIn2S4 floriated microsphere as an efficient photocatalyst for H2 production under visible-light irradiation. *The Journal of Physical Chemistry C*, *115*(13), pp.6149-6155.

[11] Xu, Z., Li, Y., Peng, S., Lu, G. and Li, S., 2012, NaCl-assisted low temperature synthesis of layered Zn-In-S photocatalyst with high visible-light activity for hydrogen evolution. *RSC advances*, *2*(8), pp.3458-3466.

[12] Adhikari, S., Charanpahari, A.V. and Madras, G., 2017, Solar-light-driven improved photocatalytic performance of hierarchical ZnIn2S4 architectures. *ACS omega*, *2*(10), pp.6926-6938.

[13] Mittal, H., Kumar, A. and Khanuja, M., 2019, In-situ oxidative polymerization of aniline on hydrothermally synthesized MoSe2 for enhanced photocatalytic degradation of organic dyes. *Journal of Saudi Chemical Society*, *23*(7), pp.836-845.

IOP Conf. Series: Materials Science and Engineering 1221 (2022) 012059 doi:10.1088/1757-899X/1221/1/012059

Investigate the Optical, Structural and electrochemical properties of PVC/PMMA/NiO blend films

M Basappa,H Ganesha, S Veeresh,Y S Nagaraju, M Vandana, H Vijeth, H Devendrappa*

Department of Physics, Mangalore University, Mangalagangothri-574199, India

Corresponding author email: dehu20210@gmail.com

Abstract. PVC/PMMA and PVC/PMMA/NiO blend films were prepared by solution casting technique. The FT-IR spectra confirms the change in the band structure and presence of hydrogen chlorine and Ni-O bond in the PVC/PMMA and NiO loaded blend film and XRD pattern reveals the formation of NiO phase in the PVC/PMMA polymer with exhibits the amorphous nature, it may increase the conductivity and stability of the material.The surface morphology shows the NiO homogeneously dispersed on the surface of PVC/PMMA matrix, and CV curve demonstrate that the rectangular shape at different scan rate from 10 mV/s to 50 mV/s. The PVC/PMMA/NiO 3% loaded polymer blend films shows the synergistic properties towards the electrochemical performance.

1. Introduction

Recently, rapid development of technology energy storage devices plays a crucial role in the modern world and necessity for human daily routines; polymer blend films are highly fashionable strategy for the development of flexible and digital portable electronics devices. Therefore, researcher focused on the polymeric materials due to long chain segment matrix, distinguished physical and chemical properties, flexible and tuneable surface morphology, easy synthesis process, and eco-friendly. Polymers derivative such as poly (methyl methacrylate) (PMMA), poly (vinyl chloride) (PVC), polyvinylidene fluoride (PVDF), poly (ethylene oxide) (PEO), and poly(vinyl alcohol) (PVA) [1-3], polymer electrolytes have been distinctly used in various fields like as, fuel cells, supercapacitor, rechargeable batteries, electrochromic devices, solar cells and Li-batteries [4].
PMMA exhaustively used in numerous applications because of light weight, transparent, low-cost, easy preparation,it PMMA plays as plasticizer, and it possessed a more mobility charge carriers to improve the electrical conductivity, poor mechanical properties, and low surface area. To overcome this deficiency by blending the PVC polymers, it is a thermoplastic polymer large scale used in industries, due to high resistance, good mechanical, stiff and rigid material, excellent insulator [5], which is attracted a more attention towards the portable electronic devices. In present work a preparation of new PVC/PMMA blend film and observed proper miscibility leads to change the interaction of chemical and physical properties, it creates a strong bonding between in blend.
Nickel oxide (NiO) is the one of the ideal candidates for various energy devices due to their different morphology structure, high surface area, excellent electrochemical window stability [6], highly active material, good ionic conductivity. NiO dispersed in a polymer blend matrix, which improve the structural, optical, electrical conductivity, and electrochemical properties [7]. Novelty of in this work the CV curve possessed a rectangular shape and maximum current found to be 33.58 μA, and these results suggests that the PVC/PMMA/NiO 3% blend film shows the more stability of potential window, compared with other blend films.

From literature, MoumitaKhutia et al [8] reported a PVC/PMMA/NiO loaded composite polymer blend films prepared via solution casting process, it shows good mechanical, dielectric, and thermal properties. A.A. Al-Muntaser et al [9] investigated optical and electrical properties of PVC/PMMA doped $Li_4Ti_5O_{12}$ blend film, N.S. Alghunaim et al [10] studied the spectroscopic analysis of PMMA/PVC/CoCl2 blend film, which suggest that the PVC/PMMA well interaction with the CoCl2, the optical band gap decreases with increasing the $CoCl_2$ amount and produces a defect in polymer matrix.
The present work illustrated the PVC/PMMA and PVC/PMMA/NiO 1%, 2% and 3% blend films, the characterised themthrough XRD, FTIR, AFM, FESEM, and optical band gap properties were investigated by using Tauc method, and electrochemical properties of the blend films evaluated by CV and LSV techniques.

2. Experimental methods

2.1 Materials used

Poly(vinyl chloride) high molecular weight, Poly (methyl methacrylate) average Mw~ 120,000, Nickel (II) oxide nanopowder<50 nm (99.8%), Tetra hydro furan (THF) solvent.

2.2 Preparation of PVC/PMMA and PVC/PMMA/NiO (1%, 2%, and 3%) blend films

The PVC/PMMA/NiO blend film was prepared by the solution casting method. The desired weight ratio of the PVC: PMMA is (60:40 wt %) was separately dissolved in a 10 ml of THF solvent and vigorously stirred at 2 hrs at 40 °C, while NiOnano (1%, 2%, and 3%) dissolved in 10 ml of THF then sonicated 1 hr, and strongly stirred at 1hr, then all the solutions were mixed together and vigorously stirred at 2 hrs at 40 °C, the obtained viscous solution poured in to petri-dish and dried in a vacuum at 50 °C about 5 hrs. The obtained polymer matrix with NiO loaded blend film was taken further characterization. PVC/PMMA blend film was prepared similar method without addition of NiO.

3. Characterization Methods

Field emission scanning electron microscopy (FESEM) images carried out by Sigma Zeiss with 5KeV, Fourier transform infrared (FTIR) spectra were obtained by Alpha-Bruker (ATR) in the scan range of 500 to 4000 cm^{-1}. The X-ray diffraction (XRD) pattern examined by using Rigaku instrument with scan range from 3° to 80°. The UV-Visible absorption spectra measured by Shimadzu UV1800, and electrochemical performance investigated by using CHI660E instrument in the window potential range -2V to 2V.

4. Results and discussion

4.1 FTIR and XRD pattern analysis of PVC/PMMA and PVC/PMMA/NiO blend films

FTIR spectroscopy one of the versatile techniques to identifying the presence of chemical bounding, shifting of band and chemical interaction. In figure 1(a) shows the PVC/PMMA blend film exhibits the prominent characteristics peak at 1712 cm^{-1} is related to stretching and bending mode of C=O element present in the polymer [11], the peak at 1375 cm^{-1} is attributed to the CH_2 wagging deformation in PVC, the band at 1127 cm^{-1}, and 813 cm^{-1} is attributed with O-CH_3 in PVC/PMMA blend film and C-Cl in PVC [26-28], peak at 949 cm^{-1}, 2967 is denoted as a (C-H) and methyl stretching group. In PVC/PMMA NiO 1%, 3%, and 5% reveals the presence of Ni-O at 608 cm^{-1} [12], the band at 949 cm^{-1}, 1136 cm^{-1} have shifted to 969 cm^{-1}, 1145 cm^{-1} in the NiO loaded composition blend films.

Figure 1(b) shows the XRD pattern of PVC/PMMA with different wt% loaded NiO blend films, PVC/PMMA blend film possessed a broad and two low intense amorphous peaks, the broad peak at 8.10° is related to the PMMA, and 17.43°, 24.62° is attributed to the PVC/PMMA polymer matrix. PVC/PMMA/NiO 1% blend film shows the similar diffraction pattern, presence of NiO peak exhibited at 62.84° is indexed with (220) plane. PVC/PMMA/NiO 3% blend film corresponding peaks at 8.18°, 17.22°, and 25.50° is ascribed to the PVC/PMMA blend and 37.29°, 62.38° confirms the NiO presence in the blend film. In case of PVC/PMMA/NiO 5% blend film PMMA peak shifted 8.10° to 7.77°, and PVC/PMMA peak at 17.38° is disappear in this blend due to increasing the NiO content in the polymer matrix.NiO characteristics peaks observed at 36.45°, 42.48°, and 62.38° is assigned with plane (111), (200), and (220) respectively [13]. These results reveal the NiO content well-interaction with PVC/PMMA, the changes in the crystallinity nature remarkably improves the electrochemical properties.

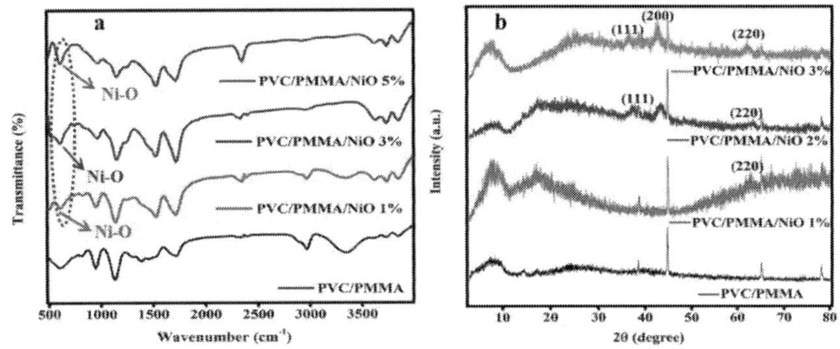

Figure 1. (a) FTIR spectra, (b) XRD pattern of PVC/PMMA, PVC/PMMA/NiO 1%, PVC/PMMA/NiO 3%, and PVC/PMMA/NiO 5% blend films.

4.2 UV-Visible absorption spectra analysis

In figure 2. The PVC/PMMA polymer matrix blend film absorption peak occurs at 277 nm, which is electronic transition ascribed to the π to π^*. The PVC/PMMA/NiO 1%, PVC/PMMA/NiO 3%, and PVC/PMMA/NiO 5% blend film absorption band shows two distinct peaks corresponding at 274 nm, 373 nm, 267 nm, 374nm, and 268 nm, 373 nm respectively.

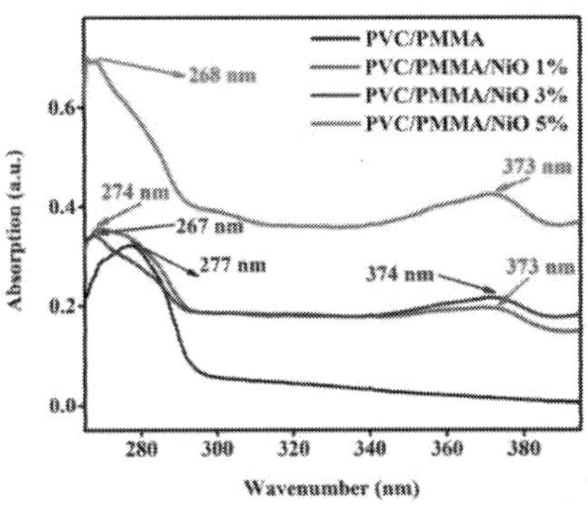

Figure 2. UV-Visible absorption spectra of PVC/PMMA, PVC/PMMA/NiO 1%, PVC/PMMA/NiO 3%, and PVC/PMMA/NiO 5% blend films.

The observed PVC/PMMA band peaks decreased in the composite films, and different weight percentage of NiO loaded in the PVC/PMMA, NiO band slightly increases towards the higher wavelength, these values agreed with previously reported results [14]. A UV-absorption spectrum clearly shows the miscibility between the PVC/PMMA loaded NiO.

4.3 FESEM analysis of PVC/PMMA and PVC/PMMA/NiO blend films

Figure 3(a) FESEM images of PVC/PMMA blend film shows the spherical and smooth surface corresponds uniformly dispersed on the surface. Figure 3 (b-d) shows the PVC/PMMA/NiO composition with different wt% loaded NiO, in PVC/PMMA matrix, it can be seen that PVC/PMMA/NiO 1% blend film indicates the NiOnano were homogeneously distributed on the PVC/PMMA blend matrix, PVC/PMMA/NiO 3% blend film reveals presence of small grains, the increasing the percentage of the NiOnano in the polymer matrix, the size of the particle also increases with agglomeration of NiO filler content [15].

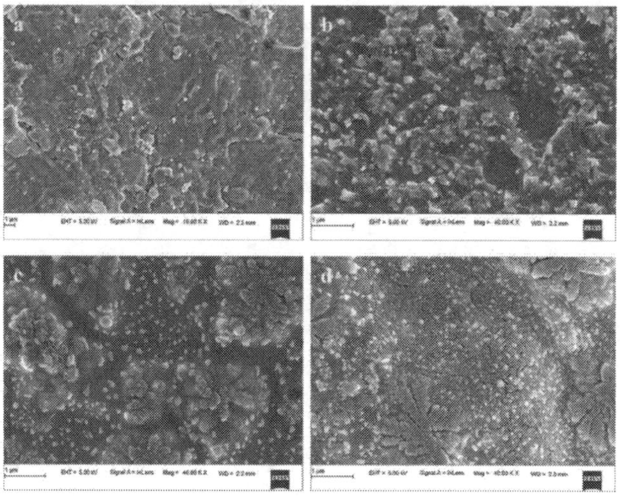

Figure 3. FESEM images of (a) PVC/PMMA (b) PVC/PMMA/NiO 1%, (c) PVC/PMMA/NiO 3%, and (d) PVC/PMMA/NiO 5% blend films.

PVC/PMMA/NiO 5% blend film shows the agglomerated with small grains structure on the surface, as addition of NiO content in the PVC/PMMA polymer matrix, which suggest that the changes the blend morphology, and it mayenhance the electrical conductivity.

5. Electrochemical characterization

5.1 CV analysis of polymer blend films

The electrochemical analysis investigated by using a cyclic voltammetry (CV) with different scan rates and applied potential window range from -2V to 2V with the help of three-electrode system Ag/Agcl is reference electrode, Glassy carbon electrode (GCE) is a working electrode and platinum electrode as an auxiliary electrode.

IOP Conf. Series: Materials Science and Engineering 1221 (2022) 012059 doi:10.1088/1757-899X/1221/1/012059

Figure 4. (a-d) CV cure of PVC/PMMA, PVC/PMMA/NiO 1%, PVC/PMMA/NiO 3%, and PVC/PMMA/NiO 5% blend film possessed a rectangular nature with pair of redox behaviour, while increasing the scan rates from 10 mV/s to 50 mV/s CV curve maintains same shape, but it changes the surface area and anodic and cathodic peaks slightly moves the positive and negative current, which demonstrate the maximum current found at 22.71 µA, 27.63 µA, 33.58 µA and 26.23 µA respectively.The PVC/PMMA/NiO 3%, blend film CV curve shows the enhanced in the current rate and surface area, then the NiO 1%, and NiO 5% blend filmsIt reveals the host polymer matrix with NiO loaded blend film blend films suitable for portable electronic devices. Electrochemical characteristics mainly depend on the surface area of the material, when morphology and crystallinity structure changes; it is evidence for improve the electrochemical properties. PVC/PMMA/NiO 3% blend film is promoting the higher electrical conductivity and electron transfer mechanism.

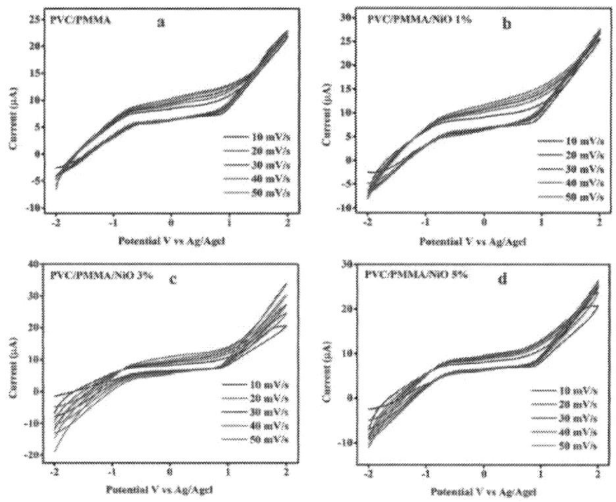

Figure 4. CV curves of blend films (a) PVC/PMMA, (b) PVC/PMMA/NiO 1%, (c) PVC/PMMA/NiO 3%, and (d) PVC/PMMA/NiO 5%. Applied potential range is -2V to 2V.

Conclusions

Polymer blend film loaded withNiO prepared via solution casting technique; the XRD pattern reveals the NiO content well- incorporated with the PVC/PMMA matrix and pattern possessed an amorphous nature, it could be enhancing the conductivity of the material due to improves the surface area. Optical properties and FTIR analysis of PVC/PMMA/NiO blend film confirms the presence of metal oxide in the composite films. Surface morphology reveals the PVC/PMMA blend film shows the spherical with smooth surface, and change in the particle size, these are evident for good miscibility between the PVC/PMMA and NiO loaded PVC/PMMA/ blend films. CV curve of the PVC/PMMA and PVC/PMMA/NiO 1%, 2%, and 3% blend film demonstrated that the rectangular behaviour with wide potential range from -2V to 2V. PVC/PMMA/NiO 3% blend film CV curve maximum current rate exhibited at 33.58 µA, which is higher than the other prepared blend films and shows the more stability in the composite film and good interface in energy devices.

Acknowledgement

The authors acknowledges to Purse Lab, Mangalore University, providing the material characterization facility.

References

[1]Mallakpour S, and Mansourzadeh S 2018, Sonochemical synthesis of PVA/PVP blend nanocomposite containing modified CuO nanoparticles with vitamin B1 and their antibacterial activity against Staphylococcus aureus and Escherichia coli. Ultrasonicssonochemistry, 43, 91-100.
[2] Kuppu S V, Jeyaraman A R, Guruviah P K, and Thambusamy S 2018, Preparation and characterizations of PMMA-PVDF based polymer composite electrolyte materials for dye sensitized solar cell. Current Applied Physics, 18(6), 619-625.
[3] Hosseinioun A, and Paillard E 2020, In situ crosslinked PMMA gel electrolyte from a low viscosity precursor solution for cost-effective, long lasting and sustainable lithium-ion batteries. Journal of membrane science, 594, 117456.
[4] Jamalpour S, Ghahramani M, Ghaffarian S R, and Javanbakht M 2020, The effect of poly (hydroxyl ethyl methacrylate) on the performance of PVDF/P (MMA-co-HEMA) hybrid gel polymer electrolytes for lithium ion battery application. Polymer, 195, 122427.
[5] Yang Q, Xia Y, Zhang D, Wang X, Oliver S, Chen X, and Lei L 2020, One-pot polyvinyl chloride preparation utilizing polyacrylate latex with tertiary amine groups for improved thermal stability, toughness, and reduced reactor scaling. Polymer Testing, 90, 106691.
[6] Yang H, Yu J H, Seo H J, Jeong R H, and Boo J H, 2018, Improved electrochromic properties of nanoporousNiO film by NiO flake with thickness controlled by aluminum. Applied Surface Science, 461, 88-92.
[7] Pepe Y, Karatay A, Donar Y O, Yildiz E A, Sınağ A, Unver H, and Elmali A, 2021, Enhanced nonlinear absorption coefficient and low optical limiting threshold of NiOnanocomposite films. Optik, 227, 165975.
[8] Khutia M, and Joshi G M, 2015, Dielectric relaxation of PVC/PMMA/NiO blends as a function of DC bias. Journal of Materials Science: Materials in Electronics, 26(7), 5475-5488.
[9] Al-MuntaserAA, Abdelghany A M, Abdelrazek E M, and Elshahawy A G (2020). Enhancement of optical and electrical properties of PVC/PMMA blend films doped with Li4Ti5O12 nanoparticles. Journal of Materials Research and Technology, 9(1), 789-797.
[10] Alghunaim N S, 2015, Spectroscopic analysis of PMMA/PVC blends containing CoCl2. Results in Physics, 5, 331-336.
[11] Devikala S, Ajith D, Kamaraj P, &Arthanareeswari M, 2019, Structural morphological and electrochemical studies on PMMA/PVP blends. Materials Today: Proceedings, 14, 630-639.
[12] Saravanakkumar D, Karthika R, Ganasaravanan S, Sivaranjani S, Pandiarajan S, Ravikumar B, and Ayeshamariam A, 2018, Synthesis of NiO doped ZnO/MWCNT Nanocomposite and its charecterization for photocatalytic& antimicrobial applications. J. Appl. Phys, 10, 73-83.
[13] Vijeth H, Ashokkumar S P, Yesappa L, Vandana M, and Devendrappa H, 2020, Hybrid core-shell nanostructure made of chitosan incorporated polypyrrole nanotubes decorated with NiO for all-solid-state symmetric supercapacitor application. ElectrochimicaActa, 354, 136651.
[14] Sheena P A, Priyanka K P, Sreedevi A, and Varghese T, 2018, Characterization of NiO/CoPcnanocomposite material synthesized by solvent evaporation route. Journal of Nanostructure in Chemistry, 8(2), 207-215.
[15] Sheng C K, Amin K A M, Hong L L, Hassan M F, and Ismail M, 2017, Investigation of morphological, structural and electrical properties of Cds/PMMA nanocomposite film prepared by solution casting method. Int. J. Electrochem. Sci, 12, 10023-10031.

Low Cost Synthesis and Characterization of Donor P3HT Polymer for Fabrication of Organic Solar Cell

D S Suresh[1], M Vandana[1], S Veeresh[1], H Ganesh[1], Y S Nagaraju[1], H Vijeth[2], M Basappa[1] and H Devendrappa[1]*

[1]Department of Physics, Mangalore University, Mangalagangothri-574199, India
[2]Department of Physics, Mangalore Institute of Technology and Engineering (MITE), Moodbidri -574225, India
*Corresponding author E-mail: dehu2010@gmail.com

Abstract: Over the years, the conjugated polymers acts as electron donor material in the active layer of Organic solar cells (OSCs). By altering means of doping, prepare a composite with suitable materials from Poly (p-phenylene vinylene) PPV polymer toPM6 polymer which gives Power conversion efficiency (PCE) can vary from ~ 2% to ~15%. But these are expensive, complicated synthetic procedure and less availability. To overcome this,the conjugated polymer Poly (3-Hexyl Thiophene)(P3HT) synthesized through oxidative coupling of 3-HT monomer and Ferric chloride ($FeCl_3$) as an oxidant which is low cost, available and very easy method of synthesis. We investigated synthesized P3HT material that shows the Fourier Transform Infrared(FTIR) peaks shows that intramolecular interactions from thiophene ring, alkyl chain and polymer chain, FESEM Images reveals morphology consists of condensed layers in micrometer dimensions and Cyclic voltammetry studies gives oxidation and reduction potentials to promote the higher ionization potential of the compound to use as donor material in OSCs.

1. Introduction

In the past few-decades, Organic photovoltaics overcome inorganic photovoltaics due to their potential to provide low cost alternatives. The conjugated polymers are used in organic solar cells in order to increase the Power Conversion Efficiency (PCE) and hence it is the most exciting topic of research due to these are easily available, low cost, versatile nature, light weight, well compatibility, good stability, eco-friendly, flexible and printable properties[1]. Among such materials Poly [3-Hexyl Thiophene] (P3HT) is a derivative of polythiophene family and one of the best photovoltaic material which is used as the donor material in active layer of organic solar cell. The conducting polymer derivatives arePoly (p-phenylene vinylene)[PPV], Poly (3-Hexyl Thiophene) [P3AT], PTB7-Th [PCE10] and PBDT-T [PM6] used as excellent donor materials in Organic solar cells. Among all the derivatives, P3HT is the best conducting polymer due to its properties as well as low cost and easy synthesis procedure. It is the excitation of the π-orbit electron in P3HT that gives the photovoltaic effect in the blend with the fullerene acceptor materials [2]. The performance of the BHJ solar cells is known to be strongly influenced by the properties of the donor polymer such as optical absorption, morphology and electrical properties. The conjugate polymer has wider bandgap than semiconductors. Due to the value of open circuit voltage(V_{oc}) depends on band gap between Higher Occupied Molecular Orbital (HOMO) and Lower Unoccupied Molecular Orbital(LUMO) levels of organic semiconductor. So that it can enhance the performance of the PCE of Organic solar cell since it is directly related with the value of V_{oc}. The active layer preparation of conjugated polymer P3HT donor material with fullerene acceptor material leads to the moderate efficiency (~4%) [3].

The Conjugated polymer consists of i. Conjugated backbone decides bandgap & intramolecular interactions, ii. Side chaincan vary solubility &mobility properties and iii.Substituents can alter HOMO &LUMO energy levels with donor and acceptor materials respectively. P3HT is the material in which contains backbone, substituents & side chains so that we can alter physical properties means of doping, adding impurities & composite with other suitable materials [4]. Hence, we can synthesis P3HT at low cost and easy procedure through oxidative coupling of 3-HT monomer in presence of $FeCl_3$ and N_2 atmosphere and different properties also studied through different characterization techniques. The P3HT is cheap, available, high molecular weight conjugated polymer which is must need future perceptive material for fabrication at low cost and also to enhance the performance of organic solar cell.

2. Experimental methods

2.1 Materials

The 3-Hexylthiopehene (3-HT) monomer, Ferric chloride ($FeCl_3$), Indium Tin Oxide (ITO) Glass substrates and Silver paste was purchased from Sigma Aldrich. Nitrogen (N_2) gas was purchased from Lakshmi cryogenics Private Ltd. Chloroform, methanol, isopropyl alcohol and distilled water were used as solvents.

2.2 Synthesis of Poly (3-hexylthiophene)

The synthesis of polymer (P3HT) using chemical reaction method in presence of nitrogen atmosphere [5]. In this case, oxidative coupling of 3-HT used as a monomer and $FeCl_3$ as an oxidant. The 3-HT monomer and oxidant $FeCl_3$ are in the ratio of 1:4 dissolved in the chloroform solvent and add solutions together, kept them stirring for 24 hours in three neck round bottom flask in presence of Nitrogen gas atmosphere and water condenser in order to maintain constant temperature and then transfer the obtained solution in 500 ml round bottom flask and extract the solvent using the Soxhlet Extraction Apparatus and resultant product is black color precipitate is formed. Then, it is filter in the Glove box at dark and vacuum conditions. Dried the sample in Tubular furnace for about 12 hours at 60°C, Collect and used for further experimental examination.

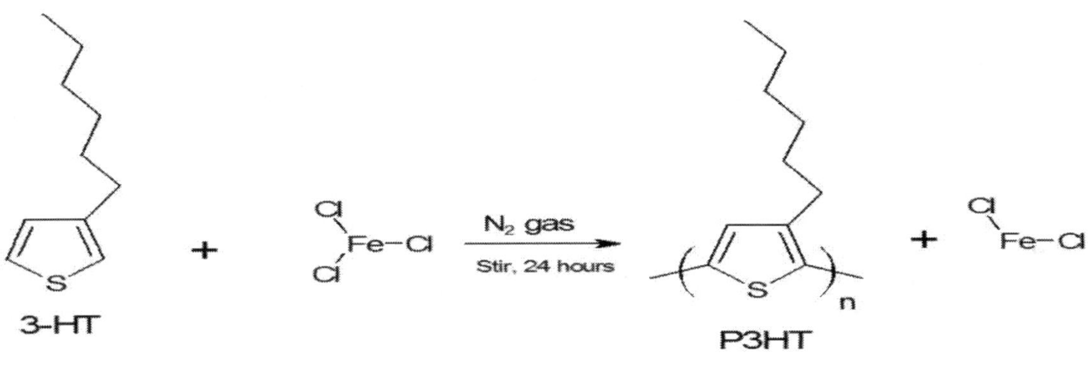

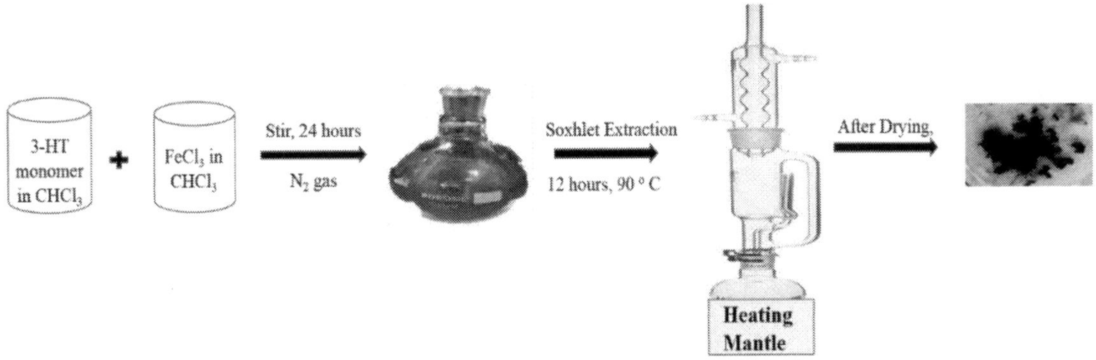

Figure 1: Representation of P3HT Synthesis

2.3 Preparation of P3HT Film

The ITO Glass substrates were cleaned using detergent, distilled water, acetone and isopropyl alcoholfor 10 minutes each respectively. The synthesized polymer (P3HT) powder was dissolved in chloroform solvent. The prepared solution was deposited onto the ITO Glass substrates through Spray pyrolysis method and dried in an oven for 30 minutes at 80 °C. Take a small piece of silver, add few drops of amyl acetate until it forms slurry, then paste it on P3HT film using small brush and dried in an oven for 1 hour with the petri dish closure.

2.4 Characterization techniques

The chemical bond formation and stretching vibration changes are studied by Fourier Transform Infrared Spectrometer (FT-IR) model ATR ALPHA BRUKER. The surface morphology and elemental analysis by the Sigma Zeiss field emission scanning electron microscopy (FESEM) and energy dispersive x-ray spectroscopy (EDX) respectively. The electrochemical characterization using a three-electrode setup such as cyclic voltammetry(CV) and linear sweep voltammetry(LSV) studies by using PC based electrochemical workstation model CHI 660E.

3. Results and Discussion

3.1 FTIR Analysis

FT-IR spectroscopy is the most important tool for the identification of chemical bonding, stretching vibrational groups and FTIR spectra of synthesized powder material (P3HT) carried out in the range 500-4000cm^{-1} as shown in figure 2. The presence of C-S single bond deformation vibrations was detected at 838 cm^{-1} and it shows the presence of C=C double bonds stretching vibration at 1446 cm^{-1}. The strong stretching vibration of aliphatic (CH_2) bonds was present at 2906 cm^{-1}[2a]. The intensity of peak increases due to larger proportion of carbon–hydrogen bonds in the polymer. The presence of these molecules was similar with the chemical structure of P3HT and well agreed with the reported results[6]. The FT-IR Spectra [2b]of P3HT film prepared from Chloroform solvent. The stretchings due to the Thiophene ring, Polymer chain and alkyl chain values are 782cm^{-1}, 1684cm^{-1} and 2955cm^{-1} respectively. The right side shifting the values from 1446 cm^{-1} to 1684 cm^{-1}& 2908 cm^{-1} to 2955 cm^{-1}due to the P3HT film degraded when exposed toatmosphere. The higher intensities of the peaks 1005 cm^{-1} and 782 cm^{-1} indicates more oxygen content in the film.

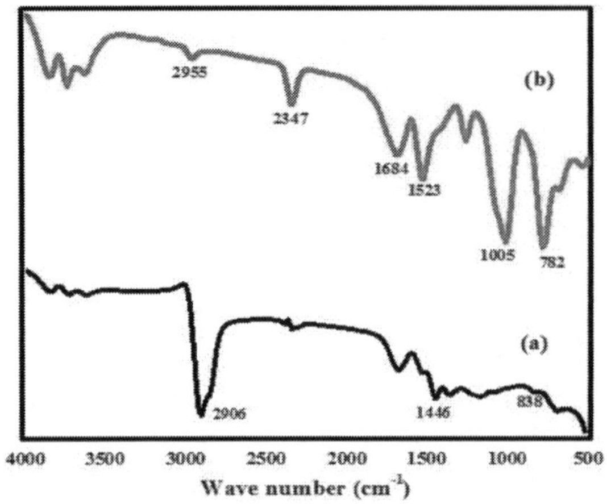

Figure 2: FT-IR Pattern of (a) P3HT Synthesized powder, (b) P3HT film.

3.2 FESEM Image & EDX analysis

The FESEM image of synthesized P3HT revealed that the morphology consists of uniformly condensed grains of several micrometers in dimension, (see figure3a& 3b).The surface of pure P3HT based film is very rough and consists of grains with different sizes. Some cracks appear in the FESEM image of powder too, (see figure3b &3c) indicating poor contact between the grains of P3HT, these cracks are filled when blend with acceptor material which increases the conductivity of the material and EDX results confirms the presence of elements Carbon & Sulphur. Some oxygen content due to the exposure of film into the atmosphere and light (see figure 3d) in the synthesized P3HT film.

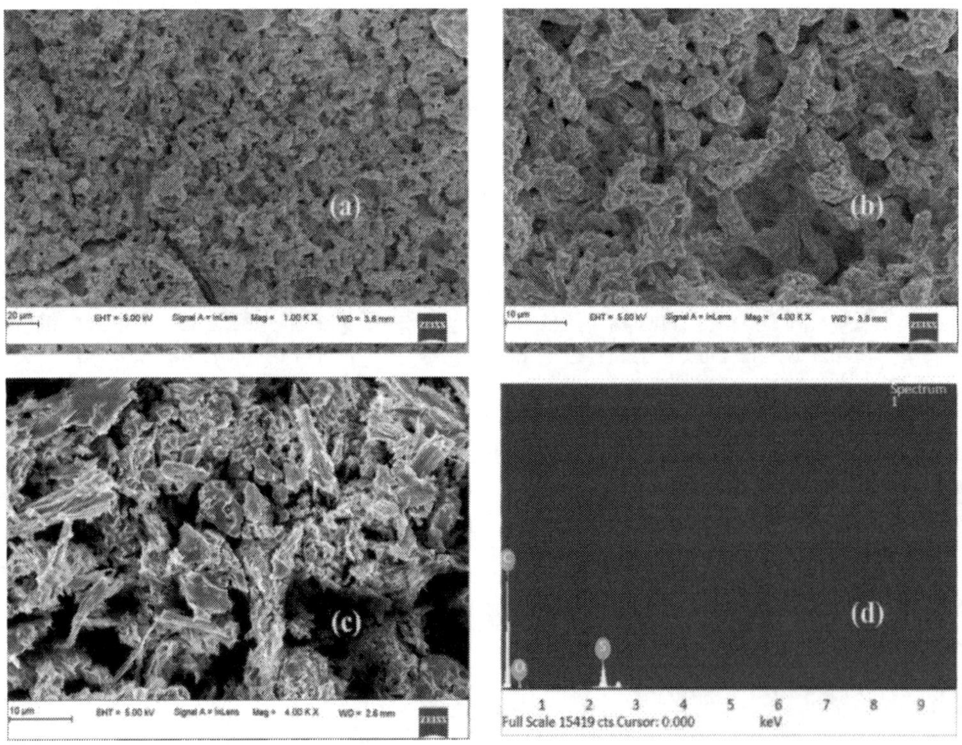

Figure 3: FESEM image of synthesized (a-b)-P3HT film, (c)-powder and (d) - EDS spectrum.

3.3 Cyclic Voltammetry and linear sweep voltammetry analysis

The Cyclic voltammetry and linear sweep voltammetry curves of three electrode system using CHI660E Electrochemical workstation shows the maximum current is observed at anodic peak potential that confirms oxidation of most of the repeat units. The onset potentials of oxidation and reduction are 0.45 V and −1.1 V versus Ag/AgCl, respectively, (see figure4b) well agreed with the reported results [7] .The obtained electrical energy band gap was 1.55 eV which is nearly equal to the energy gap of P3HT (i.e., 1.9 eV). Cyclic voltammetry curves at different scan rates shows increase in area with increase in scan rates, (see figure4a). The LSV curves of P3HT Solution shows increase in the area with increase in scan rates. The linear sweep voltammetry curves for P3HT film shows the variation of current with voltage linearly, (see figure4d).

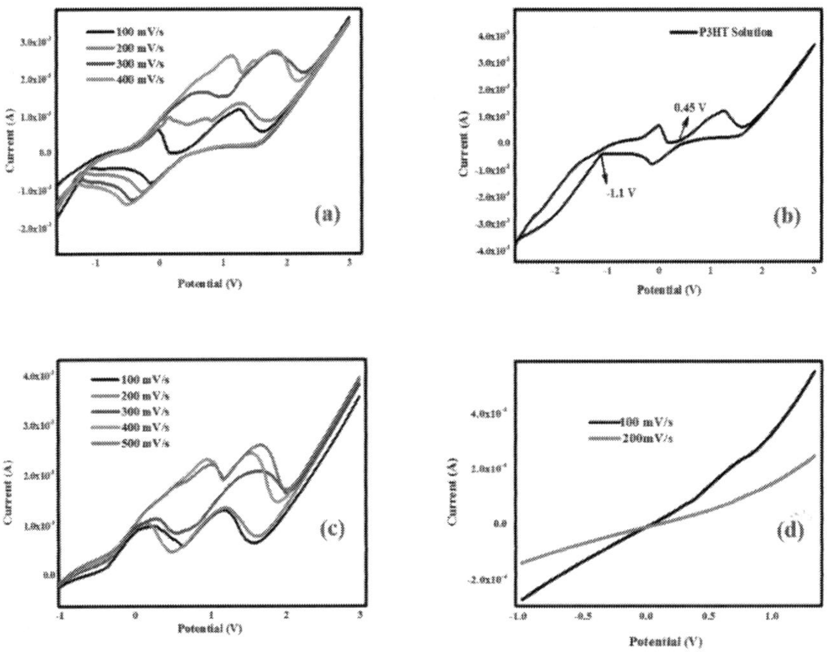

Figure 4: Cyclic voltammetry curve for (a) P3HT Solution at different scan rates, (b) P3HT oxidation and reduction potentials. & linear sweep voltammetry curves for (c) P3HT Solution at different scan rates (d) P3HT Film at different scan rates.

4. Conclusion

In summary, the synthesis of P3HT at low cost using the oxidative coupling of 3-HT monomer in presence of oxidant $FeCl_3$ and Nitrogen gas atmosphere. The FT-IR spectra confirms the synthesized material exhibits the C-S, C=C and CH_2 bonds at 838 cm^{-1}, 1446 cm^{-1} and 2906 cm^{-1} due to its thiophene ring, polymer chain and alkyl chain respectively. The morphology of P3HT reveals that FESEM images consists of uniformly condensed grains having cracks in between them, EDS results also confirms the presence of Carbon & Sulphur atoms and The cyclic voltammetry curve shows that oxidation and reduction potentials and energy gap between HOMO & LUMO is 1.55 eV and also with increase in the scan rates, increase in area of CV curves. The LSV curves of P3HT film shows the variation of current with voltage linearly. Due to this low cost synthesis and characterization properties of P3HT material is one of the most promising material used for the photovoltaic applications.

Acknowledgement

This work was financially supported by Science and Engineering Research Board, New Delhi (SERB/F/11204/2018-2019 dated 15 February, 2019) and DST-PURSE Laboratory, Mangalore University for FESEM characterization facility.

References

[1] By Qian Liu, Zunfeng Liu, Xiaoyan Zhang, Liying Yang, Nan Zhang, Guiling Pan, Shougen Yin, Yongsheng Chen and Jun Wei, 'Polymer Photovoltaic Cells Based on Solution-Processable Graphene and P3HT' 2009 *Adv. Funct. Mater.* **19**, 894–904.

[2] Ge and Weihao, 'An overview on P3HT:PCBM, the most efficient organic solar cell material so far'2009 *Solid State Phys. II*, Spring.

[3] Vanlaekea P, Swinnen A, Haeldermans I, VanhoylG and Aernoutsa T, Cheynsa D, Deibela C, D'Haena J, Heremansa P, Poortmansa J and MancaaJ V, 'P3HT/PCBM bulk heterojunction solar cells: Relation between morphology and electro-optical characteristics' 2006 *Sol. Energy.* **90,** 2150–2158.

[4] Meng Qiu, Rasmus G Brandt, Yingli Niu, Xichang Bao, Donghong Yu, Ning Wang, Liangliang Han, Liangmin Yu, Shuwei Xia and Renqiang Yang, 'A Theoretical Study on the Rational Design of CyanoSubstituted P3HT Materials for OSCs: Substitution Effect on the Improvement of Photovoltaic Performance' 2015,*J. Phys. Chem. C* .**119(16)**, 8501–8511.

[5] Muhammad Azhar Ansari, Shaikh Mohiuddin, Fatma Kandemirli and Muhammad Imran Malik 'Synthesis and characterization of poly(3-hexylthiophene): improvement of regioregularity and energy band gap' 2018*RSC Adv.,* **8(15)**,8319–8328.

[6] Kalonga G 'Characterization and optimization of poly (3-hexylthiophene-2, 5- diyl) (P3HT) and [6, 6] phenyl-C61-butyric acid methyl ester (PCBM) blends for optical absorption' 2013. *J. Mater. Sci.,* **4(7)**, 93–102.

[7] Jin Wang, Chenchen Guo,Yongqiang Yu, Huabing Yin, Xueting Liu and Yang Jiang, 'Self-doped 3-hexylthiophene-b-sodium styrene sulfonate block copolymer: synthesis and its organization with CdSe quantum dots'2015, *RSC Adv,* **5(23)**, 17905–17914.

AUTHOR INDEX